TILLS AND RELATED DEPOSITS

PROCEEDINGS OF THE INQUA SYMPOSIA ON THE GENESIS AND LITHOLOGY
OF QUATERNARY DEPOSITS / USA 1981 / ARGENTINA 1982

Tills and Related Deposits

Genesis / Petrology/Application / Stratigraphy

Edited by

Edward B.Evenson
Lehigh University, Bethlehem, Pennsylvania, USA

Ch.Schlüchter
ETH-Hönggerberg, Zürich

Jorge Rabassa
Universidad Nacional del Comahue, Neuquén, Argentina

A.A.Balkema / Rotterdam / 1983

The texts of the various papers in this volume were set individually
by typists under the supervision of each of the authors concerned.

ISBN 90 6191 511 2

© 1983 A.A.Balkema, P.O.Box 1675, 3000 BR Rotterdam, Netherlands

Distributed in USA & Canada by: MBS, 99 Main Street, Salem, NH 03079, USA

Printed in the Netherlands

Table of contents

Applied glacial geology

Glaciofluvial and glaciolacustrine deposits

Pre-Pleistocene glaciations

General

Contributions related to the field excursions:
A. United States: Wyoming and Idaho

Contributions related to the field excursions:
B. Argentina

Introduction

EDWARD B.EVENSON
Lehigh University, Bethlehem, PA, USA

CH.SCHLÜCHTER
ETH-Hönggerberg, Zürich, Switzerland

JORGE RABASSA
Universidad Nacional del Comahue, Neuquén & CIC de la Provincia de Buenos Aires, La Plata, Argentina

"A man who keeps company with glaciers comes to feel tolerably insignificant by and by." (Mark Twain after a tour of Swiss glaciers in the late 1870's)

This volume is the result of two separate INQUA Commission on the Genesis and Lithology of Quaternary symposia held in 1981 and 1982 in the United States and Argentina, respectively. One of the primary purposes for the existance of the INQUA-Commission is to provide a forum where Quaternary scientists from all parts of the globe can meet and exchange ideas. A brief glance at the authorships and editorship of this volume should demonstrate that this objective is being fulfilled.

This volume contains 43 papers, most of which were originally presented orally at either the North American or South American meeting of the Commission's Work Group I (Genetic Classification of Till and Criteria for their Differentiation). A few additional papers are included that, for various reasons, were not presented at the U.S. or Argentinian meetings.

The editors would like to express their gratitude to all those authors who labored so diligently to produce accurate, high quality, camera-ready manuscripts. All manuscripts included in this volume have been reviewed by several individuals who acted as critical and constructive reviewers and to them should go much of the credit for the scientific merit of this volume. In many cases, manuscripts were thoroughly revised and/or retyped by the editors.

Special thanks are due those authors who labored to produce articles in languages other than their "mother tongue". We hope that our readers will recognize the difficulties which faced both the authors and editors in such situations. Although we have struggled to remove or avoid them, the editors accept responsibility for all grammatic, spelling and typographic errors. There are, however, ideas, models, concepts and interpretations expressed in some papers which do not necessarily coincide with those of the editors, and their inclusion does not automatically imply approval by the Commission or the editors.

In the interest of logistic ease the editors have, in general, divided the editorship of individual papers as follows: North American authors - Evenson; European authors - Schlucter; and South American authors - Rabassa. We have organized the papers under one of seven catagories: (1) Till Genesis, (2) Till Petrology, (3) Applied Glacial Geology, (4) Glaciofluvial and Glaciolacustrine, (5) Pre-Pleistocene Glaciations, (6) General and, (7) Contributions Related to the Field Excursions. Many papers would fit as well into another catagory as they do into the one selected, which simply serves to demonstrate the breadth and interdisciplinary nature of many of the papers.

The section on Till Genesis contains twelve papers specifically related to the Commissions' aim of developing a genetic classification of tills. In order to satisfy this objective, detailed work on specific deposits, such as that presented by Shaw, Eyles, Eyles and Day, Schluchter, and Huddart is required as are the more theoretical interpretations such as those presented by Muller and some of the remaining authors in this section.

The section on Till Petrology consists of three papers and serves to emphasize the importance of understanding the changes that occur in glacial debris during transport (Haldorsen) and subsequent weathering (Hall). Brugger and others present a classic study on the use of provenance to decipher flow paths and glacial dynamics. The three papers on Applied Glacial Geology illustrate the utility and economic applications of concepts originally developed from more academic studies similar to those presented in other sections of the volume.

Four papers by Spalletti, Serat and others, Rubulis, and Cohen comprise the section entitled "Glaciofluvial and Glaciolacustrine Deposits". The intimate association of tills and waterlain deposits originating from the glacial environment is well demonstrated by the papers presented in this section.

As our understanding of modern and Quaternary glacial deposits grows, it is increasingly being applied to the investigation of Pre-Pleistocene glacial deposits and to the determination of the origin of Pre-Pleistocene diamictons. This is a particularly important extension of the work of the Commission, as only accurate interpretation of Pre-Pleistocene diamictons as "glacial" or "nonglacial" will allow us to understand and interpret the frequency, timing and distribution of paleoglaciations and the associated climatic changes which have occurred repeatedly throughout geologic time.

The section called "General" is just that. It includes twelve papers on such diverse topics as paleomagnetism of glacial deposits (Easterbrook), stratigraphy (Rabassa), rock glaciers (Seret, Ellis and Calkin), paleontology (Graf) and relative age differentiation of glacial deposits (Butler, et al., Punning and Raukas) to mention only a few.

The final section of the volume contains five papers related to the excursions that accompanied the two Commission meetings which form the basis for this volume. Eschman, et al. and Rabassa review the symposia and field excursions in North and South America, respectively. These two papers should acquaint non-participants with a summary of the main topics and controversies that were associated with the meetings and excursions and serve as valuable guides for those who might wish to duplicate all, or parts of, the field trips through these two spectacular areas. The three accompanying papers provide a detailed description of a portion of the excursion area in Idaho (Cotter, et al.), a discussion of one of the few multiple till sequences in Wyoming (Eschman, et al.) and a review of the contributions of Caldenius to the early studies of glacial geology in Argentina (Lundqvist).

As editors and field excursion leaders, we would like to thank all those who made the symposia, this volume and field trips financially and logistically possible. Without the support and cooperation of our home institutions, authors, secretaries, students, publisher, friends and families neither this volume nor the excursions would have been possible. The Department of Geological Sciences at Lehigh University provided almost unlimited logistic and financial support. The United States National Science Foundation provided an underwriting grant that allowed several foreign delegates to participate in the North American excursion.

The Institute of Foundation Engineering and Soil Mechanics at ETH-Zurich gave extended office support for the editing of the contributions from European authors.

The Geography Department of the Post-graduate School of the Universidad Nacional del Comahue provided funds and personnel during the organization of the Argentine Meeting and all necessary facilities for the Symposium. The Government of the Province of Neuquen supported the meeting with a generous grant and offered a plane of the provincial airline to overfly the largest glaciers of the province, on the slopes of Volcan Lanin.

To the above individuals and institutions and especially to those that we may have inadvertently omitted, we owe a debt of gratitude that we can only hope to fully repay.

Till genesis

Forms associated with boulders in melt-out till

JOHN SHAW
University of Alberta, Edmonton, Canada

1 INTRODUCTION

Ice is distinguished from most other transporting agencies by the virtual absence of
a competence level for large clasts. Consequently, glacial deposits are often recognised by the presence of such clasts, a
practice implicit in the term boulder clay,
formerly the pre-eminent English language
term for till. Erratics are thus used to
show the former existence and extent of ice
sheets. The spatial distributions of large
clasts have also been used to furnish information of a more detailed kind on the
flow paths in former ice sheets (Jones
1973). The shapes and surface markings of
clasts have been used to illustrate the
mode and position of their transport by ice
(Drake 1972), and also to demonstrate the
abrasion of stationary clasts beneath
sliding ice (Okko 1955 Fig. 21, Boulton
1978). Such forms as "bullet-shaped"
boulders (Boulton 1978) and associated
strongly preferred orientation of clast
long axes parallel to the direction of ice
movement (Mills 1977, Humlum 1981) evidence
deposition of till by lodgement. Where
boulders have been dragged through soft
subglacial sediment they commonly leave
trailing grooves and bulldoze mounds of
sediment before them. Elsewhere, individual
stationary boulders or clusters of boulders
give rise to lee-side cavities in which
small-scale flutings are created by soft
sediment deformation (Dyson 1952, Hoppe and
Schytt 1953, Paul and Evans 1974, Boulton
1976, Morris and Morland 1976, Åmark 1980).
Imbrication of boulders is noted on lodgement till surfaces close to modern glaciers
and has been observed in ancient tills
(Haldorsen in press). Consequently, detailed observations on the characteristics
of boulders and their relationships with

other deposits proves to be one of the most
useful tools in the recognition of lodgement
tills.

A further set of relationships between
boulders and associated deposits is found
in tills interpreted to have been deposited
by melt-out. These relationships allow some
estimate of debris concentrations in the
debris-rich ice which gave rise to the till
and permit conclusions to be drawn on the
depositional conditions and processes in
the basal part of the glacier.

Till genesis and properties are
often related without clear
definition of the genetic processes
involved. This has caused much confusion with such terms as lodgement
assuming a number of meanings. Until
agreement is reached on genetic
terminology authors should make their
use of terms clear. Lodgement and
melt-out are treated here according
to the following definitions:

Lodgement – the lodging of debris
from overlying,
sliding ice.

Melt-out – the release of debris
from ice that is
neither sliding nor
deforming internally.

Thus, lodgement involves a dynamic
prising of clasts from the wet base
of a sliding glacier. Melt-out is
a more passive process that occurs
most commonly in association with
stationary ice. In addition, the
term basal melt-out till is used
here. In this case, basal refers to
transport in a basal debris-rich
zone of a glacier or ice sheet.

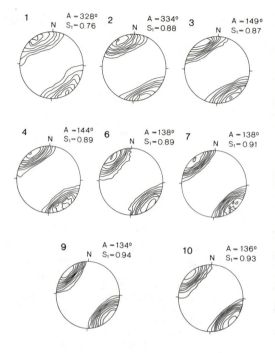

N-50 all samples   Contour interval 20

Fig. 1. Clast orientation and dip distributions in the Sveg till at Overberg (Shaw 1979). The samples were taken over a horizontal area of several hundred square metres. Ice flow direction was from northwest to southeast (Lundqvist, 1969).

Fig. 2. Draped sand layer over large clast in the Sveg till, Overberg, near Sveg, Sweden. See Shaw (1979) for site location.

2 THE SVEG TILLS AND DRAPED BEDS

Lundqvist (1969) described the structural appearance of the Sveg tills of central Sweden. These tills are characterised by intrabeds of sorted sediment which give them a stratified appearance. The tills contain a large proportion of far travelled erratics and overlie more massive tills which contain a relatively higher proportion of locally derived materials. Faceted and striated clasts are common in the Sveg tills and clast long axes show a strikingly consistent preferred orientation which parallels the former ice flow direction (Fig. 1). The sorted layers are draped over large clasts (Fig. 2).

On the basis of the above observations the Sveg tills were interpreted to be a product of melt-out from a basal debris-rich layer of the depositing glacier (Shaw 1979). In this interpretation "basal" signifies glacial transport in a basal layer of high

debris concentration and "melt-out" signifies the mode of debris release from ice. The draping of the stratified layers over large clasts is interpreted to represent volume loss with the melting of debris-rich ice adjacent to the clasts. Consequently, using the definition diagram (Fig. 3), the debris concentration of ice may be estimated in the following manner:

$$\frac{V_d + V_p}{V_d + V_i} = \frac{h_2}{h_1} \qquad 1$$

Using the void ratio $e = \frac{V_p}{V_d}$

$$\frac{V_d(1+e)}{V_d + V_i} = \frac{h_2}{h_1} \qquad 2$$

Defining debris concentration $C_d = \frac{V_d}{V_i + V_d}$

$$C_d = \frac{h_2}{h_1(1+e)} \qquad 3$$

4

A plot of h_2 vs h_1 for draped beds of the
Sveg tills at the Overberg site is presented
(Fig. 4). Assuming a void ratio of 0.4, the
estimated debris concentrations of the basal
part of the glacier have a mean of 46 percent (standard deviation 12%). The range
is from 16 to 60%. For a thickness of Sveg
till of 5 m, with a void ratio of 0.4, the
average debris concentration implies a basal
debris-rich layer of 7.8 m thickness. However, it may be that the debris concentration is overestimated and the thickness of
the debris-rich layer underestimated because of the probability that the large
clasts were located in debris-rich bands;
and thick segregated ice layers, which did
not intersect clasts, are not detectable by
the method described above.

2.1 Discussion of the debris concentration
     estimates

Debris concentration measured in a number
of glaciers shows a wide range in value.
Goldthwait (1971: 8) reported that, despite
a dirty appearance, Greenland glaciers
yielded only 0.03-3 percent debris by weight.
Boulton (1971: 46) reported a range of 10 to
>50 percent debris by volume in debris bands.
Clapperton (1975) noted that surging glaciers
in Iceland possess very high debris contents
and, although he does not give quantitative
estimates, he suggested that in some bands
the ice was merely interstitial. Lawson
(1979: 7) reported debris contents for
Matanuska Glacier, Alaska, ranging from 0.02
to 70 percent by volume with a mean of 25
percent in the debris-rich basal zone. For
Myrdalsjökull, an Icelandic temperate glacier, debris was concentrated in a basal
zone of only 20 to 50 mm thickness with
concentrations from 15 to 31 percent by
volume (Humlum 1981). Pleistocene glaciers
which carried bedrock blocks several tens of
metres in thickness (Moran 1971) or which
incorporated and transported large clasts
of subglacially derived, unlithified sediment (Shaw 1971) were clearly capable of
transporting both high concentrations and
large total loads in their basal zones.
   Although the ice/bedrock interface has
been the topic of much discussion, a second
interface between relatively clean glacier
ice and underlying debris-rich ice may be
of considerable importance (Hughes 1973).
The basal glacial zone of high debris concentration is often folded and sheared
(Boulton 1970 Fig. 3, Post and Lachapelle
1971 Fig. 31, Lavrushin 1976 Fig. 20, Shaw
1977 Fig. 8). When combined with deposition
by melt-out this folding and shearing presents a plausible explanation for the
formation of drumlins and Rogen moraines

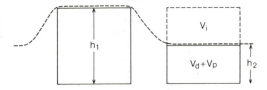

h_1 vertical diameter of clast
h_2 final thickness of material represented by h_1
      in the glacier
V_d volume of debris
V_p volume of pore space in till
V_i volume of ice

Fig. 3. Definition diagram for the estimation of debris-concentration from draped
sorted layers.

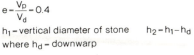

$$e - \frac{V_p}{V_d} - 0.4$$

h_1 – vertical diameter of stone $h_2 = h_1 - h_d$
where h_d – downwarp

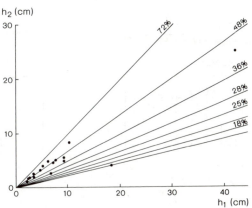

Fig. 4. h_2 vs h_1 and debris concentration
estimates assuming a void ratio in the
deposited till of 0.4. The lines on the
graph give values of C_d expressed as
percentages.

(Shaw 1979, 1980). Equally plausible is the
soft-sediment deformation model for drumlin
formation advocated, among others, by Smalley (1981). Clearly, some insight into the
actual debris loads and transport mechanisms
of the glaciers which produced ancient landforms is required before we can favour a
particular theory of their formation.
   The estimates of debris concentrations for
the Sveg tills are within the range of
values observed in modern glaciers but show

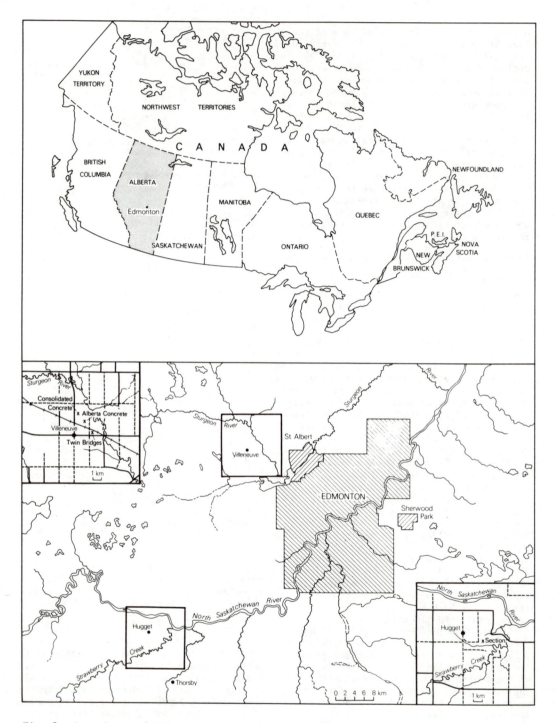

Fig. 5. Location map, Edmonton area and principal sites.

a relatively high average value. Thus the direct contribution of basal melt-out to the total thickness of tills may be considerable. In particular, stacking of debris by thrusting and folding of the basal debris-rich layer presents a powerful mechanism for the initial development of Rogen moraines and drumlins. The above conclusions on the debris content of ice in a similar geographic location and contemporary with glaciers which produced both drumlins and Rogen moraines lend support to theories which stress the role of high debris content in the formation of these landforms.

3 BOULDER SCOURS

Boulders at the base of till sheets are sometimes associated with grooves cut into the underlying strata (Westgate 1968, Ehlers and Stephan 1979). These grooves, which illustrate sliding and lodgement at the ice sediment interface have been observed in the Edmonton area, Alberta (Westgate 1968, Ramsden and Westgate 1971, Shaw in press). They are observed at the base of the so-called upper till (Westgate 1969) and are generally noted where it overlies sand. However, other boulders at or near the base of this till are not associated with grooves. Rather, a localised scour occurs beneath each boulder which is considerably wider than the boulder itself. I propose first to give a brief description of the upper till before considering the origin and significance of the boulder scours.

3.1 The Upper Till of the Edmonton area

The location of the area and the two principal locations where observations were made, Hugget and Villeneuve, are given in Fig. 5. The till is usually from 1 to 5 m thick and is normally of sandy loam texture. It contains crystalline erratics from the Canadian Shield to the northeast, carbonates from Paleozoic rocks which fringe the Shield, and quartzites which were originally transported from the Rocky Mountains by northeastward flowing rivers. These well lithified clasts are commonly faceted and striated. There is also a considerable contribution of materials from the local Cretaceous bedrock including coal, sandstone, and shale clasts. A striking and commonly noted feature is the appearance of large, intact clasts of poorly lithified bedrock or unconsolidated preglacial sands and gravels (Bayrock and Hughes 1962, Rains 1969, Kathol and McPherson 1975, Westgate et al. 1976, Babcock et al. 1978,

Shaw (1982). Westgate (1969) described a preferred northeast-southwest orientation for the long axes of clasts in the upper till. Subsequent work confirms this finding but also shows that there may be considerable deviation about the trend for the preferred orientations of individual distributions (Fig. 6). The eigenvectors of individual distributions may be analysed to obtain an estimate of the preferred orientation of the group of all distributions combined (Fig. 7). This plot of the grouped data shows clearly the highly clustered nature of the individual eigenvectors. As sole markings and surface flutings on the Upper Till show trends similar to the preferred azimuth of the grouped fabric distribution, the clast orientation is ascribed directly to glacier flow. Stratified sediments occur within and at the base of the till in the form of elongate, relatively narrow bodies

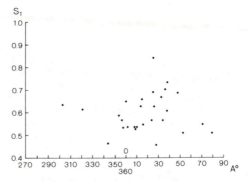

Fig. 6. Azimuth A vs normalised eigenvalue S_1 for fabric distributions in the upper till.

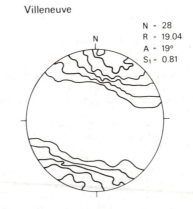

Villeneuve

N = 28
R = 19.04
A = 19°
S_1 = 0.81

Fig. 7. Distribution of grouped eigenvectors for all individual fabrics in the upper till.

often with a planoconvex cross-section
(Fig. 8). Where the stratified sediments
occur as intrabeds, that is within the till,
they are invariably faulted or otherwise
disturbed. However, faulting is rare where
they occur at the base of the till and rest
directly on subtill sediments. A strongly
unidirectional paleoflow is illustrated by
the cross-bedding and cross-lamination which
occur in the stratified sediments (Fig. 9).
This flow direction is close to the preferred
orientation of clasts in the till (Fig. 7).
From the form of these beds and their relationship to the till, Shaw (1982) concludes that they and the till were syndepositional. The implication of the large
unlithified clasts, which preclude till
deposition by lodgement, and the syndepositional sub and intratill beds, which preclude deposition by flow, is that till deposition was largely by melt-out. This conclusion is supported by the boulder scours
to be described.

From the standpoint of flow competence
the boulder of Figure 10 is unlikely to have
been transported by the flowing water which
transported and deposited the surrounding
sands. Furthermore, this and other scours

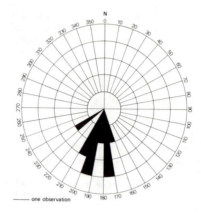

Fig. 9. Paleocurrent estimates based on
cross-bedding and cross-lamination in sub
and intratill stratified beds.

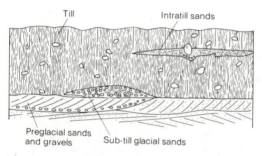

Fig. 8. Stratified sediments occurring at
the base of and within the upper till.

3.2 Description of the boulder scours

Boulder scours are erosional forms which
occur in both till and stratified sediments
and are, as their name implies, associated
with boulders. As mentioned above, the
cross-sectional widths of the scours are
greater than the diameters of associated
boulders.
 The scour troughs occur in a number of
styles regarding the relationship between
scours and the associated boulder, till, and
stratified sediment. In style 1 (Fig. 10)
the boulder is separated from the associated
till by stratified sands. Medium grade
stratified sands, conformable with the
scour surface, also intervene between the
boulder and this surface.

Fig. 10. Style 1 boulder scour in which
the boulder is separated from the associated till.

to be described are clearly related to the
presence of a boulder. It is difficult to
explain, by any normal fluvial process, how
scouring and deposition of sand occurred
beneath the boulder before it came to rest
on the bed. Both of these difficulties can
be resolved rather simply by the introduction of glacier ice. Consider that the
boulder was held in overlying ice and projected downwards into a body of subglacially
flowing water. The constriction of the flow
and the obstacle itself would have caused
both increased velocity and locally increased turbulence. These would in turn have
caused local erosion beneath and around the
boulder and thus the scour may be explained.
Provided the boulder continued to be held
in the ice, a decrease in discharge or an
increase in depth of the subglacial flow
could account for the deposition of sand
beneath the boulder. The full sequence is
explained if the boulder is then released
and sand deposited above it with debris-
rich ice finally settling to the accumulating sediment surface and till being deposited by melt-out.

In style 2 scours, the boulder remains
partly embedded in sand and partly in till
(Fig. 11). Once again the boulder is
separated from the scour floor by sands
deposited subsequent to scouring. However,
no sands intervene between the till and the
boulder.

A similar explanation to that given for
style 1 scours applies to those of style 2.
However, for style 2 it may be concluded
that the boulder was held in debris-rich ice
until the final stages of till release or,
alternatively, the boulder was not buried
by sand deposition prior to the deposition
of till. However, the problems posed by the
flow competence to transport the boulder
and its apparent suspension above the bed
during scour and fill are again overcome by
resorting to the presence of glacier ice
above a subglacial water flow.

In style 3 scours, the boulder is completely surrounded by till although a small
pocket of lag gravel occurs to its left
(Fig. 12). Sand cross-beds are truncated
abruptly by the scour surface and till rests
directly on this surface. The boulder
extends across the projection of the till/
sand contact and so is partly embedded in
the main body of the till.

Once again a boulder is considered to
have been held in the ice and erosion
occurred as a result of flow constriction
and increased turbulence beneath it. In

Fig. 11. Style 2 boulder scour in which
the boulder remains partly embedded in
till. The pronounced dark layer running
diagonally beneath the boulder is not
primary but is of diagenetic origin.
Truncation of the primary bedding can be
seen beneath this layer.

Fig. 12. Style 3 boulder scour in which
the boulder is completely surrounded by
till but for a minor lag gravel deposit
to the left. Note that the scour base
sharply truncates large-scale trough
cross-stratification in the underlying
sands.

this third case, no subsequent deposition of sand occurred. Rather, the scour was filled by till deposited by mass flow upon melting of the debris-rich ice. It may be that the small pocket of gravel represents a lag produced by winnowing of the earlier till flows.

Scours of style 4 occur in till and are associated with intratill beds. A boulder is embedded in till, which overlies an intratill sand, and rests above a scour (Fig. 13). Gravel lies between the boulder and the scour surface.

Explanation of the style 4 scours requires an englacial cavity or, alternatively, a cavity formed at the junction of already deposited till and overlying debris-rich ice. A boulder held in this overlying ice caused erosion of a scour in the underlying till or debris-rich ice. The formative flow had competence less than that required to move the coarsest clasts in the material being eroded such that a lag gravel accumulated. Sands which were also deposited in the cavity are to be seen at the level of the pick head (Fig. 13). Finally the overlying till was deposited.

Fig. 13. Style 4 boulder scour in which a boulder projects through an intratill sand bed and the scour occurs in the underlying till. A lag gravel is seen between the boulder and the scour surface just to the left of the hammer point.

3.3 Discussion on the boulder scours

The geometric relationships between the boulders and the scours are interpreted above to support a melt-out process for

till deposition. A schematic diagram illustrating the stages in the processes of scour formation and till deposition is presented (Fig. 14). As the location of the boulder was fixed during formation of the scours, sliding or internal deformation of the ice above the cavity are ruled out. The concurrence between preferred clast orientation in the associated till (Fig. 7) and other indicators of the regional ice flow direction supports the conclusion that the ice which held the boulders was of glacial rather than fluvial or lacustrine origin.

The proposed mechanism of scour formation is analogous to scour around bridge piers (Neill 1973) or boulders on the bed of a stream (Karcz 1968). It is also analogous to scours in snow caused by perturbation of wind around obstacles. It is a commonly used engineering practice to use pier width as the length scaling dimension in equations to predict scour at bridge piers (Neill 1973) and, consequently, a linear relationship is expected between obstacle width and scour width. Similarly a linear relationship is expected for wind formed scours around obstacles projecting above snow surfaces. This second expectation was tested by measuring wind scours in snow around the bases of trees, signposts and light standards on the University of Alberta campus. The relationship obtained by least squares linear regression between scour width (S) and obstacle diameter (D) is:

$$S = 0.02 + 2.47 \ D \ (r^2 = 0.93) \quad 4$$

This relationship and the data points are given in Figure 15. Combination of width measures obtained from the literature for scours produced by water flow around pebbles and scour width measures obtained for the boulder scours of this study gives the relationship (Fig. 15):

$$S = 0.03 + 1.79 \ D \ (r^2 = 0.96) \quad 5$$

where S is scour width and D is boulder or pebble diameter. Finally a regression of scour width on boulder diameter for the boulder scours of this study alone gives:

$$S = 0.07 + 1.68 \ D \ (r^2 = 0.94) \quad 6$$

The important conclusion is that each of these relationships is strongly linear and the prediction of a linear relationship for the boulder scours of this study is accurate.

4 CONCLUSIONS

Observations in Sweden and Canada show the importance of melt-out of debris from ice

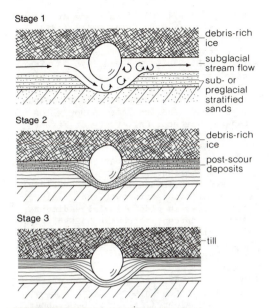

Stage 1

debris-rich ice

subglacial stream flow

sub- or preglacial stratified sands

Stage 2

debris-rich ice

post-scour deposits

Stage 3

till

Fig. 14. The stages of development of a boulder scour and till deposition by melt-out. A style 2 scour is depicted.

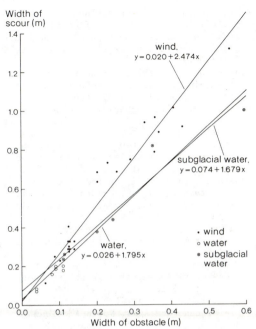

Fig. 15. Relationships between obstacle diameter and scour width for wind produced scours in snow, water produced scours around obstacles (data from Karcz 1968, 1973 and Johansson 1976) and scours produced by subglacial waters beneath boulders.

to the formation of till in the areas of observation. Clearly, this melt-out involves release of large quantities of melt-water and it is not surprising that this had significant depositional and erosional effects. The relationships between boulders and draped laminations in the Sveg tills of Sweden allow estimates of debris concentrations which illustrate the significant basal load of the glacier which deposited these tills. As the same glaciers which produced Sveg tills also produced drumlins and Rogen moraines it is reasonable to consider this high basal load to be a significant factor in the genesis of these landforms.

The boulder scours in the Edmonton area are also associated with melt-out till but with a very different structural appearance to the Sveg tills. The meltwaters associated with the tills of the Edmonton area appear to have occupied larger cavities and flowed with greater competence. Thicker stratified deposits occur and the boulders are associated with active scour rather than passive draping of sorted beds. However, both draped beds and boulder scours may be useful tools in the identification of melt-out tills.

5 ACKNOWLEDGEMENTS

This work was financed by a grant from the NSERC of Canada. Inge Wilson drafted the figures and Fran Litschko arranged the typography.

6 REFERENCES

Åmark, M. 1980, Glacial flutes at Isfalls-glaciären, Tarfala, Sweden, Geol. Fören. Stockh. Förh. 102: 251-259.

Babcock, E.A., M.M. Fenton, & L.D. Andria-shek 1978, Shear phenomena in ice-thrust gravels, central Alberta, Can. J. Earth Sci. 15: 277-283.

Bayrock, L.A. & G.M. Hughes 1962, Surficial geology of the Edmonton district, Alberta Research Council Preliminary Report 62-6.

Boulton, G.S. 1970, On the deposition of subglacial and melt-out tills at the margins of certain Svalbard glaciers, J. Glaciol. 9: 231-245.

Boulton, G.S. 1971, Till genesis and fabric in Svalbard, Spitsbergen. In R.P. Gold-thwait (ed.), Till: a symposium, p. 41-72. Ohio, Ohio State University Press.

Boulton, G.S. 1976, The origin of glacially fluted surfaces - observations and theory, J. Glaciol. 17: 287-309.

Boulton, G.S. 1978, Boulder shapes and grain-size distributions of debris as indicators of transport paths through a glacier and till genesis, Sedimentology 25: 773-799.

Clapperton, C.M. 1975, The debris content of surging glaciers in Svalbard and Iceland, J. Glaciol. 14: 395-406.

Drake, L.D. 1972, Mechanism of clast attrition in basal till, Geol. Soc. Amer. Bull. 83: 2159-2166.

Ehlers, J. & H.-J. Stephan 1979, Forms at the base of till strata as indicators of ice movement, J. Glaciol. 22: 345-356.

Goldthwait, R.P. 1971, Introduction to till, today. In R.P. Goldthwait (ed.), Till: a symposium, p. 3-26. Ohio, Ohio State University Press.

Haldorsen, S. in press, The genesis of tills from a central part of the Scandinavian ice sheet, Boreas.

Hoppe, G. & V. Schytt 1953, Some observations on fluted moraine surfaces, Geogr. Ann. 35: 105-115.

Hughes, T. 1973, Glacial permafrost and Pleistocene Ice Ages, North American Contribution, Second International Permafrost Conference Yakutsk, National Academy of Sciences, p. 213-222.

Humlum, O. 1981, Observations on debris in the basal transport zone of Myrdalsjökull, Iceland. Annals of Glaciol. 2: 71-77.

Johansson, C.E. 1976, Structural studies of frictional sediments, Geogr. Ann. 58A: 201-300.

Jones, M.J. (ed.), 1973, Prospecting in areas of glacial terrain, Instit. Mining and Metallurgy, London.

Karcz, I. 1968, Fluviatile obstacle marks from the wadis of the Negev (southern Israel), J. Sediment. Petrol. 38: 1000-1012.

Karcz, I. 1973, Reflections on the origin of source small-scale longitudinal streambed scours. In M. Morisawa (ed.) Fluvial geomorphology, p. 149-178, Binghamton, State University of New York.

Kathol, C.P. & R.A. McPherson 1975, Urban geology of Edmonton, Alberta Research Council, Bull. 32.

Lawson, D.E. 1979, Sedimentological analysis of the western terminus region of the Matanuska Glacier, Alaska, United States Army CRREL Report 79-9.

Lavrushin, Yu. A. 1976, Structure and development of ground moraines of continental glaciers, Academy of Sciences of the USSR Transactions 288.

Lundqvist, J. 1969, Beskrivning till jordartskarta över Jämtlands Län, Sver. Geol. Unders. Ca 45.

Mills, H.H. 1977, Basal till fabrics of modern alpine glaciers, Geol. Soc. Amer. Bull. 88: 824-828.

Moran, S.R. 1971, Glaciotectonic structures in drift. In R.P. Goldthwaite (ed.), Till: a symposium, p. 127-148, Ohio, Ohio State University Press.

Morris, E.M. & L.W. Morland 1976, A theoretical analysis of the formation of glacial flutes, J. Glaciol. 17: 311-323.

Neill, C.E. 1973, Guide to bridge hydraulics, Roads and Transport Association of Canada, Toronto, University of Toronto Press.

Okko, V. 1955, Glacial drift in Iceland, its origin and morphology, Bull. Commiss. Geolog. Finlande 170.

Paul, M.A. & H. Evans 1974, Observations on the internal structure and origin of some flutes in glaciofluvial sediments, Blomstrandbreen, northwest Spitsbergen, J. Glaciol. 13: 393-400.

Post, A. & E.R. Lachapelle 1971, Glacier ice, Toronto, University of Toronto Press.

Rains, R.B. 1969, Differentiation of till deposits in the Whitemud Creek Valley, Edmonton, Alberta, Albertan Geographer 5: 12-20.

Ramsden, J. & J.A. Westgate 1971, Evidence for reorientation of a till fabric in the Edmonton area, Alberta. In R.P. Goldthwait (ed.), Till: a symposium, p. 335-344. Ohio, Ohio State University Press.

Shaw, J. 1971, Mechanism of till deposition related to thermal conditions in a Pleistocene glacier, J. Glaciol. 10: 363-373.

Shaw, J. 1977, Till body morphology and structure related to glacier flow, Boreas 6: 189-201.

Shaw, J. 1979, Genesis of the Sveg tills and Rogen moraines of central Sweden: a model of basal melt-out, Boreas 8: 409-426.

Shaw, J. 1980, Drumlins and large-scale flutings related to glacier folds. Arctic and Alpine Res. 12: 287-298.

Shaw, J. 1982, Melt-out till in the Edmonton area, Alberta, Canada, Can. J. Earth Sci. 19: 1548-1569

Smalley, I.J. 1981, Conjectures, hypotheses, and theories of drumlin formation, J. Glaciol. 27: 503-505.

Westgate, J.A. 1968, Linear sole markings in Pleistocene till, Geol. Mag. 105: 501-505.

Westgate, J.A. 1969, The Quaternary geology of the Edmonton area, Alberta. In S. Pawluk (ed.), Pedology and Quaternary research symposium. National Research Council of Canada.

Westgate, J.A., L. Kalas, & M.E. Evans 1976, Geology of Edmonton area, Field Trip C-8 Guide Book, Annual Meeting of the Geological Association of Canada.

Dewatering during lodgement of till

ERNEST H.MULLER
Syracuse University, NY, USA

1 SCOPE AND OBJECTIVES

For geotechnical considerations, as well as for purely academic reasons, major efforts in the study of glacial till have been focused upon the processes and environments of till deposition. Properties of unweathered glacial till are, to large degree, relics of the conditions of emplacement. Problems inherent in the handling of till in construction and engineering stem from the manner in which it was emplaced and modified during glaciation. Conversely, interpretation of past glacial environments and conditions of deposition depends largely upon correct understanding of features and characteristics distinguished in ancient tills.

Emplacement of till beneath a glacier involves lodgement of basal load and collapse of the dilatant structure that characterized the material while in transport. Emplacement involves the change from the dilatant condition of basal debris to a firm and compacted state, immobilized as part of the substratum. In ultimate analysis, water plays a role in the deposition of all glacial till. The timing and the manner of escape of basal and interstitial water are crucial in determining the character of the till.

This paper examines the lodgement process, focuses upon dewatering in the lodgement of till, and discusses features of tills which may relate to the dewatering process during lodgement.

2 PREVIOUS WORK

A considerable body of literature deals with dewatering phenomena in hydrodynamically deposited sediments, but little attention has been given to water expulsion during deposition of glacial till.

Many studies bear on the escape of channelized subglacial meltwater, for instance, as it relates to the rapid drainage of "self-dumping" glacial lakes (Nye 1976; Rothlisberger 1972; Shreve 1972; Björnsson 1974). The role of unchannelized subglacial meltwater in glacier flow and surging has been examined in papers by Lliboutry (1979) and Weertman (1979) among others. Based on a drilling program, Hodge (1979) infers that most of the glacier bed of South Cascade Glacier is hydraulically inactive and isolated from subglacial conduits. On the other hand, Walder and Hallet (1979) recognize a nearly continuous network of cavities and channels in the mammilated limestone bedrock recently uncovered by shrinkage of Blackfoot Glacier. In all these papers, the stress is hydrologic. Little consideration is given to relationships of sediment water and pore pressure upon dynamic conditions in the subglacial water layer.

Particularly in the work of the INQUA Commission on Genesis and Lithology of Glacial Deposits, notable progress has been made toward interpretation of depositional conditions on the basis of till characteristics (Dreimanis 1980). Lavrushin's monograph (1976) examines in detail the structure and development of ground moraine, distinguishing till facies and subfacies. Boulton (1968; 1970; 1971) applied information on processes of till deposition by modern glaciers in Spitzbergen and Iceland to interpretation of both lodgement and melt-out tills.

Insight into the lodgement process derives from examination of lodgement tills. Implicit in interpretation of consolidation pressures of glacial sediments is the assumption that pore water escaped freely during compaction and loading. Boulton and

Dent (1974) and Boulton et al. (1974) describe field evidence from Iceland which indicates the role of interstitial water pressure in determining till characteristics. Smalley and Unwin (1968), Muller (1974) and others invoked collapse of the dilatant subglacial debris-water system in the origin of drumlins. Virkkala (1952) conceived the characteristic laminar structure of many tills as resulting from the melting out of intercalated ice during lodgement. Alternatively, Boulton et al. (1974) suggest that platy structure may originate through "consolidation at depths where the glacier stress has been attenuated" below the threshold necessary for continued deformation of the till. Bjelm (1976) describes attempts to simulate the compaction and structures of lodgement till. Sitler and Chapman (1955), Kresl (1965) and Evenson (1970) analyzed microfabrics and microstructures of till - features that record the intimate dynamics of final emplacement as well as direction of glacier flow. Hallet (1976a; 1976b; 1979) recognized in subglacial precipitation of calcium carbonate a sensitive indicator of subglacial environments and of conditions during meltwater escape. These diverse investigations notwithstanding, no systematic treatment has yet been given to evidence bearing on dewatering in the lodgement of tills.

3 LIMITING CASES

Two contrasting situations represent limiting extremes in the range of conditions under which lodgement occurs.

At one extreme, isolated water films develop briefly and microscopically at pressure points between debris particles and contiguous ice grains, but are generally refrozen and transported further. Deposition under these conditions must be considered extraordinary, for prompt and complete regelation is a condition approached only beneath a dry-soled glacier with significant reservoir of cold beyond that required to maintain freezing.

At the opposite extreme, confined glacial meltwater persists interstitially and as a more or less continuous layer during deposition beneath most wet-soled glaciers. The resulting deposit may range from characteristic firm lodgement tills, the compactness of which is attained by subsequent dewatering while still heavily loaded by overriding ice (Boulton et al. 1974), to "soft till" with obscure bedding. Such lodgement tills grade, with increased fluidity of the debris-water system, into

subglacial melt-out tills (Boulton 1970), leeside tills (Hillefors 1973) and subaquatic flow tills (Evenson et al. 1977).

Most lodgement takes place under conditions intermediate between the two extremes characterized above. It involves neither prompt and complete regelation beneath a dry-soled glacier nor total confinement of basal meltwater beneath a wet-soled glacier. Rather, lodgement is a progressive process completed as water is eliminated and the dilatant structure of the sediment collapses.

4 LODGEMENT, A MULTIDIMENSIONAL PROCESS

In proposing the term "lodgment till" (in the present paper spelled "lodgement" following the consistent practice of the INQUA Commission), R.F.Flint (1971:171) implied that "slow pressure-melting of the flowing ice frees the particles and allows them to be plastered one by one and under pressure on the subglacial floor". The "smearing-on" process involves particle by particle lodgement as frictional resistance against the substrate exceeds stress imposed by the overriding ice. A sharp contrast is implied between deforming ice and immobile substratum. The view is essentially two-dimensional, in that the lodgement process is conceived as being planar and instantaneous. While this process suffices for emplacement of clasts, it is a simplistic model for immobilization of matrix material.

Increasingly, the lodgement process is conceived as being multidimensional in that it takes place progressively through the interval of time and the thickness of the zone of deformation in which the dilatant debris-water system is transformed into lodgement till. The value of conceiving the lodgement process in all its dimensions becomes apparent in considering the familiar phenomenon of fissility in till.

Many tills display fissility (Fig. 1) that ranges from a weak, platy structure to a bedded aspect suggestive of lamination, though generally less regular, and typically deformed around clasts (Dreimanis 1976). Lineation akin to microslickensiding can be occasionally distinguished on horizontal fracture surfaces. Although fissility becomes increasingly apparent in the weathered profile of tills, it is generally recognized as reflecting primary textural and structural arrangements that developed during lodgement.

In one view, such fissility has been attributed to "accretion" of layers of drift from the base of a glacier in a kind of

14

Figure 1. Planar fissility in lodgement
till near Constableville, New York. This
platy character, though made apparent by
weathering, reflects inherent sedimentary
structure imposed during deposition.

layer by layer lodgement. Virkkala (1952)
interpreted these layers as being "made by
the successive stagnation of thin, overloaded basal layers of ice in the terminal
zone of the glacier. The base of the moving
glacier shifted upward in jumps, leaving
beneath it drift with interstitial ice."
(Flint 1971:159). According to this view,
fissility in till originates as a vestige
of a structure that was already present in
the active glacier. Elimination of interstitial ice in the form of meltwater is
all that is necessary for the structure to
achieve its final form. Clearly, the lodgement process in such situations is multidimensional, involving both changes with
depth and with time in a zone of lodgement
rather than on a planar surface.

5 CONSOLIDATION, DEWATERING AND PLANAR
  STRUCTURE IN TILLS

Some, but by no means all, of the surfaces
which define fissility in till may have
originated as indicated in the previous
discussion. Alternative hypotheses have
been suggested. Of these, the suggestion
by Boulton et al. (1974) is instructive
particularly because it involves a four
dimensional approach. In accounting for
relationships observed in freshly exposed
lodgement till of Breidamerkurjökull in
Iceland, these authors show lodgement to
be a progressive rather than an instantaneous process.

Till in process of lodgement beneath a
wet-soled glacier is in an expanded or dilatant condition. Water fills the voids
and bears part of the superincumbent load.
Immobilization of this deformable medium
results either from decreased shear stress
or collapse of the dilatant structure due
to escape of pore water, or to a combination of both processes. Boulton et al. suggest that downward attenuation of shear
through the deforming medium diminishes
stress below the threshold necessary to
maintain dilatance. Just such a juxtaposition of fluid till over immobile and consolidated till was observed in the Breidamerkurjökull exposures.

It is here suggested that immobilization
may be induced by loss of pore pressure
and structural collapse due to escape of
interstitial water as the primary factor,
rather than downward attenuation of shear
stress as suggested by Boulton et al. Some
fissile tills are characterized by thin
planar zones that demonstrate differential
erosivity, either with, or as in Figure 2,
without apparent difference in mechanical
composition of the matrix. Where no vertical change in texture is apparent in a
till, differential resistance to erosion
may be ascribed to differences in bulk
density and compactness of the till layers
resulting presumably from differences in
confinement and escape of pore water

Figure 2. Planar structure etched into
relief on weathered till exposure, Chim-
ney Point, on south shore of Lake Ontario,
New York. Coherence varies as a function
of compactness and microstructure.

15

during lodgement. Such changes in the abundance of meltwater in the depositional zone beneath a glacier and of pore pressure and water content within the till are functions of the shifting balance between basal melting and escape of meltwater from the system.

Preconsolidation of ancient subglacial sediments has been interpreted as indicative of the thickness of overriding ice at time of deposition. Although informative, this is risky and may be deceptive because of the assumptions involved (Boulton 1975:296-297). Because load imposed by overriding ice is expressed as a combination of lithostatic and hydrostatic pressure, preconsolidation pressures of subglacial sediments measure total load only if pore water escapes freely from the system. At the time of deposition of basal till, a portion of the load may still be borne by relatively incompressible pore water. This condition persists until the confined water can escape during waning of the overlying glacier and ordinarily in a narrow marginal zone (Clayton and Moran 1974; Boulton et al. 1974).

Basal meltwater is commonly assumed to escape primarily between the glacier sole and the substratum. Such is the case over impermeable bedrock. Where glacier dynamics are the object of investigation, this simplifying assumption is justified (Weertman 1966; Shreve 1975). In an investigation of subglacial lodgement, however, such an assumption is unjustified. Whether transmissivity of the substratum is exceeded by meltwater escape at the glacier sole, or not, the movement of water through and out of the sediment during lodgement significantly affects till properties.

6 SILT CAPS AND CLAY SKINS

Clay skins are familiar pedologic developments, coating clasts and fracture surfaces in illuviated portions of the weathering profile. The existence of related features in unweathered till indicates a possible primary origin during lodgement as well.

Silt films (cutans) on horizontal fracture surfaces are associated with fissility, and with platy or laminar structure in some tills. They range in prominence up to the distinct bedding of basal meltout and glaciolacustrine tills, but commonly are too thin to be considered sedimentary interbeds. Some such occurrences are believed to record variations in the rate at which meltwater accumulated or escaped at

the glacier sole during deposition involving a continuous ice-water contact at the base of the glacier. Other similar occurrences may, instead, represent surfaces of water escape from within the collapsing till structure during lodgement (Fig. 3).

Figure 3. Stratification in till northwest of Sveg, Sweden. Thin layers of sorted sediment intercalated between diamicton units are inferred to have resulted during relatively free movement of meltwater beneath the glacier sole.

Determining the origin of such intercalation in till is often difficult and the conclusions, ambiguous. Sveg till involves intercalation that is too intimate and undeformed to have resulted from repeated oscillatory retreat and readvance. The diamicton clasts appear to be in characteristic attitudes of lodgement and not imbricate as in traction or mud-flow deposition. Deposition beneath continuous and uninterrupted ice cover with varying conditions of meltwater or pore water escape is suggested.

Distinguishing between these latter two origins may be possible on the basis of the relationship of planar structure to enclosed cobbles. Expulsion of water after lodgement is suggested by continuity of planar structures over clasts even though thinned and slightly domed as a result of differential compaction. In contrast, washing by a water film at the glacier sole may result in interruption of clastic layers where they abut against enclosed clasts.

Thin interbeds and clastic films that coat horizontal fracture surfaces are commonly spaced rather uniformly in vertical dimension in manner suggestive of rhythmic repetition of depositional conditions. An hypothesis of seasonal control may be tempting in situations immediately adjacent to the ice margin. The control was, however, dynamic and related either to dewatering or to glacier flow and basal meltwater escape. The controlling changes in glacier flow regime and basal conditions probably were not closely tied to seasonal cycles.

Water passes through a clastic medium by means of interconnecting pore channels. The coarser the particles and the larger the interstices the less is the resistance to flow. In a poorly sorted medium, the favored lines of flow converge against coarse particles. The result may be a washing and sorting of the matrix immediately in contact with enclosed clasts. Thus Boulton (1975:297), for instance, noted an enrichment in fines immediately adjacent to large clasts in undisturbed lodgement till at Breidamerkurjökull. This relationship may be ascribed to favored lines of flow during the dewatering process, though indeed subsequent water movement through the material may also be a factor.

7 MICROSTRUCTURES AND CHEMICAL EVIDENCE

Macrostructures indicative of plastic deformation of till have been described extensively in the literature. Such features range from folds and deformed shear structures to clastic dikes of magnitude readily observed in the field (e.g. Lavrushin 1970). Far less adequate is the attention given to microstructures in till. Yet it is at this scale that details of the lodgement and dewatering processes should be best recorded.

Sitler and Chapman (1955) described characteristic microfabrics of tills from Ohio and Pennsylvania as involving an horizontal microfoliation which they accounted for by rotation and packing of flaky mineral particles. This they attributed to intergranular movement in the till, greatly facilitated by the initial presence of interstitial water. Presumably, the stresses involved were those of lodgement prior to dewatering.

The same authors further described as characteristic a "vein structure" involving more perfect orientation of silt grains than in the microfoliation. This "vein structure" was ascribed by Sitler

and Chapman to dynamic differentiation due to shear. It is here suggested that the imposition of shear after development of microfoliation may have resulted directly from collapse during dewatering.

McGinnis (1968) pointed to the possible role of meltwater in mineralization. More recently, Hallet (1976a; 1976b; 1979) has analyzed the conditions under which subglacial precipitation of calcium carbonate takes place. Calcium carbonate dissolved from bedrock on the stoss sides of subglacial obstacles is concentrated on the lee sides where refreezing takes place. The finely laminated structure of such incrustation suggests repeated episodes of precipitation interrupted periodically by changes in the through-flow or chemistry of subglacial waters. For small cirque glaciers, Hallett (1979;329) suggests that seasonal variation provides the rhythm responsible for this lamination. A dynamic control related to glacier flow velocities, basal melting and meltwater release is suggested as an alternative explanation.

8 CONCLUSION

This discussion has touched upon a few of the kinds of features which may be informative as to the role of dewatering during lodgement of till. Some of the features are common and characteristic of many tills; others are more unusual. All can be better understood when viewed as products of lodgement processes that are multidimensional rather than instantaneous and local. These processes take place through an interval of time and in a discrete till zone beneath the ice for lodgement involves not only embedding of clastic particles in a plastic substratum, but also the immobilization and collapse of a dilatant system through expulsion and escape of interstitial water.

9 REFERENCES

Bjelm, L. 1976, Deglaciation of the Smaland Highland, with special reference to deglaciation dynamics, ice thickness and chronology. Thesis 2, Univ. of Lund, 77p.

Björnsson, H. 1974, Explanation of Jökulhlaups from Grimsvötn, Vatnajökull, Iceland, Jokull 24:1-24.

Boulton, G.S. 1968, Flow tills and related deposits on some Vestspitsbergen glaciers, Jour. Glaciology, 7:391-412.

Boulton, G.S. 1970, The deposition of subglacial and melt-out tills at the margins of certain Svalbard glaciers, Jour. Glaciology, 9:231-245.

Boulton, G.S. 1971, Till genesis and fabric in Svalbard. In R.P.Goldthwait (ed) Till: a Symposium, Ohio State Univ Press p. 41-72.

Boulton, G.S. 1975, Processes and patterns of subglacial sedimentation: a theoretical approach. In Wright & Moseley (eds.), Ice Ages Ancient and Modern, Geol. Jour. Spec. Issue 6:7-42.

Boulton, G.S. & D.L.Dent 1974, The nature and rates of post-depositional changes in recently deposited till from Southeast Iceland, Geog. Annaler 56A:121-134.

Boulton, G.S., D.L.Dent & E.M.Morriss 1974, Subglacial shearing and crushing and the role of water pressures in till from southeast Iceland, Geog. Annaler 56A 125-145.

Clayton, L. & S.R.Moran 1974, A glacial process-form model. In D.R.Coates (ed.) Glacial Geomorphology, Proc. 5th Ann. Geomorphology Symp., S.U.N.Y. Binghamton p. 89-119.

Dreimanis, A. 1976, Tills, their origins and properties. In R.F.Lettett (ed.), Glacial Till, Roy, Soc, Canada Spec. Pub. 12:11-49.

Dreimanis, A., 1980, Terminology and classifications of material transported and deposited by glaciers. In W.Stankowski (ed.), Tills and Glacigenic Deposits, Univ. im. Adama Mickiewicz w. Poznaniu Seria Geografia 20:5-10.

Evenson, E.B., 1970, A method for 3-dimensional microfabric analysis of tills obtained from exposures or cores, Jour. Sed. Pet., 40:762-764.

Evenson, E.B., A. Dreimanis & J.W.Newsome 1977, Subaquatic flow tills: a new interpretation for the genesis of some laminated till deposits, Boreas 6:115-133.

Flint, R.F. 1971, Glacial and Quaternary Geology, New York, John Wiley & Sons.

Hallet, B. 1976a, The effect of subglacial chemical processes on glacier sliding, Jour. Glaciology 17:209-222.

Hallet, B. 1976b, Deposits formed by subglacial precipitation of CaCO3, Geol. Soc. America Bull. 87:1003-1015.

Hallet, B. 1979, Subglacial regelation water film, Jour. Glaciology 23:321-334.

Hillefors, A. 1973, The stratigraphy and genesis of stoss- and lee-side moraines, Bull. Geol. Inst. Univ. of Uppsala 5:139-154.

Hodge, S.M 1979, Direct measurement of basal water pressures: progress and problems, Jour. Glaciology 23:307-319.

I.A.S.H., 1973, Symposium on the Hydrology of Glaciers, Cambridge, 7-13 Sept., 1969, Pub. 95, 262p.

Kresl, R.J., 1964, A preliminary thin-section analysis of glacial till from east-

ern Wells County, North Dakota, Proc. North Dakota Acad. Sci. 18:156-161.

Lavrushin, Y.A., 1970, Reflection of the dynamics of glacier movement in the structure of ground moraine, tr. Consultants Bur. from Litologiya i Poleznye Iskopaemye 1:115-120.

Lavrushin, Y.A. 1976, Structure and development of ground moraines of continental glaciation, Acad. Sci. U.S.S.R. Trans. 288, 237p.

Lliboutry, L. 1979, Local friction laws for glaciers: a critical review and new openings, Jour. Glaciology 23:67-95.

McGinnis, L.D., 1968, Glaciation as a possible cause of mineral deposition, Econ. Geol. 63:390.

Muller, E.H. 1974, Origins of drumlins. In D.R.Coates (ed.) Glacial Geomorphology, Proc. 5th Ann. Geomorphology Symp. S.U.N.Y. at Binghamton, p. 187-204.

Rothlisberger, H. 1972, Water pressure in intra- and subglacial channels, Jour. Glaciology 12:2-18.

Shreve, R.L. 1972, Movement of water in glaciers, Jour. Glaciology 11:205-214.

Sitler, R.G. & C.A.Chapman 1955, Microfabrics of till from Ohio and Pennsylvania, Jour. Sed. Pet. 25:262-269.

Smalley, I.J. & D.J.Unwin 1968, The formation and shape of drumlins and their distribution and orientation in drumlin fields, Jour. Glaciology 7:377-390.

Virkkala, K. 1952, On the bed structure of till in eastern Finland, Comm. Geol. de Finlande Bull. 157:97-109.

Walder, J & B. Hallet 1979, Geomoetry of former subglacial water channels and cavities, Jour. Glaciology 23:335-346.

Weertman, J 1966, Effect of a basal water layer on the dimensions of ice sheets, Jour. Glaciology 6:191-208.

Weertman, J 1979, The unsolved general glacier sliding problem, Jour. Glaciology 23:97-115.

Till genesis and the glacier sole

ERNEST H.MULLER
Syracuse University, NY, USA

1 INTRODUCTION

Knowledge of what goes on at the sole of a glacier is as difficult to obtain by direct observation as it is essential for understanding of the genesis of basal tills. This paper is intended to examine the concept of the glacier sole, to focus on processes at the base of a glacier during deposition, and to suggest implications for till classification.

The recent literature on subglacial processes is too extensive to permit of brief review. It will suffice instead to call attention to a few volumes containing relevant reports. Papers presented at the International Glaciological Society's "Symposium on Glacier Beds: the Ice-Rock Interface" (Glen et al. 1979) center particularly on the role of basal meltwater in sliding and the glacier-surge mechanism. Subglacial deposition is the focus of landmark papers by Boulton (1970, 1975) and by Boulton et al. (1974a, 1974b). Till symposium volumes (Goldthwait 1971; Legget 1976) and numerous papers stimulated by activities of the INQUA Commission on Lithology and Genesis of Quaternary Deposits have further developed the subject (Schlüchter 1979; Stankowski 1980). It was, in fact, the difficulty in reaching agreement on terminology during Commission discussions which stimulated the present effort at clarification and resolution of conflicting views.

2 THE GLACIER SOLE

Further discussion requires clarification as to just what is meant by the glacier sole, a term frequently used but seldom defined. The "Glossary of Geology" (Bates and Jackson 1980) offers several alterna-tives for the geological use of the word "sole".

According to the first definition, the sole is "the under surface of a rock body" (Fig. 1). Clearly, a glacier is a large rock body and the definition is therefore simple and straightforward. The sole of a glacier is the underside of the glacier.

But, reading on, one finds that the sole is "the middle and lower parts of the shear surface of a landslide". Although a glacier is not usually thought of as a landslide, the role of basal slip in glacier movement certainly justifies analogy to a slowly moving slide mass. If one questions how a shear surface, by definition a planar feature, can have a middle

Figure 1. Sole of San Rafael Glacier, austral Chile. Minor moraine ridges and patches of planar substrate result from oscillatory thrusting of terminus with sole shearing up and over subsole debris.

and a lower part, this definition is, at
best, ambiguous; at worst it is impossible.

The third definition offered by the
"Glossary of Geology" reads that the sole
is "the lower part, or basal ice of a glacier, which often contains rock fragments,
appears dirty, and is separated from clean
ice by an abrupt boundary."

We are left in uncertainty as to whether
the sole of a glacier is really a surface
of contact, or a zone at the base of the
ice. If the former, where is it? If the
latter, where does it begin and end?

For the glaciologist, distinction between
debris-laden near-basal ice and the overlying relatively clean ice is of importance. In such case terminology such as
that employed by Lawson (1979) is useful.
In investigation of Matanuska Glacier,
Alaska, Lawson distinguished between the
"dispersed" and "stratified facies" of a
"basal zone" beneath the overlying "englacial zone". Such usage is clear, informative and eliminates the need for two definitions of the glacier sole, one of which
is planar, while the other implies a vertical dimension. The term "glacier sole"
ought not to be required to fill double
duty in these dissimilar roles.

For the glacial geologist, the surface
that separates glacier from substrate is
the sole of the glacier, its under surface
(Fig. 2). By delivery across this basal
surface, material that had been in transport ceases to be "glacial debris" and becomes part of the underlying sediment.
The glacier sole, then, may be defined as
the contact between still-frozen and plastically deformed ice with or without contained debris, and an underlying stratum

which either has not been glacially entrained or having undergone glacial transport has been subsequently melted.

Acceptance of the definition
proposed above affords clear criteria for
distinction between sole and substrate for
modern glaciers, except for stagnant, cold
or dry-based glaciers over frozen ground.

3 SOLE ON ICE

In exceptional and somewhat ephemeral
situations, temperate glaciers may expand
over permanently frozen ground. The condition may have been more extensive in
times past. If, for instance, continental
ice sheet generation was more ponderous
than the spread of cryogenic climates,
permafrost might be developed ahead of the
expanding ice margin. Even in such case,
the slow rise of geothermal heat coupled
with effective insulation of glacial cover
must have promoted gradual thawing of the
overridden permafrost.

Where average annual air temperature is
sufficiently low, or where glacier cover
is inadequately thick, the freezing iso-
therm may penetrate deeply into the underlying earth. This is typically the situation beneath cold, dry-based glaciers as
beneath submarginal portions of high latitude ice sheets and ice caps.

Ice flow may cease and glacier stagnation
occur long before deposition of the entrained sediment. Ice and sediment which
have undergone glacial transport remain
glacial during stagnation and until melting occurs. Where stagnating glacier ice
overlies frozen ground, the simple criterion for distinguishing the glacier sole as
the base of the frozen material is inade-
quate. Distinction must then be made between material which has been glacially
transported and that which has not.

4 DEFORMATION TILL AND DEFORMED SUBSTRATE

Till containing marked development of
folded, faulted, sheared and brecciated
structures has been called deformation till
(Dreimanis, 1976). Some of the structures
in deformation till result from aborted
entrainment involving only incipient homo-
genization of the basal load; other structures are produced by differential stresses
during transport or subsequent deposition.

Stresses induced by overriding ice may
well deform the substrate to depths of tens
of meters below the glacier sole. Simple
deformation of bedrock does not constitute
entrainment and does not convert it into

Figure 2. Sole of San Rafael Glacier comprising fluted ceiling of ice cave. The
ice is severely stressed and freed of basal
sediment where it has sheared up and over
a bed obstruction.

till. Ice-thrust features do not by the simple act of dislocation become till. Some very large rock and drift masses may be contained within till, but they are properly part of the till sheet only if they themselves were frozen into and transported as part of the glacier.

The distinction between subglacially brecciated bedrock and overlying till is sometimes difficult. The position of the lowest exotic, presumably glacially transported component is often taken as marking the contact between crushed and broken rock below and till above. The characteristic fissility of loamy tills may in other situations afford a criterion even where the till incorporates only local rock material.

Because till is sediment transported and deposited by glacial ice, use of the term "deformation till" implies that the material was for some interval a part of the glacier, frozen into and moving with the deforming ice.

5 GLACIER SOLE IN DEPOSITIONAL MODE

Defining the glacier sole as the basal surface of still frozen ice and debris which have undergone glacier flow carries the further interesting and fundamental implication that in ultimate analysis essentially all till is water deposited. The role of meltwater is essential to the movement of material through the glacier sole to the substrate. Whereas in some cases this involves only the briefest melting and regelation of ice at pressure points between clasts, much more often it results in a slurry, a dispersed sediment-water system the subsequent development of which determines the nature of the subglacial till.

Those who have worked around the terminus of a warm glacier during early melt season may need no further introduction to the slurry which can be found at the ice margin. Boulton et al. (1974) demonstrated the existence of this slurry as a transition between glacial debris and consolidated lodgement till. The fluidity of subglacial drift enables it to deform into flutes and ridges and fracture fillings in the glacier sole. Confined under glacier loading, such slurry has been credited with the polishing and scouring of hard rock. The role of pore pressure in sub-glacial thrusting is generally accepted. Even though we cannot directly observe what goes on at the base of a glacier without disrupting the system, we must recognize and seek to understand the role of the basal slurry in the glacial depositional process.

In glacial erosion, an initially immobile rock structure, subjected to stress enters an expanded state as a result of microfracturing (dilatation) before yielding. During transport it continues in dilatant state maintained by the ice matrix. The depositional process has been simplistically conceived as taking place instantaneously at a point in time and space. Rather, basal deposition is a multidimensional process, taking place through a discrete interval of time in a three-dimensional zone and involving a number of steps (Boulton et al 1974).

The depositional process involves a) the movement of glacial debris into vulnerable positions at or projecting through the glacier sole, b) detachment, meaning transfer from the enfrozen to the unfrozen state, c) subsequent disturbance or flow in the dispersed system, and d) final positioning that usually results from compaction and dewatering.

Movement into position vulnerable for detachment occurs while the materials are still englacial and the sediment remains glacial debris as long as it is contained in glacial ice, whether the ice is still flowing or has become stationary.

Detachment takes place as the resisting stresses imposed by the substrate exceed the containing forces exerted on the clastic particle by partially enclosing ice. This is the lodgement process as generally conceived and as described by Flint (1971). It implies the lodgement of a clast into a relatively unyielding bed and it probably takes place. The displacement of debris through the glacier sole into the subglacial sediment-water system probably involves this kind of lodgement only in the broad sense. Typically, however, the bed in depositional glacial mode is still mobile but involves "a layer of active subsole drift ... which intervenes between bedrock and ice sole, is partially to completely ice free, and is mechanically and visibly distinct from the overlying debris-laden ice." (Kamb et al. 1979).

The subsole drift, in dispersed state and mobile may exist beneath either flowing or stationary ice. Its final immobilization and positioning take place when overriding ice ceases to flow, or when the shear strength of the sediment exceeds the imposed stresses. This latter may result either when accumulating drift thickens enough to attenuate stress deeper in the drift, or when the dispersed sediment-water system collapses through elimination of interstitial water.

6 IMPLICATIONS FOR TILL CLASSIFICATION

Fundamentally, glacial debris is detached by being melted out, or by being restrained by the substrate more strongly than it is contained by the ice. The action of one of these processes to the total exclusion of the other must be extraordinary. Accordingly, the resulting tills fall somewhere in the spectrum between end members which involve either pure lodgement or pure meltout. Pure meltout should be sought only where detachment took place after glacier flow had ceased. At the other end of the spectrum, pure lodgement involves only mechanical emplacement and is probably applicable only to individual clasts, not to a sediment mix such as comprises till.

Palimpsest structure reflecting the distribution of debris while it was still englacial may be preserved after deposition. Shaw (1977) has suggested that such preservation is favored in arid, polar environments by the relative importance of sublimation.

Among till structures to which such palimpsest character is ascribed, perhaps the most common is planar fissility. Recalling a photograph of the debris-charged basal ice of a Greenland outlet glacier, Virkkala (1952) ascribed this fissility to a vestigial englacial structure preserved during the melting of basal ice. This hypothesized relationship carries the further implication that planar fissility should be associated with meltout till. On the contrary, however, planar fissility has been considered an identifying characteristic of lodgement till. Hypotheses that account for fissility in terms of lodgement (sensu lato) relate it either to downward attenuation of stress through accumulating drift (Boulton et al. 1974) or collapse of the dispersed sediment-water system.

In the continuum between pure meltout and pure lodgement till, boundaries can be only artificially drawn and based upon arbitrary criteria. On such basis it has proven informative to distinguish as meltout tills those which retain inferred palimpsest englacial structures and evidence of basal meltwater action. Similarly lodgement tills are those for which dominantly mechanical emplacement and delayed dewatering of the dispersed system are inferred.

Although boundaries are recognized as artificial and criteria as arbitrary, the increased understanding of the processes of deposition and the resulting characteristics of till is adequate justification for the effort involved in their adaption.

7 REFERENCES

Bates, R.L. & J.A. Jackson 1980, Glossary of Geology, 2nd ed. American Geological Institute, Falls Church, Md, 749.

Boulton, G.S. 1970, The deposition of subglacial and melt-out tills at the margins of certain Svalbard glaciers, Jour. Glaciology 9:231-245.

Boulton, G.S. 1975, Processes and patterns of subglacial sedimentation: a theoretical approach, in Wright, H.E. and Moseley, (eds.), Ice Ages Ancient and Modern, Geol. Jour. Spec. Issue No. 6:7-42.

Boulton, G.S. & D.L. Dent 1974a, The nature and rates of post-depositional changes in recently deposited till from southeast Iceland, Geog. Annaler 56A:121-134.

Boulton, G.S., D.L. Dent & E.M. Morriss 1974, Subglacial shearing and crushing and the role of water pressures in tills from southeast Iceland, Geog. Annaler 56A:135-145.

Dreimanis, A. 1976, Tills: their origins and properties, pp. 11-49 in Legget, R.F., Glacial Till. Royal Society of Canada Spec. Pub. 12, 412p.

Flint, R.F. 1971, Glacial and Quaternary Geology. John Wiley & Sons, N.Y. 892p.

Glen, J.W., R.J. Adie, D.M. Johnson, D.R. Homer & A.D. MacQueen 1979, Symposium on Glacier Beds: the Ice-Rock Interface, Ottawa, 15-19 August 1978. Jour. Glaciology 23(89):441p.

Goldthwait, R.P. 1971, Till, a Symposium. Ohio State Univ. Press, 402 p.

Kamb, W.B., H.F. Engelhardt & W.D. Harrison 1979, The ice-rock interface and basal sliding process as revealed by direct observation in bore holes and tunnels; Jour. Glac. 23:416-419.

Lawson, D.E. 1979, Sedimentological analysis of the western terminus region of the Matanuska Glacier, Alaska, Cold Regions Research Engineering Laboratory Report 79-9, Hanover, N.H. 112 p.

Legget, R.F. 1976, Glacial Till. Royal Society of Canada Spec. Pub. 12, 412 p.

Schlüchter, Ch. 1979, Moraines and Varves. A.A. Balkema, Rotterdam, 441 p.

Shaw, J. 1977, Tills deposited in arid polar environments, Canadian Jour. of Earth Sciences 14:1239-1245.

Stankowski, W. 1980, Tills and Glacigenic Deposits. Universytet im Adama Mickiewicz w. Poznaniu Seria Geografia No. 20.

Virkkala, K. 1952, On the bed structure of till in eastern Finland. Comm. Geol. de Finlands Bull. 157:97-109.

Sedimentologic and palaeomagnetic characteristics of glaciolacustrine diamict assemblages at Scarborough Bluffs, Ontario, Canada

NICHOLAS EYLES & CAROLYN H.EYLES
University of Toronto, Ontario, Canada

TERENCE E.DAY
University of Western Ontario, London, Canada

1. INTRODUCTION AND RATIONALE

This is a progress report on process-oriented sedimentological research on the status and genesis of Quaternary tills. Recent experiences with three supposed Early and Middle Wisconsin glacial units (the Sunnybrook, Seminary and Meadowcliffe Tills) and intervening sands at the classic Late Quaternary site of Scarborough Bluffs, Ontario are described.

With the rapid expansion of process sedimentology since the late 1960's the sub-discipline of glacial sedimentology has been strengthened by observations of currently active depositional processes at glacier margins of a wide variety of glacier thermal characteristics and settings. Till classifications have developed rapidly based both on field observations and a priori theoretical considerations. These classifications are not easily applied to older sequences as the latter have overwhelmingly been described in the past by reference to laboratory-derived physical and mineralogical analytical data reflecting concern with stratigraphic correlation rather than with genesis and a depositional systems approach to stratigraphy (Mitchum et al., 1977). Rarely have appropriate field lithofacies descriptions been made creating a situation where for many Quaternary tills and pre-Quaternary tillites there is considerable confusion as to what is being described or sampled.

This paper outlines an approach to the description and interpretation of sequences based on lithofacies description employing lithofacies codes, analysis of stratigraphic sequence (vertical profile analysis) and derivation of remanent magnetism.

Employing this approach at the Scarborough Bluffs demonstrates that the multiple 'tills' there are glaciolacustrine diamicts formed by pelagic rain-out and ice rafting separated by sandy deltaic lithofacies.

2. LITHOSTRATIGRAPHY OF THE SCARBOROUGH BLUFFS

The Scarborough Bluffs sequence along the northern Lake Ontario shoreline is widely recognised to contain a major North American Late Quaternary succession and has even been referred to as an 'ideal world type section for the continental, glacial chronostratigraphy of the Last Ice Age' (Morner 1973 p. 200). Sections over 100 m high are considered to exhibit the most complete record yet known of last glaciation deposits (Dreimanis 1977) with sequences below current lake level that may be correlative with the so-called interglacial (Sangamon) Don Formation and possible Illinoian till in the adjacent Don Valley (Williams et al., 1981). The Scarborough sequence lies on a bedrock suface, varying from 6 to 12 m below current lake level along the Bluffs sloping gently to the southeast with a mean gradient of about 1:125 (Lewis and Sly, 1971).

Karrow (1967, 1969) identified the Bluffs lithostratigraphy as being composed of lowermost Scarborough Formation extending up to 45 m above lake level, some 60 m of Early and Middle Wisconsin tills and interstadial sands capped by the Late Wisconsin Halton Till (Fig. 2). This paper is concerned with the character and genesis of the Early and Middle Wisconsin sequence.

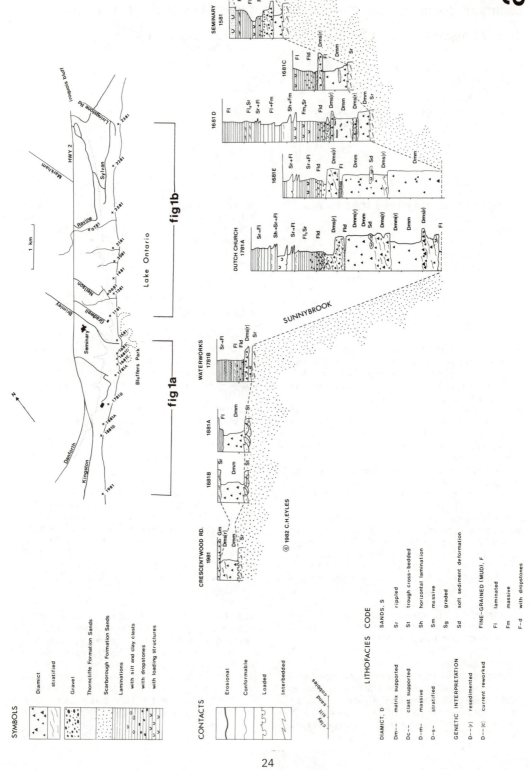

a

SYMBOLS

Diamict
stratified
Gravel
Thorncliffe Formation Sands
Scarborough Formation Sands
Laminations
with silt and clay clasts
with dropstones
with loading structures

CONTACTS

Erosional
Conformable
Loaded
Interbedded

clay
silt
sand
cobbles

LITHOFACIES CODE

DIAMICT, D
Dm-- matrix supported
Dc-- clast supported
D-m- massive
D-s- stratified

GENETIC INTERPRETATION
D--(r) resedimented
D--(c) current reworked

SANDS, S
Sr rippled
St trough cross-bedded
Sh horizontal lamination
Sm massive
Sg graded
Sd soft sediment deformation

FINE-GRAINED (MUD), F
Fl laminated
Fm massive
F-d with dropstones

© 1982 C.H. EYLES

24

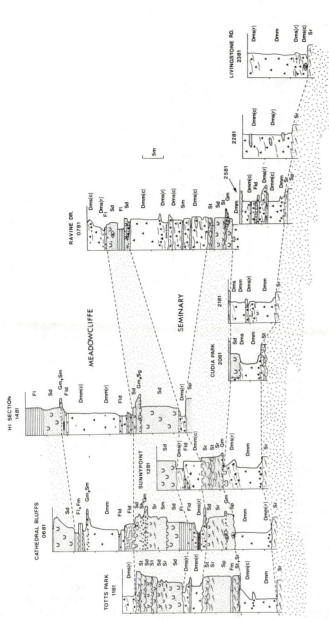

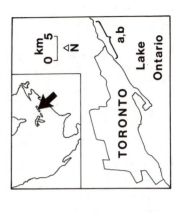

Figure 1a,b. Vertical profiles along Scarborough Bluffs employing the diamict lithofacies code. Lithostratigraphic divisions of Karrow (1967) are depicted alongside the logs; detailed logging shows that these unit boundaries, as previously defined, are extremely difficult to place emphasing possible continuity of deposition. Vertical profiles west of Totts' Park are truncated by the bench of the Late Wisconsin Lake Iroquois. Note the absence of sheared and graded diamicts. The full lithofacies code for diamict description is reviewed in Eyles et al., In Press.

25

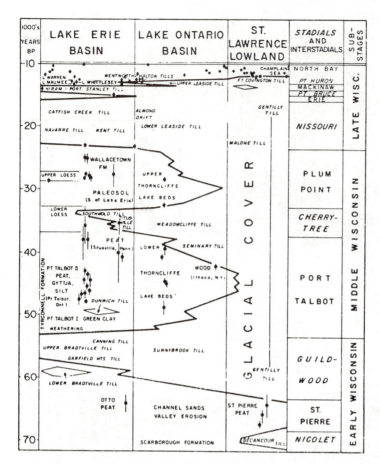

1000's YEARS BP	LAKE ERIE BASIN	LAKE ONTARIO BASIN	ST. LAWRENCE LOWLAND	STADIALS AND INTERSTADIALS	SUB-STAGES	

Figure 2. Time-distance diagram from Dreimanis (1977). Glacial sediments are depicted by slanted lettering, nonglacial events, sediments and lake stages by vertical lettering. The ice margin is shown as a heavy line, radiocarbon dates by black dots, with standard deviations.

The Scarborough Formation consists of a 30 m thick lower unit, of laminated silts and clays probably deposited as a pro-delta sequence in deeper lake water. The unit is overlain sharply by a prograding deltaic sand unit, with an average thickness of 15 m, deposited by a low-sinuosity fluvial system flowing southward into a lake basin. Deformation and injection structures resulting from subaqueous slumping and dewatering are ubiquitous (Yagishita et al., 1981). The height of the delta surface indicates a lake water body at least 45 m deeper than at present. Terasmae (1960) inferred a boreal climatic regime from pollen and plant macrofossils with a mean annual temperature some 5.5°C lower than the present value of 7.5°C whilst a sub-arctic climate has been inferred for the upper part of the Scarborough Formation by Morgan (1972). The top of this deltaic wedge shows broad channels up to 60 m deep such as at the Dutch Church Section (Fig. 1). It has been suggested that these channels result

from lowered lake levels (Dreimanis, 1977), but channel avulsion on the surface of the delta body or subaqueous erosion by strong traction currents provide other possible explanations (Weimer, 1981). Above the Scarborough Formation delta Karrow (1967) described three units which he identified and formalized as tills (the Sunnybrook Till, Seminary Till and Meadowcliffe Till) separated by so-called interstadial lacustrine sands of the Thorncliffe Formation. The succession was interpreted to be the result of three advance/retreat phases of grounded ice (Fig. 2).

The Meadowcliffe and Seminary units are assigned to the Middle Wisconsin on the basis of radiocarbon dates on resedimented organics in the lacustrine Thorncliffe sands. The Sunnybrook is assigned to the Early Wisconsin (Fig. 2) as is the Scarborough Formation (Karrow, 1974; Berti, 1975; Dreimanis, 1977; Dreimanis, 1982) on the basis of infinite radiocarbon ages.

26

Figure 3. Massive structureless diamict (pebbly mud: Dmm).

Figure 4. Matrix-supported massive diamict (Dmm) overlain by matrix-supported stratified and resedimented diamict (Dms(r)) within the Sunnybrook unit at the Dutch Church (Section 1681D Fig. 1a).

Figure 5a,b. Structure of the
Sunnybrook diamict assemblage at
Sylvan Park (Sect. 2581 Fig.
1b). Note boulder concentrate
and interbedded contact with the
Scarborough Formation at the
base (Fig. 13), large flow nose
and stratified resedimented
diamict (Dms(r)). See Fig. 1
for lithofacies code.

3. LITHOFACIES IDENTIFICATION AND CODES

Indiscriminate use of the term 'till' for poorly sorted admixtures of gravel, sand and mud is a major obstacle to environmental interpretation. A consensus among workers familiar with modern depositional complexity suggests that the term till should be reserved strictly for diamicts deposited directly by glacier ice and which experience no disaggregation by normal non-glacial sedimentary processes either during or after deposition. Any identification of till in the field should ideally be accompanied by a list of criteria by which that till is identified and what genetic type is present. Till classifications are evolving rapidly but are unfortunately composed of intimate blends of direct field observation and theoretical constructs and are further developed than the establishment of rigorous field criteria by which till types can be identified through some progress is being made in this direction. (INQUA: Comm. Gen. Lith. Quat. Dep. Workgroup No.9). Birkeland et al., (1979) stressed the need for objective description of glacigenic sequences and a standard lithofacies code has been developed for this purpose. This code, which is described in detail elsewhere (Eyles et al., In Press), allows objective appraisal and notation of both field sequences or drill-core. The term till is not employed in the code as it is a genetic term; in this objective lithofacies scheme the term diamict is used to refer to any poorly sorted gravel-sand-mud admixture **regardless of depositional environment**. The code can thus be employed for any sequences that contain poorly-sorted sediments. The code is based on that of Miall (1977, 1978) who showed that the majority of fluvial deposits, including glaciofluvial outwash, are satisfactorily described using standard lithofacies types each with its own coding. The next step, of identifying lithofacies assemblages, vertical lithofacies sequences and lateral lithofacies relationships for the purpose of interpreting depositional environments and erecting facies models, can then proceed. Vertical lithofacies assemblages (or associations) are particularly diagnostic in the interpretation of glacial diamict sequences and can be examined statistically in several forms of Markov Chain analysis.

The structure and content of the lithofacies code is depicted in Fig. 1b. The code designator D represents diamict.

The second part of the code, used as a mnemonic, refers either to matrix-support (m) or clast-support (c). The third letter refers to massive (m), stratified (s) or graded (g). A frequently encountered situation with regard to glacial marine and lacustrine diamicts is that where massive units show subsidiary stratification such as minor textural inhomogeneities; a problem has then arisen in defining the limit at which the extent of such interbeds justifies use of the term stratified. A guideline for discrimination is that stratified units can be defined as possessing marked textural differentation over greater than 10% of unit thickness.

The last letters of the code, emphasizing aspects of diamicts that are useful in environmental reconstruction, are set in parentheses to clearly identify their genetic connotation. The term sheared (s) relates to evidence of shearing or glacier traction during deposition and commonly takes the form of shear banding, oriented glacially-shaped clasts, bedding plane shears and incorporation of substrate materials (Eyles et al., 1982). The term resedimented (r) is used for diamicts that indicate deposition by sediment gravity flows or slides having several support mechanisms and resulting in variable upward grading characteristics (Lowe, 1979; Nardin et al., 1979). Current reworking (c) is also a significant sorting process in many diamict sequences, e.g. those that accumulate on lake or sea floor by pelagic fall-out coupled with ice rafting of coarser clastics (Kurtz and Anderson, 1979; Link and Gostin, 1981). Depending on the scale of observation, current bedded lamellae can be recorded either separate from the diamict or, two intimately related lithofacies can be depicted by employing a comma (,) (e.g. Dmm,Sr). The scale of analysis clearly will dictate where to place bed limits; in our experience when logging many hundreds of metres of drill core, commas are used but given more detailed work individual beds can be depicted and logged. The type of bed contact is of crucial importance (Fig. 1b).

Note in Fig. 1b that hyphens are employed to indicate that diamicts show several possible combinations of internal characteristics. A diamict lithofacies code used in combination to identify 18 major lithofacies types coupled with the fluvial lithofacies code of Miall (1977, 1978) is found to be adequate for most sedimentological investigations. Additional information that should be noted on field logs is colour (soil colour chart) clast

Figure 6 a) Resedimented stratified diamict (Dms(r)) at Sylvan Park (Section 2581; Fig. 1b). Knife handle for scale. b) cut slab of same. Label is 1.5 cm long. Note flow breccias of imbricated pebbly mud and silt clasts and floating angular silt clasts.

Figure 7. Massive resedimented diamict lithofacies (Dmm(r)). Edge of knife hilt for scale. Note subtle textural differentiation, abundant angular silt clasts. There has been limited mixing of the two pebbly muds during flow. Sunnybrook unit; Dutch Church section 1681E, fig. 1a. This structure is only seen after detailed cleaning and logging of small section thicknesses. More rapid logging can only identify Dmm (fig. 1a).

lithology, clast shape and orientation, packing, jointing, weathering and other criteria reflecting the particular investigation in question e.g. simple geotechnical properties or the precise location of sample sites.

4. VERTICAL PROFILE ANALYSIS OF THE SCARBOROUGH BLUFFS

Fig. 1 shows representative lithofacies logs of the Scarborough Bluffs employing diamict and fluvial lithofacies coding. A vertical profile analysis of these sequences consists of an examination of the following:

4.1 Diamict Lithofacies

Considerable lithofacies variability exists within the Sunnybrook, Meadowcliffe and Seminary units. Massive diamict (Dmm), massive diamict masses with evidence of resedimentation (Dmm(r)) or current reworking (Dmm(c)), and stratified diamicts (Dms(r)/Dms(c)) predominate. Diamict units indicating deformation accompanying deposition at an ice base by lodgement

processes (Dmm(s); lodgement till) are conspicuously absent (Fig. 1). Detailed logging of the units shows that a pebbly clayey-silt texture is most common; the units cannot be characterized by averaged textural data given considerable lithofacies variability (Fig. 1; Eyles and Eyles, 1983).

The Sunnybrook unit exhibits massive facies (Dmm) with variable clast component (Fig. 3), sandwiched between stratified resedimented units (Fig. 4; Dms(r)) with frequent flow noses of various size (Figs. 5,6) and good clast fabrics as a result of flow; two such cycles are exhibited at the Dutch Church section (Fig. 1a). Massive resedimented units (Dmm(r)) are common (Fig. 7). Included lenses of starved rippled sands, frequently loaded, within massive diamict indicate traction current reworking during diamict deposition (Dmm(c)). Note the irregular boulder concentrate at the base of the Sunnybrook at the Sunnypoint and Sylvan Park sections (Figs. 1b,5). This is not a boulder pavement of oriented, striated and shaped clasts; instead clasts are variably

31

oriented and commonly well rounded.

The Seminary unit is frequently stratified as a result of resedimentation (Dms(r)) and at some localities (e.g. Section 0781, Fig. 1b), it has been reworked by traction currents to form crudely stratified sandy diamicts with interbeds of rippled and cross-bedded sands. At no point along its outcrop is the Seminary unit entirely massive (Dmm).

The Meadowcliffe unit is predominantly massive (Dmm) with minor current reworking (Dmm(c)) and resedimentation (Dmm(r)), and exhibits fewer large clasts, compared to the other units.

A particularly valuable approach to the study of these diamicts has been the use of X-ray and macro-photography on thin (<4 cm) slabs prepared from moist samples. Consequently we would emphasize the need for careful attention to detail in order to distinguish truly massive diamict units (Fig. 3) from apparently massive units (Fig. 7) that show fine structure resulting from either resedimentation (Dmm(r)) and episodic current activity (Dmm(c)).

4.2 Sandy Lithofacies

Intervening sand lithofacies are predominantly fine-grained with either ripple lamination, horizontal or trough cross-bedding. The most frequent facies is massive, or crudely-bedded silty-sands with abundant dish structures (Sd) and ubiquitous evidence of deformation accompanying rapid dewatering in a subaqueous environment.

4.3 Laminated Lithofacies

Rhythmically laminated units grading from find silty sand to mud are a conspicuous component of the Bluff's stratigraphy and occur within, on top and at the base of diamict units (Figs. 1,8,9). The occurrence of undeformed laminae within the diamict units is critical (Fig. 9a). These laminations frequently contain ice-rafted clots of diamict and dropstones, with locally abundant silt clasts. Laminations appear to be of turbidite origin and are thickest in basin infillings draping the Sunnybrook diamict assemblage in the Dutch Church Valley (Fig. 1a) and within the Seminary unit at Sections 1181, 0681,1281 (Fig. 1b). These laminations have been described by Antevs (1928) as varves. Logging indicates that the term

'varve' as applied to these units is incorrect; these rhythmites are largely composed of single graded units from fine sand, silt to clay (C, D, E components of a Bouma sequence) and do not show the sharp division between two texturally distinct parts (summer silt, winter clay) nor the consistent winter clay layer thickness reported by Ashley (1975). The penodicity of deposition is under further investigation employing secular variation curves of natural remanent magnetism (e.g. Verosub, 1979) and pollen content.

4.4 Contacts

Conformable contacts between diamicts and laminated lacustrine silty-clays, (Fig. 9a) are significant. Particularly so is the soft sediment deformation, in the form of loaded diamict tops and bases at the contact with associated sands (Fig. 11). Between Cudia Park and the Sunnypoint Section (Fig. 1b) large pillows of silty-sand, up to 3.5 m deep and 10 m wide are loaded into the top of the Sunnybrook diamict (Fig. 12a). The loaded sands show ubiquitous dewatering structures (Fig. 12b) and upward narrowing diamict dykes are intruded into the sands. Note the frequent interbedded character of the contact 'zone' at the base of the Sunnybrook and the underlying Scarborough Formation (Figs. 1, 5, 13a,b). Local erosion at the base of resedimented diamicts is evident (Fig. 10).

4.5 Lithofacies Associations and Geometry

The intimate, conformable and frequently interbedded relationship between diamict units and laminated lacustrine silty-clays is striking. The geometry of this association, revealed in three dimensions along the Bluffs and in ravines, shows diamict units, along with laminated silts and clays, largely as planar tabular strata with an absence of cross-cutting erosional geometry. The lowermost diamict, the Sunnybrook, fills up and flattens out the irregular topography of the underlying deltaic sand body of the Scarborough Formation and it is in the Sunnybrook that resedimented diamict lithofacies are most significant, e.g. the Dutch Church Section (Fig. 1a).

5. PALAEOMAGNETIC INVESTIGATIONS

To the sedimentologist the measurement of natural remanant magnetism is a rapid technique offering particularly useful information as to genesis of diamicts. A major contrast can be established betweeen

32

Figure 8 a). Turbidite sequence
grading from horizontally-lamin-
ated fine sands to rippled (loaded) silty sands and weakly
laminated muds representing overall waning density current flow
with minor higher energy pulses
(Sh→Sr→Fl; Dutch Church section
1681D, fig. 1a).

b) Graded turbidite unit with
massive muds overlying horizontally laminated silty sands
and muds (Fl→Fm). Discrete
rippled silty sand laminae have
been loaded (Dutch Church section
1681D, fig. 1a). The bar is
2 cm long.

Figure 9 a). Undeformed
laminated silty-clay breccias
grading to mud occuring within
matrix supported massive
diamict of the Sunnybrook unit
at the Dutch Church (Section
1681D: fig. 1a).

b) Matrix-supported stratified
resedimented diamict (Dms(r))
of the Meadowcliffe unit passing
up into laminated silty-clays of
probable turbidite origin (Hi
section; 1481 Fig. 1b).

Figure 10. Contact between resedimented weakly stratified diamict (Dms(r)) of the Meadowcliffe unit and underlying laminated silty-clays (Fl). Note the trail or cloud of silty-clay picked up by the flow and subtle textural banding within the diamict.

diamicts that are deposited as dense fluids (i.e. debris flows, mud flows or where diamict accumulates by settling through a water column) and those units accumulating by the lodgement of sedimentary particles against the underlying substrate at a glacier base (lodgement till). Lodgement tills are reported to have a natural remanent magnetism that is sheared-out and dispersed as a result of glacial traction and shear rotation of magnetic particles whereas unhindered orientation of magnetic particles occurs in a dense fluid that slowly dewaters and compacts (Dreimanis,

1976). The rate of dewatering and compaction determines the age difference between deposition and development of a fixed natural remanent magnetism.

5.1 Sampling Techniques

Sampling consists of removing oriented sediment blocks (20x10x10 cm) at natural moisture contents (Fig. 14) to the laboratory and the cutting out, with a high speed band saw, of sediment cubes that are then fitted in to plastic boxes (2x2x2 cm) for measurement on a spinner

Figure 11. Loading at the contact of Lower Thorncliffe Sands and Sunnybrook diamict (Sect. 0781 Ravine Drive; Fig. 1b).

35

Figure 12 a) Large silty-sand pillow loaded into Sunnybrook diamict at Cudia Park (Fig. 1b). Figure to left. b) detail of pillow margin.

Figure 13. Interbedded contact of Sunnybrook diamict and Scarborough Formation Sands (Fig. 5a,b) at section 2581 (Fig. 1b).

Figure 14. Paleomagnetic sample site section 1681D (Dutch Church Fig. 1a). Block samples were collected from freshly excavated sediment. Figure at right of trench for scale. The bottom of the trench penetrated the laminated silt-breccias (arrowed) found within the Sunnybrook diamict (Fig. 15). Interbedded contact of Sunnybrook unit and Scarborough sands at X.

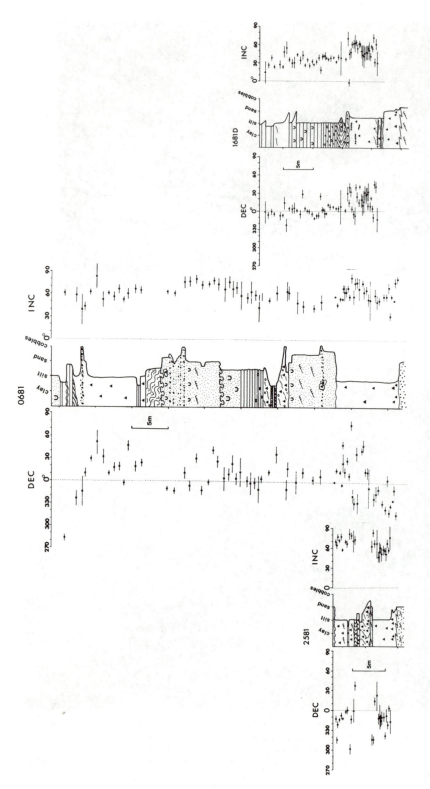

Figure 15. NRM's of sediments at Scarborough Bluffs (Figs. 1,14). Data points are vectorial means of three to five sub-samples from single blocks. Horizontal lines are confidence limits based on Fisher's (1953) cone of 95% confidence for data on a sphere. Directions to magnetic north. The base of each profile is the base of the Sunnybrook. Work is still in progress to clarify the dependence of paleomagnetic data on sediment facies. Note for example increased confidence limits in inclination data upward in section within the laminations sitting on the Sunnybrook unit at Section 1681D associated with loading structures (Fig. 8). Lower inclination values compared with the Sunnybrook may result from sediment compaction.

magnetometer. Sampling techniques employing pushing or hammering sample boxes directly into fine grained sediments may re-set the remanent magnetism to the ambient magnetic field as a result of shock thixotropy (Symons et al., 1980).

Sediment units that show evidence of resedimentation or loading structures were specifically avoided during the sampling programme e.g. the large flow nose at the Sylvan Park site (Figs. 1b, 5, 15) and disturbed laminated silty clays loaded into sands at the base of Meadowcliffe diamict at Cathedral Bluffs. The base of the Sunnybrook unit at the Dutch Church section could not be reached (Figs. 1a,15).

5.2 Measurement and Results

652 Samples (3 to 5 subsamples per sampling site) were measured on a Schonstedt DMS-1 spinner magnetometer to obtain the direction of natural remanent magnetism (NRM). NRM in most sediments is variably overprinted by an unstable viscous magnetism resulting from the ambient earth's magnetic field; this unstable element is usually removed by step demagnetisation in an alternating field (AF). Demagnetisation characteristics are illustrated for four representative samples in Fig. 16. Demagnetisation shows an initial stable remanent magnetisation with an unstable laboratory induced rotational component at high alternating fields (Wilson, 1980). These results indicate that the mean NRM of the subsamples approximates to the remanent magnetisation at or close to the time of deposition. Mean NRM and confidence limits are plotted against vertical profiles in Fig. 15.

Preliminary results indicate the following:
(1) Within sample block scatter is low with no evidence of shear-induced dispersion by glacitectonism.
(2) Broad corelations are apparent between sections but detailed correlation is not possible; erratic directions relate to resedimentation and loading.
(3) The excursion reported by Stupavsky et al., (1979) from the Meadowcliffe 'Till' and correlated with the Lake Mungo excursion of Barbetti and McElhinny (1976) can be related not to paleofield variation but to resedimented diamict horizons. The 'disturbing lack of correlation of NRM's over short interprofile separations' noted by Stupavsky et al. (1979, p. 271) also results from the sampling of resedimented diamict and not supposed homogenous

'basal tills' (Stupavsky and Gravenor, 1974).

5.3 Interpretation

Mean natural remanent magnetism declination and inclination values plotted against vertical profile sequences show a smooth data series and generally high within block precision that is inconsistent with deposition by lodgement processes at the base of a glacier (Fig. 15). The ability of magnetic particles to align themselves with the earth's palaeofields indicates at saturated mud-like consistency for the diamicts at the time of deposition.

6. DISCUSSION OF SEDIMENTOLOGICAL AND PALAEOMAGNETIC DATA

Vertical profile analysis shows assemblages of massive or resedimented diamicts having conformable, frequently loaded upper and lower contacts with turbiditic lacustrine silty-clays and sands. There is no evidence for direct glacitectonic disturbance such as occurs at an ice base in the form of incorporation, rafting and shearing of underlying sediments and sharp erosive contacts (Brodzikowski and Van Loon, 1980; Eyles et al., 1982).

The evidence for a subaqueous origin for the diamicts is considerable.
1) The palacomagnetic remanence characteristics indicate a mud-like consistency at time of deposition.

2) Evidence for a high water content in the diamicts at time of deposition is upheld by the loading of diamict upper surfaces as a result of reverse density gradients (Figs. 11,12).

3) Starved ripples occur within the diamicts (Dmm(c)). The ripples are frequently loaded to form pseudonodules.

4) The base of the Sunnybrook diamict is interbedded with the underlying Scarborough Sands (Fig. 13).

5) Undeformed bands of silt-clasts and turbiditic silty-clay laminae with ice-rafted clasts and diamict clots are a ubiquitous component of the diamicts (Figs. 1,9a).

6) Irregular boulder concentrates at the base of diamicts and vertical stones suggest settling of ice-rafted clasts through mud (Fig. 11b) (Spjeldnaes, 1973; Gibbard 1980).

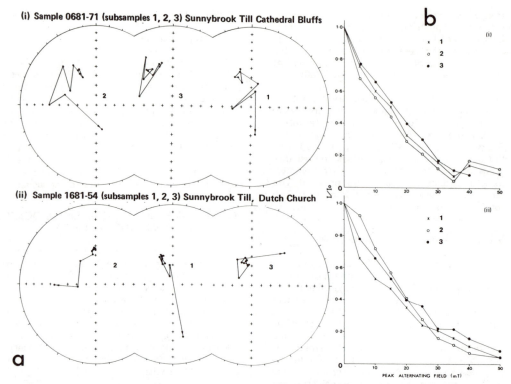

(i) Sample 0681-71 (subsamples 1, 2, 3) Sunnybrook Till Cathedral Bluffs

b

(ii) Sample 1681-54 (subsamples 1, 2, 3) Sunnybrook Till, Dutch Church

a

PEAK ALTERNATING FIELD (mT)

Figure 16a. Equal area plots of direction of remanent magnetisation for selected samples (Fig. 15) to show stability of direction after cleaning by alternating field demagnetisation. If the remanence direction moves in low fields then the samples are magnetically unstable and require demagnetisation or rejection. The directions are quite stable at low fields, so the undemagnetised NRM approximates the direction of the earth's field at the time of deposition. The instability in higher fields is thought to be a laboratory induced effect (Wilson, 1980), due in part to the very low magnetic intensities exhibited by the samples, known as rotational remanent magnetism. The rapid decay of intensity upon demagnetism (b), suggests that the remanent magnetism is carried by silt size magnetite and is not due to chemical precipitation.
I/I_o = ratio of intensity upon demagnetisation to original intensity. mT = milliteslas.

7) The massive or crudely-bedded silty sand with abundant dish structures of the Thorncliffe Formation closely resembles fine-grained subaqueous glacial outwash described by Rust and Romanelli (1975), Rust (1977) and Cheel and Rust (1980) from Late Wisconsin glaciomarine sequences in the Ottawa Valley.

8) Diamicts are intimately and conformably associated with laminated lacustrine silty-clays containing drop stones and dropped diamict masses (Fig. 9b).

An environmental interpretation based on vertical profile analysis and palaeomagnetic remanence directions of the three major diamict units (the Sunnybrook, Seminary and Meadowcliffe Tills) and intervening lacustrine sands and laminated silty clays indicates a glaciolacustrine origin for the entire package. Significantly ther would appear to be no major erosional hiatusses in the sequence (Fig. 1).

7. DEPOSITIONAL MODEL

Depositional models of diamict deposition in the glaciolacustrine environment

40

involving subaqueous gravity flows from ice margins (Evenson et al., 1977; May, 1977), accumulation in subglacial openings below a partially floating ice shelf where basal debris melts-out and falls through the water column without size softing (May, 1977; Gibbard, 1980) and deposition from grounded ice bergs (Dreimanis, 1979) have been carefully considered. These mechanisms are rejected as models for the deposition of diamicts exposed at Scarborough Bluffs given the lack of evidence for grounded ice over more than 8 km of section. We would instead emphasise deposition of a variable clastic component below floating ice, be it an ice shelf ramp, bergs or seasonal ice, coupled with the rain-out of suspended pelagic fines; a model that is commonly described from glacial marine environments with a wide variation in deposition rates being reported (Miller, 1957; Spjeldnaes, 1973; Barrett, 1975; Anderson et al., 1980; Link and Gostin 1982; Orheim and Elverhoi 1981).

The possibility of extensive thick and fine grained diamict sequences originating as stony lake muds below floating ice in lacustrine basins has been discussed in geotechnical description of problematical waterlain 'tills' in the Sarnia district of southwestern Ontario and around Detroit and Chicago (Wu, 1958; Soderman et al., 1961; Quigley, 1980) and in a brief report from Manitoba by Matile and Neilsen (1981). No detailed lithofacies descriptions are available however. A review of the extensive glacial marine literature and consideration of diamict lithofacies exposed at the Scarborough Bluffs suggests that diamict assemblages are generated by the interaction of three processes: (1) accumulation of massive diamicts (Dmm) on the lake floor by pelagic rain-out with ice rafted clasts.
(2) resedimentation by subaqueous slope processes and infilling of topographic lows (Dmm(r),Dms(r)) in association with turbidite activity
(3) weak traction current activity (Dmm(c),Dms(c)). The relationship between these processes can be shown on a ternary diagram and used to characterize diamict assemblages in vertical profile (Eyles and Eyles, 1983). Subaqueous slope processes appear to have been active, particularly at the Dutch Church valley (Fig. 1), over the irregular surface of the underlying Scarborough Formation delta. Slumping is a common sea and lake floor process in rapidly deposited fine-grained sediments

subject to the development of excess pore water pressures (Harris 1977; Carlson 1978; Johnson 1980). Resedimentation accounts for the close association of resedimented diamict and laminated silty clay of turbidite orgin in vertical profile (Figs. 1, 5; see also Morgenstern, 1967; Quigley, 1980). The overlying Seminary and Meadowcliffe diamict assemblages show a relatively reduced resedimented component due to infilling of topographic irregularities on the basin floor. Overconsolidation of the diamict assemblages has probably been a response to isostatic recovery, falling lake levels, underdrainage through the intervening sands (cf. Quigley and Ogunbadejo, 1976; Quigley, 1980) and erosion of an unknown thickness of overlying sediments.

8.0 IMPLICATIONS OF THE DEPOSITIONAL MODEL

The major lithological breaks on the Bluffs are traditionally assigned to alternating incursions of grounded ice margins (tills) and the formation of frontal lakes as the ice margin withdrew (Thorncliffe Formation; Karrow, 1967, 1969; Dreimanis, 1969, 1977; Fig. 2). The data presented here demonstrates, in the absence of erosional breaks indicating low lake levels, that lacustrine conditions were maintained throughout. The top of the Meadowcliffe diamict assemblage is 95 m above present lake level. Rather than alternating ground ice margins and interstadial lakes evidence suggests basinward delta progradation (sandy lithofacies) over stony lake bottom sediments. Highly significant in this regard is the widespread evidence for loading of diamict upper surfaces (e.g. Figs. 8,11,12) and, at sections 1781A to 1581 (Fig. 1), the gradual upward transition from stratified resedimented diamict (Dms(r)) to thin-bedded turbidites with dropstones (Fld) to thick bedded turbidites (Sh→Sr→Fl) and starved ripples in mud (Fl,Sr) prior to the incursion of sandy lithofacies with palaeocurrents to the south and west. A sequence such as this demonstrates delta progradation over glaciolacustrine diamicts; a model that also described the relationship of sandy lithofacies with the two other diamict assemblages upsection.

It is not possible to specify the type of floating ice, below which stony muds accumulated, be it a floating glacier ice shelf, ice bergs, lake ice cover or a combination of all or several. Having identified that the Scarborough sequence is not the product of repeated grounded ice incursions and reformed interstadial lakes the significance of the lithostratigraphy for interbasin litho and chronstratigraphic correlation (e.g., Fig. 2) can be seriously questioned since controls on glaciolacustrine deposition are likely to be intrabasinal (Eyles and Eyles, 1983).

In view of the paleogeography of the Great Lakes Basins during glaciation, with large deep lakes at the margin of the Laurentide ice sheet lobes, (Prest, 1970; Andrews, 1973; Denton and Hughes, 1983; Hilaire-Marcel et al., 1981) and glacier retreat in the direction of increased isostatic depression the frequently reported interbedded association of 'basal tills' with lake sediments in Pleistocene lake basins (e.g. Peck and Reed, 1954; Wu, 1958; Soderman et al., 1961; Farrand 1969; Sly and Thomas, 1974; Lineback et al., 1974; Teller, 1976; Wickham et at., 1978; Quigley, 1980; Desaulniers, et al., 1981) is of interest. A depositional model for diamict assemblages emphasizing glaciolacustrine mud deposition with coarser clastics derived from floating ice (sensu lato) coupled with non-glacial lake floor processes may be a useful model for other 'tills' in the Great Lake Basins either mapped onshore or identified subaqueously by reference to seismic profiling and coring.

ACKNOWLEDGEMENTS

We are grateful to J.B. Anderson, J.T. Andrews, G.S. Boulton, E.W. Domack, P. Fralick, B. Greenwood, A. Miall, R. Ostry, B.R. Rust, J. Shaw, R.M. Slatt, J.A. Westgate and R. Wright for discussion and comments. J. Clunas, B. Clark and B. Kaye provided excellent field assistance. A research grant from the University of Toronto for palaeomagnetic investigations in glacial sediments is gratefully acknowledged as is discussion with H. Halls and W. Morris. Nancy Kaye assisted with the manuscript. Fieldwork was supported by the National Science and Engineering Research Council of Canada and the Ontario Ministry of Colleges and Universities.

REFERENCES

Antevs, E., 1928. The Last Glaciation. Amer. Geogr. Soc. Res. Ser. 17, 292 pp.

Anderson, J.B., Kurtz, D.D., Domack, E.W. and Balshaw, K., 1980. Glacial and glacial marine sediments of the Antarctic continental shelf. Jnl. Geology, 88 399-414.

Andrews, J.T., 1973. The Wisconsin Laurentide ice sheet; dispersal centers, problems or rates of retreat and climatic implications. Arctic and Alpine Research 5, 185-199.

Barrett, P.J., 1975. Textural characteristics of Cenozoic preglacial and glacial sediments at site 270, Ross Sea, Antarctica. In Hayes, D.E. and Frakes, L.S. (Eds) Initial reports of the Deep Sea Drilling Project. 28, 757-766.

Barbetti, N.F. and McElhinney, M.W., 1976. The Lake Mungo geomagnetic excursion. Phil. Trans. Soc. Lond. 281, 515-552.

Brodzikowski, K., and Van Loon, A.J., 1980. Sedimentation deformations in Saalian glaciolimnic deposits near Wlostow. Geol. En. Mijnbouw, 59, 251-272.

Berti, A.A., 1975. Paleobotany of Wisconsinan Interstadials, Eastern Great Lakes Region, North America, Quat. Res. 5, 591-619.

Birkelund, P.W., Colman, S.M., Burke, R.M., Shroba, P.R., Meserding, T.C., 1979. Nomenclature of alpine glacial deposists or, what's in a name? Geology, 7, 532-36.

Boulton, G.S., 1972. Modern Arctic glaciers as Depositional Models for Former Ice Sheets. J. Geol. Soc. Lond., 182, 361-393.

Carlson, P.R., 1978. Holocene slump on continental shelf off Malaspina Glacier, Gulf of Alaska. A.A.P.G. Bull. 62, 2412-2416.

Cheel, R.J. and Rust, B.R., 1980. A sequence of soft sediment deformation structures near Ottawa, Ontario. Geol. Ass. Can. Prog. Abst. 5, 45.

Denton, G. and Hughes, T.J., 1981. The Last Great Ice Sheets. Wiley Interscience.

Desaulniers, D.E., Cherry, J.A., and Fritz, P., 1981. Origin, age and movement of porewater in argillaceous Quaternary deposits at four sites in southwestern Ontario. Jnl. Hydrology. 50, 231-257.

Dreimanis, A., 1969. Late Pleistocene Lakes in the Ontario and Erie basins. Proc. 12th Conf. Great Lakes Res. 170-180.

Dreimanis, A. 1976. Tills: Their origin and properties. In Glacial Till. Ed. by R.F. Legget., Roy. Soc. Can. Spec. Pub. 12, 11-49.

Dreimanis, A. 1977. Correlations of Wisconsin glacial events between the Eastern Great Lakes and the St. Lawrence Lowlands, Geogr. Phys. Quat. XXXI, 37-51.

Dreimanis, A., 1979. The problems of waterlaid tills. In; Moraines and Varves. Ed by C. Schlüchter, A.A. Balkema, Rotterdam, 167-177.

Dreimanis, A., 1982. Middle Wisconsin substage in its type region, the eastern Great Lakes and Ohio River basin, North America Quaternary Studies in Poland 3 pt.2. 21-27.

Evenson, E.G., Dreimanis, A., and Newsome J.W., 1977. Subaquatic flow tills: a new interpretation for the genesis of some laminated till deposits. Boreas 6, 115-133.

Eyles, C., and Eyles, N., 1983. Sedimentation in a large lake; a reinterpretation of the Late Pleistocene stratigraphy at Scarborough Bluffs, Ontario, Canada. Geology, In Press.

Eyles, N., Sladen, J.A. and Gilroy, S., 1982. A depositional model for stratigraphic complexes and facies superimposition in lodgement tills. Boreas, 11, 317-333.

Eyles, N., Eyles, C. and Miall, A. (In Press). Lithofacies types and vertical profile models; an alternative approach to the description and environmental interpretation of glacial diamict and diamictite sequences. Sedimentology.

Fisher, R.A., 1953. Dispersion on a sphere. Proc. Roy. Soc. Lond. Ser. A, 217, 295-305.

Gibbard, P. 1980. The origin of Stratified Catfish Creek Till by basal melting. Boreas, 9, 71-85.

Gustavson, T.C., Ashley G.M., and Boothroyd J.C., 1975. Depositional sequences in glaciolacustrine deltas. In A.V. Jopling and B.C. McDonald (Eds)

Glaciofluvial and Glaciolacustrine sedimentation. S.E.P.M. Spec. Pub 23 Tulsa. pp. 264-280.

Harris, P.G., 1977. Sedimentation in proglacial lake Jokulsarlon, southeast Iceland, Unpub. Ph.D. Thesis, Univ. East Anglia, U.K.

Hilaire-Marcel, C., Occhietti, S. and Vincent, J.S., 1981. Sakami moraine, Quebec. a 500 km long moraine without climatic control. Geology, 9, 210-214.

Johnson, T.C., 1980. Late-glacial and post-glacial sedimentation in Lake Superior based on seismic-reflection profiles. Quat. Res., 13, 380-391.

Karrow, P.F., 1967. Pleistocene geology of the Scarborough Area, Ont. Dept. Mines Geol. Rep. 46.

Karrow, P.F., 1969. Studies in the Toronto Pleistocene. Proc. Geol. Ass. Can., 20, 4-16.

Karrow, P.F., 1974. Till stratigraphy in parts of Southwestern Ontario. Geol. Soc. Am. Bull., 85. 761-768.

Kurtz, D.D. and Anderson, J.B., 1979. Recognition and sedimentologic description of recent debris flow deposits from the ross and Weddell Seas, Antarctica. J. Sed. Pet. 49, 1159-1170.

Lewis, C.F.M. and Sly, P.G., 1971. Seismic profiling and geology of the Toronto waterfront area of Lake Ontario. Proc. 14th. Conf. Great Lakes Res. 303-354.

Lineback, J.A., Gross, D.L. and Meyer R.P., 1974. Glacial tills under Lake Michigan. Illinois State Geol. Surv. Env. Geol. Note 69. 48 pp.

Lowe, D.R., 1979. Sediment gravity flows: their classification and some problems of application to natural flows and deposits. In: Geology of Continental Slopes (Ed. by L.J. Doyle and O.H. Pilkey) 75-95. S.E.P.M. Spec. Pub. 27.

Matile, G. and Nielsen, E., 1981. Proglacial lake sedimentation in southeastern Manitoba. Geol. Ass. Can. 1981 meeting. Abstracts. P. 38.

May, R.W., 1977. Facies model for sedimentation in the glaciolacustrine environment. Boreas 6, 175-180.

43

Miall, A.D., 1977. A review of the braided river depositional environment. Earth Sci. Rev. 13, 1-62.

Miall, A.D., 1978. Lithofacies types and vertical profile models in braided rivers: A Summary. In: Fluvial Sedimentology. Can. Soc. Pet. Geol. Mem. 5, 597-604.

Miall, A.D. In Press. Glaciomarine sedimentation in the Gowganda Formation (Huronian), Northern Ontario. J.Sed. Pet.

Miller, D.J., 1957. Lake Cenozoic marine glacial sediments and marine terraces of Middleton Island, Alaska. J. Geology, 61, 17-40.

Mitchum, R.M., Vail, P.R. and Thompson S., 1977. The depositional sequence as a basic unit for stratigraphic analysis. In Payton, A. (Ed) Seismic Stratigraphy-Applications to Hydrocarbon Exploration. A.A.P.G. Mem. 26, 53-62.

Morgan, A., 1972. The fossil occurence of Helophorus Articus Brown (Coleoptera; Hydrophilidae) in Pleistocene Deposits of the Scarborough Bluffs, Ontario Ca. J. Zool. 50, 555-558.

Morgenstern, N.R., 1967. Submarine slumping and the initiation of turbidity currents. In Richards, A.F. (ed) Marine Geotechnique. Univ. Illinois Press. 189-230.

Morner, N.A., 1973. Till stratigraphy in North America and Northern Europe. bull. Geol. Inst. Univ. Uppsala 5, 199-208.

Nardin, T.R., Hein, F.J., Grosline, D.S. and Edwards, B.D., 1979. A review of mass movement processes, sediment and acoustic characteristics and contrasts in slope and base-of-slope systems versus canyon-fan-basin floor basins. In: Geology of continental Slopes (Ed. by L.J. Doyle and O.H. Pilkey) 61073. S.E.P.M. Spec. Pub. 27.

Orheim, O., and Elverhoi, A., 1981. Model for submarine glacial deposition. Annals of Glaciology, 2, 123-128.

Peck R.B., and Reed, W.C., 1954. Engineering properties of Chicago subsoils Univ. Illinois. Eng. Exp. Station. Bull. 423 62 pp.

Prest, V.K., 1970. Quaternary Geology. In: Geology and Economic Minerals of Canada. Geol. Surv. Can. Ec. Geol. Rep. 1.

Quigley, R.M. and Ogunbadejo, T.A. 1976. Till geology, mineralogy and geotechnical behaviour, Sarnia, Ontario. In: Legget, R.F. (Ed) Glacial Till. Roy. Soc. Can. Spec. Pub. 12. 336-345.

Quigley, R.M., 1980. Geology, Mineralogy and geochemistry of Canadian soft soils: A geotechnical perspective. Can. Geotech. J., 17, 261-285.

Rust, B.R. and Romanelli, R., 1975. Late Quaternary sub-aqueous deposits near Ottawa, Canada. In: Glaciofluvial and Glaciolacustrine Sedimentation. (Ed. by A.V. Jopling and B.C. McDonald) 177-192. S.E.P.M. Spec. Pub. 23.

Rust, B.R., 1977. Mass flow deposits in a Quaternary succession near Ottawa, Canada: diagnostic criteria for subaqueous outwash: C.J. Earth Sci. 14, 175-184.

Sly, P.G. and Thomas, R.L., 1974. Review of geological research as it relates to an understanding of Great Lake Limnology. J. Fish. Res. Board Canad., 31 795-825.

Soderman, L.G., Kenney, T.C., and Loh, A.K., 1961. Geotechnical properties of glacial clays in Lake St. Clair region of Ontario. In 14th Can. Soil. Mech. Conf. Proceedings. NSERC Tech. Memo 69, 55-90.

Stupavsky, M., and Gravenor, C.P., 1974. Water release from the base of active glaciers. Geol. Soc. Amer. Bull. 85, 433-436.

Stupavsky, M., Gravenor, C.P., and Symons, D.T.A., 1979. Paleomagnetic stratigraphy of the Meadowcliffe Till, Scarborough Bluffs, Ontario; A late Pleistocene excursion? Geophys. Res. Lett. 6, 269-272.

Symonds, D.T.A., Gravenor, C.P. and Stupavsky, M., 1980. Remanence resetting by shock induced thixotropy in the Seminary Till, Scarborough, Ontario, Canada. Geol. soc. am. Bull., 91, 593-98.

Teller, J.T., 1976. Lake Agassiz deposits in the main offshore basin of southern Manitoba. Can. J. Earth Sci. 13, 27-43.

Terasmae, J., 1960. A palynological study of pleistocene interglacial beds at Toronto, Ontario. Geol. Surv. Can. Bull., 56, 23-41.

Verosub, K.L., 1979. Paleomagnetism of

varved sediments from western New England: secular variation Geophys. Res. Lett. 6, 245-248.

Weimer, R.J., 1981. Deltaic and shallow marine sandstones; sedimentation, tectonics, and petroleum occurences. Ant. Ass. Petrol. Geol. Ed. Course Note Series No. 2. Tulsa.

Wickham, J.T., Gross, D.L., Lineback, J.A. and Thomas, R.L., 1978. Late Quaternary Sediments of Lake Michigan. Illinois State Geol. surv. Env. Geol. Note 84, 26 pp.

Williams, N.E., Westgate, J.A., Williams, D.D., Morgan, A. and Morgan A.V., 1981. Invertebrate fossils (Insecta: Trichoptera, Diptera, Coleoptera) from the Pleistocene Scarborough Formation at Toronto, Ontario and their palaeo environmental significance. Quat. Res. 16, 146-166.

Wilson, R.L., 1980. Gyromagnetism: Rocks in a spin. Nature, 284, 12-13.

Wu, T.H., 1958. Geotechnical properties of glacial lake clays. Inl. Soil. Mech. and Found. Div. Proc. Am. Soc. Civ. Eng. 84, 1732-33.

Yagishita, K., Westgate, J.A., and Pearce, G.W., 1981. Remanent magnetization in penecontemporaneous structures of the Pleistocene Scarborough Formation, Ontario, Canada. Jl. Geol. Soc. 138, 549-557.

Modern Icelandic glaciers as depositional models
for 'hummocky moraine' in the Scottish Highlands

NICHOLAS EYLES
University of Toronto, Ontario, Canada

1 INTRODUCTION

This paper describes the application of glacial depositional models derived from Icelandic ice caps to deglaciation features of the lateglacial Scottish Highlands. It is argued here that deglaciation of the Lateglacial Stadial Loch Lomond Ice Cap was by frontal retreat not areal deglaciation as is generally assumed. Widespread 'hummocky moraine' from which the stagnation of the ice cap has been inferred, can be identified as resulting from both uncontrolled and controlled supraglacial deposition by actively retreating glaciers. These ice margins transported substantial supraglacial loads as a result of severe periglacial weathering of exposed valleysides.

1.1 The Scottish Lateglacial

According to the classic model of the Scottish Lateglacial, the ice-free Windermere Interstadial, of c. 1300 radiocarbon years duration (12,500-11,200 y.b.p.), separates the final decay of Devensian ice from a vigorous but geographically limited and short lived lateglacial build-up of ice during the Loch Lomond Stadial (Coope, 1977; Coope and Pennington, 1977; Gray and Lowe, 1977; Pennington, 1977; Sissons, 1979a; Lowe and Walker, 1980). This stadial is commonly equated with the Younger Dryas (Pollen zone III) of the Scadinavian Lateglacial and therefore with the Ra, Central Swedish and Salpausselka moraines of Scandinavia.

The regeneration of ice caps and corrie glaciers on the British mainland during the Loch Lomond Stadial, commenced sometime prior to 11,200 y.b.p. and is thought to have terminated around 10,000

y.b.p. (the Loch Lomond Advance; Sissons, 1979b,c). By between 9,500 and 9,640 ± 180 y.b.p. a 'thoroughly thermophilous assemblage of insect species was widespread . . . as far north as southwest Scotland' (Coope, 1981 p. 219). The Loch Lomond Advance is identified by fresh glacial terrain and ice limits that are inferred from the distribution of 'hummocky moraine' (Clapperton, 1977; Cornish, 1981; Gray and Brooks, 1972; Gray and Lowe, 1977; Sissons, 1967, 1974a,b, 1976, 1977a,b, 1980; Sissons and Grant, 1972; Thompson, 1972). The latter is a distinctive terrain of irregular bouldery mounds apparently lacking any systematic lineation or orientation with respect to former ice flow direction, underlain by a wide range of sediments from coarse rubbly 'till' to outwash. From such hummocky, uncontrolled, so-called 'stagnation' topography it is generally concluded that the stadial glaciers and ice cap stagnated *in situ* and dissipated by areal deglaciation; all areas were thus deglaciated simultaneously (Gray and Lowe, 1977; Sissons, 1974b, 1979a). This scenario is thought to be supported by rapid post-glacial climatic amelioration at the end of the stadial that is inferred from beetle faunas (Coope, 1977).

The concept of areal deglaciation and simultaneous stagnation conflicts with the earlier work of J.K. Charlesworth (1955) who identified a succession of undated ice marginal positions indicating recession of ice back into parent corries. A particular problem is that there has been no detailed work on the sedimentology and genesis of 'hummocky moraine'. Brief references to the sedimentology and internal structure of

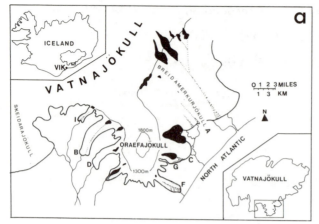

Figure 1a.   Southeast Iceland.
The Oraefajokull and Vatnajokull
Ice Caps provide a suitably scaled
analog for Scottish Lateglacial
glaciers (See Fig. 6).   Valley
glaciers draining Oraefajokull,
transport supraglacial diamict and
deposit it as latero-frontal and
medial moraine ridges.   Nunataks
are shown in black.   The large
expanded-foot glaciers draining
the Vatnajokull ice cap deposit
the subglacial landsystem (Boulton
and Paul, 1976; Eyles, 1979).
Hummocky supraglacial diamict at
Hrutajokull (G), Fjallsjokull (C)
and Morsarjokull (I) is shown in
Fig. 5.

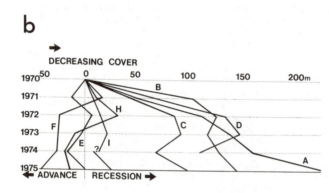

Figure 1b.   The glaciological
significance of supraglacial
diamict covers.   Frontal positions
of southeast Icelandic ice margins
(1970-75) relative to their 1970
position.   Ice margins A, B, C and
D (Fig. 1a) transport little
supraglacial cover and have
receded since 1970 in response to
ameliorating mean annual air
temperature.   Ice margins E, F, H
and I carry a substantial
supraglacial cover and show
erratic frontal activity and
frequent advances beyond the 1970
position.   The supraglacial cover
reduces ice melt in the ablation
zone thus promoting a more
positive mass-balance equation and
frequent bulldozing of
supraglacial diamicts along the
ice margin.

this terrain type indicate that a wide range of materials underlies such topography (Clapperton 1977; Clapperton and Sugden, 1977; Young 1978; Sugden, 1980). The model of wholesale stagnation (or areal deglaciation) used by workers after the fashion of Sissons (1967) cannot be used with any certainty in the absence of detailed work on the sedimentology and depositional history of 'hummocky moraine'.

2   MODERN ANALOGUES FOR THE DEPOSITION OF LATEGLACIAL 'HUMMOCKY MORAINE'

The southeast coast of Iceland (Fig. 1) is a close morphological replica of conditions that would have occured in Scotland during times of substantial ice cover. The

juxtaposition of confined valley glaciers and large (30 km wide) expanded-foot outlet glaciers affords a great variety of glacial sedimentary processes (Boulton and Paul, 1976).

The large outlet glaciers such as Breidamerkurjokull (Fig. 1a) drain the southeast margins of the Vatnajokull Ice Cap (8,410 km^2) which lies adjacent to the ice-capped quiescent volcanic caldera of Oraefajokull (elev. 2,000 m). Narrow valley glaciers, entrenched into easily eroded basaltic extrusives, drain the volcano (Fig. 1a). The larger glaciers are fringed by extensive sandur plains, proglacial lakes and drumlinized lodgement till plains crossed by push ridges and eskers (Price, 1973), an association of landforms and sediments that is referred to as the subglacial landsystem

Figure 2. Deposition of supraglacial diamicts; after Eyles (1979). Ice-cores melt under thinner covers creating topographic lows that are then filled by slides. In this fashion the ice-cored topography is inverted (1,2,3). Mechanical mixing of diamict and kettle hole sediments, frost action, surface washing and soil formation, result in a wide range of included lenses of matrix-rich and matrix-poor material within the unconsolidated diamict (Fig. 4).

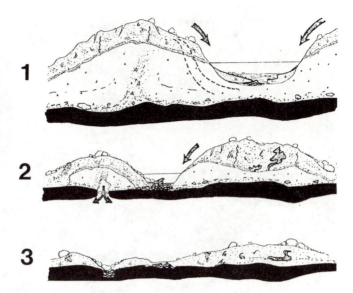

by Boulton and Paul (1976) in their classification of glaciated terrains.

Along the confined valleys draining Oraefajokull the subglacial landsystem is obscured over large areas by a hummocky drape of coarse-grained supraglacial diamicts derived by periglacial weathering of the valleysides and transported in conveyor-belt fashion on the glacier surface. The term supraglacial morainic till has been used for these diamicts where deposited directly by glacier ice (Eyles, 1979). Given however, the importance of non-glacial resedimentation processes (slides, flows, etc.) operating during supraglacial deposition (e.g. Lawson 1981, 1982) and the resultant wide range of glaciofluvial and lacustrine sediments incorporated within the final hummocky landform (Fig. 2) the term supraglacial diamict is used in this paper in keeping with a strict genetic definition of 'till' (Eyles et al., this vol.).

2.1 The Importance of Climatic Regime

In general, wet-based sliding glaciers fail to incorporate much englacial load as a result of continued basal melting. On valley glaciers in southeast Iceland, englacial debris loads are increased substantially as a result of erosion prone valleysides in accumulation areas and the consequent dumping of bedrock debris onto the firn surface. Another important process leading to larger volumes of englacial debris is the overriding and

subglacial shear of decaying ice-cores at the contact between active clean ice, devoid of much supraglacial cover, and well-insulated ice-cores along the outer ice margin. Many valley glacier termini thus exhibit englacial debris bands (Fig. 3a).

In southeast Iceland, a related manifestation of easily weathered bedrock types and the production of large volumes of debris from rockwalls is the occurrence of rockglaciers on nunataks within the Vatnajokull Ice Cap. Rockglaciers are rarely found in maritime areas (Eyles, 1978).

Decaying ice-cores below an active, mobile cover of supraglacial diamicts contain englacial debris that is progressively lowered onto the substrate. In southeast Iceland where the mean annual air temperature is + 1°C, ice-cores are short lived and karstic destruction (Fig. 2) ensures that a distinct basal melt-out unit is not a marked feature of proglacial stratigraphies. Typical melt-out accumulations are thin (< 1 m), overlie lodgement till and are crudely bedded with a wide variation in matrix content as a result of matrix loss under conditions of rapid ice melt. Ice-cores in southeast Iceland disappear by both top and bottom melt with englacial and subglacial openings progressively enlarged in karstic fashion. The term 'drop-out' perfectly describes the deposition of much englacial debris (Fig. 3b). Thicker melt-out sequences may develop where ice-cores are buried below substantial thicknesses of outwash. Valley

Figure 3a. The terminus of Glacier de la Brenva, Italian Alps. A slide has stripped supraglacial diamict from the ice front to reveal outcropping englacial debris. Ice front about 20 m high.

Figure 3b. Overconsolidated lodgement till, with an oriented glacially shaped boulder, overlain by loose, clast-supported diamict layers dropped from the roofs of a subglacial opening as a result of karstic development in dead ice. These diamicts are overlain by matrix-supported units having flowed into the subglacial cavity as it opened to the surface.

Figure 4a. Supraglacial diamict in Iceland deposited as in Fig. 2. Notebook for scale. The bouldery matrix-deficient lens results from size-sorting on the steep hummock slopes and subsequent mixing during topographic inversion.

Figure 4b. Crudely stratified supraglacial diamict in Glen Torridon Scotland. Multiple units of coarse, clast-supported layers record the infilling of an ice-cored trough (Fig. 2). Note the clast angularity. See Fig. 6 for location.

glaciers under more continental conditions in areas of discontinuous or continuous permafrost produce thicker melt-out sequences that are slowly lowered onto the substrate (Shaw, 1977, 1979; Lawson, 1981 a,b).

Driscoll (1980) determined from Klutlan Glacier in the Yukon Territory, that a 950 year period was required to melt down 180 m of ice with a 1% debris content - this increased to 1200 years with a 1.5% debris content. The mean annual temperature at Klutlan Glacier is -8°C, that of southeast Iceland is +1°C. Coope (1977) suggested that average annual temperatures during the Scottish Lateglacial Stadial were below

-8°C. It must be emphasized therefore that whilst the processes of supraglacial deposition on the margins of Icelandic Valley glaciers can be used as a genetic model for Scottish sequences, excavations through Scottish 'hummocky moraine' deeper than at present available, may in the future reveal a more substantial englacial and basal melt-out component reflecting the known greater severity of the Scottish Lateglacial stadial climatic regime.

The predominant landforms that characterize deposition of coarse-grained supraglacial diamicts are latero-frontal moraine ridges and medial moraines with a hummocky bouldery topography that frequently covers the entire valley floor. Collectively this glacigenic association is referred to as the glaciated valley landsystem (Boulton and Paul, 1976; Boulton and Eyles, 1979) and it is this association that is extensively developed in the Scottish Highlands.

2.2 Supraglacial diamict deposition in Iceland

Supraglacial diamicts on valley glaciers are extremely variable in character since they are aggregated principally from valleyside debris ranging from bedrock masses, valleyside fans and soils to fluvial sediments transported by supraglacial streams. These are mixed during deposition (Fig. 2). Sedimentological characteristics are summarized in Eyles (1979). Fig. 2 in Boulton and Eyles (1979) shows the typical distribution of supraglacial diamicts on the surface of a valley glacier and as deposited landforms along the valley. There are two principal locales of deposition - as medial moraines along the valley floor and as lateral moraine ridges along the valleyside.

As medial moraines along the valley floor supraglacial diamict is deposited as:
(i)   a thick hummocky non-compact cover that is ice-cored during deposition (Figs. 2,4,5) and
(ii)   as a dispersed bouldery cover that is not ice-cored at the time of deposition but which may be locally thickened in the form of dump moraine ridges.

(i)   As a result of severe compression of ice in the glacier's terminal zone, surface diamict thicknesses increase as a result of enhanced englacial debris concentrations. As the glacier recedes upvalley diamict covered ice-cores, isolated from cleaner ice as it melts back, are left along the valley floor with a hummocky high-relief surface resulting from subsurface melt. Deposition is generally analogous to that described from Spitsbergen and Alaskan subpolar glaciers where basal debris is exposed in a supraglacial position along the ice margin (Boulton, 1972; Boulton and Paul, 1976; Lawson, 1981a,b, 1982). Supraglacial diamicts on Icelandic glaciers are more coarse-grained however because they are derived from the valleysides. Thus whilst the workers cited above have identified sediment gravity flow as an important process in the deposition of supraglacial diamicts, in Iceland flowage typically only

occurs during periods of intense rainfall when low liquid limits are exceeded and diamicts liquefy. Loss of fines quickly arrests movement. In Iceland the most important mass movement process is sliding (Fig. 2). Saturation of the mud layer that forms on the ice surface at the base of the diamict cover, as a result of downward washing of fines, assists in the sliding process.

Following several cycles of topographic inversion, that may take a few tens of years in a maritime climate and much longer under continental conditions (Wright, 1980), an irregular topography consisting of a steep-sided bouldery mounds and ridges and dead ice hollows is produced (Fig. 5). A detailed lithofacies log showing the range diamicts and related sediments exposed below such terrain is shown in Eyles et al., (In Press).

(ii)   Where the supraglacial cover on the glacier surface is thinner, diamict is shed directly from the icefront as a bouldery drape on top of recently exposed subglacial surfaces. Crevasse fills and dump moraine ridges are common. Such end-dumped material frequently forms the feather edge of thicker, hummocky accumulations of supraglacial diamict preserved as one or more medial moraines along the valley floor. In large basins of complex geology each medial moraine may be composed of different bedrock lithologies; a feature used for mineral prospecting and mapping in mountain areas (Evenson et al., 1979).

Thick sequences of diamict and outwash accumulate in the trough between the glacier and valleysides because in the ablation zone there is a net component of flow towards the valley side. With destruction of landforms by meltstreams along the valley floor these lateral moraine ridges survive as the dominant element in many glaciated valleys.

2.3 Lateglacial Supraglacial Deposition in Scotland

Hummocky moraine has been examined at a number of localities in Scotland where according to J.B. Sissons and his co-workers it is most clearly expressed. The 'valley of a hundred hills' is a recurring and expressive local name in these areas (Sissons, 1976; Clapperton, 1977). These are Glens More and Forsa on the Island of Mull, affected by the separate Lateglacial Mull Ice Cap (Gray and Brooks, 1972), the Pass of Drummochter between Edendon

Figure 5. Supraglacial diamicts in Iceland and Scotland.
a) Kviarjokull, Iceland; the lateral moraine ridge in the background is 100 m high. So-called 'stagnation topography' along the valley floor (foreground) has been generated incrementally during frontal retreat since c. 1960.

Figure 5b) Glen Torridon, Scotland, at Coire a Cheud-chnoic with Lochan Neimhe in background (O.S. 1:25,000 map, The Torridon Hills) See Fig. 6 for location.

Fig. 5c) The floor of Coire a Cheud-chnoic; note the peat or water filled dead ice hollows and uncontrolled hummocky moraine. The area of view is about 1.5 km wide.

Figure 5d) Bouldery hummocky drape of supraglacial diamict being let down onto a lodgement till surface at Morsarjokull. The section at bottom left is 5 m high.

Figure 5e) Supraglacial diamict up to 2.5 m thick exposed on the medial moraine between Hrutajokull (**G**, Fig. 1a) and Fjallsjokull (**C**, Fig. 1a). Note the steeply dipping foliation at **A** as a result of intense compression between the two glaciers (see also Eyles and Rogerson, 1977). As a result of such compression the basal debris layer is folded upwards between the two ice masses and fine-grained basal debris is exposed at the ice surface. In this way glacially-shaped 'flat-iron' boulders are found in a supraglacial position.

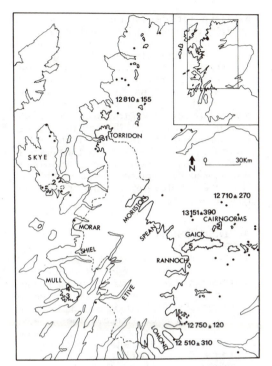

Figure 6. The Loch Lomond Ice Cap
and other ice masses at their maximum
extend during the Lateglacial stadial.
The dashed lines are tentative ice limits.
Dots are pollen sites, with selected
radiocarbon dates, demonstrating that most
of Scotland may have been ice-free during the
Lateglacial interstadial (c. 12,500-11,200
ybp) that preceded the Loch Lomond Ice Cap
(After Sissons, 1976). 1) Coire a Cheud-
chnoic (Figs. 4b, 5b, c), 2) Sligachan Inn,
3) Glen More, 4) Glen Forsa, 5) Pass of
Drummochter.

Bridge and Crubenmore and the adjacent valleys that drained the plateau Gaick Ice Cap (Sissons, 1974a), and along Glen Torridon (Fig. 6). More general examinations of hummocky moraine terrain have been made on Skye, in the area of Sligachan Inn, and along Glens Shiel and Morriston referred to by Clapperton (1977) and Sissons (1977a,b).

The depositional facies and landforms of supraglacial diamict outlined from the margins of modern Icelandic glaciers can be recognized in Scotland (Figs. 2,4,5).

The distribution of hummocky moraine along the valley floor is of great significance in any genetic interpretation of hommocky moraine in Scotland. In many valleys, close to the down valley limit of the Loch Lomond ice lobes, hummocky moraine typically extends right across the valley floor (Figs. 5b,c) reflecting radial flow of ice in the terminal zone and the transport of supraglacial material to laterofrontal positions. Upvalley the distribution of hummocky moraine is more controlled in the form of linear medial moraines along the valley centre-line or as lateral moraines. This is most clearly expressed along the valleys draining the former Gaick Ice Cap and in the area of Glen

Torridon (Fig. 6) where a distinct medial and lateral distribution of hummocky moraine can be mapped (e.g. Fig. 2 in Sissons, 1979a and Figs. 5,6,8,9 in Sissons, 1977b). Medial moraines originate upglacier on the backwalls of the firn basin or downglacier at glacier confluences, their size usually being a reflection of the susceptibility of different rock types to periglacial weathering on the valleysides, and area of bedrock exposed (Eyles and Rogerson, 1978; Dawson, 1979). Many of the smaller former corrie glaciers mapped by Sissons (1977a,b; 1980) are delineated by a dispersed bouldery cover ('a sea of boulders'; Sissons, 1979c, p. 73) often transversely lineated with respect to ice flow direction by both push and dump moraine ridges.

Lateral moraine ridges are a dominant feature in many Scottish Highland valleys frequently damming up lakes in sidevalleys such as Loch Ailsh in Glen Oykel and Loch a'Gharbhrain (Sissons 1977b). Boulton and Eyles (1979) have described the formation and internal stratigraphy of such moraine ridges; they frequently contain buried soils and organics (Rothlisberger and Schneebeli, 1979) and are thus worthy of further detailed investigation.

Boulton and Eyles (1979) also describe the pattern of glaciofluvial deposition adjacent to glaciers transporting large volumes of supraglacial diamict; complex stratigraphies result from diamict slides moving into troughs between ice-cored highs. These trough fills show a wide variety of sediments from resedimented diamicts, crudely bedded coarse grained outwash to lacustrine units. With the melt of ice-cores, a hummocky kamiform terrain composed of a wide variety of churned, often mechanically mixed sediments results

(Fig. 2) which may account for the variable sediments reported below hummocky moraine in Scotland (e.g. Young, 1978; Clapperton, 1977; Clapperton and Sugden, 1977; Sugden, 1980).

3. THE STYLE OF DEGLACIATION OF THE LOCH LOMOND ICE CAP

The widespread distribution of supraglacial diamict facies in Highland valleys may reflect intense periglacial weathering processes during the Lateglacial stadial (Sissons, 1979a) when large volumes of frost weathered bedrock debris were dumped onto the Loch Lomond glaciers. A suitable model would appear to be offered by Little Ice Age Norwegian glaciers when climatic deterioration was associated with increased weathering of valleysides and greater supraglacial debris loads (Grove, 1972).

In Scotland the hummocky depositional surfaces that result from supraglacial deposition have been interpreted as uncontrolled stagnation topography. From such 'dead ice forms that abound within the Loch Lomond limit' (Sissons, 1967, p. 143) the wholesale in situ stagnation and areal deglaciation of these glaciers have been inferred (Thompson, 1972; Sissons and Grant, 1972; Sissons and Sutherland, 1976; Young, 1974, 1975, 1978; Gray and Lowe, 1977; Sissons, 1977a,b, 1979a,b,c; Clapperton, 1977).

In Iceland, tracts of hummocky supraglacial diamict with abundant 'dead ice forms' are being deposited by active glaciers (Okko, 1955; Eyles, 1979; Vikingsson, 1978; Kaldal, 1978). The mechanism is one of 'incremental marginal stagnation' whereby a diamict covered and insulated rim of the ice margin (from 10 to 500 m wide) is isolated as cleaner ice upglacier thins and retreats. This process continues as the glacier recedes and diamicts become subsequently exposed on the ice surface by ice melt. Repeated incremental marginal stagnation induced by the supraglacial diamict cover gives rise to a hummocky ice-cored surface topography that is transversely lineated as a result of bulldozing by the clean, more active, ice upglacier (Fib. 1b). Surging glaciers exhibit this controlled topography very clearly (Wright, 1980) in response to intense compression of active ice against the dam of slow moving or stagnant ice at the terminus.

With the karstic decay of ice-cores

evidence for deposition by an active frontally-receding glacier is either partially or completely obliterated. The final result is uncontrolled 'stagnation' topography. The controlled distribution of supraglacial diamicts (hummocky moraine) as medial and lateral moraine ridges in Highland valleys, the presence of push and dump moraine ridges among tracts of hummocky moraine (Sissons, 1977b) and the picture from modern Icelandic environments, indicates that the Scottish Lateglacial Stadial glaciers, whose extent is demarcated by hummocky moraine, did not stagnate wholesale but dissipated by frontal retreat.

4. DISCUSSION

The interpretation of the genesis and glaciological significance of hummocky moraine in many Scottish Highland valleys supports the conclusions of more extensive field mapping in Scotland by Charlesworth (1955). Charlesworth traced the general recessional positions of Scottish Lateglacial ice by reference to regional patterns of moraine ridges. Whilst the chronology established by Charlesworth in the absence of radio-metric dating has been demonstrated to be suspect, and his ice marginal positions are generalized, his overall model of active glacier recession has been ignored by later studies that emphasise the mapping of hummocky moraine as a basis for delimiting Lateglacial ice cover and its interpretation as uncontrolled stagnation topography.

Within this scenario of active frontal retreat and incremental marginal stagnation it is to be expected that the rate of recession by individual Lateglacial valley glaciers in response to climatic amelioration at the end of the stadial was variable from one ice margin to another by virtue of different volumes of supraglacial diamict being transported (e.g. Fig. 1b); this may partially help to explain recent pollen-stratigraphic evidence for complex time-transgressive Lateglacial deglaciation across the Scottish Highland (Lowe and Walker, 1981).

ACKNOWLEDGEMENTS

This work was supported by a grant from the University of Newcastle Upon Tyne. Carolyn Eyles and Andy Heald assisted in the field. Nancy Kaye assisted with the manuscript.

REFERENCES

Boulton, G.S., 1972, Modern Arctic glaciers as depositional models for former ice-sheets. J. Geol. Soc. Lond. 128, 361-393

Boulton, G.S., and Paul, M.A., 1976, The influence of genetic processes on some geotechnical properties of glacial tills. Q.J. Engng Geol. 9, 159-194

Boulton, G.S., and Eyles, N., 1979, Sedimentation by valley glaciers; a model and genetic classification, In Schlucter, C.H. (ed): Moraines and Varves, 11-23 Balkema, Rotterdam.

Charlesworth, J.K., 1955, The late-glacial history of the Highlands and Islands of Scotland. Trans. R. Soc. Edinb. 62, 769-928.

Clapperton, C.M., 1977, The northern highlands of Scotland. INQUA. X Congress 1977. Guidebook for Excursions A10 and C10. Geo. Abstracts Ltd. Norwich.

Clapperton, C.M., and Sugden, D., 1977, The Late Devensian Glaciation of north-east Scotland. In Gray, J.M., Lower, J.J. (eds): Studies in the Scottish Late-glacial Environment, 1-14. Pergamon, Oxford.

Coope, G.R., 1977, Fossil coleopteran assemblages as sensitive indications of climatic changes during the Devensian (Last) cold stage. Phil. Trans. R. Soc. Lond. B. 280, 313-337.

Coope, G.R., 1981, Episodes of local extinction of insect species during the Quaternary as indicators of climatic change. In, Neale, J., Flenley, J. (eds) The Quaternary in Britain, 216-221. Pergamon, Oxford.

Coope, G.R., Pennington, W., 1977, The Windermere Interstadial of the Late Devensian. Phil. Trans. Roy. Soc. Lond. B. 280, 337-339.

Cornish, R., 1981, Glaciers of the Loch Lomond Stadial in the western Southern Uplands of Scotland. Proc. Geol. Ass. 92, 105-114.

Dawson, A.G., 1979, A Devensian medial moraine in Jura. Scot. J. Geol. 15, 43-48.

Driscoll, F.G., 1980, Wastage of the Klutlan ice-cored moraines, Yukon Territory, Canada. Quat. Res. 14, 31-49

Evenson, E.B., Pasquini, T.A., Stewart, R. A., Stephens, G., 1979, Systematic provenance investigations in areas of alpine glaciation: applications to glacial geology and mineral exploration. In Schlucter, H. (ed): Moraines and Varves, 25-42. Balkema, Rotterdam.

Eyles, N., 1978, Rock glaciers in Esjufjoll Nunatak area southeast Iceland. Jokul, 28, 53-56.

Eyles, N., 1979, Facies of supraglacial sedimentation on Icelandic and Alpine temperate glaciers. Can. J. Earth Sci. 16, 1341-1361.

Eyles, N., Eyles, C.H. and Miall, A.D., (In Press), Lithofacies types and vertical profile models; an alternative approach to the description and environmental interpretation of glacial diamict and diamictite sequences. Sedimentology.

Eyles, N. and Rogerson, R.J., 1977, Glacier movement, ice structures and medial moraine form at a Glacier confluence, Berendon Glacier, British Columbia, Canada. Can. J. Earth Sci., 14, 2807-2816.

Eyles, N., and Rogerson, R.J., 1978, A framework for the investigation of medial moraine formation: Austerdalsbreen, Norway and Berendon Glacier, British Columbia. J. Glaciol., 20, 99-114.

Gray, J.M. and Brooks, C.L., 1972, The Loch Lomond Readvance moraines of Mull and Menteith. Scottish J. Geol. 8, 95-103.

Gray, J.M. and Lowe, J.J., 1977, The Scottish Lateglacial environment: a synthesis. In Gray, J.M., Lowe, J.J. (eds): Studies in the Scottish Lateglacial Environment, 163-181. Pergamon, Oxford.

Grove, J.M., 1972, The incidence of landslides, avalanches and floods in western Norway during the Little Ice Age. Arctic and Alpine Res. 4, 131-138.

Kaldal, I., 1978, The deglaciation of the area north and northeast of Hofsjokull, central Iceland. Jokull, 28, 18-31.

Lawson, D.E., 1981a, Distinguishing characteristics of diamictons at the margin of Matanuska Glacier, Alaska. Annals of Glac. 2, 78-84

Lawson, D.E., 1981b, Sedimentological characteristics and classification of depositional processes and deposits in the glacial environment. CRREL Rep. 81-27.

Lawson, D.E., 1982, Mobilization, movement and deposition of active subaerial sediment flows. Matanuska Glacier, Alaska. J. Geol. 90, 279-300

Lowe, J.J., and Walker, M.J.C., 1980, Problems associated with radiocarbon dating the close of the Lateglacial period in the Rannoch Moor area, Scotland. In Lowe, J.J., Gray, J.M., Robinson, J.E. (eds.) The lateglacial of North-west Europe. 123-138. Pergamon, Oxford.

Lowe, J.J., and Walker, M.J.C., 1981, The early postglacial environment of Scotland: evidence from a site near Tyndrum, Perthshire. Boreas, 18, 281-293.

Meier, M.F., 1961, Mass budget of south Cascade Glacier, 1957-60. U.S. Geol. Surv. Prop. Paper, 424B, 206-211

Okko, V., 1955, Glacial drift in Iceland, its orgin and morphology. Comm. Geol. de Finlande, Bul. 170, 130 p.

Pennington, W., 1977, The late Devensian flora and vegetation of Britain. Phil. Trans. R. Soc. Lond. B. 280, 247-71

Price, R.J., 1973, Glacial and Fluvioglacial Landforms. Oliver and Boyd.

Rothlisberger, F., and Schneebeli, W., 1979, Genesis of lateral moraine complexes, demonstrated by fossil soils and trunks; indicators of postglacial climatic fluctuations. In: Schluchter, H. (ed) Moraines and Varves, 387-420. Balkema, Rotterdam.

Shaw, J., 1977, Tills deposited in arid polar environments Can. J. Earth. Sci. 14, 1239-1245

Shaw, J., 1979, Genesis of the Sveg Tills and Rogen Moraines of central Sweden: a model of basal melt-out. Boreas, 8, 409-426.

Sissons, J.B., 1967, The Evolution of Scotland's Scenery. Edinburgh.

Sissons, J.B., 1974a, A lateglacial ice cap in the central Grampians, Scotland. Trans. Inst. Br. Geogr., 62, 95-114

Sissons, J.B., 1974b, The Quaternary in Scotland: a review. Scot. J. Geol. 10, 311-337.

Sissons, J.B., 1976, The Geomorphology of the British Isles: Scotland, Methuen, London. 150 pp.

Sissons, J.B., 1977a, The Scottish Highlands. INQUA X CONGRESS 1977. Guidebook for excursions All and Cll. Geo. Abstracts Ltd. Norwich.

Sissons, J.B., 1977b, The Loch Lomond Readvance in the northern mainland of Scotland. In Gray, J.M., Lowe, J.J. (eds): Studies in the Scottish Lateglacial Environment, 45-60. Pergamon, Oxford.

Sissons, J.B., 1979a, The Loch Lomond Stadial in the British Isles. Nature. 280, 199-203.

Sissons, J.B., 1979b, The limit of the Loch Lomond Advance in Glen Roy and vicinity. Scot. J. Geol., 15, 31-42.

Sissons, J.B., 1979c, The Loch Lomond Advance in the Cairngorm Mountains. Scot. Geog. Mag., 95, 66-82.

Sissons, J.B., 1980, The Loch Lomond Advance in the Lake District, Northern England. Trans. Roy. Soc. Edinb. Earth Sciences, 71, 13-27

Sissons, J.B., 1981a, Ice-dammed lakes in Glen Roy and vicinity: a summary. In Neale, J., Flenley, J. (eds): The Quaternary in Britain, 174-183. Pergamon, Oxford.

Sissons, J.B., and Grant, A.J.H., 1972, The last glaciers in the Lochnagar area, Aberdeenshire. Scot. J. Geol., 8, 85-93.

Sugden, D., 1980, The Loch Lomond advance in the Cairngorms, Scot. Geo. J. 96,

Sutherland, D.G. 1980, Problems of radiocarbon dating deposits from newly deglaciated terrain: examples from the Scottish Lateglacial. In Lowe, J.J., Gray, J.M., Robinson, J.E. (eds): Studies in the Lateglacial of Northwest Europe. 139-150. Pergamon, Oxford.

Sissons, J.B., Sutherland, D.G., 1976, Climatic inference from former glaciers in the south-east Grampian Highlands, Scotland. J. Glaciol. 17, 325-346.

Thompson, K.S.R., 1972, The last glaciers in western Perthshire. Unpublished Ph.D. thesis: University of Edinburgh.

Vikingsson, S., 1978, The deglaciation of the southern part of the Skagafjordur District, Northern Iceland. Jokull 28, 1-17.

Wright, H.E., 1980, Surge moraines of the Klutlan Glacier, Yukon Territory, Canada. Quat. Res. 14, 2-19.

Young, J.A.T., 1974, Ice wastage in Glenmore upper Spey valley, Inverness-shire. Scot. J. Gol. 10, 147-157.

Young, J.A.T., 1975, Ice wastage in Glen Feshie, Inverness-shire, Scot. Geog. Mag. 91, 91-101.

Young, J.A.T., 1978, The landforms of Upper Strathspey. Scot. Geog. Mag. 94, 76-94.

Different till types in North Germany and their origin

J.EHLERS
Geologisches Landesamt Hamburg, Germany

1 INTRODUCTION

Recently a lot of research work has been done on the genesis of Pleistocene deposits, and within this field of work most effort has been concentrated on the formation of till. Since the formation of the INQUA Commission 'On Genesis and Lithology of Quaternary Deposits', symposia on this subject have been held in Columbus 1969 (Goldthwait 1972), in Ottawa 1975 (Leggett 1976), in Warsaw 1975 (Stankowski 1976), in Stockholm 1976 (Johansson 1977), in Birmingham 1977, in Zurich 1978 (Schlüchter 1979), in Trondheim 1979 (Sollid 1980), in Orono 1980 and in Wyoming 1981. At these meetings and field excursions many problems concerning till genesis have been discussed, including results from the recently glaciated areas.

The contribution of German authors to these activities has been remarkably small. Much work has been done on north German tills, but most of it is published in German with only short English summaries. This paper briefly summarizes the different till types which can be found in North Germany.

The Pleistocene stratigraphy of North Germany can be seen in fig.1. During the Elsterian and Saalian, North Germany was covered by a continental ice sheet from the Baltic Sea to the Central German Uplands (fig. 2). During the Weichselian the ice did not cross south of the Elbe River. Its margin lay in Schleswig-Holstein, just north of Hamburg. During the Saalian and the Weichselian several separate ice advances can be distinguished. The same is assumed for the Elsterian, but as the Elsterian deposits are almost completely covered by younger sediments and are mostly known only from borings, no subdivision of the Saalian has been possible yet. An earlier attempt by Richter (1962) had to be discarded; only in the GDR have two different ice advances been distinguished so far (Eissmann 1975, Eissmann & Müller 1979).

2 DIFFERENCES DUE TO THE AGE OF THE DEPOSITS

During the first half of this century it was assumed that tills of different age could be distinguished by their different degree of weathering (Mückenhausen 1939). No evidence of that has been found in the field. Tills of Elsterian age often contain pieces of chalk and limestone which are completely unweathered and look as fresh as clasts from the youngest Weichselian till. Weathering intensity is closely related to the soil formation which took place mainly during the interglacials and - to a lesser degree - during the interstadials (Felix-Henningsen 1979, 1982). The pedogenetic processes depend largely on the composition of the substratum. It will take much more time to decalcify a till, which is rich in calcium carbonate than another that is poor in it. The differences Mückenhausen found in the field

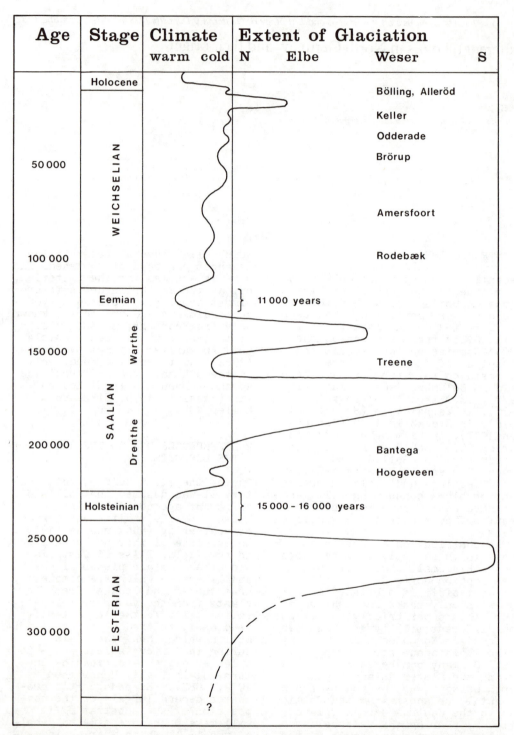

Figure 1. Pleistocene stratigraphy in North Germany. Based on Woldstedt & Duphorn 1974, Müller 1974a, b, K.-J. Meyer 1974, Zagwijn & Doppert 1978, Berggren et al. 1980, Berglund & Lagerlund 1981, Birchfield, Weertman & Lunde 1981, Woillard 1978.

and attributed to the age of the sediments were, in reality, primary compositional differences.

Apart from pedogenic alterations, the tills show some age-dependant differences. During the first major ice advances in the Elsterian the pre-Pleistocene surface was most easily accessible to glacial erosion. Consequently glaciers reworked a great deal of the Tertiary quartz sands and clays and incorporated them in the till. Therefore the Elsterian tills in many places have a very local appearance. Whereas in Hamburg the typical Elsterian till is of a black and clay-rich type, in the Schwanewede area (north of Bremen) it is light grey to greenish and extremely sandy (Grube 1967, Höfle 1977, 1979).

During the later glaciations the Tertiary surface was increasingly covered by younger sediments. Therefore the Saalian and the Weichselian tills contain less quartz and Tertiary clay and are less altered by local influences than the older ones.

3 DIFFERENCES DUE TO CHANGES IN ICE MOVEMENT

In Germany the Pleistocene stratigraphy is traditionally based on indicator pebble counts. It was recognized very early that different tills seemed to contain different assemblages of indicator pebbles (Milthers 1934, Hesemann 1934). Consequently, the pebble assemblages were used as stratigraphical tools. It was assumed that tills of a certain age were characterized by a certain indicator pebble assemblage (Lüttig 1958). Accordingly, the North German Pleistocene stratigraphy was based mainly on the indicator pebble countings. The following rules were established:
1. The Elsterian tills are characterized by a rather high percentage of Norwegian erratics (Rhomb Porphyries, Larvikites, etc.).
2. The tills of the Drenthe substage (Saalian I) contain mainly material from Central and Southern Sweden (Dala Porphyries, Stockholm Granites, etc.),

whereas the Warthe deposits (Saalian II) are rich in Eastern Baltic material (Åland Granites, Rapakivis, Brown Baltic Porphyries).
3. For the Weichselian deposits no strict rules could be established.

But in recent years geologists have found it increasingly difficult to fit their results into the above-mentioned scheme. Tills that are rich in Eastern Baltic material occur not only in Warthe deposits, but also in Elsterian, Drenthe and Weichselian till sequences, mainly in the upper parts (Meyer 1976, Stephan 1980). These irregularities were explained in most cases as floes by those who, in spite of some critical remarks by foreign (Marcussen 1978, Schuddebeurs 1980/81) and East German scientists (Eissmann 1975) about the stratigraphical value of indicator pebbles, did not want to abandon the scheme: one ice advance = one indicator pebble assemblage. Others, however, looked for a glaciodynamic solution of the problem.

From research work in Scandinavia, we know that the ice divide did not remain stationary, but that it changed its position during the course of a glaciation. At the beginning of the glaciation it was situated near the main Scandinavian water divide. With a growing accumulation of ice in the lee of the high mountain ranges the ice divide shifted towards the east. It was finally centred near the Baltic Sea coast (Lundqvist 1982). When the ice started to melt down at the end of the glaciation, the ice divide returned slowly to its position near the main water divide.

This shifting of the ice divide during the course of each glaciation had consequences for the ice movement. In Scandinavia the changing directions of ice movement could be proved by measurements of the striations or by analyzing the fabric and composition of thick till sequences in some valleys in Central Norway (Bergersen & Garnes 1972, 1981) for example. But the changing direction of ice movement was important in the more marginal parts of the

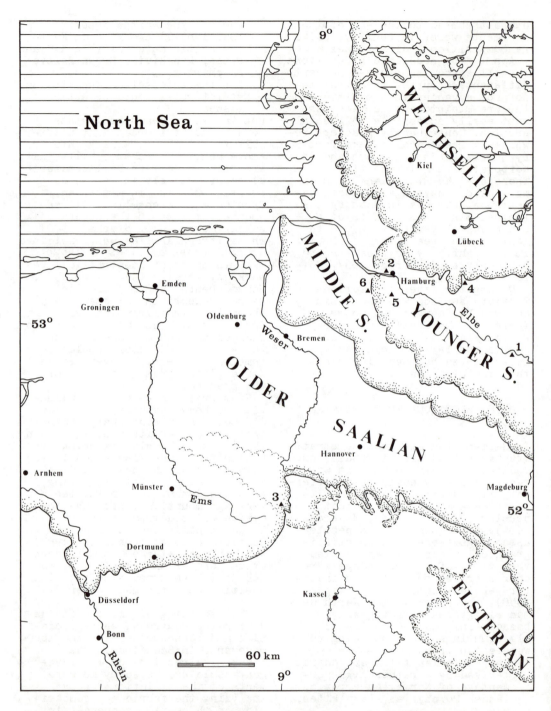

Figure 2. Limits of the Pleistocene glaciations in North Germany (after Liedtke 1981, Ehlers & Stephan 1982) and locations mentioned in the text. 1 = Höhbeck, 2 = Hamburg-Bahrenfeld, 3 = Detmold, 4 = Segrahner Berg, 5 = Beckedorf, 6 = Eilendorf.

glaciated areas as well.

Till fabric measurements in the Hamburg area (Ehlers 1978a) revealed that during the Saalian glaciation there were two cycles of till deposition with ice advances from north to south, then from the northeast and finally from the east. Stephan (1980, 1982) came to a similar conclusion for the Saalian tills of Schleswig-Holstein. He stated that during the Saalian there had been two major ice advances (the Drenthe and the Warthe substages, see fig. 1). During each of these substages the ice movement changed in a clockwise direction. There are a few exceptions to the rule, but in general this changing pattern of ice movement seems to reflect the migration of the main ice divide. The return of the ice divide to the mountain areas had no consequences for North Germany because during the decline of the glaciation the marginal parts of the ice sheet became inactive and thus no longer reflected the dynamics of the centre. Much of this later migration took place after deglaciation in North Germany.

Which course did the ice take from Scandinavia to North Germany? Normally a glacier - like water - tends to flow downslope. The same is true of a continental ice sheet. That means that major obstacles were avoided rather than crossed. This can be illustrated by an example from central Germany. Recently the idea that the Saalian ice had crossed the Teutoburger Wald mountains was rejected (Thome 1980). It was proved that the glaciers instead invaded the Münsterland basin 'round the corner' from the lower areas further to the west. The flow pattern of the Saalian ice was reconstructed by Ehlers & Stephan (1982a).From this it can be assumed that an ice sheet flowing from Scandinavia to North Germany was strongly influenced by the pre-existing topography.

There are three main courses the ice could take from Scandinavia to North Germany (fig. 3):

1. The westernmost course (stage I of a glaciation) leads from Norway and Central Sweden via the Kattegat and the Danish Isles to Schleswig-Holstein. On this path the glaciers bring along indicator pebbles from the Oslo area and from Southern Sweden. The tills of this stage contain little chalk and flint. Only in the Elsterian did the ice of this phase reach as far south as North Germany. In the early Saalian and in the early Weichselian these advances terminated in Denmark (Sjørring 1982).

2. During the second stage of the glaciation the ice advanced across the Swedish peninsula and approached Germany from between Bornholm and Rügen islands. The tills of this ice advance contain indicator pebbles from Central and Southern Sweden. As the ice had to cross the Cretaceous chalk areas at the bottom of the Baltic Sea, the tills are very rich in chalk and flint.

3. Finally, the ice followed the course of the depression of the Baltic Sea and reached North Germany flowing in an east-west direction. During this third stage the glaciers brought abundant Eastern Baltic rocks from the Åland Islands and from the bottom of the Baltic Sea.

This model of the course of a glaciation easily explains the fact that the uppermost till layers of all known ice advances in North Germany are extremely rich in Eastern Baltic material (Gauger & Meyer 1970, Meyer 1976). Although this idea was discussed by several authors (Eissmann 1967, Duphorn 1978, Stephan 1980, Ehlers 1981), it is still not generally accepted. The reasons for this are as follows:

1. First of all it must be stated that not all of the mentioned till types form a complete cover over all of North Germany. In particular, the remnants of the East Baltic ice advances are rather seldom found. Only during the last few years has it been shown that these tills are more than just a few small floes in the large, uniformly composed till sheets of Swedish origin.

2. In many places some of the different tills are separated from each other by thin layers of sand. This outwash was deposited during minor oscillations of the

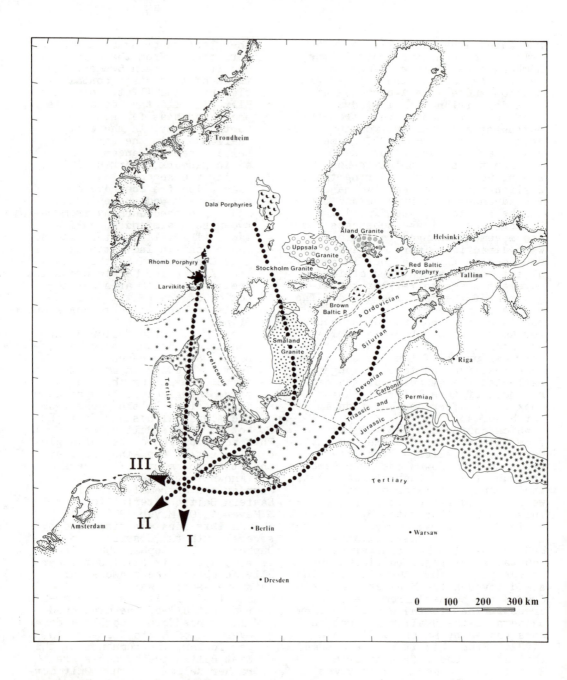

Figure 3. Changing directions of ice movement from Scandinavia to Hamburg during the course of a glaciation and source areas of important indicator rock types. Based on Hesemann 1975, Hucke & Voigt 1967, Martinsson 1979.

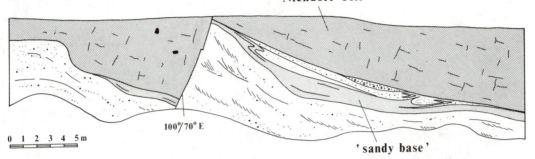

SE   Niendorf Till   NW

100°/70° E

0 1 2 3 4 5 m

'sandy base'

Figure 4. Exposure south of Neu Wulmstorf, showing the Niendorf Till and its 'sandy base', thrusted by the Fuhlsbüttel advance.

ice front. Till strata that appear to be different, separated by sand layers, used to be interpreted as tills of different, independent ice advances. A typical example is the Fuhlsbüttel Till of the Hamburg area (Grube 1967, 1981a).

3. Because in each of these till layers there was only one fabric maximum and a uniform lithological composition, there seemed to have been no change in the direction of ice movement during one glaciation.

All of these points seemed to support the hypothesis that each glaciation was represented by one till, characterized by one fabric, one direction of ice movement and one pebble assemblage. All deviations from this pattern were interpreted as floes.

To illustrate the above-mentioned model of changing directions of ice movement, let us have a closer look at the sediments of the Middle and Younger Saalian in the Hamburg area (the Warthe substage).

The sequence starts with a very sandy till of yellowish-brown colour. The clast fabric shows a maximum around 0 to 5°. The till contains many Swedish indicator pebbles, but not too much chalk and flint. It seems to represent a direction of ice movement somewhere between phase I and II of a glaciation (see above).

This sandy till, when it was first described by Meyer (pers. comm.) was assumed to be the sandy base of the overlying clay-rich Middle Saalian till, formed by

partial reworking of the underlying outwash material that was incorporated by shearing into the till. Grube (1981a) regards this type of sediment as 'sole till'. But reworking cannot explain the differences in composition and fabric which occur between this sandy till and the overlying clay-rich one (see Ehlers 1978a). Therefore both tills have to be regarded as different members of the same Middle Saalian till unit.

This sandy till is overlain by a clay-rich, grey till which together with its content of Swedish indicator pebbles contains a lot of chalk and flint. Its calcium carbonate content is about 20 % (Baermann, Iwanoff & Wilke 1982). The fabric shows an orientation maximum at about 35° (phase II).

In an exposure south of Neu Wulmstorf the two Middle Saalian tills can be seen (fig. 4). The grain-size differences between both tills are obvious in the field. Whereas the lower till resembles a sand rather than a till, the clay-rich Niendorf Till is much more poorly sorted and shows a grain-size distribution typical of till. Because of thrusting by the Younger Saalian Fuhlsbüttel advance no fabric measurements could be made in this exposure, but several measurements have been published from neighbouring exposures, all showing the same results (Ehlers 1978a).

The Younger Saalian ice advance came almost directly from the east (fig. 7). It is separated from the Middle Saalian tills only by a minor layer of mostly fine-grained

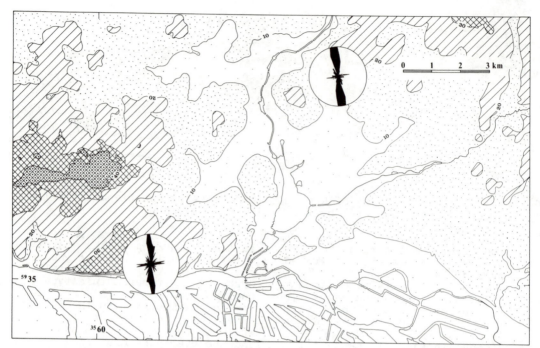

Figure 5. Fabric measurements in the 'sandy base' of the Niendorf Till in Hamburg. Ice movement from the north.

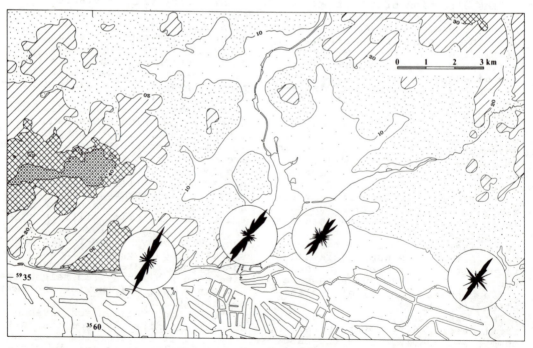

Figure 6. Fabric measurements in the Niendorf Till in Hamburg. Ice movement from the northeast (stage II in fig. 3)

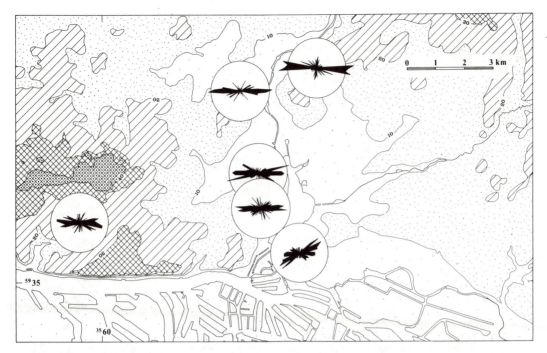

Figure 7. Fabric measurements in the Fuhlsbüttel Till in Hamburg. Ice movement from the east (stage III in fig. 3).

outwash. It is assumed that this interval represents only a minor oscillation of the ice margin within the Warthe substage, and that the Middle and Younger Saalian tills stratigraphically belong closely together (Ehlers 1981).

The Younger Saalian Fuhlsbüttel Till is rather sandy again. Its pebble assemblage does not differ markedly from that of the Niendorf Till, but fabric measurements show a long axis orientation of about 90 - 100°. Therefore this till may well be the last deposit of the Warthe substage in the Hamburg area.

Some tens of kilometres east of Hamburg this till is partially replaced by the reddish coloured Vastorf Till, which contains a vast amount of East Baltic indicator pebbles (Gauger & Meyer 1970, Gauger 1978, Meyer 1976). In the Höhbeck area both tills occur side by side (pers. comm., Kabel). Fabric analyses there show an E - W orientation, too.

All of these compositionally different tills represent changes in facies and not in the mode of deposition. Therefore it should be stressed that a different composition does not necessarily mean a different genesis. Fabric, composition, and the sedimentary environments of the tills in question have to be taken into consideration, before genetical interpretations can be made.

4 DIFFERENCES IN TRANSPORT

Studies in Scandinavia have shown that till material is not just picked up by the glacier in one place, carried along in the ice and deposited somewhere else, but that it rather undergoes a permanent process of reworking and redeposition. Studies by Lindén (1975), Perttunen (1977) and Haldorsen (1977) have shown that the Scandinavian tills closely reflect the composition of the local bedrock.

With some restrictions the same is true of the tills in North Germany, although we are much closer to the ice margin and in an area of net deposition. It has been widely disregarded that the striking boulders and pebbles of Scandinavian

69

origin that are found in our tills form only a 'minor' component surrounded by a differently composed matrix. The > 2 mm fraction normally makes up for less than 2 % of the material. The sand, silt and clay fractions of the tills are to a high degree composed of material derived from the local bedrock, which is Tertiary quartz sands and mica clays. As the Upper Tertiary formations at the base of the Pleistocene deposits have no coarser particles, they are rare in the fine gravel fraction, and practically absent in the medium and coarse gravel fractions. These are the fractions which are mostly investigated, and that leads to the idea of tills completely consisting of far-travelled Scandinavian material.

That the local component plays an important role can be seen quite easily, if we go further to the south, where the glaciers could rework fluvial gravel of southern origin. Here the composition of the tills undergoes a sudden change, and the 'southern' components start to dominate. This is illustrated by the fine gravel analysis of two samples of Older Saalian Drenthe till taken near Detmold (Kater gravel pit) at the northern slope of the Teutoburger Wald mountains (fig. 8).

Thus, it can be seen that all till material that was transported in the basal parts of the ice (as is typical for glacial transport) must have a mixed composition of all sorts of material the ice has overridden on its way to the area of deposition, and that to a certain degree the tills always reflect the local material. If this material is an older till, this will influence the composition of the younger till, too. Therefore our tills usually consist of a mixture of reworked material, a good part of which has already undergone Pleistocene reworking.

But there are some tills that are strikingly different from this concept. The red tills of the late ice advances are all similar in that they consist of rather pure far-travelled material and that they

Hamburg Bahrenfeld

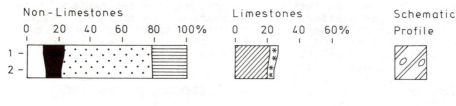

Detmold

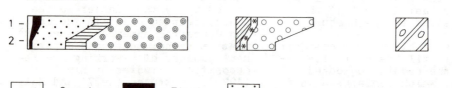

Figure 8. Fine gravel analyses (3 - 5 mm fraction) of Older Saalian Drenthe till in Hamburg-Bahrenfeld (construction site) and in Detmold (Kater gravel pit). The enormous reworking of local material in the Detmold samples has largely changed the original composition of the till.

Segrahner Berg

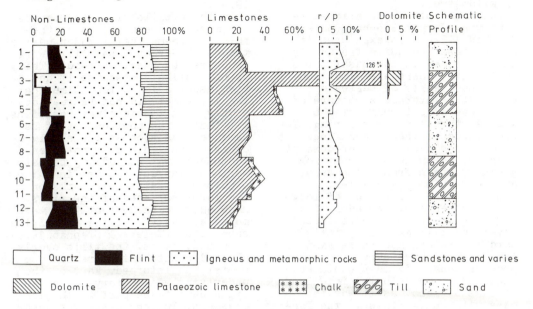

Figure 9. Fine gravel analysis of samples from the Segrahner Berg sand pit (Schleswig-Holstein). The composition of the red till (sample 3) is different from the other samples – possibly due to englacial transport.

contain almost no material of local or regional origin (Meyer 1976, 1982, Ehlers 1980). This can be shown by indicator pebble counts as well as by fine gravel analyses. In the fine gravel fraction the local component of North German tills is quartz (from the Upper Tertiary lignite sands) and, in a broader sense, chalk and flint (from the Upper Cretaceous in the Southern Baltic Sea area).

The fine gravel diagram of the Segrahner Berg section in eastern Schleswig-Holstein (fig. 9) shows that among the tills in this thrusted sequence of Saalian sediments there is one thin layer of reddish till, the composition of which is completely different from the rest. This layer (sample 3) contains almost no flint and chalk and no quartz, whereas its content of Palaeozoic limestones is extremely high. It is assumed that this till is the red facies of the Younger Saalian till, the Vastorf member (Gauger 1978).

To explain how this till could be transported unchanged from its source area somewhere in the East-

ern Baltic to the area of deposition, an englacial transport has to be envisioned. The alternative would be basal transport over a ground, which was unaccessible to erosion, either because it was completely frozen, or because the glaciers had a wet base and could only deposit all the way long. But the very long distances of transport (several hundred kilometres; see fig. 3) weighs against this assumption.

This means that generally two different modes of transport have to be considered for the North German tills, a basal one and an englacial one.

5 DIFFERENCES IN DEPOSITION

Until recently it was assumed that almost all North German tills were deposited as basal tills, mainly as lodgement tills. This theory was supported by the massiveness of most of the tills, by their strong fabric and by the fact that till sheets of a very uniform composition could often be followed at nearly the same altitude over tens of kilometres.

This concept was questioned recently by Jürgen Stephan (Ehlers & Stephan 1982b). In exposures it can be observed quite frequently that imbedded in the basal parts of the till there are almost undisturbed lenses of meltwater sands, the internal structure of which was almost unaffected by glacial thrusting or shearing. The occurrence of these sand lenses is unconformable with the process of lodging, which means deposition grain by grain from the base of an actively moving glacier.

Even if the definition of lodging is taken a bit broader, including lodging of centimetre-thick layers of till material, it is hard to explain how these sand lenses could have survived this process unharmed. They would be expected to be deformed into long streaks of sand, as it was the case with chalk floes in the Dan-

ish cliff sections (Berthelsen 1979).

If these tills with sand lenses are more than exceptions of the rule, it seems that a major part of the North German basal tills were deposited by basal melt-out. This is not unlikely, because it has been assumed for a long time that the ice sheet did not retreat actively from its marginal positions, but that large portions of it became dead as soon as the supply decreased. But this does not mean that there was no lodging beneath North German ice sheets. In quite a number of cases sole marks on the lower surfaces of the tills show quite clearly that they were moved actively over the underlying substratum (Ehlers & Stephan 1979).

In many cases, the tills overly older deposits which are not completely undisturbed. Part of the underlying sedimentary sequences often have taken part in the ice movement. In many cases the shear planes run parallel to former bedding planes, which makes them hard to detect. Only where the underlying layers have been pushed or folded (fig. 11) can the influence of the overriding glacier be clearly proved.

It has been proposed that these layers be called 'deformation till' (Dreimanis 1976). Grube (1979) speaks of 'sole till'. But it seems doubtful whether these strata should be called 'till' at all. Some geologists argue that everything that was moved by the ice, even if it was for a few centimetres, should be called till. But as the sediments in question are hardly deformed and have neither the structure nor the composition of tills (diamicton), we prefer to call them 'subglacially deformed sediments' (Ehlers & Stephan 1982b).

The North German tills are hardly ever overlain by anything which could be called an ablation till. This may be partly due to the fact that the uppermost parts of the tills were either eroded by later ice advances or altered by periglacial action and soil forming processes in such a way that they cannot be identified as a distinct type of till anymore.

The same is true of supraglacial 'flow till', which so far has been

Figure 10. Reworked chalk floes in till in a cliff section on the island of Møn. The chalk forms centimetre thick lenses within the till.

Figure 11. Glaciofluvial sands deformed by the overriding glacier. The deformations are invisible where they run parallel to the bedding planes (left); only where folding occurs (centre) the pushing becomes obvious. The photograph was taken in a sand pit on the island of Falster.

hardly found in North Germany. Stephan recently mentioned some examples from Schleswig-Holstein (in Sollid 1980: 99). Most of the 'flow tills' deposited at the snout of advancing glaciers would have been reworked immediately when they were overridden. 'Flow tills' deposited during the retreat phase of the ice sheet were mostly altered and changed by post-depositional processes. Only those that were deposited in a considerable thickness in depressions in front of the advancing ice had a chance to be preserved. Therefore most of the 'flow tills' of Saalian age which have been found until now are situated beneath the basal till of the main advance rather than on top of it (Ehlers 1978b).

In a gravel pit west of Beckedorf 'flow till' of Middle Saalian age was exposed (figures 12 and 13). Rather fine-grained proglacial outwash is overlain by about 1 m of a till-like material, intercalating with fine sands and silts (figs. 12 and 13). This sequence of glacigenic and glaciofluvial material is interpreted as a 'flow till' of the Middle Saalian ice advance. It is overlain by several metres of fine sands and silts, probably deposited during a stagnation phase, when the ice margin retreated slightly to the east.

When the ice started to readvance, the whole sequence of 'flow till' and glaciolacustrine sediments was covered with thick meltwater sands. The basal till of the Middle Saali-

Figure 12. 'Flow till' interbedded with sand layers in the Beckedorf sand pit, south of Hamburg.

an advance is missing; on top of the chalk-rich Middle Saalian outwash the Younger Saalian Fuhlsbüttel Till is to be found, with its characteristic fine gravel composition (little quartz and flint) and with its E - W pebble orientation (Grube & Ehlers 1977, Ehlers 1978a). In the eastern part of the pit beautiful folds, thrusted from the east, are exposed.

The Beckedorf 'flow till' is a very sandy till (fig. 16). Its grain-size distribution comes very close to that of a meltwater sand, but the sediment has a completely different structure. Although some kind of sorting (or reworking of sorted material) has taken place, the 'flow till' shows no signs of grading.

In spite of its till-like appearance the 'flow till' is not till in a true sense. When Gripp (1929) described these till flows on the snouts of Spitsbergen glaciers, he called them 'mud flows' (Schlamm-

ströme) and not 'tills'. Although for practical reasons (mapping) it seems suitable to call these sediments 'till', from a genetical point of view they are sediment flows, not deposited by the glacier. Therefore, according to Lawson (1979), the term 'till' for these deposits should be discarded.

The same problem arises with the so-called 'waterlain tills', i.e. tills, deposited from the base of a floating ice sheet into stagnant or flowing water. From a genetical point of view these 'waterlain tills' resemble much more lacustrine than glacigenic sediments. In North Germany so far they are rarely found. Big ice-dammed lakes were formed only during the Elsterian, and their deposits are at a considerable depth now and are hardly accessible for detailed studies.

An Exposure in Eilendorf SW of Hamburg (figs. 14 and 15) shows a 'waterlain till' in Saalian deposits, probably formed in a small ice-dammed lake in front of the ad-

Figure 13. 'Flow till' in the Beckedorf sand pit.

SAND PIT EILENDORF

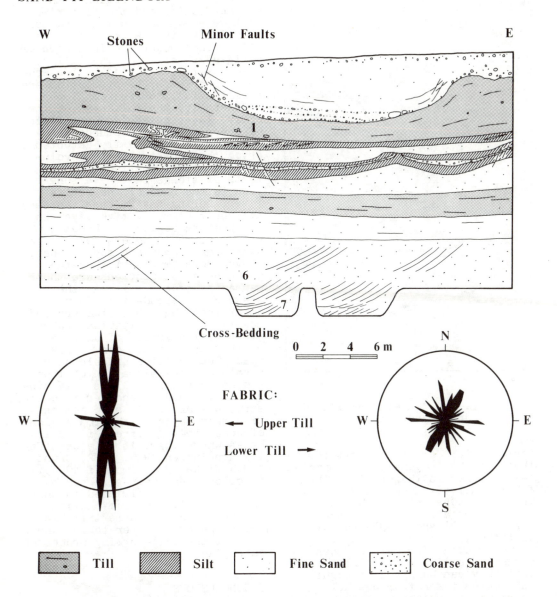

Figure 14. The Eilendorf sand pit, southwest of Hamburg. The upper till
is a basal till with a strong fabric (left diagram), whereas the lower
till is waterlain (see fig. 15) and shows only a weak orientation of the
long axes. The numbers in figs. 14 and 15 indicate from where the samples for the grain-size analyses were taken (fig. 16).

vancing ice. The interfingering
with glaciolacustrine silts and
sands gives evidence of the lacustrine environments under which the
deposition took place.

The cumulative grain-size curves
(fig. 16) show the difference between the pure glaciolacustrine deposits and the 'waterlain till'.
The grain-size curves of the 'wa-

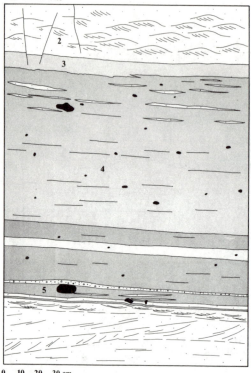

EILENDORF SAND PIT
The Lower Till

0   10   20   30 cm

Figure 15. The 'waterlain till' in the Eilendorf sand pit.

terlain till' are rather similar to those of the lodgement till sample, taken higher in the profile. The fabric analysis shows, however, that the lodgement till has a good N – S orientation, whereas the 'waterlain till' shows only random distribution of the long axes.

One argument for calling these rather rare till types 'till' is that they can hardly be recognized in samples taken from boreholes. Where the sedimentary environments of the deposits cannot be clarified, it will not be possible for the mapping geologist to distinguish the different genetic till types.

6 POST-DEPOSITIONAL ALTERATIONS

Many of the differences between tills, as we find them in the field today, are not primary but are due to post-depositional alterations.

The older tills, especially those of the Saalian and the Elsterian that were exposed on the land surface during the Weichselian Glaciation, came under the influence of periglacial climate. The till sheets were fragmented by ice wedges, reworked by cryoturbation and solifluction occurred on slopes.

Generally, the thickness of the active layer determined the depth of the structural disturbances. These range from easily detectable cryoturbations to single involutions or the formation of pavements of stones that have sunk down to the base of the active layer. During the Weichselian, the thickness of the active layer in the Hamburg region was about 2 m (see fig. 36 in Ehlers 1978a). Ice wedges may penetrate much deeper.

Even more important than the periglacial alterations are those caused by pedogenesis. As these are highly dependant on factors other than the substratum (relief, vegetation, precipitation) on every till a series of different weathering profiles can be developed. The solution of calcium carbonate and the migration of iron and manganese as well as clay minerals all lead to changes in colour, structure and grain-size composition that may give the impression of completely different tills in cases where there is only one (Felix-Henningsen 1979, 1982, Grube 1981b, Wilke 1981, Iwanoff 1981). As colour and grain-size composition are routinely used to characterize till, special caution is advised in cases where till lies above the groundwater table, where most of the soil-forming processes take place.

7 CONCLUSIONS

In North Germany a number of different till types can be distinguished. The most important differences are in facies due to changes in the direction of ice movement. These differences are also age-dependant because some of the source areas got buried and others got excavated during the course of later glaciations. The mode of transport led to further compositional differences. Some

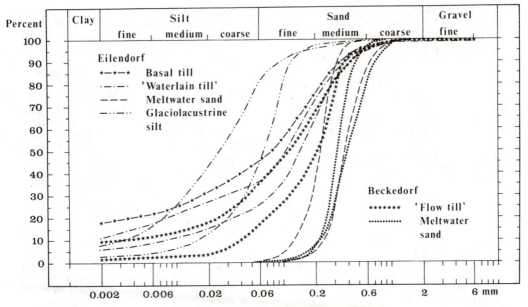

Figure 16. Grain-size analyses of samples from Beckedorf and Eilendorf, showing the difference in sorting between the 'flow till' and 'waterlain till' samples and the glaciofluvial and glaciolacustrine deposits.

tills seem to have been transported englacially, without being mixed with any local material.

The differences in mode of deposition, on which much interest has focused in recent studies, does not play such an important role in North Germany. The vast majority of all tills is of a massive, uniformly composed type with a strong fabric and was most likely deposited as basal till, partly by lodging and partly by melt-out processes. Supraglacial tills are rarely preserved. 'Waterlain tills' occur, but they are rare as well.

Grain-size composition, structure and colour of tills has often been changed by post-depositional periglacial or pedogenic processes. In this way a number of new, secondary till types was created. These post-depositional changes must be taken into consideration in any till study.

8 ACKNOWLEDGEMENTS

I wish to thank Professor Dave Mickelson, University of Wisconsin, who kindly helped me to translate the manuscript.

9 REFERENCES

Baermann, A., A. Iwanoff & H. Wilke 1982. The calcium carbonate content of North German tills, in print.

Bergersen, O.F. & K. Garnes 1972. Ice movements and till stratigraphy in the Gudbrandsdalen area. Preliminary results, Norsk Geogr. Tidsskr. 26: 1-16.

Bergersen, O.F. & K. Garnes 1981. Weichsel in central south Norway: the Gudbrandsdal Interstadial and the following glaciation, Boreas 10: 315-322.

Berggren, W.A., L.H. Burckle, M.B. Cita, H.B.S. Cooke, B.M. Funnell, S. Gartner, J.D. Hays, J.P. Kennett, N.D. Opdyke, L. Pastouret, N.J. Shackleton & Y. Takayanagi 1980. Towards a Quaternary Time Scale, Quaternary Res. 13: 277-302.

Berglund, B.E. & E. Lagerlund 1981. Eemian and Weichselian stratigraphy in South Sweden, Boreas 10: 323-362.

Berthelsen, A. 1979. Recumbent folds and boudinage structures formed by subglacial shear: an example of gravity tectonics, Geologie en Mijnbouw 58 (2): 253-260.

Birchfield, G.E., J. Weertman & A. Lunde 1981. A Palaeoclimate Model of Northern Hemisphere Ice Sheets, Quaternary Res. 15: 126-142.

Dreimanis, A. 1976. Tills: their origin and properties. In Legget, R.F. (ed.), Glacial till. An inter-disciplinary study. Royal Soc. Can. Spec. Publ.12: 11-49.

Duphorn, K. 1978. Diskussionsbeitrag in: Der Geschiebesammler 12 (2/3).

Ehlers, J. 1978a. Die quartäre Morphogenese der Harburger Berge und ihrer Umgebung. Mitt. Geogr. Ges. Hamburg 68: 181 pp.

Ehlers, J. 1978b. Vor dem Eisrand abgelagerte Sedimente - Beispiele aus dem nördlichen Niedersachsen, Mitt. Geol.-Paläontol. Inst. Univ. Hamburg 48: 17-32.

Ehlers, J. 1980. Feinkieszählungen im südlichen Geestgebiet Dithmarschens, Schr. Naturw. Ver. Schleswig-Holstein 50: 37-55.

Ehlers, J. 1981. Problems of the Saalian Stratigraphy in the Hamburg Area, Meded Rijks Geol. Dienst 34 (5): 26-29.

Ehlers, J. & H.-J. Stephan 1979. Forms at the base of till strata as indicators of ice movement, J. Glaciol. 22 (87): 345-355.

Ehlers, J. & H.-J. Stephan 1982a. Till fabric and ice movement, in print.

Ehlers, J. & H.-J. Stephan 1982b. North German till types, in print.

Eissmann, L. 1967. Rhombenporphyrgeschiebe in Elster- und Saalemoränen des Leipziger Raumes, Abh. u. Berichte d. naturkundl. Museums „Mauritianum" Altenburg 5: 37-46.

Eissmann, L. 1975. Das Quartär der Leipziger Tieflandsbucht und angrenzender Gebiete um Saale und Elbe. Schriftenreihe f. Geol. Wissenschaften 2: 228 pp.

Felix-Henningsen, P. 1979. Merkmale, Genese und Stratigraphie fossiler und reliktischer Bodenbildungen in saalezeitlichen Geschiebelehmen Schleswig-Holsteins und Süd-Dänemarks. Kiel (Thesis): 219 pp.

Felix-Henningsen, P. 1982. Palaeosols and their stratigraphical interpretation, in print.

Gauger, W. 1978. Zehn Jahre Forschung in den Kiesgruben des Raumes Vastorf (10 km östlich von Lüneburg), ein Abschlußbericht, Jahrbuch Naturw. Verein Fürstentum Lüneburg 34: 65-84.

Gauger, W. & K.-D. Meyer 1970. Ostbaltische Geschiebe (Dolomite, Old Red-Sandsteine) im Gebiet zwischen Lüneburg und Uelzen, Der Geschiebesammler 5 (1): 1-12.

Goldthwait, R.P. (ed.) 1971. Till - A Symposium. Columbus, Ohio State University Press: 402 pp.

Gripp, K. 1929. Glaciologische und geologische Ergebnisse der Hamburgischen Spitzbergen-Expedition 1927, Abh. Naturw. Verein Hamburg XXII: 147-249.

Grube, F. 1967. Die Gliederung der Saale-(Riß-)Kaltzeit im Hamburger Raum. Fundamenta B 2: 168-195. Köln/Graz, Böhlau.

Grube, F. 1979. Zur Morphogenese und Sedimentation im quartären Vereisungsgebiet Nordwestdeutschlands, Verh. naturwiss. Ver. Hamburg (NF) 23: 69-80.

Grube, F. 1981a. The Subdivision of the Saalian in the Hamburg Region, Meded. Rijks Geol. Dienst 34 (4): 15-22.

Grube, F. 1981b. Postsedimentäre Veränderungen von Gletscherablagerungen, Verh. naturwiss. Ver. Hamburg (NF) 24 (2): 103-112.

Grube, F. & J. Ehlers 1977. Landschaft und Geologie des Landkreises Harburg. Heimatchronik des Kreises Harburg: 7-22. Köln, Archiv f. deutsche Heimatpflege.

Haldorsen, S. 1977. The petrography of tills - a study from Ringsaker, south-eastern Norway, Norg. Geol. Unders. 336: 36 pp.

Hesemann, J. 1934. Ergebnisse und Aussichten einiger Methoden zur Feststellung der Verteilung kristalliner Leitgeschiebe, Jb. preuß. geol. L.-A. 55: 1-27.

Hesemann, J. 1975. Kristalline Geschiebe der nordischen Vereisungen: 267 pp. Krefeld, Geologisches Landesamt Nordrhein-Westfalen.

Höfle, H.-C. 1977. Die Geologie des Elbe-Weser-Winkels. Führer zu vor- und frühgeschichtlichen Denkmälern, Das Elbe-Weser-Dreieck 1: 31-41.

Höfle, H.-C. 1979. Klassifikation von Grundmoränen in Niedersachsen, Verh. naturwiss. Ver. Hamburg (NF) 23: 81-92.

Hucke, K. & E. Voigt 1967. Einführung in die Geschiebeforschung

(Sedimentärgeschiebe): 132 pp. Oldenzaal, Verlag Nederlandse Geol. Vereniging.

Iwanoff, A. 1980. Zur Quartärgeologie im Raum Langenhorn-Glashütte. Hamburg (unpublished Diplomarbeit).

Iwanoff, A. 1981. Zur Profildifferenzierung einer saalezeitlichen Moräne in Hamburg-Rissen, Verh. naturwiss. Ver. Hamburg (NF) 24 (2): 113-122.

Johansson, H.G. 1977. Till Sweden - 76, Boreas 6: 71-227.

Lawson, D. 1979. Sedimentological analysis of the western terminus region of the Matanuska Glacier, Alaska, CRREL Report 79-9: 112pp.

Legget, R.F. (ed.) 1976. Glacial Till - An Inter-disciplinary Study: 412 pp. Ottawa, The Royal Society of Canada.

Liedtke, H. 1981. Die nordischen Vereisungen in Mitteleuropa, 2nd. ed., Forschungen zur deutschen Landeskunde 204.

Lindén, A. 1975. Till petrographical studies in an Archaean bedrock area in southern central Sweden, Striae 1: 57 pp.

Lüttig, G. 1958. Methodische Fragen der Geschiebeforschung, Geol. Jb. 75: 361-418.

Lundqvist, J. 1982. The Glacial History of Sweden, in print.

Marcussen, I. 1978. Über die Verwendbarkeit von Geschieben in Grundmoränen als Hilfsmittel der Stratigraphie, Der Geschiebesammler 12 (2/3): 13-20.

Martinsson, A. 1979. The Pre-Quaternary substratum of the Baltic. In Gudelis, V. & L.-K. Königsson (eds.), The Quaternary History of the Baltic. Acta Universitatis Upsaliensis, Annum Quingentesimum Celebrantis 1: 77-86.

Meyer, K.-D. 1976. Studies on ground moraines in the northwest part of the German Federal Republic. In Stankowski, (ed.), Till - its Genesis and Diagenesis, Zeszyty Naukowe Uniwersytetu Im. Adama Mickewicza w Poznaniu, Geografia 12: 217-221.

Meyer, K.-D. 1982. Indicator pebbles and stone count methods, in print.

Meyer, K.-J. 1974. Pollenanalytische Untersuchungen und Jahresschichtenzählungen an der holstein-zeitlichen Kieselgur von Hetendorf, Geol. Jb. A 21: 87-105.

Milthers, V. 1934. Die Verteilung skandinavischer Leitgeschiebe im Quartär von Westdeutschland, Abh. preuß. geol. L.-A. N.F. 156:74pp.

Mückenhausen, E. 1939. Die Böden des Warthe-Stadiums in Nordhannover im Vergleich zum westlichen Alt- und zum östlichen Jung-Diluvium, Abh. naturw. Ver. Bremen 31: 335-346.

Müller, H. 1974a. Pollenanalytische Untersuchungen und Jahresschichtenzählungen an der holsteinzeitlichen Kieselgur von Munster-Breloh, Geol. Jb. A 21: 107-140.

Müller, H. 1974b. Pollenanalytische Untersuchungen und Jahresschichtenzählungen an der eem-zeitlichen Kieselgur von Bispingen/Luhe, Geol. Jb. A 21: 149-169.

Perttunen, M. 1977. The lithologic relation between till and bedrock in the region of Hämeenlinna, southern Finland, Geol. Surv. Finl. Bull 291: 68 pp.

Richter, K. 1962. Geschiebekundliche Gliederung der Elster-Kaltzeit in Niedersachsen, Mitt. Geol. Staatsinst. Hamburg 31: 309-343.

Schlüchter, Ch. (ed.) 1979. Moraines and Varves: 455 pp. Rotterdam, Balkema.

Schuddebeurs, A.P. (1980/81). Die Geschiebe im Pleistozän der Niederlande, Der Geschiebesammler 13 (3/4): 163-178, 14 (1): 33-40, 14 (2/3): 91-118, 14 (4): 147-198, 15 (1/2): 73-90, 15 (3): 137-157.

Sjørring, S. 1982. The Glacial History of Denmark, in print.

Sollid, J.L. 1980. INQUA Till Norway 1979, Norsk Geogr. Tidsskr. 34: 97-106.

Stankowski, . (ed.) 1976. Till - its Genesis and Diagenesis, Zeszyty Naukowe Uniwersytetu Im. Adama Mickewicza w Poznaniu, Geografia 12.

Stephan, H.-J. 1980. Glazialgeologische Untersuchungen im südlichen Geestgebiet Dithmarschens, Schr. Naturw. Ver. Schleswig-Holstein 50: 1-36.

Stephan, H.-J., G. Schlüter & Ch. Kabel 1982. Stratigraphical problems in the glacial deposits of Schleswig-Holstein, in print.

Thome, K.N. 1980. Der Vorstoß des nordeuropäischen Inlandeises in das Münsterland in der Elster- und Saale-Eiszeit - Strukturelle, mechanische und morphologische

Zusammenhänge, Westfälische Geogr.
Studien 36: 21-40.

Wilke, H. 1981. Postsedimentäre Veränderungen kaltzeitlicher Sedimente im Hamburger Raum, Verh. naturwiss. Ver. Hamburg (NF) 24 (2):
185-198.

Woillard, G.M. 1978. Grande Pile
Peat Bog: A Contnuous Pollen Re-
cord for the Last 140,000 Years,
Quaternary Res. 9 (1): 1-21.

Woldstedt, P. & K. Duphorn 1974.
Norddeutschland und angrenzende
Gebiete im Eiszeitalter: 500 pp.
Stuttgart, Koehler.

Zagwijn, W.H. & J.W.Ch. Doppert
1978. Upper Cenozoic of the
Southern North Sea Basin: Palaeoclimatic and Palaeogeographic
Evolution, Geol. en Mijnb. 57 (4):
577-588.

Flow tills and ice-walled lacustrine sediments, the Petteril valley, Cumbria, England

D.HUDDART
I.M.Marsh College, Liverpool Polytechnic, UK

1 INTRODUCTION

During motorway construction south of
Carlisle excellent sections in glacigenic
sediments were exposed at Carrow Hill and
at other sites in the Petteril valley
(fig. 1). Together with these exposures,
the borehole stratigraphy made available
by Cumberland County Council, the 1:10,600
Geological Survey maps (1970) and the
field mapping of the author provided a
basis for a study of the mode of ice decay,
flow till deposition and the sedimentation
associated with ice-walled lake basins.
The latter, although known in North America
and Poland, have not been described before
in the British Isles. Several well defined
facies are described and their origins
discussed.

2 ICE-WALLED LAKE ENVIRONMENTS

Ice-walled lake sediments and associated
flow tills have been located throughout the
Petteril Valley but were best exposed at
Carrow Hill (Grid Reference NY 436516).

Topographically this is a flat topped hill
at about 57.0m.O.D., overlooking the
Petteril valley, (fig. 2). It falls away
steeply to the west, north and north-east
but rises gradually in elevation to the
south and south-east. Dixon et al (1926)
described Carrow hill as a delta deposited
in a lake ponded up by ice in the Carlisle
plain. Morphologically this was probably
the most reasonable explanation and there
is evidence for a series of proglacial lakes
between 61.0m and 30.5m in the Carlisle
plain, which were deposited during the
retreat of the Main Glaciation ice sheet
(Huddart 1970,1981). However, when the
evidence from the sediments exposed during
the construction of a motorway interchange
is taken into account the delta explanation

is no longer tenable and illustrates the
danger of utilising morphology alone in
studying the genesis of fluvioglacial and
glacial landforms.

Motorway interchange excavations gave an
opportunity to make an intensive study and
to build up a three-dimensional picture of
the sediment distribution. From a prelim-

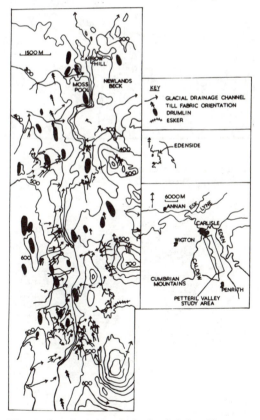

Figure 1. Location of glacial landforms
in the Petteril valley.

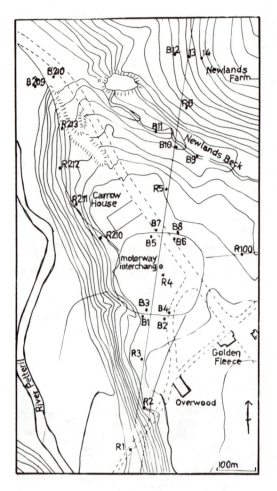

Figure 2. Morphology and borehole location, Carrow Hill.

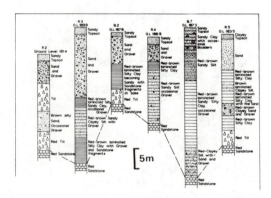

Figure 3. Borehole stratigraphy, Carrow Hill.

inary study of the borehold records, illustrated in fig. 3, it was thought probable that the hill was an example of an ice-walled lake plain. An ice-walled lake is a lake bottomed on bedrock or basal till and not on ice, but it is surrounded by stagnant ice (Clayton and Cherry, 1967). If this proved to be the case, it would be an important indicator of the mode of ice decay in the Petteril valley and would be the first landform in Britain to be interpreted as an ice-walled lake plain.

2.1 Stratigraphy

The bedrock is red, Triassic St. Bees sandstone, with interbedded shale units and a depression in its surface acted as the basin where the laminated silts and clays accumulated. Above bedrock is a hard, compact, red-brown till, packed with angular sandstone and shale fragments. At its maximum development it is 2m thick but a more usual thickness is 1m. Macrofabric study of the till particles was impossible because of the angularity of the majority of the gravel sized clasts and the till compactness. However, four thin sections indicated a consistent southeast to north-west orientation and a section cut in the vertical plane parallel with this orientation indicated a marked imbrication of the sand-size particles to the south-east. This is consistent with basal till fabric orientations in the Petteril valley shown in fig. 1 and seems to indicate the influence of Scottish ice to the north which resulted in a diversion of the Lake District ice to the north-west.

Collapsed superglacial sediment shown in figs. 4-6 above the basal till has a maximum thickness of 2m and is located near the southern margin of Carrow Hill. It has several distinguishing characteristics:

i) heterogeneity of sediment grain size, with all grades from clay to pebble gravel.

ii) large-scale disturbances are very common.

iii) greatly deformed clasts of laminated clay and silt, with included sand lenses and occasional gravel. These clasts can be up to 54cm long and 15cm thick.

iv) irregular stringers of laminated clay, up to 50cm long and 3cm thick.

v) oval clasts of red till, with the largest being 40cm long and 14cm thick, set in a green silt matrix.

vi) occasional pebble gravel bands and isolated pebbles in a sandy matrix. Macrofabric analyses show poorly marked east-west orientation maxima, with a tendency for a westerly dip of the pebbles.

vii) thin sections show clasts of sand-size particles in a silt matrix.

viii) sand clasts, faintly cross stratified, up to 20cm long and 10cm thick, in a green silt matrix.

Cut into the surface of this unit are sand-filled channels, with the largest measuring 3.7m wide and 70cm deep. In depressions in the surface of unit B are basins of laminated clays and silts which are undisturbed in general but folded and contorted in their upper layers. These basins are usually small, up to 1.5m wide and 20cm thick but one could be traced for 15m and was 50cm thick. The basal layer of these basins is composed of a red, sandy clay, 4cm thick, which contains small pebbles and shows a preferred orientation of the sand-sized particles. The laminated silts and clays consist of couplets of green silt and red clay, with occasional till clasts, pebbles and sand partings.

It is thought that this unit represents a marginal collapsed superglacial facies which would account for the range in grain size, the disturbances, the fabric pattern of the pebble gravel and the clast development in sand (which must have been frozen to preserve the cross stratification), till and laminated clay. Superglacial sediments, which would be liable to collapse on the melting of the underlying ice, would be deposited in many environments on the debris covered ice. These would include superglacial lakes (laminated silts and clays), superglacial streams (sand and gravel), and the melt-out of debris bands along marginal shear planes (till). After a period of ice melting and collapse of superglacial sediment small subaerial ponds would develop on the irregular surface and streams would begin to dissect the deposit and produce the channel fills.

Unconformably cutting across the collapsed superglacial sediment were several thin till units which could be differentiated by slight colour changes through shades of red and brown; changes in composition, some being very pebbly and others predominantly sandy clays and by sandy, washing lines which divided the till units. Where several units were superimposed the lower units tended to be sandier and where till overlay laminated clay the latter was disturbed, although there was little evidence of incorporation. These till-laminated clay junctions are illustrated in fig. 7. The thickest till unit was 94cm. with the total flow till thickness being 1.5m. The lateral

extent was 75m-100m. Both the macro- and microfabric of these till units were studied at several sites as the excavation progressed and the areal and two and three-dimensional distributions are presented in Huddart (1970) and the three-dimensional statistics in tables 1 and 2. The most important points to be noted from the fabric work are as follows:

i) the till particles have a southeast to north-west trend which changes to a more westerly trend in the southerly sites and a more northerly trend in the northern sites.

ii) the particle dip is mainly to the west and north-west.

iii) the general parallel alignment of macro- and microfabric distributions at any one site.

iv) the concentration of particles in the 0-10° dip class, with mean dips between 5.5° and 11.4°.

v) most of the microfabric distributions have one main mode, with several subsidiary peaks, usually one of these at right angles to the main peak. In the macrofabric distributions transverse peaks were found at sites 2, 7 and 18 but are absent at 21, 28, 30, 33 and 34.

To try and find out why there are only occasional transverse peaks; to see if there were any vertical changes in fabric pattern and to find more evidence for the origin of these till units, the thickest unit (94cm) was divided into eight equal zones, with twenty-five stones sampled from each. The till unit showed no colour changes, sandy washing lines or any other heterogeneities which might have suggested that the whole unit had not been deposited under uniform conditions. The results of the macrofabric analyses are illustrated in figs.8&9 and the three-dimensional statistics in table 2. Significant trends from these results are:

i) the similarity of the resultant orientations in the lowest five zones, with the slight divergence in the top three zones explained by the small transverse peaks.

ii) an increase in the percentage of stones with dips to the north-west quadrant as they are traced from the top of the unit to the base, and a corresponding decrease in the importance of transverse stones.

iii) an increase in the percentage of stones with dips in the 0-10° class as they are traced from the top of the unit to the base, with a corresponding decrease in mean dip from 11.2° in A to 5.2° in H.

The stratigraphical position, fabric pattern and vertical changes in pattern and basal contacts suggest that these till units are the product of flow tills which moved down a gentle palaeoslope which was orientated in a south-east to north-west

Figure 4. Structures within the collapsed superglacial sediment. Sand and till clasts within a heterogeneous matrix (E), disturbed clay and silt at the scale (D) and till (A).

Figure 5. Large scale disturbances within collapsed superglacial sediment.

Figure 6. Disturbed sand with till clasts, scale 20cm.

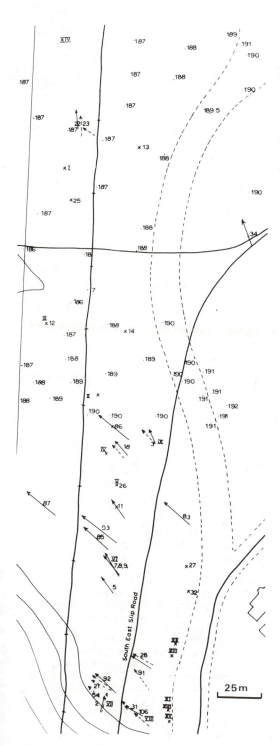

Figure 7. Flow till orientations,
Carrow Hill.

Figure 8. Flow till laminated sediment
contact, Carrow Hill a) flow till 94cm
thick, b) thin unit of laminated sediment,
c) flow till 30cm thick, faintly laminated
with occasional pebbles, d) laminated,
lacustrine sediment 10cm thick, e) sand
4cm thick.

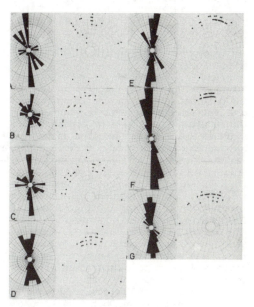

Figure 9. Flow till macrofabric distributions, Carrow Hill.

Table 1. Three dimensional statistics (after Steinmetz 1962), flow tills, Carrow Hill

Site Number	N	$\bar{A}°$	R	$D°$	$\theta5\%°$	K	Mean dip°	% dipping			% dipping		
								U.	D.	T.	0-10°	10-20°	20°
24	25	10	14.6	12	25	2.3	7.0	16	76	8	60	40	-
7	25	334	9.2 (s.5%)	21	39	1.5	7.6	32	60	12	84	16	
5	50	320	38.6	11	11	4.3	8.4	8	77	18	84	8	8
2	25	11	4.8 (not/s)	64	63	1.2	10.2	28	40	32	60	24	16
30	50	331	34.0	8	14	3.1	5.5	12	88	-	94	4	2
21	50	311	35.9	12	13	3.5	8.3	12	80	8	82	18	-
9	50	314	18.7	20	27	1.6	7.5	28	66	6	90	10	-
11	50	319	34.9	16	13	3.2	11.4	10	88	2	68	26	6
18	50	319	23.2	24	22	1.8	11.0	22	74	4	64	32	4
22	50	349	24.7	17	21	1.9	8.5	10	74	16	76	20	4
33	25	337	22.8	9	9	11.0	8.4	-	100	-	88	12	-
34	25	314	17.3	12	19	3.1	7.5	16	84	-	88	12	-

Where U - upslope, D - downslope and T - transverse particles lying between 60° and 120° on either side of the mean

Table 2. Three dimensional statistics, one flow till unit, Carrow Hill

	$\bar{A}°$	R	$D°$	$\theta5\%$	K	Mean dip	% dipping			0-10°	10-20°	20°
							U.	D.	T.			
A	338	14.7	19	25	2.3	11.2	12	68	20	64	20	16
B	323	14.4	19	26	2.7	11.0	16	64	16	68	24	8
C	331	14.9	15	24	2.4	8.8	12	72	16	76	24	-
D	356	18.1	14	18	3.5	9.8	8	80	12	72	24	4
E	355	18.2	11	18	3.5	8.0	8	92	-	96	4	-
F	353	24.0	9	6	23.7	8.6	-	100	-	92	8	-
G	347	19.2	9	16	4.2	6.6	4	92	4	92	8	-
H	351	23.7	5	7	18.5	5.2	-	96	4	100	-	-

direction. The changes in fabric through the unit reflect a decrease in velocity gradient from the base to the surface of the mudflow (Bagnold 1956, Lindsay 1968). Mudflows are usually described as travelling by viscous fluid flow in a turbulent manner and the Carrow Hill till units are homogeneous as would be expected from turbulent flow. However, Lindsay (1966) and Stauffer (1967) noted that some mudflows travel in a manner closely approximating to laminar flow and Bagnold (1956) stated that when the load of dispersed grains was progressively increased in turbulent flow the turbulence was damped down by the overall shear resistance so that laminar flow resulted. As these two modes of flow lie on a continuum with decreasing velocity, which was described by Dott (1963), flows which travelled with initial turbulence will eventually pass below the critical

velocity as the slope decreases and will flow in a laminar manner. Lindsay (1968) thought that clast fabric in a mudflow developed in this short interval before coming to rest, with the modes parallelling the flow direction. He saw fabric development as cyclic, building up and then degenerating, a feature expected because of the periodic motion of the particles. The mode was associated with a weak girdle which dipped up or downstream depending on the point in time at which the mudflow was arrested. After the flow came to rest he thought gravity settling might be important as the clasts would have a slightly higher density than the mudflow matrix. However, the Carrow Hill units show a uniform clast dispersal throughout the tills indicating rapid gelling after movement had ceased and all flow tills show a marked downcurrent dip. This is in contrast to the work of Cohen (1979) who suggested that up and down current dips can form within a single flow till unit during its deposition. This could be due to the basal part of the flow being deposited first and retaining different dip directions to the later deposited overlying part of the unit. This would imply that either the unit is composed of multiple subunits indistinguishable by eye or that the flow froze from the base up such that separate instances of fabric formation would occur. At Carrow Hill the sub-units can be distinguished and all have strong downslope orientations and dips. From what has been reported in the literature we can expect flow tills to show varying degrees of up and down current dips but most have strong orientations parallel with flow direction.

The sand/metrix ratios (Sitler 1967, Huddart 1971) for the tills and diamictons at Carrow Hill confirm the visual impression of a wide range of grain sizes for the flow tills, ranging from 0.15 to 0.88 (mean 0.57, S.D. 0.19, n=16), with the basal till giving more consistent values from 0.4 - 0.6 (mean 0.49, S.D. 0.07, n=4).

These tills at Carrow Hill have many of the characteristics of flow tills reported by Boulton (1967, 1968) and Marcussen (1973, 1975).

Stratigraphically above the flow till in the southern part of the hill is a unit of laminated silts and clays, with minor quantities of diamicton. This unit increases in thickness from 1m in the south-west to about 6m east of the Carrow House hotel. The silts and clays consist of couplets of grey silt and red/brown clay which have been described by Huddart (1970). The silts exhibit several sedimentary structures similar to those described by Banerjee (1973) which include parallel lamination, small-scale cross stratification, convolute lamination, mudball conglomerates, thin gravel and diamicton units, with bedding plane marks (parting lineation, bounce and prod marks) along some silt divisional planes fig.10-13. Many of the clay divisions show small erosional channels in their upper surfaces. These silts are thought to have been deposited from density underflows and the diamictons from subaqueous mudflows or subaquatic flow tills (Evenson et al 1977). They did not form by the remobilisation process described by Cohen (1979). It was estimated that the lake had a minimum life of eighty seven years from couplet counts.

To the north of the old Durdar - A.6 link road were exposed 3.25m of sand and gravel, overlying the laminated sediment. Grading laterally into the latter unit were small deltaic sequences, 1.5m thick, which

exhibited the classic tripartite division. The bottomset sediments were composed of 20cm of silty fine sand with ripple formsets, dipping at 2-5°. Overlying are foresets, 1.02m thick, composed of type 'a' ripple cross lamination and fine sand, with thin sets of sinusoidal ripple lamination in sandy silt, dipping at 15° to the north. The succeeding units are formed of pebble gravel and horizontal stratification in coarse sand, dipping at 5°. The deltaic sequence is overlain by a fluvial sequence of horizontally stratified medium to coarse sand with clay balls (derived from both till and laminated sediments), trough cross stratified medium sand and imbricate pebble gravel units with clay balls. Palaeocurrent analysis indicates deposition from the south-east between 144-158°. These gravels and sand give a minor rim topography to the south-eastern part of the hill. Near the Carrow House hotel the sands overlying the laminated sediment are much thinner and are composed mainly of sinusoidal ripple cross lamination and ripple formsets in silty sand dipping at 4° to the north.

2.2 Stages in the development of Carrow Hill

The stratigraphy exposed both in the interchange excavations and in the borehole logs gives rise to the following interpretation of the depositional stages which resulted in Carrow Hill (see fig.14.). In stage 1 the Late Devensian ice decayed in situ in the deep Petteril valley, with basal till deposition. The stagnant ice would be loaded with ablation debris which would cover the surface of the ice and facilitate the development of a complex system of super-glacial environments. This insulating, super-glacial debris facilitated the differential melting of the ice where the cover was thin. This would be accentuated along crevasses above till/bedrock highs and gave rise to superglacial depressions and lakes. As downwasting continued the basal till/St. Bees sandstone high at Carrow Hill appeared as a nunatak at the ice surface. Melting would be rapid around this nunatak and meltwater became ponded. The lake was bottomed on basal till and surrounded by a wide range of superglacial fluvial and lacustrine environments. Where ice-walled lake plains are found nearly at the same height as adjacent ground moraine it is possible that the lakes formed in lows on the ice surface that initially contained less superglacial debris and so melted faster than the adjacent ice. These lows would be filled with sediment which would increase their final elevation. At Carrow Hill this latter process is thought to have occurred because

Figure 10. Parallel laminated silts, succeeded by rip-up clasts, clay and a mudball conglomerate.

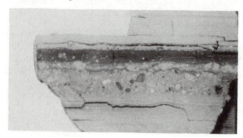

Figure 11. Thin diamicton succeeded by clay, both above and below parallel laminated silts.

Figure 12. Mudball conglomerate.

Figure 13. Laminated silts and clays showing couplets.

although the hill is above the surrounding post-glacial river valley it is below or at approximately the same elevation as the surrounding ground moraine.
In stage 2 when the lake had been initiated meltwater streams would discharge their debris load into it producing small deltaic

sequences and the lake would grow both in length and width as the surrounding ice melted. Marginal melting would result in the collapse of the overlying supraglacial sediment and the production of the complex collapsed supraglacial facies. Where hollows developed on the surface of this collapsed zone small ponds would form and small supraglacial streams would start to erode the collapsed sediment.

In stage 3 after further extension of the lake through melting a widespread phase of flow till deposition was initiated which covered the collapsed supraglacial facies and produced thin subaqueous flow till/density underflows in the proximal lake area. The source of the till was supraglacial moraine which had been concentrated by ablation to the south and south-east of the lake site.

In stage 4 supraglacial streams flowing from the south and south-east eroded part of the lacustrine sediment, deposited small

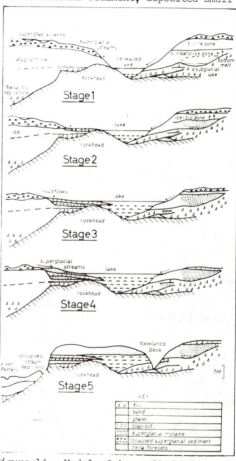

Figure 14. Model of ice-walled lake plain deposition, Carrow Hill.

88

deltas and finally filled up the lake basin with a thin fluvial capping. There is no evidence to suggest that the ice-walled lake was destroyed by ice-wasting which drained the lake through ice-walled valleys or stream trenches as has been reported by Parizek (1969).

Finally in stage 5 the surrounding ice melted with the production of collapse structures at the western margin of the plain and an ice-contact slope. Small detached ice-blocks slowly melted in the sands and gravels to produce occasional contorted sands. The post-glacial river Petteril has not been able to erode into the lake plain and only slight modification has taken place by the cutting of the Newlands Beck glacial drainage channel.

2.3 Concluding remarks on the Carrow Hill site.

Clayton and Cherry (1967) recognised two main types of ice-walled lake plain (moraine lake plateau of Parizek 1969) in Dakota: unstable lakes where there was only a thin superglacial debris cover and stable lake sites where there was a thick cover of superglacial drift on the surrounding ice. In the former case the ice topography underwent continual change and mass movement of the superglacial drift was common. This type of ice-walled lake environment was formed at the Carrow Hill site. A comparison with stable environment lakes in Dakota is given below in table 3.

In the case of Carrow Hill the sediment exposure was so good that the reconstruction of the depositional environment can be attempted with a high degree of accuracy and its recognition as an ice-walled lake plain contributes greatly to the understanding of the method of ice decay in the Petteril valley. This type of landform is not thought to have been an isolated example but elsewhere in the Petteril valley the exposure was not as good and the interpretation of the depositional environments are not as accurate. Nevertheless there are several examples of flat-topped hills which could be interpreted as ice-walled lake plains. However, without sedimentological evidence this can only be conjecture. At Moss Pool described in Huddart (1981) and at Calthwaite Beck described in the following section, there is borehole evidence for lacustrine sedimentation and field evidence for flow tills although whether this deposition was in a completely ice-surrounded, ice-walled environment or ice-marginal is difficult to judge from the available field evidence.

3 FLOW TILL DEPOSITION AT CALTHWAITE BECK-LOW OAKS (G.R.NY 472406 to 466413)

Sections revealed during motorway construction work at Calthwaite Beck showed complex interdigitation of till and sand, although the stratigraphy illustrated in Huddart (1970) is based largely on borehole evidence. Field observations were made on a sand unit and the succeeding till/sand unit above the

Table 3. A comparison between ice-walled lakes in the Petteril valley and Dakota.

Carrow Hill - unstable ice-walled lake	Stable ice-walled lakes in Dakota
superglacial drift probably thin (probably under 5m)	superglacial drift thick (over 5m)
mudflows common	mudflows uncommon
superglacial rivers carried sand and gravel into the lake, deposition by density underflows and suspension lake plain concave upwards, the result of coarse-grained sediment in the marginal area and compaction of the mid-lake sediment	relatively tranquil rivers carrying silt/clay into the lake, deposition by suspension lake plain convex upwards because of the relatively uniform clay deposition over the entire lake
gravel rim	no rims
ice topography changed quickly in a low area, although above the adjacent Petteril valley; a short existence, estimated at around ninety years	ice topography changed slowly perched above the surrounding dead ice moraine; lasted for a much longer time period, with thicker sediment accumulation.
non-existent fossil assemblage	more abundant fossil assemblage as the water warmer, more silt free and more time for establishment of life forms.

89

beck level. At Low Oaks the stratigraphy was not well exposed and only the upper till was revealed.

The stratigraphy in the field showed complex, interpenetrated units of till and sand. The lower part of the upper till unit has many sand stringers alternating with 1mm thick till laminae. In the sand unit there are complex lateral interpenetrations of till with included sand lenses and stringers which emanate from a 38cm thick till unit. Above the contact with the sands in the upper part of the upper till unit the sand pockets and stringers die out and the till is more homogeneous. Often the underlying sands are folded and sand pockets and clasts are very clear in the lower part of the till. The till is usually compact and hard, as at site T.158, where it dips at 25° to the east. Here the main till unit is 27cm thick with thin till layers, up to 2cm thick, interbedded in the sands below. At T.162 the till is only 18cm thick, dipping at 10° to the east, overlying 75cm of sand which in turn is succeeded below by another till unit. The steepest dips measured in the till and underlying sand were 45°, although a more average figure is around 15-20°, with all the dips to the east, towards the Petteril valley.

Fabric analysis was carried out on both a macro- and microscale, with the distributions illustrated in figures 15 & 16. The resultant orientations show that although there is some scatter the general transport direction was to the east. This is the present day downslope direction. Tables 4 and 5 show the three-dimensional statistics and the marked pronounced dip of the particles.

The texture of the tills was studied using the sand-matrix ratio and three samples gave values of 0.39(T.163), 0.39(T.160) and 0.42(T.164), with a mean value of 0.4 and a standard deviation of only 0.013. These values are less than for basal tills in the Petteril valley where at Henley Sike, Low Street, Meadow Sike and Carrow Hill, values of 1.02, 0.75, 0.5 and 0.6 were respectively obtained.

The local topography in the area is a poorly marked, fan-shaped sediment body which has a very gentle gradient down to the Petteril and has been incised by the Calthwaite Beck. This area is situated to the east of higher, drumlinised land which reaches 170.7m O.D. west of Calthwaite village. Regional, basal ice movement in the area is indicated by the parallelism of macrofabric and drumlin long axis at Robinson House, just to the south of Calthwaite Beck. Here the resultant orientation was 130°, with a mean dip of 10.2° and a preferred upglacier dip of the pebbles.

Clearly the till exposed at the Calthwaite Beck section is not basal in origin and its origin seems to be related to mass movement of sediment from the higher, drumlinised land to the west. Evidence supporting this genesis comes from the many individual till units; their thinness; their interpenetration with sand, which shows evidence of incorporation and disturbance; the general parallelism between the macro- and micro-fabrics in an easterly direction; the high dips of the particles and their low scatter in orientation both in the micro- and macrofabrics. The stratigraphic succession derived from borehole records suggests that

Table 4. Three dimensional statistics and direction of pebble dip, Calthwaite Beck and Low Oaks.

	Number of Stones	R	$\overline{A}$°	D°	Θ5%°	K
A	50	29.5	60	31	17	2.4
B	25	18.7	107	21	17	3.8
C	25	23.6	358	29	7	17.7
D	25	24.2	32	22	5	30.4
Low Oaks	40	31.9	62	21	12	4.8

	% Dips to E	% to W.	% Transverse	0-10°	10-20°	20°	Mean Dip°
A	72	8	20	52	18	-	17.6
B	80	4	16	56	16	-	15.8
C	100	-	-	28	72	-	27.6
D	100	-	-	60	40	-	21.6
Low Oaks	94	2	4	52	22	-	17.0

Table 4 continued

Comparison of Precisions

Homogeneity of variance $\dfrac{30.4}{3.8} = 8.0$,

F max ratio with 2 (N–1) d.f. = 14.8

therefore homogeneity indicated.

Comparison of mean directions

$$\dfrac{(\Sigma N_i - s)}{2\,(s-1)} \cdot \dfrac{\Sigma R_i - R}{\Sigma N_i - R_i} = 8.47,$$

5% F ratio = 2.4

therefore the hypothesis is rejected and one vector mean cannot be calculated.

Table 5. Microimbrication, Calthwaite Beck.

	Number of Particles	% Dip to E.	% Dip to W.	Mean Dip
T.159	156	72.5	27.5	17.4
T.160	143	90.0	10.0	23.2
T.163	90	59.0	41.0	14.8
T.164	90	72.0	28.0	18.8
T.169	137	79.0	21.0	21.0

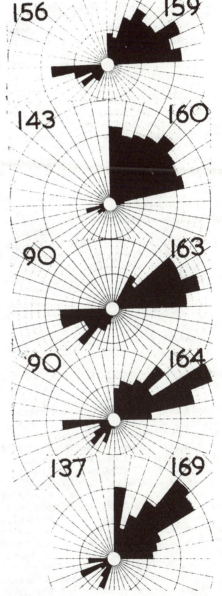

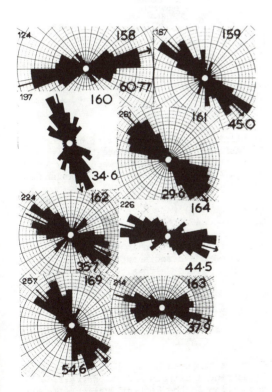

Figure 15. Till microfabric distributions, Calthwaite Beck. Figure in top left: population of grains counted; top right: sample number, bottom right: chi-square value.

Figure 16. Till microimbrication, Calthwaite Beck.

much of the upper till unit in this section
of the Petteril valley formed by flowage of
sediment as the till at Low Oaks (chainage
676) gave a resultant orientation of 62°
and had a high mean dip of 17°in a downslope direction.

From the stratigraphy it is clear that
much of the sequence at this site could have
accumulated during deglaciation in an ice
marginal environment. In the stratigraphic
succession there is a rapid change in
sediment type, although till predominates,
both laterally and vertically and in borehole B.50 a mixture of till and sand and
gravel is found both at the top and base of
the borehole log. If the lower of these
two units is interpreted in the same way as
the upper unit, that is as flow till interpreted with fluvial sands, then there is at
least 15m of sediment in this borehole which
can be attributed partially to mass movement
in an ice-marginal environment. The interpretation of the associated laminated clays,
as around chainage 750 and at Calthwaite
Beck, can also be best explained as deposition in a marginal or ice-walled position.
During the early stages of deglaciation mass
movement of already deposited sediment would
be a feature between the stagnant ice, which
gradually occupied lower positions in the
valley and the higher, drumlinised topography to the west. Figure 17 summarises
hypothetically the development of the glacigenic sediment suite in the Calthwaite area,
but this type of succession seems to have
been common throughout the Petteril valley
during the later stages of deglaciation.
Unfortunately extensive exposures are non
existent and only motorway borehole records
are available for other parts of the
Petteril valley.

4 OTHER LAMINATED CLAY LOCALITIES IN THE PETTERIL VALLEY

In the motorway borehole logs laminated
sediments appear frequently and occur interpenetrating within till and sand, or as
distinct units. Some of the localities are
given below from south to north: around
Dentons G.R. 492338 (chainage 449 to 456)
in the lee of a drumlin, interpenetrating
with sand; around Beck Cottage G.R. NY445338
(chainage 799 to 790); north of Roughton
Gill G.R. NY442463 (chainage 861 to 866),
interpenetrating with till, sand and silt;
around Howfield G.R. NY437472 (chainage 890
to 893) within till and north of Gillhead
Bridge G.R. NY 431478 (chainage 945 to 970),
interpenetrating with tills above Penrith
sand stone. Topographically the laminated
clays seem to be situated in basins where
the drift is thick, to the lee of the higher

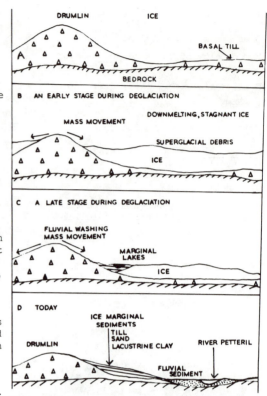

Figure 17. Hypothetical model of glacigenic
sedimentation, Calthwaite Beck – Low Oaks.

drumlinised ground to the west and the river
Petteril to the east. Where the drift cover
is relatively thin, as between Kitty House
G.R. NY481381 (chainage 575) and Ravensgill
Beck G.R. NY475392 (chainage 605), and
between west of Petteril Hill G.R. NY445435
(chainage 750) and the access to Toppinhill
G.R. NY440447 (chainage 825), laminated
clays appear to be very rare. Where the
sediments can be seen in the field the
origin of the associated till units, as at
Carrow Hill and Calthwaite Beck, should
give a clue to the origin of these laminated
clays.

5 CONCLUSION: ICE-WALLED LAKES AND THE MODE OF DEGLACIATION

Evidence has been presented for ice-walled
lacustrine sediments and extensive flow till
deposition in the Petteril valley. In the
same area ice-marginal fluvial and subglacial depositional environments which were
associated with the decay in situ of the
last glaciation ice sheet (Late Devensian)
have been described by Huddart (1970).

During this decay much melt-water was liberated and although subglacially deposited eskers are not unknown, for example around Morton and in the Low Plains col, (Huddart 1981), subglacial meltwater erosion was more important and produced many typical suites of glacial drainage channels. As can be seen from figure 1 most of these channels focus on the Petteril valley and this valley must have been a major subglacial meltwater exit to the north. However, at certain stages meltwater escaped subglacially from the Petteril system to the Eden, for example between south of Wallnook and Baronwood, via the channel and esker complex, where there must have been a region of lower subglacial hydrostatic pressure.

Many of the complex, stratigraphic sequences revealed in borehole logs in the Petteril valley have been explained as deposition in ice-marginal or in superglacial situations where there was a relatively thick drift cover. Till was deposited both subglacially and commonly as subaerial flow tills, laminated silts and clays in ice-marginal, ice-walled, superglacial and subglacial lakes and sand and gravel in subglacial, ice-marginal and superglacial environments. The landforms produced in these environments include ice-walled lake plains, ice-marginal kame terraces, sub-glacial eskers and ice-marginal deltas but the correct genetic interpretation of any landform must be primarily based on detailed sedimentological evidence.

Conclusive proof has been given which suggests that the ice sheet was stagnant during deglaciation and is not thought to have retreated in the manner described by Trotter (1929) and Hollingworth (1931) where they envisaged an active ice front retreating back to the ice source, although this does seem to have been the case in the Carlisle plain to the north (Huddart 1970, 1981). The major factor explaining this difference in ice decay is the topography in the two areas. The pre-glacial Edenside rivers had produced deep depressions where the ice during deglaciation would be cut off from its supply and would stagnate in situ. Ice in the lee of the Pennines, in the South Tyne drainage basin would be cut off from its source once the ice sheet had downmelted below the crestline of the Pennines. In Edenside the flow lines of the basal ice layers would rise to counteract the obstruction of the Pennines and so, although the ice was not at first cut off from its source, shear planes would bring debris to the surface continually during the early stages of deglaciation. Ablation would concentrate this debris and a thick, superglacial, debris cover would result. At any one time only a relatively narrow, stagnant marginal zone is thought to have developed to the west of the Pennines, but with continual downwasting this zone would widen and a zone up to 12km wide would be stagnant (Huddart 1981). Once the ice downmelted below the crest of the Penrith sandstone ridge which separates the Petteril and Eden valleys individual stagnant ice lobes would form in the Caldew, Petteril and Eden valleys. The drumlins of the Caldew-Petteril divide, Thiefside Hill and Blaze, Barrock and Lazonby Fells would project as nunataks above these lobes and the superglacial debris would become reworked in marginal, subaerial and subglacial, fluvio-glacial environments. The ice-walled lake with its associated flow tills was one characteristic environment in the Petteril valley and indicates the mode of deglaciation in Edenside during the decay of the Late Devensian ice.

6 ACKNOWLEDGEMENTS

The author gratefully acknowledges receipt of a Natural Environment Research Council Research Studentship at the University of Reading and thanks Drs. J.R.L. Allen and J.B. Whittow for their critical reading of an earlier manuscript. He acknowledges the help given by Cumberland County Council Highways and Bridges department, in particular the former County Surveyor and Bridgemaster, Mr. F.M. Broughton, who gave permission to study the motorway excavations and borehole records and provided copies of maps, logs and soil mechanics data. Thanks are due also to the officials and workmen of Tarmac and Dowsett Engineering companies who never objected to the author working on sections exposed by their excavators.

7 REFERENCES

Bagnold, R.A. 1956, The flow of cohesionless grains in fluids, Phil.Trans. R.Soc.A249, 235-97.

Banerjee, I. 1973, Sedimentology of Pleistocene glacial varves in Ontario, Canada, Bull.geol.Surv.Can. 226, 59pp.

Boulton, G.S. 1967, The development of a complex supraglacial moraine at the margin of Sørbreen, Ny Friesland, Vestspitzbergen, J.glaciol.6, 717-35.

Boulton, G.S. 1968, Flow tills and related deposits on some Vestspitzbergen glaciers, J.glaciol.7, 391-412.

Clayton, L. and J.A. Cherry 1967, Pleistocene superglacial and ice-walled lakes of west-central North America, N.Dak.geol. Surv.Misc.Ser.30 47-52.

Cohen, J.M. 1979, Aspects of glacial lake

sedimentation, Unpubl.Ph.D.thesis
Trinity College, University of Dublin.

Dixon, E.E.L., J. Maden, F.M. Trotter,
S.E. Hollingworth and L.H. Tonks, 1926,
The geology of the Carlisle, Longtown and
Silloth districts, Mem.geol.Surv.U.K.

Dott, H.R.T. 1963, Dynamics of sub-aqueous
gravity depositional processes, Bull.Am.
Assoc.Petrol.geol.47, 104-28.

Evenson, E.B., A. Dreimanis and J.W. Newsome
1977, Subaquatic flow tills: a new interpretation for the genesis of some laminated till deposits, Boreas 6, 115-33.

Huddart, D. 1970, Aspects of glacial sedimentation in the Cumberland lowland,
Unpubl.Ph.D. thesis, University of Reading.

Huddart, D. 1971, Textural distinction of
Main Glaciation and Scottish Readvance
tills in the Cumberland lowland, Geol.
Mag.108, 317-24.

Huddart, D. 1981, Fluvioglacial systems in
Edenside, In Boardman, J.ed. Eastern
Cumbria Field Guide, Quat.Res.Assoc.,
81-103.

Lindsay, J.F. 1966, Subaqueous mass movement, New South Wales, J.sedim.Petrol.36,
719-32.

Lindsay, J.F. 1968, The development of clast
fabric in mudflows, J.sedim.Petrol.38,
1242-53.

Marcussen, I. 1973, Studies on flow till in
Denmark, Boreas 2, 213-31.

Marcussen, I. 1975, Distinguishing between
lodgement till and flow till in Weichselian deposits, Boreas 4, 113-23.

Parizek, R.R. 1969, Glacial ice-contact
rings and ridges, Geol.Soc.Am.Spec.Pap.
123, 49-102.

Sitler, R.F. 1968, Glacial till in orientated thin section, 23rd Int.geol.Congr.25,
262-69.

Steinmetz, R. 1962, Analysis of vectorial
data, J.sedim.Petrol.32, 801-12.

Trotter, F.M. 1929, Glaciation of Eastern
Edenside, the Alston Block and the
Carlisle plain, Q.J.geol.Soc.Lond.88,
549-607.

The readvance of the Findelengletscher
and its sedimentological implications

CH.SCHLÜCHTER
ETH-Hönggerberg, Zürich, Switzerland

ABSTRACT: Since winter 1979/80 the Findelengletscher in the Zermatt area, Switzerland, has been readvancing. The spectacular glacial-dynamic activity results in a surprisingly low production of "new till" but in a voluminous remobilisation of older glacial deposits at the lateral and frontal ice margins. The remobilisation processes are predominantly glacitectonically controlled and meltwater influence is minimal. The glacier has been moving by 1 cm/hour at the ice front and by 2 cm/hour at 220 m in elevation above the terminus (Bezinge 1982). This has caused considerable stress on the frontal parts of the icetongue. Stress release is accomplished not only by downvalley sliding but also by strong lateral extension of the ice which brings basal till to the lateral ice margins. The obstacle to the noncompressive "steady downvalley flow" of the ice is probably the cross-valley bedrock ridge at the present ice front.

1 INTRODUCTION

The INQUA-Commission on Genesis and Lithology of Quaternary Deposits visited the Findelengletscher area in 1978 during its annual fieldmeeting (Schlüchter 1978). For almost a hundred years the position of the glacier front has been surveyed every autumn but, ironically enough, not in 1979 - and during the following winter of 1979/80 was a readvance of the glacier observed. Since then, the glaciers' dynamics have been carefully investigated by the glaciologists Dr. A. Iken and PD Dr. H. Röthlisberger of the Federal Institute of Technologys' Glaciology Section. In addition to the glaciological studies the sedimentological aspects of the readvance are recorded. And this report gives a summary of the first results and an introduction to further investigations.

Findelengletscher belongs to the draining system of the glaciated area of the Southern Swiss Alps (Walliseralpen). The highest peaks of the Alps occur in this region with, for example, the Dufour Spitze in the Monte Rosa Massif reaching 4634 m.

As a consequence, the most spectacular alpine morphology is found in this area (Fig. 1 and 3). Findelengletscher is a medium size alpine glacier with a total length of 7 km. Its catchment area reaches to altitudes of 3800 m. The glacier terminates at 2500 m and its snout rests there on a cross-valley bedrock ridge (Fig. 1 and 2). The catchment basin is 24.5 km^2 and 19.5 km^2 thereof are glaciated (Bezinge 1982). The accumulation area is given by Bezinge (1982) as 11.5 to 16.5 km^2 and the ablation area averages 3 to 8 km^2. At the end of September the snowline is, on the average, between 3000 and 3250 m.

The bedrock lithology of the glacier bed is complex. The southern part, including the cross-valley ridge at the ice front, is in the gneisses of the Monte Rosa Nappe. The northern part comprises the ultramafic series of the "Ophiolit-Zone Zermatt-Saas Fee" (Bearth 1953). The large-scale topography of the glacier bed is controlled by bedrock lithology.

Figure 1: The Findelengletscher near Zermatt, descending between Rimpfisch- and Strahl-horn (left) and the Monte Rosa Massif (right). Photo taken Sept. 19, 1964, looking east. Note the well defined lateral moraines and the initial stage of the "pyramids" along the right frontal ice margin. - Photo by Swissair Photo + Vermessungen AG, Zurich.

2 GLACIAL GEOLOGY

The dominant glacial geological features in the Findelen area are the prominent left and right lateral moraine ridges (Fig. 2). A complex lithostratigraphy demonstrates that these voluminous lateral moraines contain deposits of multiple glacier oscillations. Radiocarbon dates on fossil soils within the lateral moraine accumulations (Röthlisberger 1976) show that the build-up of the moraines is due to multiple readvances of the glacier since 2500 y BP. The most important advance - in ice volume and downvalley extension - occurred between 1848 to 1865 and all other Post-glacial oscillations have remained within the limits of this advance.

The terminal zone of the Findelengletscher exhibits an interesting facies-association: a zone of gravel- and sandcovered "ice-pyramids" marks the front of the glacier. This feature is better developed along the right frontal part between the main meltwater discharge and the lateral moraine (Fig. 2). - The "pyramids" are completely covered by sandy gravel with occasional boulders. The thickness of the sediment cover varies from 10 to 40 cm and exceeds 50 cm only in cavities and in places of supraglacial gravity flows. The material is sandy gravel with a sub-rounded to well rounded gravel fraction. The glacigenic characteristics are minimal to lacking. Their occurance on ice does not imply glacigenic origin. They can be exposed by simple melting. Of interest is the geometry of the frontal "pyramids": their up-glacier dip is attributed to shear planes in the frontal zone as schematically illustrated and discussed in the Guidebook for the 1978 field excursion. The origin of the sandy

Figure 2: Frontal zone of the retreating Findelengletscher. Airphoto taken Sept. 14, 1977, by Swiss Topographic Survey. Note the well defined but genetically complex lateral moraine ridges. The distance between the very pronounced parts of the moraines accross the glacier front is 750 m. B: cross-valley bedrock ridge, G: Gletschertor, P: "Pyramids". - Published with permission by Swiss Topographic Survey (Bundesamt für Landestopographie) of May 19, 1983.

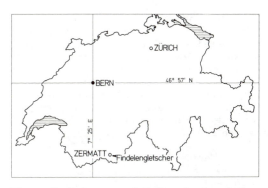

Figure 3: Index map of Switzerland and the location of the Findelengletscher in the Southern Swiss Alps.

gravel remains speculative: it is either reworked material from fluvial/fluvioglacial sediments in an overdeepened basin behind the cross-valley bedrock ridge and sheared upwards to the present position (Schlüchter 1978) or - as PD Dr. H. Röthlisberger has suggested - it is a deformed seimentfilled intraglacial channel system. The accumulation of material in intraglacial channels is dependant on the "flowtopography" complicated here by the cross-valley bedrock ridge. Such an intraglacial channel system is progressively closed and deformed when the ice movement is reaching an area of compressive flow as is the case here at Findelen behind the cross-valley bedrock ridge. In favour of a meltwater origin for the sands and gravels in

Figure 4: Forefield of the Findelengletscher in 1978. - S: Sandur and main meltwater discharge. T: Position of artificial trench through annual moraine ridge for 1978 Field-excursion. 1: artificially modified moraine ridges. 2, 3: reference boulders, see Figures 5 and 6. - At this time the glaciers' snout is to the left of the photo.

intraglacial channels are bedding structures. Such structures have been observed in the intraglacial sediment beds along the downvalley slopes of the pyramids before they collapse through melting. Both genetic hypothesis point to the existence of an overdeepened basin upglacier from the cross-valley bedrock ridge.

The initial formation of the sediment-covered ice "pyramids" is related to the ice retreat before 1979 and, more precisely, to the considerable downwasting in this part of the glacier during the last 20 years. Between 1925 and 1978 backwasting caused a retreat of the ice front by 1450 m (Bezinge 1982) and left the glacier just on the cross-valley bedrock ridge. Further retreat of the ice front upvalley from the bedrock ridge makes glaciodynamically initial downwasting necessary - but right now the glacier is still advancing!

Before the present readvance the lateral ice margins have been zones of "quiet downwasting" with little visible traces of ice movement. The ice margin has been covered by slope deposits and wash-off debris from the lateral moraines.

3 THE READVANCE

During the winter of 1979/80 a well defined shearzone of 2 km in legth has been discovered along the right lateral margin of the glacier. Subsequent measurements of the ice-movement by Dr. A Iken have revealed an advance of 1 cm/hour at the ice front (= altitude of 2500 m) and of 2 cm/hour at an altitude of 2720 m. This velocity has been maintained by the ice to the end of 1980 (Bezinge 1982).

The difference in speed from 1 cm/hour at the front to 2 cm/hour at 220 m in elevation above the glacier front must produce considerable stress-fields (and compressive flow regime) in the frontal parts.

The actual readvance of the ice front has been measured as follows (Bezinge 1982):

Figure 5: Forefield on August 3, 1980. - Note the glacier front with the gravel and sand covered "pyramids" (P) and the arch of pressed scales (SC) in the active frontal zone of advance. T, 2, 3: see Fig. 4.

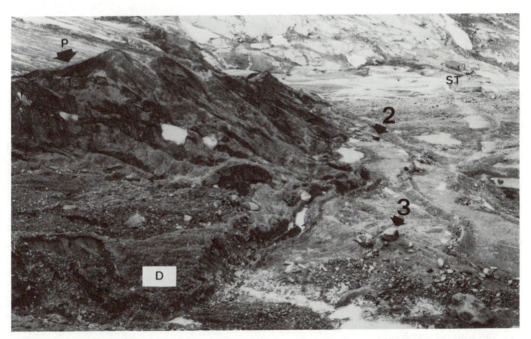

Figure 6: Forefield in July 1981. - P, 2, 3: see Fig. 4. D: reactivated dead ice at the right frontal margin. ST: water intake station by Grande Dixence SA, now modified for subglacial functioning.

Figure 7: Overriding of a pressed scale at the active ice front. The icicles reaching the till are bent due to the movement of the ice front. Length of icicles is appr. 1 m. Photo: Febr. 27, 1981.

from October 1979 to October 1980: 84 m
from October 1980 to October 1981: 52 m.

Three pictures are given for comparison to illustrate the forefield in 1978, 1980 and 1981 (Fig. 4, 5 and 6).

4 SEDIMENTARY PROCESSES

The frontal zone of the Findelengletscher consists of three distinct seimentary environments:

(a) the actual forefield, (b) the gravel and sand - "pyramids" between the melt-water stream and the right ice margin, and (c) the lateral ice margins and moraine ridges. In the following, the active sedimentary processes of these three zones during the recent readvance of the ice shall be discussed.

4.1 The glacier forefield

Figures 1 and 4 to 6 illustrate the pos-ition of the advancing ice tongue on the cross-valley bedrock ridge. It is evident, that in such a morphodynamic position very little sediment is accumulated: on the left side of the ice front almost bare bedrock extends upslope towards the lateral moraine ridge and characteristic erosional marks are abundant there. In the immediate fore-field, the main meltwater discharge is re-lated to a broad bedrock channel where a small sandur has accumulated (Fig. 4 to 6). In the main forefield bedrock is partially covered by supraglacial and basal melt-out till. The average thickness is less than 50 cm but may reach 2 m in the more pro-nounced annual moraine ridges (Fig. 4). Towards the right margin of the forefield the bedrock drops off and the till accumu-lation increases rapidly reaching a con-siderable thickness at the right lateral moraine.

In 1978 a trench was dug through an un-disturbed annual moraine: the internal structure of this ridge revealed evi-dence suggesting bulldozing of material to produce the landform (Schlüchter 1978).

Figure 8: Pressed scales and folds in the winter snow cover in front of the advancing ice due to bulldozing by the pushed boulder. Height of boulder is appr. 2 m. Photo: Febr. 27, 1981.

The same bulldozing and pushing mechanisms as well as their modifications can be actively observed during the present re-advance. Especially with the winter snow-cover the glacitectonic activity was on spectacular display at the ice-front (Fig. 8). One aspect of the glacitectonic pushing mechanism is the occurrence of pressed scales (in german = Press-Schuppen) in the old till immediately in front of the active ice. These pressed scales (identical of those described by Gripp 1979: 160) develop along the glacier front wherever there is a noticeable till cover on bed-rock.

Considering the amount of till left by the former retreat of the ice and observing the formation of pressed scales but no push-endmoraine one simple assumption remains: there must be a glacidynamic process active at the advancing ice front which causes the sediments at the ice margin to disappear. The observations during the winter months of rapid advance (1979/80, 1980/81) demonstrate how previously pushed and pressed material is overridden by the advancing front and incorporated to the base of the ice (Fig. 7). This "eating-up" mechanism is not directly observed in summer time as delicate structures at the ice front collapse or are mantled by supraglacial melt-out till. This mechanism of active incorporation of older sediments ("re-mobilisation") helps to explain why no "fresh" sediment production due to the advance has taken place in the frontal zone so far.

4.2 The "pyramids" at the right-frontal/ lateral ice margin

In Figure 1 (taken in 1964) an initial stage in the evolution of the "pyramids" is visible and in Figure 2 - a picture taken 13 years later - the arch of the "pyramids" is clearly shown. Just prior to the readvance the relative height of the "pyramids" above the adjacent clean ice averaged 10 to 12 m and reached a maximum of appr. 20 m at the highest ridge. With the readvance of the ice front a considerable thickening of the glacier is combined which has decreased the relative height of the "pyramids" with regard to adjacent up-glacier ice but increased its absolute height with regard to the ice-free forefield.

With the readvance neither the position nor the structure of the "pyramids" has changed; they were moved as part of the frontal zone. At the right lateral edge of the ice front only a large slab of dead ice became reactivated during 1981 (Fig. 6) and the debris-covered part was broadened that way.

With increased activity in frontal and lateral parts of the glacier debris enriched shear planes became visible also along the left margin of the ice. There, the same granulometric composition of the debris is evident: sandy gravel with little fines. So far, during the readvance the "pyramids", which developed during the phase of downwasting, have not changed, either in form or in position.

4.3 The lateral margins of the ice

As mentioned above, the readvance was initially detected in the winter of 1979/80 when an appr. 2 km long well defined displacement was noted at the right lateral margin. This formerly inactive debris covered part of the glacier became very active during the following months and a clearcut lateral shear plane developed. The ice thickness has also increased considerably and this has resulted in voluminous glacitectonic remobilisation of the tillcomplex along the ice margin and the slope of the lateral moraine ridge.

The most spectacular phenomena are visible in winter, when, in zones of compressive flow, the well defined shear plane doesn't collapse due to continuous melting of the ice (Fig. 10). As a consequence the geometry of the movement can be recorded at this place. Qualitatively important is the observation of strong lateral and "upward" (= vertical) flow components combined with the main downvalley movement of the ice in the zones of compressive flow regime. This is demonstrated (Figure 10) by well preserved "striae" at the glacier sole which dip upglacier at appr. 30°. This clear-cut lateral shear plane reactivates the lateral ice margin. This has previously been covered by slope-debris and exhibits now an active supraglacial till complex (Fig. 9).

The production of fresh till related to the readvance is restricted to a thin basal till layer frozen to the zone of displacement (Fig. 10). This fresh till is produced by the advance of the glacier and is, with regard to the position of transport, basal till; with regard to its deposition it is melt-out till as it is released as soon as the temperatures (daily and/or seasonal) are rising above freezing. Before the displacement reaches the surface some of this basal material is lodged as basal lodgement till to the stable part of the glacier's bed. The texture of this basal till is identical of what is known from "Grundmoräne" (= basal lodgement till) outside the recent glaciers (Schlüchter 1980).

The debris - rich basal zone of sliding is extremely thin and averages, with the exception of attached pebbles and boulders, only a few centimetres (Fig. 10). Till production in this glaciodynamically active zone is surprisingly low. Furthermore, the influence of meltwater as sedimentological agent is, at the right lateral surface, almost lacking apart from being linked to the thawing processes. The main sedimentary process in the active marginal zone of the glacier is therefore glacitectonic and glacidynamic reworking and remobilising of older till. Textural characteristics of the remobilised sediments are, therefore, inherited and may not reflect the actual glacidynamic process.

5 CONCLUSIONS AND FURTHER STUDIES

The following sedimentological conclusions can be drawn from the recent readvance of the Findelengletscher:

a) The autochthonous production of new till directly related to the advance is restricted. The main active sedimentary processes in lateral and frontal zones are reworking and remobilisation of older till.

b) Meltwater activity has influenced the sedimentary processes related to the readvance very little so far. Reworking and remobilisation are primarily of glacitectonic nature in a broad sense.

c) Pressed scales (Press-Schuppen sensu Gripp 1979) at the ice front are formed

Figure 9: The right margin of the glacier: view downglacier along the well defined lateral zone of movement with glaciodynamically reworked older till to form a supraglacial till ridge. Photo: Febr. 27, 1981.

Figure 10: Active lateral shear plane with a thin layer of basal till attached to the base of the moving ice. Note "striae" and orientation of some clasts. Height of total visible "outcrop" is appr. 2.20 m. Photo: Febr. 27, 1981.

within th older till above bedrock up
to 10 m in distance from the actual ice
front. The lack of a prominent push-
end-moraine and observations during the
winter months (Fig. 7 and 8) demonstrate
subsequent overriding of the pressed
scales by the advancing ice front.

d) The well defined zone of lateral shearing demonstrates that the movement of
the glacier has a considerable vertical
and lateral component. This is caused
by the difference in ice velocity
(2 cm/hour at 220 m in elevation versus
1 cm/hour at the ice front: The compressive stresses are therefore released not only by downvalley sliding
but also by lateral and vertical extension ("thickening").

Further studies will include the following topics:

- installation of pressure gauges on the
  cross-valley bedrock ridge

- relationship between flow patterns of
  the advancing glacier and the resulting
  till facies (also in comparison with the
  older tills in the lateral moraine ridges
  and in the forefield).

- small-scale till facies variations at
  the ice front and the behaviour of the
  "pyramids"

- the origin of glacially sculptured
  clasts.

6 ACKNOWLEDGEMENTS

Most cordial thanks are due to Dr. A.
Iken and PD Dr. H. Röthlisberger, ETH-
Zurich, for stimulating discussions and
for their companionship and encourage-
ments. Best thanks are transmitted to
A. Bezinge of Grande Dixence SA who has
helped with logistic support and to Prof.
H.J. Lang, Director of the Institute of
Foundation Engineering and Soil Mechanics,
ETH-Zurich, who has given permission to
extend my studies on tills to the recent
glacial environment. Very special thanks
are due to those colleagues and friends,
especially to Dr. W. Haeberli, ETH-Zurich,
who have helped to carry sediment samples
from the glacier area for laboratory tests.

7 LITERATURE

Bearth, P. 1953, Geologischer Atlas der
  Schweiz 1:25 000, Atlasblatt 29 (Zermatt).
  - Schweiz. Geol. Kommission (Kümmerly &
  Frey AG) Bern.
Bezinge, A. 1982, Glacier de Findelen,
  avance rapide et construction d'une
  nouvelle prise sous glaciaire. - Société
  Hydrotechnique de France, Section de
  Glaciologie, Journées de Paris, les
  4 et 5 mars 1982 (manuscript).
Gripp, K. 1979, Glazigene Press-Schuppen,
  frontal und lateral. - In: Schlüchter,
  Ch. (Ed.), Moraines & Varves (Balkema)
  Rotterdam: 157-166.
Röthlisberger, F. 1976, Gletscher- und
  Klimaschwankungen im Raum Zermatt,
  Ferpècle und Arolla. - Die Alpen, 3./4.
  Quartal 1976, 52. Jg.: 59-150.
Schlüchter, Ch. 1978, INQUA-Commission on
  Genesis and Lithology of Quaternary
  Deposits, Symposium 1978, Switzerland.
  - Guidebook for the Excursion, 112 p.,
  ETH-Zurich.
Schlüchter, Ch. 1980, Bemerkungen zu eini-
  gen Grundmoränenvorkommen in den
  Schweizeralpen. - Zeitschr. f. Gletscher-
  kde. u. Glazialgeologie, 16/2: 203-212.

A recent drumlin with fluted surface

in the Swiss Alps

J.J.M.VAN DER MEER
University of Amsterdam, Netherlands

ABSTRACT: In the "Vorfeld" of the Biferten glacier in E.Switzerland the only inner-alpine drumlin known occurs. In this paper the drumlin, surface features like annual moraines and flutes, and the composition of the drumlin are described. The origin of drumlin and flutes is discussed.

INTRODUCTION

In 1979 German, Hantke & Mader published a paper on a drumlin in the proglacial area of the Biferten glacier in the Swiss Alps. Since this is the only one of its kind in the Alps it seemed well worth visiting this object. After comparison of the drumlin with the observations of German et al. it seemed that there was more to say about this drumlin. Besides, it seems worthwhile to give more publicity to this feature.

It must be mentioned that drumlins commonly described as occurring in the Alps (e.g. N of the Bodensee; Embleton & King, 1975, p. 416) actually are located in the Alpine foreland. The drumlin described here is found in an alpine valley. De Quervain & Schnitter (1920) mention the occurrence of a similar feature in the proglacial area of the Steinglacier (Sustenpass area, C.Switzerland). These authors indicate however, that this was already partly destroyed by meltwater and furthermore that it was actively overridden by the glacier (the 1920 advance). In this study of the glaciers of the Sustenpass area, King (1974) does not mention the occurrence of a drumlin in the proglacial area of the Steinglacier. Contrary to the Biferten glacier, the Steinglacier had the same extent during the 1920 advance as during the late 1890's and remodeled its proglacial area completely. Drumlin-like mounds are further mentioned by German (1972, in German et al., 1979) as occurring near the Theodulglacier (near Zermatt, Switzerland; see also Hantke, 1978). Menzies (1979) mentions a few more authors who found drumlins in highland areas outside the Alps. Although the occurrence of drumlins is usually mentioned in relation to ice sheets (Embleton & King, 1975; Menzies, 1979) it must be concluded that they are not completely unknown from the valley glacier environment. But in the latter their chances of survival must be considered small as exemplified by the case of the Steinglacier mentioned above.

DESCRIPTION

The drumlin described here is located in the upper reaches of the Biferten-valley (fig. 1), a tributary of the Sandbach, which later becomes the Linth River and empties in the Walenseen in E.Switzerland. It is easily detected on the 1:25,000 Swiss National Map, sheet 1193, Tödi.

Upon entering the rather flat proglacial area the drumlin is clearly visible (fig.2). It occupies part of the area enclosed by a large moraine, which dates from the mid-nineteenth century (German et al., 1979). A few hundred meters to the S of the drumlin there is a rocky step in the valley. The glacier nowadays just reaches the upper part of this step and is located about 300 m higher than the drumlin (fig. 1).

The drumlin itself is about 230 m long and 120 m wide, giving a ratio of ca. 2:1 which is about the average for drumlins in general (Menzies, 1979). The height of the drumlin is about 15 m and it trends almost exactly S-N. The drumlin is not completely elliptical in shape, but shows some indentations, while the SE part is actively eroded by the Bifertenbach. The surface of the drumlin is fairly flat (figs. 2 and 3),

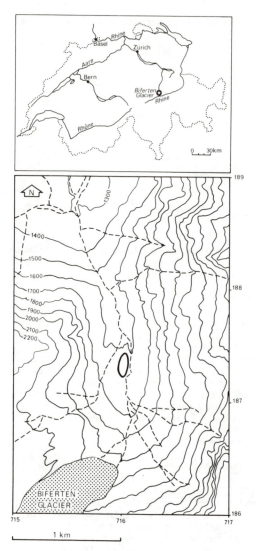

Fig. 1. Location of the drumlin

it does not show the typical egg-shape described so often. At the surface of the drumlin several minor forms can be observed. Upon approaching the drumlin several low ridges oriented more or less parallel to its short axis can be observed (figs. 2, 3). These ridges are only a few dm high and are, in accordance with German et al. (1979), interpreted as annual moraines (see also Rutishauser, 1972). When walking over the drumlin a number of straight ridges can be observed. De Quervain (1918) counted about 12 of these. Probably they showed up better in those days, due to the less well developed vegetation. These ridges are completely

straight (fig. 4) and ran almost exactly S-N. Their maximum dimensions are: width 2.5 m, height 1 m, and length 70 m. Near the up-glacier end a large block (mainly limestones) can be observed, which is of the same width as the ridge. The width of the ridges remains fairly constant in the down-glacier direction. Up-glacier from this block there is also an extension of the ridges. These extensions are at most 10 m long and have a pointed end. The down-glacier end of the ridges is in most cases rather vague, the height of the ridges diminishes gradually and the exact end is difficult to indicate. The surface of the ridges is littered with blocks of all sizes, while the swales in between the ridges show fewer blocks. The ridges are oriented parallel to the long axis of the drumlin, only at the down-glacier end they slightly diverge (De Quervain & Schnitter, 1920). The ridges consist of till. They are interpreted as flutes, and as such they are positive, constructive forms (Sugden & John, 1976). Flutes are not uncommon in the Alps, see e.g. Boulton (1976) and Haeberli et al. (1979), or in general, in the valley glacier environment (Boulton & Eyles, 1979).

COMPOSITION OF THE DRUMLIN

According to De Quervain & Schnitter (1920) the drumlin consists of till, while German et al. (1979) interpreted the lower part as consisting of lateral moraine sediments (Randmoränensedimente), covered by two meters of till. Investigation of the eroded SE part of the drumlin only showed till however. Granulometric analyses of a sample of the matrix (< 2 mm) of the till is presented in table 1. It shows that there are two modes, the first mode in the silt fraction and a second mode in the sand fraction. Sediment parameters (according to Folk & Ward, 1957) are also included in the table. These are fairly typical of lodgement till (Schlüchter, 1981). As for the petrographic composition, only limestones were observed in the eroded SE part of the drumlin. De Quervain & Schnitter (1920) pointed out that at the surface of the drumlin also dolomites, oolites and cristalline rocks occur. These authors considered these blocks to be part of the - thin - ablation till. From the exposed till in the SE part of the drumlin one sample of 100 limestones (fraction 2-7 cm) was collected for roundness measurements. The Cailleux-indices for this sample are: roundness 53 and flatness 202. These indices are typical of till (Schlüchter, 1981; van der Meer, 1982). Hardly any of these stones were striated

Fig. 2. View of the drumlin and the enclosing end moraine. To the right a lateral moraine can be observed.

Fig. 3. The drumlin viewed from the N. In the centre several annual moraines can be seen.

Table 1. Granulometry of the matrix (< 2 mm) of the till.

2000-1200 µm	9.2 %	M_Z	+ 3.48 ϕ
1200- 850 "	10.2 "	τ_I	+ 3.03 ϕ
850- 600 "	7.3 "		
600- 420 "	6.3 "	Sk_I	+ 0.13
420- 300 "	5.4 "		
300- 210 "	4.7 "	K_G	+ 0.70
210- 150 "	3.6 "		
150- 105 "	3.3 "		
105- 75 "	1.6 "		
75- 50 "	1.7 "		
50- 32 "	6.8 "		
32- 16 "	10.6 "		
16- 8 "	12.8 "		
8- 4 "	8.1 "		
4- 2 "	3.8 "		
2- 0 "	4.5 "		

however, but all of them show abraded edges and corners. This is interpreted as typical of stones in tills of a uniform lithology (Van der Meer, 1982; p. 58).

DISCUSSION

The fact that hardly any drumlins are known from alpine areas seems to indicate that in this environment they only form under special conditions. Because none of the necessary data, like ice velocity or the existence of secondary flows, are known, one can only hint at the possible processes that are thought to be responsible for the origin of the drumlins.

As pointed out by Boulton (1982) stream-lined forms can - among others - be produced by lodgement or by subglacial sediment deformation. In this case only till was found in the drumlin and thus both processes can be responsible for the formation of the drumlin. If the first

process is responsible then the rather flat surface of the drumlin points to either large ice velocity or small effect-ive pressure (Boulton, 1982). The possibility that the drumlin formed around a rock protuberance cannot be excluded. In this case subglacial sediment deformation can be the process responsible. Especially if German et al. (1979) are right in their observation of lateral moraine sediments underneath the till. The till was then plastered on these sediments during or after deformation.

German et al. (1979) considered the nucleus of the drumlin to consist of lateral (o.c., p. 97) or medial (o.c., p. 102) moraine material. The latter was derived from the Biferten glacier and the Hinter Rötifirn. As the western lateral moraine clearly swings around the drumlin to join the frontal moraine (fig. 2) the location of a medial or a lateral moraine in the centre of the valley is not convincing.

According to De Quervain & Schnitter (1920) and as indicated on the topographic map of this area the Biferten glacier is partly covered by debris. The eastern (debris-covered) part of the glacier also shows some ridges. This might indicate that the process proposed by Shaw (1980) is responsible for the formation of the drumlin. In this process threedimensional secondary flows are considered to produce agglomerations of subglacial material. The debris-covered ridges on the glacier are thought to be the surface expression of these flows.

As for the origin of the flutes, De Quervain & Schnitter (1920) give an excellent description of the process which has later been termed "the till squeeze theory" (Menzies, 1979; see also Boulton,

Fig. 4. View up-glacier of the largest flute on the drumlin.

1976). De Quervain & Schnitter state: "*The pressure of overlying ice can laterally squeeze material in the pressure-relieved cavities*" (developed behind boulders; translation by the author). They based this idea on field observations near the Obere Grindelwald glacier (o.c., note 3 on p. 146). According to Menzies (1979, p.331) this till squeeze theory was proposed by Alden in 1918.

Most authors consider flutes to be positive constructive forms. It is remarkable that German et al. (1979) consider the flutes on the drumlin of the Biferten glacier as erosion remnants. The swales between the flutes are thought to be eroded by the glacier, while the till behind large boulders was protected against erosion. During deglaciation the swales further deepened by meltwater. Especially the emphasis they put on the role of melt-water in reworking glacial deposits is not supported by the facts. For instance, the fact that the flutes are straight and of constant width over a length of some 70 m suggests otherwise. Further, the fact that the surface of the flutes is littered with blocks while the swales in between contain fewer blocks does not support this reasoning. If meltwater had been important one would expect most blocks in the swales,

where they would then have concentrated as a lag. The fact that at some places shallow rills are found in the swales only indicates that after deglaciation of the area, running water has slightly reshaped part of the surface of the drumlin.

ACKNOWLEDGEMENTS

I am very grateful to Messrs. F.Wälti Sr. and Jr. for support in the field. I further wish to express my thanks to Mrs. I.I.Y.Castel for preparing the figures, to Drs. M.Rappol for the granulo-metric analyses and for critically reading the manuscript, and to Mrs. M.C.G.Keijzer-v.d.Lubbe for the typing. Discussions with Dr. Ch.Schlüchter and Dr. G.S.Boulton are gratefully acknowledged.

REFERENCES

Boulton, G.S., 1976. The origin of glacially fluted surfaces - observations and theory. J.Glac. 17-76: 287-309.
Boulton, G.S., 1978. Boulder shapes and grain-size distributions of debris as indicators of transport paths through a glacier and till genesis. Sedimentology 25, 773-799.
Boulton, G.S. & N.Eyles, 1979. Sedimentation

by valley glaciers; a model and genetic classifications. In: Ch.Schlüchter (ed.), Moraines and varves, p. 11-23, Rotterdam, Balkema.

Boulton, G.S., 1982. Subglacial processes and the development of glacial bedforms. In: R.Davidson-Arnott, W.Nickling & B.D.Fahey (eds.), Research in glacial, glacio-fluvial, and glacio-lacustrine systems, p. 1-31. GeoBooks, Norwich.

Embleton, C. & C.A.M.King, 1975. Glacial geomorphology. London, Arnold.

Folk, R.L. & W.C.Ward, 1957. Brazos river bar: a study in the significance of grain size parameters. J.Sed.Petr. 27, 3-26.

German, R., R.Hantke & M.Mader, 1979. Der subrezente Drumlin im Zungenbecken des Biferten-Gletschers (Kanton Glarus, Schweiz). Jh.Ges.Naturkde. Württemberg 134, 96-103.

Haeberli, W., L.King & A.Flotron, 1979. Surface movement and lichen-cover studies at the active rock glacier near the Gruben-gletscher, Wallis, Swiss Alps. Arc. & Alp..Res. 11, 421-441.

Hantke, R., 1978. Eiszeitalter, Band 1. Thun, Ott.

King, L., 1974. Studien zur postglazialen Gletscher- und Vegetationsgeschichte des Sustenpassgebietes. Basler Beitr. Geogr. 18.

Meer, van der, J.J.M., 1982. The Fribourg area, Switzerland. A study in Quaternary geology and soil development. Ph.D. thesis, Univ. Amsterdam.

Menzies, J., 1979. A review of the literature on the formation and location of drumlins. Earth.-Sc. Rev. 14, 315-359.

Quervain, de, A., 1917. Ueber einen rezenten Drumlin. Ecl.geol.Helv. XIV, 684.

Quervain, de, A. & E.Schnitter, 1920. Das Zungenbecken des Bifertengletschers. Denkschr.Schweiz.Naturf.Ges. LV-II, 137-149.

Rutishauser, H., 1972. Beobachtungen zur Bildung von Jahresmoränen am Tschingel-gletscher (Berner Oberland). Ecl.geol. Helv. 65, 93-105.

Schlüchter, Ch., 1981. Genetic relation-ships between outwash deposits and basal till. Quaest.Geogr., in press.

Shaw, J., 1980. Drumlins and large scale flutings related to glacier folds. Arc. & Alp.Res. 12, 287-298,

Sugden, D.E. & B.S.John, 1976. Glacier and landscape. London, Arnold.

Die Grundmoränen des durch Schub
und des durch Zerrung fliessenden Eises

K.GRIPP
Lübeck, Germany

ZUSAMMENFASSUNG

S c h u b f l i e s s e n d e Gletscher
werden durch Schmelzwasserbäche an der
Oberfläche und im Tunnel entwässert. In ihrer eisgefüllten Grundmoräne werden die
feinsten Gesteinspartikel zurückbehalten. -
Regen und Schmelzwasser zirkulieren in den
durch Z e r r u n g fliessenden Gletscher
in Spalten und es kommt so zu einem Auswa-
schen der Gletscherbasis und zu einer Sor-
tierung der dortigen Sedimente. In dieser
Grundmoräne treten vor allem Sand und Kies
auf.

ABSTRACT

Textural characteristics of groundmoraines
in relation to compressing or extending
flow of the ice.

Glaciers under compressive flow regime are
drained on the surface and in tunnels.
Their icefilled groundmoraine preserves
the finest grained material. - On extending flowing glaciers rain and meltwater is
circulating in crevasses and is washing
the base of the ice thus sorting the sediments there. In the groundmoraine sand and
gravel are prevailing.

Auf Anregung von Professor Eduard Richter
(Graz) kamen im August 1899 die namhaften
Gletscherforscher in Gletsch (Kt. Wallis,
Schweiz) zusammen. Es wurden nicht nur die
Fragen der Nomenklatur im Saale diskutiert,
sondern auch Rhone- und Unteraargletscher
begangen.

Hinsichtlich der Benennung der Moränen
wurden nach der Lage unten im Eis die
Untermoräne (moraines inférieures) und unter dem Eis die Grundmoräne (moraines de
fond) getrennt. Diese Benennung hat
A. von Böhm (1901) eingehend kritisiert.
Da aber jene Untermoräne nach Schwund des
Eises zur Grundmoräne wird, und nur solche
im ehemals vereisten Flachland anzutreffen
ist, und zudem in Grönland und auf Spitzbergen an rezenten Gletscherränden eisfreie Grundmoräne nicht aufgefunden wurde,
spielt der Begriff Untermoräne keine be-
deutende Rolle mehr.

Festzuhalten aber ist der Unterschied in
der Ausbildung der Grundmoräne. In den
Alpen und den skandinavischen Gebirgen ist
sie sandig-kiesig, daher locker, in den
Flachländern jedoch durch Gesteinsmehl
oder aufgenommenen Ton gebunden und in
trockenem Zustand fest. - Da die Erdwärme
nicht ausreicht, um bei den Gebirgsglet-
schern im basalen Teil der Untermoräne das
Eis schmelzen zu lassen, und da in Dänemark
und Norddeutschland Grundmoräne mit Schliff-
Fläche auf Grundmoräne ruht, also vormals
eisfreie sandig-kiesige Basislagen fehlen,
muss für die unterschiedliche Ausbildung
der Flachlands- und Gebirgsgrundmoräne
nach einer Deutung gesucht werden.

J.F. Nye (1952) hat auf die zwei ver-
schiedenen Arten vom Fliessen des Eises
hingewiesen. Auf schwach geneigtem und
besonders auf ebenem Untergrund wird das
Eis durch den Druck hinzugekommenen Eises
nach allen Seiten geschoben; es fliesst
also durch Schub (compressing flow). Hier-
bei können keine Spalten offen bleiben.
Niederschlag und Schmelzwässer sammeln
sich daher auf der Oberfläche des Eises,
z.T. zunächst in Seen oder Eissümpfen, und

fliessen in Bächen hangabwärts, bis sie in einer Bachschwinde in das Eis gelangen. Durch ein Tunnelsystem gelangen sie zunächst bis an die Basis und weiter bis an den Rand des Eises.

H.W. Ahlmann (1935) traf bei der Durchquerung der flachen Insel Nordost-Land (Spitzbergen) auf 350 km Weges keine Gletscherspalte. Am Rande des Skeidarar Jökull liefen nach Jewtuchowicz (1973) von der Oberfläche des Eises 120 Bäche herab. - Die basale, mit Gesteinsschutt erfüllte Lage des vorher geflossenen Eises bleibt in Ruhe solange erhalten, bis durch Tieftauen von der Erdoberfläche her das Eis schwindet. Da eine Durchspülung ausbleibt, enthält jene Grundmoräne den gesamten vom Eis aufgenommenen und in ihr entstandenen Feinstschutt (Gesteinsmehl).

Die zweite Art des Eisfliessens ist nach Nye das gravitative Fliessen (extending flow), also das Abgleiten auf geneigtem Untergrund. Da die Bezeichnung Zerfliessen phonetisch missverstanden werden könnte, wird Zerrungs-Fliessen als deutsche Bezeichnung vorgeschlagen. - Die Folge dieser Zerrung sind die zahlreichen Gletscherspalten. Diese bedingen das Fehlen von Sümpfen und Bächen und auf der Oberfläche der Gletscher. Vielmehr steht dem gesamten Niederschlag sowie dem Eis-Schmelzwasser nur der Weg unter das Eis zur Verfügung. Zwar reichen Gletscherspalten zumeist nicht mit erheblicher Breite bis auf den Untergrund; aber da Wasser von 4°C die höchste Dichte hat, sind im Sommer genug Kalorien im Wasser vorhanden, um auf engen Spalten das Wasser bis unter das Eis gelangen und dieses schmelzen zu lassen.

In den im Zehrgebiet liegenden, durch Zerrung fliessenden Gletschern tritt daher durch das Talab-Wandern der Spalten weitflächig unter dem Eis ein Schmelzen der tiefsten Eislagen und Ausspülen des Feinkornanteils des basalen Schuttes ein. - Die Art des Fliessens des Eises bedingt die Art von dessen Entwässerung und diese die Unterschiede in der Ausbildung der Grundmoräne. - Diese Korngrössenunterschiede sind schon von K. Richter (1937) für Grossräume dargelegt worden (siehe auch H. Liedtke, 1975, S. 18).

Dieser Wechsel hängt nicht, wie bislang angenommen, mit der Entfernung vom Moränenschutt liefernden Felsgestein zusammen und nicht mit der Möglichkeit, feinkörniges Sediment aufzunehmen, sondern weitgehend von der basalen Auswaschung der Untermoräne im Bereiche des gravitativ fliessenden Eises.

LITERATUR

Ahlmann, H.W., 1935, Contribution to the physics of glaciers, Geogr.Journ. 86.
Böhm, A. von, 1901, Geschichte der Moränenkunde, Abh. geogr.Ges.Wien 3, Nr. 4.
Jewtuchowicz, St., 1973, The present-day marginal zone of Skeidararjökull, Scient.Results Polish Geograph. Expedition to Vatnajökull (Iceland), Geographia Polonica 26, 115-137, Warszawa.
Liedtke, H., 1975, Die nordischen Vereisungen in Mitteleuropa, Forsch.dtsch.Landeskunde 160, 1-204, Bonn.
Nye, J.F., 1952, The mechanics of glacier flow, J. Glaciol.2 (12), 82-93
Richter, K., 1937, Die Eiszeit in Norddeutschland, Berlin.

Lodgement tills and syndepositional glacitectonic processes
Related to subglacial thermal and hydrologic conditions

HANNA RUSZCZYŃSKA-SZENAJCH
Warsaw University, Poland

INTRODUCTION

The most commonly known interpretation of the lodgement process undoubtedly is that published by Flint (1971) which says: "... pressure melting of the flowing ice frees drift particles and allows them to be plastered ... on to the subglacial floor." This interpretation has been confirmed by observations on active glaciers (Boulton 1971, and others). This process requires particular subglacial thermal conditions which, when combined with pressure, "allow" the melting of lowermost ice and the deposition of entrained basal debris.

PRE-DEPOSITIONAL PROCESSES

Thermal conditions more severe than those which allow melting of subglacial debris-rich ice usually cause freezing of the substratum material onto the base of the glacier (Boulton 1972, Sugden 1976). This freezing-on process (Ruszczyńska-Szenajch 1980) is restricted to the zone of glacial erosion. A comparatively long prevalence of these conditions in the subglacial zone, followed by an abrupt change into ice-depositional conditions, may result in the creation of an abrupt contact between lodgement till and the underlying sediments. Commonly stratification, and even a delicate lamination, of the underlying sediments may be perfectly preserved up to the bottom surface of the till (Ruszczyńska-Szenajch 1976a: Plate III).

However, the above situation is comparatively rare, and disturbed sediment beneath the till is much more common. The average thickness of the disturbed zone in the Polish Lowlands varies from a few centimetres to several tens of centimetres.

The disturbed sediments are usually with obliterated or completely destroyed sedimentary structures soft sands, silts, and/or clays (Fig. 1 and 2). The disturbed zone is attributed to shearing of the unfrozen underlying material (Ruszczyńska-Szenajch 1976a:102-103). The existence of those sheared sediments, composed of pure displaced substratum with obliterated sedimentary structures, points to the fact that mechanical (i.e. glacitectonic) transport underneath the moving ice was not yet accompanied by glacial deposition. Comparatively thick horizons of this kind also indicate a lack of erosion during the overriding process. In the above cases subglacial thermal conditions favour neither freezing-on of the substratum nor melting down of basal debris. These conditions allow only glacial transport - both within the ice (i.e. glacial transport sensu stricto) and below (i.e. subglacial glacitectonic transport). The necessity of existence of a subglacial zone, characterised by the above conditions, has previously been discussed by the author and described as a "transitional zone underneath the ice between the zones of prevailing glacial erosion and prevailing glacial deposition" (Ruszczyńska-Szenajch 1976a: 105 and Fig. 35).

OCCASIONAL LODGEMENT

The displaced sediments underlying till in lowland areas often contain thin layers, lenses or streaks of till (Fig. 3). This points to an occasional lodgement of basal debris accompanying the subglacial glacitectonic processes, and (since the soft substratum in lowlands causes much weaker friction than a hard bedrock does) the lodgement must have been mainly caused by positive temperature changes - high enough to produce melting of the basal ice but short-lasting and again being followed by cooler conditions allowing only transport.

In some places displaced sediments below till also contain irregular, usually small, lumps of till. Some of the lumps are elongated in the direction of ice flow and join the lowermost part of till (Fig. 4).

The writer has interpreted such features as syndepositional load casts (Ruszczyńska-Szenajch 1976a:103). Their formation must have required considerable water saturation of the substratum, and the basal debris which melted down must have also been saturated. The friction was in these cases minimal and the (occasional) melting must have been caused by the positive thermal changes, as pressure (the other factor controlling subglacial melting) could hardly display such delicate variations in a vast lowland area where the thickness of ice sheet was comparatively uniform and was controlled by larger-scale balance changes.

Horizons composed of deformed substratum changing upwards into deformed substratum with inclusions of till reflect, in the writer's opinion, a trend of temperature-and-pressure conditions in the subglacial environment, with a gradual change from "indifferent" (allowing glacial transport only) to "depositional". The smaller-scale thermal changes, superimposed on the general trend, are examined by studying the till inclusions which are the result of yet occasional, ice-depositional conditions.

INTERMITTENT LODGEMENT

The basal parts of lodgement tills often show interbeddings of two different kinds of substratum material: (a) comparatively regular layers separating individual till beds, and (b) irregular layers which are strongly deformed - together with till beds within which they occur.

(a) The comparatively regular interbeddings of substratum material occur in the basal parts of tills with considerable consolidation - accumulated as well defined beds, more or less horizontal, with the flattened clasts lying conformably with the bedding planes. The substratum material separating the till beds forms regular layers, often very thin - not

exceeding a few centimetres (Fig. 5). These layers are interpreted as a result of mechanical, glacitectonic, shifting of the substratum underneath the ice sheet, i.e.: "... process of till deposition was interrupted by episodes of non-deposition accompanied by movement of (substratum) material, which was spread thinly by the moving ice over already accumulated till beds" (Lindner and Ruszczyńska-Szenajch 1979:251). Therefore, the temperature (and pressure) conditions causing lodgement were interrupted by periods of a more severe thermal regime which allowed transport only. Such breaks of non-deposition are recorded also by clearly marked regular beds of till, with sharp boundaries, and not necessarily separated by displaced substratum.

The consolidation of these tills, the regularity of their bedding, and the position of flattened clasts point to the "classic" lodgement process, interpreted by Flint (1971), and observed in the recently glaciated areas. This points also to a wet but not oversaturated subglacial environment.

(b) During the examination of basal parts of lodgement tills plastically deformed horizons of substratum and till are commonly found. In places of cavities in the moving debris rich basal ice (caused e.g. by overriding of a boulder) deformed substratum and till layers are squeezed-up into them (Fig. 6).

Figure 1: Sands with obliterated sedimentary structures, underlying till at Wytowno.

Figure 2: Index map and localities mentioned in the text from Poland.

Figure 5: Thin sandy interbeddings of Jurassic material (white), separating till beds in the lower part of till horizon at Rozwady. Photo from color slide by K. Zielińska.

Figure 3: Thin till layer within displaced sands underlying till horizon at Czerwinsk (Ruszczynska-Szenajch 1976a).

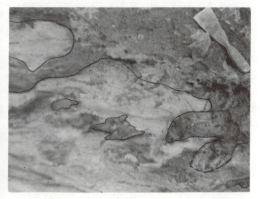

Figure 4: Lumps of till within displaced sands underlying till horizon at Wytowno (after Ruszczyńska-Szenajch 1976a).

This is evidence for lodgement of the lowermost irregular till layers combined with a displacement of the substratum material. The plastic deformation of both, substratum and till suggests saturation of the subglacial environment. These facies associations prove the glacitectonic melange of the till and substratum below the moving ice. This process has already been reported by Charlesworth (1957:380), and evidenced by Pleistocene data (Ruszczyńska-Szenajch 1976a:102-103 and Fig. 35). At present it is studied in a systematic way at existing glaciers (Boulton: paper presented at the glacitectonic symposium in Denmark, 1981). The zone where substratum and till just being deposited is shifted by an overriding glacier is called "mobile zone" by Boulton.

The (a) or (b) - type of the basal part of lodgement till units may occur within individual till horizons, but may exhibit also spatial variations within the same till (Ruszczyńska-Szenajch 1976a). Both reflect thermal conditions causing lodgement though yet interrupted by more severe thermal periods (seasons?). To produce the (a)-type the friction must have been important - the structure of the till points to a "hard" lodgement - where the (b)-type needed little friction (but water saturation), and thus was formed at higher ice temperatures. Both types of subglacial environments mentioned were also characterised by active glacitectonic processes.

115

Figure 6: The displaced substratum sands
(light) and till layers (dark) filling-in
a "hollow" in the lowermost part of till
horizon at Wytowno. The section is about
1 m high. Photo from color slide by K.
Zielińska.

STEADY LODGEMENT

Outcrops often display massive lodgement
till structures without traces of breaks
in deposition and are characterised by a
considerable thickness.

However, careful studies reveal that a
pure massive structure is not so common
as it was thought before. Apart from such
features, as reflecting cooler regime
during deposition – for example "sheet-
like accumulation of cobbles" within a
till– (cf. Dreimanis and Reavely 1953:
242), thick massive tills very often show
strong deformation. The deformations are
caused either by "hard" shearing processes
recorded by perfectly marked shear planes
with slickensides or by plastic deformation
recorded by folds and diapirs. Both features usually reflect the direction of
ice movement and are often truncated by
till of the same age as the deformation.
This proves the syn-depositional glacitectonic action affecting the whole (or
almost whole) lodgement till beds. The
difference in deformation (shearing or
diapirism) must have been controlled by
the amount of subglacial water. The
occurence of different types of deformation in different segments of a till (e.g.
at Wyszogród – Ruszczyńska-Szenajch 1976b)

may point to local differences in water
saturation, and this in turn may have
been related to the lithology of substratum.

Lodgement tills with no traces of breaks
during deposition most probably reflect
a continuous lodgement, resulting from
steady thermal conditions, often accompanied by glacitectonic processes. The considerable thickness of such tills may
point tó a considerable length of time of
prvalence of the given thermal conditions.

FINAL REMARKS

The differences in lodgement tills may
result from:
    (1) continuity or discontinuity of the
lodgement process,
    (2) amount of water in the subglacial
environment,
    (3) syn-depositional glacitectonic pro-
cesses.

The thermal conditions in the subglacial
environment, as a main factor "allowing"
lodgment to occur may also cause an increase in water content due to more
effective melting and, in turn, influence
the character of glacitectonic subglacial
deformation (hard shearing or plastic
deformation).

The occasional and intermittent lodgement
is referred by the writer to "cold
deposition" (connected with cooler
thermal conditions), while the continuous
steady lodgement is attributed to comparatively "warm subglacial deposition".

The evidence discussed here also shows that
there is no particular sedimentary process responsible for deposition of the so
called "deformation till", which is just
a result of combination of glacitectonic
action and a regular lodgement process.

REFERENCES

Boulton, G.S. 1971, Till genesis and
  fabric in Svalbard, Spitsbergen. In:
  Goldthwait R.P. (Ed.): Till. A symposium:
  41-72.
Boulton, G.S. 1972, The role of thermal
  regime in glacial sedimentation. Inst.
  Br. Geogr. Spec. Publ., 4:1-19.
Charlesworth, J.K. 1957, The Quaternary
  Era. E. Arnold Publ. Ltd., London.
Dreimanis, A. & Reavely, G.H. 1953,
  Differentiation of the lower and the
  upper till along the north shore of Lake
  Erie. Journ. of Sedimentary Petrology,
  23, 4:238-259.
Flint, R.F. 1971, Glacial and Quaternary
  Geology. J. Wiley and Sons, Inc.,
  New York.
Lindner L. & Ruszczyńska-Szenajch H. 1979,
  Changing conditions of glacial erosion
  and deposition reflected by differen-
  tiation of glacial deposits at Rozwady
  (Świętokrzyskie Mountains). In: Schlüchter
  Ch. (Ed.): Moraines and varves:249-255.
  Rotterdam.
Ruszczyńska-Szenajch, H. 1976a, Glaci-
  tektoniczne depresje i kry lodowcowe na
  tle budowy geologicznej południowo-
  wschodniego Mazowsza i południowego
  Podlasia (Glacitectonic depressions and
  glacial rafts in mid-eastern Poland).
  Studia Geol. Polon., 50:106. (Engl.
  summ.).
Ruszczyńska-Szenajch, H. 1976b, Examples
  of differentiation of tills due to bed-
  rock conditions prevailing in the place
  of deposition. In: Stankowski W. (Ed.):
  Till - its genesis and diagenesis:81-89.
  Poznań.
Ruszczyńska-Szenajch, H. 1980, Glacial
  erosion in contradistinction to glacial
  tectonics. In: Stankowski W. (Ed.):
  Tills and glacigene deposits:71-76.
  Poznań.
Sugden, D.E. 1976, Glacial erosion by the
  Laurentide ice sheet and its relationship to ice, topographic and bedrock
  conditions. Dep. of Geography, University of Aberdeen:45-75, Aberdeen.

Load deformations in melt-out till and underlying laminated till
An example from northern Poland

EUGENIUSZ DROZDOWSKI
Polish Academy of Sciences, Toruń, Poland

ABSTRACT

Synsedimentary and metasedimentary load
deformation structures found in a till
section are presented and discussed, example being drawn from the second glacial-
drift horizon on the lower Vistula River,
the deposition of which was associated
with large-scale stagnation of the Scandinavian ice sheet. A laminated till lying
below a massive supraglacial melt-out till
is explained as having been redeposited
due to liquefaction of subglacially produced melt-out till and local changes of
effective normal pressure at the ice base.
Load deformations comprise graben-like
displacement of the sediment, faults,
overconsolidation of the sediment expressed by a closer spacing of the laminae,
and downwarping. The last three types of
the structures were brought about by the
weight of a large boulder after its release from ice and subsequent subsidence.
The significance of these deformations
for the environmental interpretation of
melt-out till and some terminologic questions are discussed in the conclusions.

1 INTRODUCTION

The most discussable genetic type of till
is possibly melt-out till. The term was
coined by Boulton (1970) to describe the
till deposited by slow melt-out of glacial
debris from masses of dead ice. However,
the existence of such kind of till was
suggested much earlier, for example, by
Goodchild in 1875 (cf Shaw 1980) or by
Gravenor and Ellwood (1957) when they discussed the difficulty in differentiating
between ablation till and lodgement till.
They stated (op.cit., p.9): "It is,

however, generally assumed that there is
a considerable amount of englacial drift
in-transport near the base of the glacier.
This in-transport material may never appear
in the superglacial environment, but during
the final stage of melting it is let down
on the underlying lodgement drift. It is
quite likely that in-transport drift would
be protected from extensive melt-water
action and from fabric disruption. Is this
material classified as lodgement drift or
ablation drift?"

According to the modern interpretation,
melt-out till is produced progressively
from the top of debris-rich dead ice downwards and from its base upwards. The resultant tills are then termed supraglacial
or subglacial melt-out till with respect
to the position of their formation.

Melt-out tills have been subsequently
identified by many researchers in the
field of glacial geology and geomorphology,
they have been also included in the modern
till classifications based on the mode of
deposition of the sediment (Boulton 1976,
Dreimanis 1976, 1978, Drozdowski 1979,
Shaw 1980), nevertheless, relatively few
descriptions of their lithology and interpretations of their origin have been yet
published, and hence the diagnostic characteristics of melt-out tills, thereby the
environmental conditions and processes
which led to their formation are actually
very little known.

The criteria used for identifying melt-
out till discussed here provided analyses
of several exposures situated along the
valley walls of the lower Vistula River,
northern Poland (Fig. 1). They show in
vertical sections the subsurface, second
(counting from the surface of the morainic
plateau) till stratum, deposited by the

Fig. 1: Location of the study area

early Middle-Vistulian ice sheet in that area, which is known from a large-scale stagnation during its recession. Considerable attention has been paid to the section at Strzemiecin, where peculiar type of melt-out till and deformation structures have been found. The investigation of these features has led not only to an explanation of their origin, but allowed to establish with some confidence that the tills formed by in situ melting of debris from a sheet of dead ice.

2 DESCRIPTION OF THE SECTION

The section is located at Strzemiecin (town-quarter of Grudziadz), near the culmination of a morainic plateau "island" called Strzemiecinska Kepa. Since the surface of this island has been strongly affected by denudation processes during the Late Vistulian and Holocene time, the exposed second till stratum (Fig. 2) is not overlied by any younger stratigraphic member, including also supraglacial varved clays which commonly cover directly the second till stratum. The stratum is here 3 m thick (the maximum thickness observed in the study area is 11 m), at its base shows a large boulder, 0.6 m in diameter, and below a layer of distinctly laminated till, enriched in sandy and silty fractions, 0.2 m thick, resting immediately on a permeable substratum composed of fine-grained sand (Fig. 3). In some places, lenses and wedging out layers of till interfingering with sand were found further below the laminated till. In other places,

Fig. 2: Section of the second till stratum at Strzemiecin. In the lowermost position laminated till is visible.
The length of the spade: 1.2 m

graben-like structures that disturb the stratification of the underlying sand have been encountered (Fig. 3).

The laminated till has been traced only over a short distance of 17-19 m within the exposure and, as a closer examination showed, it is associated with an alternation of till and thin laminae of silty sand. The thickness of the silty sand laminae grows downwards from a tenth of millimetre to one millimetre or so near the base of the till stratum. Once partings are created, they produce rectangular blocks of till of various height and length, depending on the spacing of the laminae and internal cohesion of the sediment.

Fig. 3: Deformation structures below the
boulder. Scale 0.2 m

3 ORIGIN OF THE LAMINATED TILL

The question arises how the laminated structure of the till originated? At least three
alternatives seem plausible:

1. The laminated structure is inherited
from the former foliation structure of the
glacier ice. This means that former debris
bands converted into the till laminae whereas the ice between them produced the silty
sand laminae.

2. The laminated till represents subaquatic flow till which originated on the
surface of the ice margin and flowed down
into a proglacial lake.

3. The laminated structure formed as a
result of local displacement of subglacial
melt-out till being deposited under conditions of abundant meltwater and high
pore-water pressure.

Alternative No. 1, assuming glacial origin
of the laminated structure from a stratified
basal ice appears to be supported by a coherent structural bond of the laminated till
with the overlying portion of the till stratum, and by their similarities in many lithological parameters, like petrographical
composition of the gravel fraction, abrasion
of quartz grains, carbonate content in the
till matrix etc., related to the section

discussed here and numerous other sections
of the same till stratum on the lower Vistula River (Olszewski 1974, Drozdowski 1974,
1979). However, if one accepts that the laminated structure was inherited from the
glacier ice, then it is not unreasonable
to suspect that the original glaciogenic
structure would have been disturbed partly
or completely by meltwater produced during
the release of debris from ice. Another
alternative that the ice decayed by sublimation (cf Shaw 1977) may be eliminated for
the reason that the depositional processes
took place at the base of a sheet of debris-
laden dead ice, being dense and thick enough
to prevent removal of ice by its full con-
version into gasous phase. Arguments against
the inheritence of the lamination structure
from glacier ice seem to deliver the sandy-
silty laminae themselves; they grow in
their thickness downwards and pass gradually
into the underlying fine-grained sand without any visible hiatus. In addition, there
exist differences in the fabric pattern between the laminated till and the overlying
massive till (Fig. 4). It appears therefore
more likely, in contradiction to the author's previous interpretation (Drozdowski
1979), that the laminated structure does
not represent original glacial structure
but resulted from some kind of secondary
depositional process.

Alternative No. 2, interpreting the laminated till as subaquatic flow till deposited in proglacial lake has supporting
arguments based on numerous examples from
the study area and other glaciated areas
(Hartshorn 1958, Boulton 1968, Drozdowski
1977), particularly those from the north
shore of Lake Erie, Ontario, described by
Evenson and others (1977). The similarities
concern, above all, the interbeddings and
interfingerings of till with sandy-silty
sediments, and also different fabric patterns in individual till layers. Nevertheless, there are several lines of evidence
which are not in accord with the paleogeographic aspect of this alternative.
They are related to the glacial history
of the study area which does not fit a
model of steep-sloping snout of the retreating ice sheet. The discussed till
stratum represents a typical terrestrial
till deposited mainly, if not entirely,
during stagnation and slow decay of a
broad marginal zone of the ice sheet (Drozdowski 1979). No evidence of proglacial
lakes which were developing during the deglaciation period was found. Worthy of note
is also the restricted areal extent of the

laminated till and its location at the height of the top surface of the underlying fine-grained sand. All the evidence makes gravity flow of ablation till from the glacier surface into a proglacial lake hard to realize.

Alternative No. 3, suggesting local displacement of subglacially deposited melt-out till, seems to be more relevant. As shown by the asymmetric rose-diagrams of fabric relating to the laminated till and the overlying melt-out till (Fig. 4), the a-axes of prolate clasts display different orientations: toward north-north/east in the laminated till and toward north-east in the overlying massive melt-out till. These differences appear to indicate different flow directions of the depositing media. In addition, the a-axes in the overlying till have a steep up-glacier dip whereas in the laminated till they dip at lower angles and into both opposite directions which suggests a sediment flow (cf Boulton 1968, Lawson 1979).

Although the fabric pattern suggests sediment flow, the mechanism which led to the formation of the laminated structure is still not clear enough. It may be related to the lack or impeded drainage of the melt-water at the base of down-wasting ice sheet. Under such conditions, a rebound occurring upon removal of ice load or seasonal confinement of basal meltwater, as suggested by Müller (1977), may have caused a rise of pore-water pressure at the ice base and – as a consequence – a decrease of effective normal pressure (normal ice overburden pressure minus pore-water pressure at the base of the ice sheet), leading to the liquefaction of till and its subsequent displacement due to local changes of the pressure.

A problem still remains how originated the graben-like structures below the laminated till (Figs. 3 and 4 – N.1)? In appearance, they are similar to the early loadings or gravifossa in their initial phase of formation (Brodzikowski and Van Loon 1980), however the instability of the

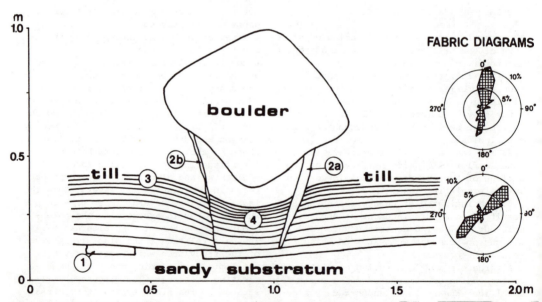

Fig. 4: Structural relationships between the boulder and the deformation structures. The fabric diagrams are shown opposite their respective sites in the section. 1. Graben-like structure, 2a. fault enlarged by plastic strain and filled with sand and occasionally gravel grains, 2b. fault deformed plastically and filled with fine sand, 3. closer spacing of the laminae (overconsolidation), 4. downwarping under the load

depositional surface, being responsible for such deformations, seems not to be applicable to the discussed case. More likely may have been local transient changes of the effective normal pressure generated, for example, by differentiated weight of debris in ice. About the transience of the deformational process gives evidence to the synsedimentary nature of the deformations inasmuch as they do not continue upwards into the overlying till.

4 LOAD DEFORMATION STRUCTURES

There are three kinds of deformations that were induced by the release of the large boulder from the ice and its subsequent subsidence (see Fig. 4): faults, closer spacing of the laminae, and downwarpings. The nature of these strains and their interrelations may play a key role in interpreting the depositional history of the melt-out tills discussed here.

The faults are exposed in a cross-section running approximately parallel to the direcion of the ice movement. They are 30-35 cm long and show vertical displacement of 2-4 cm. Fault 2a expanded laterally up to 5 cm, and was filled with fine-grained sand in which occasionally grains of gravel are found. Both faults die out at a distinct sandy lamina before reaching the base of the till. The walls of the expanded fault 2a have undergone some plastic modifications. Apart from the plastic modifications of the fault, there are two kinds of plastic strains: local overconsolidation of the laminated till fixed in a closer spacing of the laminae near the top of the till, particularly in the axial part of the buckling structure displaced down between the faults, and the downwarping strain itself. The plastic downwarping resembles, within certain limits, concentric-like folds in elastic multilayers obtained experimentally by Johnson and Honea (1975). Like in the experiment, the downwarping shows different wave lenghts and amplitudes at different laminae.

5 CONCLUSIONS

Different strains brought about by the weight of the boulder seem to reflect quite well the slowly proceeding formation of a layer of melt-out till and associated with this process changes of the mechanical properties of the material. Two melting zones operating within a sheet of dead ice should be envisaged in this process: an upper zone moving downwards, affected by the heat from the air, and a lower zone moving upwards under the influence of the geothermal heat flux. The described cracking and faulting developed presumably before the upper melting zone contacted with the lower melting zone, that is before the formation of melt-out till in the ice directly underlying the boulder. Therefore, the material tended to crack rather than to warp when a rapid compressive stress has been applied. This situation differs from that existing during the deformation within the underlying laminated till which provides evidence for ductile behaviour of the material, thereby suggesting that the deformational process post-dated here the deposition of till.

Thus, based on the data presented, a general conclusion can be reached that tills occurring in the section at Strzemiecin represent two varieties of melt-out till. The upper massive till, containing the large boulder, formed englacially, whereas the laminated till beneath formed as a result of basal melting and redeposition at the ice base due to liquefaction and local changes of the effective normal pressure. The mechanism of process operating in the deposition of melt-out till and particularly the laminated till is far from simple, and at the present deficiency of information based on direct measurements and observations it is not possible to examine the matter at this point.

Considering the question from the terminilogic standpoint, two types of melt-out till can be distinguished in the discussed section: supraglacial melt-out till and a laminated subglacial melt-out till that can be termed tentatively liquefied subglacial melt-out till or liquefied subglacial ablation till, following the terminology proposed by Elson (1961).

6 REFERENCES

Boulton, G.S. 1968, Flow tills and related deposits on some Vestspitsbergen glaciers. Jour. Glaciology, 7, no. 51: 391-412.

Boulton, G.S. 1970, On the deposition of subglacial and melt-out tills at the margins of certain Svalbard glaciers. Jour. Glaciology, 9, no. 56: 231-45.

Boulton, G.S. 1976, A genetic classification of tills and criteria for distinguishing tills of different origin. In: Stankowski W. (Ed.) Till - its genesis and diagenesis, Geografia 12: 65-80.

Brodzikowski, K. and Van Loon, A.J. 1980, Sedimentary deformations in Saalian glacilimnic deposits near Wlostow (Zary area, western Poland). Geologie en Mijnbouw, 59/3: 251-72.

Dreimanis, A. 1976, Tills: their origin and properties. In: Legget, R.F. (Ed.) Glacial till, Ottawa: 11-48.

Dreimanis, A. 1978, Till and tillite. In: Fairbridge, R.W. and Bourgeous, J. (Eds.) The Encyclopedia of Sedimentology: 805-09.

Drozdowski, E. 1974: Zmiennosc facjalna glin morenowych w profilu drugiego pokladu morenowego w Sartowicach/sum.: Facial variation of tills in the profile of the second moraine stratum at Sartowice (Lower Vistula Valley) Zesz. Nauk. UAM, Geografia 10: 121-36.

Drozdowski, E. 1977: Ablation till and related indicatory forms at the margins of Vestspitsbergen glaciers. Boreas, 6, no. 2: 107-14.

Drozdowski, E. 1979, Deglacjacja dolnego Powisla w srodkowym Würmie i zwiazane z nia srodowiska depozycji osadow (sum.: Deglaciation of the Lower Vistula region in the Middle Würm and associated depositional sedimentary environments). Prace Geograficzne IGiPZ PAN no. 132, 132 p.

Elson, J.A. 1961, The geology of tills. Proc. 14th Canadian Soil Mech. Conf., Natl. Rs. Counc. Canada. Com. Soil. Mech., Techn. Mem. 69: 5-36.

Evenson, E.B., Dreimanis, A. and Newsome, W. 1977, Subaquatic flow tills: a new interpretation for the genesis of some laminated till deposits. Boreas, 6, no. 2: 115-33.

Gravenor, C.P. and Ellwood, R.B. 1957, Glacial geology of Sedgewick district, Alberta. Research Counc. Alta, Prel. Rpt. 57-1, 42p.

Hartshorn, J.H. 1958, Flowtill in southeastern Massachusetts. Geol. Soc. A. Bull., 69: 477-82.

Johnson, A. and Honea, E. 1975, A theory of concentric, kink and monoclinal flexuring of compressible elastic multilayers. Part III - Transition from sinusoidal to concentric-like and chevron folds. Tectonophysics, 27, no. 1: 1-38.

Lawson, D.E. 1979, Sedimentological analysis of the western terminus region of the Matanuska Glacier, Alaska. CRREL Rpt 79-9, 112 p.

Müller, E.H. 1977, Dewatering during lodgement of till. INQUA Commission on Genesis and Lithology of Quaternary Deposits, Programme and Abstracts of Commission Meetings: 18.

Olszewski, A. 1974, Jednostki litofacjalne glin subglacjalnych nad dolna Wisla w swietle analizy ich makrostruktur i makrotekstur (sum.: Lithofacial units of subglacial boulder clays on the lower Vistula in the light of the analysis of their macrostructures and macrotextures). Stud. Soc. Sci. Torunensis, sectio C, v. 8, no. 2, 145 p.

Shaw, J. 1977, Tills deposited in arid polar environments. Can. Jour. Earth Sci., 14, no. 6: 1239-45.

Shaw, J. 1980, Melt-out till. INQUA Commission on Genesis and Lithology of Quaternary Deposits. Appendix to Circular no. 19.

Till petrology

The influence of till petrology on weathering of moraines
in southwestern Montana and northwestern Wyoming

R.D.HALL
Indiana University-Purdue University, Indiana, USA

1 ABSTRACT

Relative age dating techniques used in interpretations of glacial chronology are based on the assumption that the primary factor influencing the weathering of a deposit is the time since deposition. Other factors are assumed to be only secondarily important or are controlled by the selection of study sites. Investigations of glacial deposits in the Tobacco Root Range in southwestern Montana, and the Wind River Range in northwestern Wyoming, demonstrate that variations in the petrology of the parent materials may result in differences in weathering that are greater than those attributed to relative ages.

In deposits derived primarily from silicic igneous and gneissic sources, clasts of the latter rock type will usually be less weathered. Mafic clasts, if present, will be less weathered still. The larger grain size of the silicic igneous clasts appears to result in higher permeability and thus more weathering than that characteristic of finer-grained gneissic or mafic clasts, although the minerals of the latter lithologies are more chemically unstable in a weathering environment.

An abundance of the silicic igneous component also often results in more weathering of the deposit as a whole. Soils in such deposits are compact and have well-defined horizons. The solum is likely to be more enriched in clay and free iron and more depleted in hornblende. Deposits derived from mixed silicic igneous and gneissic sources are usually less weathered. Clay minerals form that reflect the petrology of the parent material. The presence of calcareous sedimentary materials in till promotes development of $CaCO_3$-rich horizons. Loess added to moraines also promotes more weathering, especially the development of pedogenic clay.

2 INTRODUCTION

A variety of relative age dating techniques has been used in many attempts to interpret the chronology of Quaternary glacial and periglacial events in the Rocky Mountains. The basic premise is that, the earlier the event, the more weathered will be its deposits. Techniques have then been designed to assess the degree of weathering of till and periglacial deposits, assuming the primary factor influencing weathering is the time since deposition. Other factors such as topography, climate, vegetation, and parent material, are either assumed to be only secondarily important or are controlled in the selection of sites in the study area. Among these techniques, many focus on assessing 1) the weathering of clasts on and within a deposit, and 2) the degree of weathering of the deposit as a whole, including the soil development. In the latter case the assessment often includes the description of soil morphology and the collection of geochemical and mineralogical data.

Clearly, the petrology of the parent material must significantly influence the results of these studies, particularly where weathering has occurred in arid and semi-arid climates (Birkeland, 1974). Although some control on the parent material factor is usually exerted by investigators through the selection of study sites with similar petrologies, so much attention is often directed at establishing chronosequences that the effects of petrologic variations on weathering may not be adequately addressed. Yet such effects can sometimes mask the differences in the data intended to indicate relative age. The purpose of this paper is to examine the influence of variations in petrology on the weathering of moraines in

and around the Tobacco Root Range in southwestern Montana and the Wind River Range in
Wyoming (Fig. 1). The paper addresses
three petrologic factors: 1) the weathering of clasts of different lithologies, 2)
the influence of till petrology on weathering of the deposit as a whole, and 3) the
influence on weathering of loess added to
and mixed down into deposits.

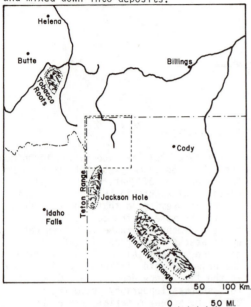

Figure 1. Location of Tobacco Root Range
and Wind River Range.

The Wind River Range in northwestern
Wyoming (Fig. 1) has been the location of
many previous studies of glacial geology
(e.g., Blackwelder, 1915; Holmes and Moss,
1955; Richmond, 1948, 1962, 1964, 1976;
Murphy, 1965; Currey, 1974; Miller and
Birkeland, 1974; Mahaney, 1978). In this
paper some data are included from studies
by the author at Bull Lake along the northeastern flank of the range (Hall and Roy,
1979) and near Boulder Lake along the
southwestern flank (Roy, 1980).

However, the primary sources of data for
this paper are studies by the author and
his students of the late Quaternary glaci-
ations and periglacial activity in the
Tobacco Root Range, located in southwestern
Montana about 100 km northwest of
Yellowstone National Park (Fig. 1). Inves-
tigations have been completed recently in
the South Boulder River drainage basin
(Roy, 1980; Roy and Hall, 1981) and the
Cataract Creek and North Willow Creek basins
(Hall and Heiny, in press) and are in

progress in the South Willow Creek basin,
all of which are along the eastern flank of
the range (Fig. 2). Also in progress is a
study of the Bear Gulch basin on the west
side of the range (Fig. 2).

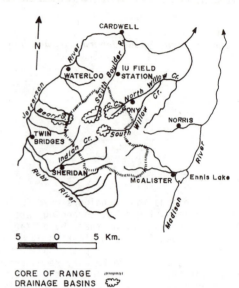

Figure 2. Map of the Tobacco Root Range
study area.

These study areas contain deposits whose
petrologies reflect contributions, in varying degrees, from silicic igneous(batho-
lithic), metamorphic (predominantly gneissic), and sedimentary bedrock sources.
Similarly, loess has been added to the till
and periglacial deposits in variable amounts.
The influence of parent material on weathering will be assessed by comparing deposits
of variable petrology, but quite similar
relative age. To the extent possible, data
have also been chosen from sites with
similar topography, climate, and vegetation.

3 WEATHERING OF CLASTS

3.1 Silicic Igneous Clasts

The core of the Tobacco Root Range is composed of the silicic Cretaceous Tobacco
Root Batholith and Precambrian metamorphic
rocks, predominantly gneisses, but including schists and amphibolites, with the whole
metamorphic complex in places cut by mafic
dikes. Paleozoic and younger sedimentary
rocks flank this crystalline core. The
outcrop pattern of these rock types within
the drainage basins that have been studied
resulted in a considerable variation in the
abundance of silicic igneous, gneissic, and
mafic clasts (the latter including fine- to
medium-grained amphibolites and diabases)

128

Table 1. Effects of Grain Size on Weathering of Silicic Igneous Clasts

| Drainage Basin | Deposit | Weathering Index (R^a) | | | Mean Weathering Rind[b] | | |
		CS[c]	MED[c]	FN[c]	CS[c]	MED[c]	FN[c]
South Boulder	SBR-1A[d]	.0.56	0.30	0.13	1.75	0.50	0.66
	SBR-11A[f]	0.75	NA	0.19	NA	NA	NA
Cataract Creek	CC-6[e]	0.31	0.33	0.20	1.14	1.33	0.60
North Willow Creek	NWC-2[e]	0.41	0.22	0.12	3.12	1.99	0.97
South Willow Creek	SWC-2A[d]	0.40	0.31	0.24	5.60	3.68	2.18
	SWC-9A[f]	0.31	0.28	0.00	3.30	1.90	NA
	MEAN	0.46	0.29	0.15	2.98	1.88	1.10

[a] $R = \dfrac{\text{\% very weathered clasts} + \frac{1}{2}\text{\% moderately weathered clasts}}{\frac{1}{2}\text{\% moderately weathered clasts} + \text{\% unweathered clasts}}$

Weathering of clasts based on degree of grussification as witnessed by the degrees of etching into relief of individual mineral grains, disintegration with a hammer blow, and staining by iron and manganese oxides along fractures.

[b] Thickness in mm.
[c] Grain sizes are: fine=<2mm, medium=2mm to 1.5cm, coarse=>1.5cm.
[d] Terminal moraine.
[e] Lateral moraine.
[f] Cirque moraine.

within the glacial and periglacial deposits along the east side of the range. In addition, sedimentary sources have contributed to the deposits in Bear Gulch on the west side of the range.

The Tobacco Root Batholith includes rocks ranging from diorite to granite, with intermediate facies of tonalite, granodiorite, quartz monzonite, and monzonite (Smith, 1970; Vitaliano and others, 1979). However, no attempt has been made to subdivide these lithologies in Quaternary deposits, and for purposes of this paper all such rocks are designated simply as silicic igneous. The grain size ranges from almost aphanitic to coarse porphyritic, bearing phenocrysts of K-feldspar as long as 3 cm.

Birkeland (1974) has indicated that coarse-grained igneous rocks commonly weather more rapidly than finer-grained varieties, even if the mineralogy of the latter is more susceptible to chemical weathering. The range in grain size of the silicic igneous clasts in the Quaternary deposits of the Tobacco Root Range provides an opportunity to assess the influence of grain size on the weathering of clasts of very similar mineralogy.

When the silicic igneous clasts from the surface of selected deposits in the Tobacco Root Range are placed in three categories based on ranges in grain size (fine = < 2mm; medium = 2mm to 1.5cm; coarse = > 1.5cm) it becomes clear that the coarse-grained clasts are likely to be significantly more weath-

ered (Table 1). Indeed, as assessed by both the weathering index R (for which higher values indicate more weathering) and the mean rind thickness, the coarse-grained clasts on these deposits are about 3 times more weathered (Fig. 3).

3.2 Gneissic Clasts

The weathering of gneissic clasts has been compared to that of silicic igneous clasts in surficial deposits of the Tobacco Root Range study area for which both lithologies are significantly abundant. In this comparison neither group has been subdivided by grain size. As assessed by the weathering index, gneissic clasts, both on the surface and in the subsurface, are less weathered than silicic igneous clasts in most deposits (Fig. 4). Gneissic clasts in the subsurface also often have a thinner weathering rind (Fig. 5). However, gneissic clasts on the surface often have a weathering rind of equal or greater thickness in comparison with silicic igneous clasts (Fig. 5, 6). The overall lesser weathering of gneissic clasts on the surface may result in better retention of weathering rinds on a significantly large number of clasts. In contrast, the greater tendency for disintegration of silicic igneous clasts may result in partial or complete loss of weathering rinds.

A consideration of the weighted means of clast weathering data for the Tobacco Root Range deposits (Table 2, Fig. 7) again demonstrates the lesser weathering of gneissic clasts in comparison with silicic igneous clasts. In general, the latter

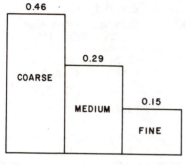

WEATHERING INDEX (R)

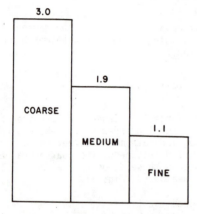

MEAN WEATHERING RIND THICKNESS (mm)

Figure 3. Weathering of silicic igneous clasts on Tobacco Root Range moraines as a function of grain-size.

lithology is about 1.5 times more weathered. Both quartzofeldspathic gneiss and hornblende gneiss are abundant in the Precambrian metamorphic assemblage of the Tobacco Root Range (Vitaliano and others, 1979) and in the surficial deposits of the study area. The mineralogic composition of the quartzofeldspathic gneiss is very similar to that of the silicic igneous rocks, although some of the gneiss is rich in plagioclase and/or hornblende and biotite. The hornblende gneiss also contains abundant plagioclase. Thus, in general, the mineralogy of the gneiss is potentially less stable in the chemical weathering environment than that of the silicic igneous rocks. The greater weathering of the latter lithology is attributed to a greater mean grain size rather than a less stable mineralogic composition. Birkeland (1974) has suggested that, because intergranular surface area increases with decreasing grain size, more energy will be required to disintegrate

finer-grained rocks. Furthermore, the smaller grain size of many of the gneissic clasts examined in the Tobacco Root Range study is believed to result in a low

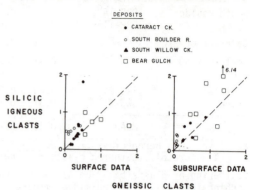

Figure 4. Weathering ratio (R) for silicic igneous and gneissic clasts in deposits of Tobacco Root Range.

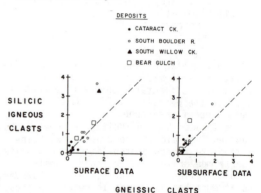

Figure 5. Mean weathering rind thickness (in mm) for silicic igneous and gneissic clasts in deposits of Tobacco Root Range.

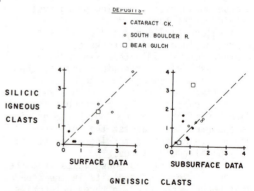

Figure 6. Mean weathering rind thickness (in mm) of ten thickest rinds for silicic igneous and gneissic clasts of Tobacco Root Range.

Table 2. Clast Weathering Data for the Deposits of the Tobacco Root Range

Drainage Basin	No. of Sites	IGN/GN[a] Clasts	Rock Type	Weathering of Clasts[b] U	M	V	R	Weathering Rind Thickness[c] Maximum	Mean	Mean$_{10}$[d]
SURFACE CLASTS										
Cataract Creek	8	33:67	IGN	42	40	8	0.59	1.0	0.3	0.4
			GN	55	40	5	0.34	1.3	0.3	0.6
S. Boulder River	6	30:70	IGN	43	47	10	0.51	3.8	1.3	1.9
			GN	73	26	1	0.18	4.2	1.0	2.3
S. Willow Creek	1	95:5	IGN	53	41	6	0.36	23.3	3.3	9.6
			GN	58	33	8	0.32	6.0	1.7	1.7
Bear Gulch	5	27:73	IGN	19	80	1	0.73	2.0	0.8	NA[e]
			GN	17	75	8	0.92	2.1	0.6	1.0
Weighted Mean	20	34:66	IGN	41	52	7	0.59	3.2	0.9	1.6
			GN	51	44	5	0.44	2.6	0.7	1.3
SUBSURFACE CLASTS										
Cataract Creek	6	43:57	IGN	42	53	5	0.56	1.9	0.5	0.9
			GN	46	48	6	0.45	1.2	0.3	0.7
S. Boulder River	6	38:62	IGN	68	24	8	0.26	3.8	1.0	1.3
			GN	83	17	0	0.09	2.4	0.7	1.2
S. Willow Creek	1	96:4	IGN	74	17	9	0.21	NA[e]	NA	NA
			GN	50	25	25	0.60	NA	NA	NA
Bear Gulch	7	30:70	IGN	15	62	23	1.80	2.3	0.7	1.8
			GN	16	73	9	0.92	1.5	0.3	0.6
Weighted Mean	20	40:60	IGN	42	46	12	0.89	2.6	0.7	1.4
			GN	47	47	6	0.51	2.0	0.4	0.8

[a] Ratio of silicic igneous to gneissic clasts
[b] U=unweathered clasts, M=moderately weathered clasts, V=very weathered clasts, R=weathering ratio. See Table 1
[c] Measured in mm
[d] Mean thickness for 10 thickest clasts
[e] Not available

A. SURFACE CLASTS

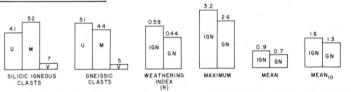

WEATHERING OF CLASTS   WEATHERING RIND THICKNESS

B. SUBSURFACE CLASTS

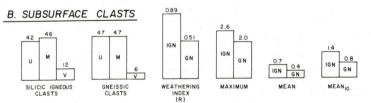

WEATHERING OF CLASTS   WEATHERING RIND THICKNESS

Figure 7. Comparison of weathering of silicic igneous and gneissic clasts on and in Tobacco Root Range moraines. All values are weighted means.

Table 3. Clast Weathering Data for Selected Deposits in the South Boulder River Basin

Site	Rock Type	Weathering Index (R)	Weathering Rind Thickness[a] Mean	Mean$_{10}$[b]
SURFACE CLASTS				
Terminal	IGN	0.47	1.1	1.3
	GN	0.10	0.9	1.9
	MAFIC	0.03	0.6	1.2
Spring	IGN	0.59	0.8	1.2
	GN	0.27	0.8	1.9
	MAFIC	NA	NA[c]	NA
Curly Ck	IGN	0.64	0.8	1.8
	GN	0.41	1.1	2.7
	MAFIC	0.11	0.9	0.9
Aspen	IGN	0.49	0.6	0.6
	GN	0.04	0.9	1.5
	MAFIC	0.06	0.5	NA
S. Aspen	IGN	0.49	1.1	2.2
	GN	0.14	0.8	1.9
	MAFIC	NA	NA	NA
Sailor	IGN	0.40	3.7	4.0
Lake I	GN	0.10	1.6	3.8
	MAFIC	0.00	0.6	NA
Mean	IGN	0.51	1.3	1.9
	GN	0.18	1.0	2.3
	MAFIC	0.05	0.7	NA
SUBSURFACE CLASTS				
Terminal	IGN	0.43	0.6	NA
	GN	0.08	0.2	0.5
	MAFIC	0.00	0.3	1.4
Spring	IGN	0.12	0.5	1.3
	GN	0.10	0.5	1.3
	MAFIC	0.18	0.4	0.4
Aspen	IGN	0.20	0.7	1.1
	GN	0.05	0.4	0.9
	MAFIC	0.02	0.2	0.6
S. Aspen	IGN	0.25	0.7	1.4
	GN	0.09	0.6	1.6
	MAFIC	0.00	0.3	0.3
Sailor	IGN	0.40	2.7	3.6
Lake I	GN	0.10	1.8	2.0
	MAFIC	NA	NA	NA
Sailor	IGN	0.10	1.3	NA
Lake II	GN	0.10	1.1	3.5
	MAFIC	0.08	0.7	1.4
Mean	IGN	0.24	1.0	1.8
	GN	0.09	0.7	1.6
	MAFIC	0.05	0.4	0.8

[a]Measured in mm
[b]Mean thickness for 10 thickest rinds
[c]Not available

permeability, which, in turn, reduces grussification through processes of physical weathering and, ultimately, chemical weathering. However, these same characteristics may enhance the preservation of weathering rinds. In some clasts weathering of the gneiss has proceeded more rapidly along foliation surfaces, but in many others weathering appears to have taken place isotropically.

3.3 Mafic Clasts

Previous investigators have noted that fine-grained mafic clasts within surficial deposits are commonly much less weathered than clasts of coarse-grained silicic rocks (Birkeland, 1964, 1974) and that mafic inclusions of finer grain size stand in relief above the surface of granitic outcrops or boulders (Blackwelder, 1931; Birkeland, 1964, 1974; Burke and Birkeland, 1979). Mafic clasts are uncommon in the surficial deposits of most of the basins of the Tobacco Root Range, but in several deposits of the South Boulder River basin such clasts constitute up to 30 percent of the total number of clasts. Such clasts have been eroded from fine- to medium-grained orthoamphibolites that occur as sills and dikes within the Archaean complex or from Proterozoic diabase dikes that cut the metamorphic rocks. Because such rocks do not crop out over much area in the South Boulder River basin (Vitaliano and others, 1979), their abundance in moraines suggests that they are more resistant to attrition during glacial transport than is true for the silicic igneous and gneissic clasts.

Although these rocks contain hornblende, clinopyroxene, calcic plagioclase, and other minerals that are potentially unstable in the chemical weathering environment, mafic clasts in the moraines of the South Boulder River basin are less weathered than either silicic igneous or gneissic clasts (Table 3, Fig. 8). In fact, depending upon which criterion is used to assess weathering, and whether comparisons are made for surface or subsurface clasts, silicic clasts are 2 to 10 times, and gneissic clasts are 1.5 to 3.5 times, more weathered than mafic clasts.

The resistance to weathering of the mafic clasts seems due to their fine grain size and perhaps high density. The resulting low permeability reduces the possibility of grussification through physical weathering processes operating along grain boundaries, and the climate is insufficiently warm and moist to permit much chemical weathering to rapidly affect the exterior of the clasts without the increase in specific surface area

provided by physical weathering. Thus, the mafic clast often has an advantage in surviving both attrition during glacial transport and post-depositional weathering. In contrast, silicic igneous clasts may be mechanically shattered with only slight chemical weathering (Birkeland, 1974), particularly if disintegration is aided by the expansion of weathered biotite (Birkeland, 1974; Isherwood and Street, 1976).

A. WEATHERING INDEX RATIO (R)

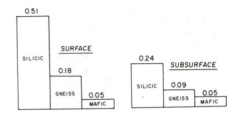

B. MEAN WEATHERING RIND THICKNESS

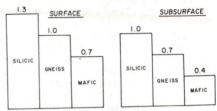

Figure 8. Comparison of weathering of silicic igneous, gneissic, and mafic clasts on and in selected moraines in the South Boulder River basin.

4 INFLUENCE OF TILL PETROLOGY ON WEATHERING OF DEPOSITS

4.1 Silicic Igneous vs. Gneissic Petrology

The petrology of the parent material is usually accepted as one of the primary factors affecting weathering and soil development (Jenny, 1941). An assessment of the parent material factor is made herein by assuming that the petrology of the deposit as a whole is reflected in the composition of the clasts. Although the composition of the matrix of the deposit may differ somewhat from that of the clasts, it should still be possible to compare the weathering of deposits derived from primarily either silicic igneous or gneissic sources or from a mixture of the two lithologies.

The glacial deposits of the North Willow Creek and Cataract Creek basins in the Tobacco Root Range provide an opportunity to assess the parent material factor by comparing deposits of similar age and topographic expression in which weathering has proceeded under similar climate and vegetation (Hall and Heiny, in press). The composition of deposits in the North Willow Creek basin is 100 percent silicic igneous, whereas in the Cataract Creek basin the petrologic ratio of silicic igneous to gneissic ranges from 100:0 to 10:90 and averages about 40:60.

The petrology of the till affects soil morphology, especially some of the characteristics of the B horizon. The B horizons developed in deposits of the North Willow Creek basin with a completely silicic igneous source are generally better defined in terms of both sharper boundaries and more distinct color differences in comparison with the overlying and underlying horizons. Such horizons also tend to be more compact.

The unweathered parent materials below the sola of the deposits of the North Willow Creek basin differ in grain size from those of the Cataract Creek basin primarily in containing somewhat more sand and less silt, although the clay content is the same. However, in general, the sola of the deposits of the North Willow Creek basin are somewhat more enriched in clay than is characteristic of the sola of the deposits of the Cataract Creek basin (Table 4), supporting the contention of greater weathering in the deposits with a completely silicic igneous source. Similarly, in the South Boulder River basin, more clay enrichment has taken place in post-glacial soils whose parent material includes a greater silicic igneous component (W. R. Roy, written communication).

Table 4. Comparison of Weathering Characteristics of Glacial Deposits of the North Willow Creek(NWC) and Cataract Creek(CC) Basins

PROBABLE AGE	BASIN	% CLAY IN SOLUM	CLAY ^{a} ENRICHMENT	FREE IRON ^{b} ENRICHMENT	HORNBLENDE ^{c} DEPLETION
EARLY TO MIDDLE PINEDALE	NWC	9	4	0.2	13.5
	CC	8	4	0.2	9
EARLY HOLOCENE	NWC	13	7	0.4	33
	CC	8	1	0.2	4
NEOGLACIAL	NWC	8	3	0.4	0
	CC	8	2	0.0	0

^{a} % clay in solum - % clay in C horizon.
^{b} % free iron in upper enriched zone - % free iron in lower unweathered zone.
^{c} % hornblende in unweathered part of profile - % hornblende in depleted zone.

The solum of the deposits of the North Willow Creek basin is also somewhat more enriched with iron (Table 4) and is substan-

tially more depleted in hornblende. Thus, it appears that weathering and soil development have not been as extensive in the deposits of the Cataract Creek basin derived from mixed silicic igneous and gneissic sources. Furthermore, there is far more variability in the weathering of such deposits (Hall and Heiny, in press), probably controlled by local variations in the abundance of the two components.

In the South Boulder River basin a series of four recessional moraines of the Middle Pinedale advance (Roy, 1980) have petrologies ranging from 29 to 42 percent silicic igneous, the rest gneissic (Table 5). However, a comparison of weathering characteristics among the sites indicates either insignificant differences in weathering or differences that cannot be related to the abundance of the silicic igneous component. Apparently the range of variation in petrology is not enough to significantly affect weathering in this case. The effects of small variations in petrology are probably masked by other factors that influence weathering and soil development. Larger variations, such as those noted between the North Willow Creek and Cataract Creek basins, are more likely to affect weathering to an extent that can be documented.

4.2 Clay Mineralogy

In addition to the weathering and soil characteristics discussed previously, the clay mineralogy of a soil profile is often closely related to the petrology of the parent material. In a study of the clay minerals produced through weathering of clasts of different composition in a boulder conglomerate, Barnhisel and Rich (1967) found that kaolin minerals were derived from granites and gneisses low in bases, whereas montmorillonite was the clay mineral weathering product derived primarily from basic rocks such as gabbros. Similarly, in a study of Quaternary deposits of the eastern Sierra Nevada, Birkeland and Janda (1971) noted associations of kaolin minerals with deposits rich in granitic rocks and of montmorillonite with deposits rich in intermediate or basic rocks.

In the Tobacco Root Range, only the glacial deposits of the South Boulder River basin contain a significant mafic component (Table 6), which may explain the occurrence of more expandable clay minerals, predominantly montmorillonite, in the sola of deposits in this basin in comparison with the deposits of other basins in the study area. However, the deposits of the South

134

Table 5. Comparison of Weathering Characteristics of Middle Pinedale Moraines in the South Boulder River Basin

DEPOSIT	% SILICIC IGNEOUS CLASTS	CLAY[a] ENRICHMENT	FREE IRON[b] ENRICHMENT	E/NE[c] CLAYS	HORNBLENDE[d] DEPLETION
SOUTH ASPEN	42	7	NA	NA	12
SPRING	38	6	0.4	0.6	4
CURLY CREEK	30	7	0.3	0.5	6
ASPEN	29	6	0.1	0.2	5

[a]% clay in solum - % clay in C horizon.
[b]% free iron in upper enriched zone - % free iron in lower unweathered zone.
[c]ratio of expandable to nonexpandable clay minerals in solum.
[d]% hornblende in unweathered part of profile - % hornblende in depleted zone.

Table 6. Comparison of the Clay Mineralogy of the Sola of Glacial Deposits in the Tobacco Root Range

BASIN	NUMBER OF SITES	PETROLOGIC[a] RATIO	PERCENT CLAY MINERALS		
			EXPANDABLE	ILLITE	KAOLINITE + CHLORITE
N. WILLOW CK.	3	100:0:0	1-8	69-82	14-24
CATARACT CK.	4	38:62:<1	12-33	44-60	23-29
S. BOULDER R.	4	24:58:18	26-36	33-40	25-36

[a]Silicic igneous:gneissic:mafic

Boulder River basin also contain a much greater abundance of kaolin minerals than might be expected from a consideration of their relatively sparse silicic igneous component (Table 6). Kaolin minerals may form from deposits with a significant mafic component if the drainage is adequate, and all localities in the South Boulder River basin are well-drained. These deposits also contain the least abundance of illite in the study area. Deposits in the Cataract Creek basin, which contain only a very minor mafic component but are dominated by gneisses, have a clay mineralogy that is intermediate between that of the deposits of the South Boulder River basin and that of the deposits of the North Willow Creek basin, in which only silicic igneous materials are present. In the latter case, the sola contain the smallest amount of expandable clays and kaolin minerals and the most illite (Table 6).

The data presented by Barnhisel and Rich (1967, Table 1) also indicate somewhat less production of both kaolin minerals and montmorillonite from the weathering of a granitic clast in comparison with a gneissic clast. The data presented by Birkeland and Janda (1971, Fig. 4) indicate that kaolin minerals are abundant wherever granitic parent materials are abundant, even if mixed with metamorphic materials. Their

data also show that montmorillonite is not present in deposits derived almost entirely from a granitic source but occurs where granitic and metamorphic materials are mixed.

Thus, based on the data from the study of the glacial deposits of the Tobacco Root Range and from previous investigations, it appears that abundant kaolin minerals are likely to be present in the solum wherever the parent materials contains a significant silicic igneous component (perhaps 25 percent), even if gneissic and/or mafic materials are also abundant. Expandable clay minerals, particularly montmorillonite, are likely to be sparse or absent if the parent material is exclusively silicic igneous but should occur if gneissic materials are also present and should be abundant where mafic parent materials are included. The amount of illite may be the greatest where the parent material is dominated by the silicic igneous component and the least where mafic parent materials are dominant. Furthermore, the effects of variations in the petrology of the parent material on clay mineralogy are likely to obscure any trends in clay mineral distribution as a function of time since deposition and render studies of clay mineralogy useless as a technique for the relative age dating of mountain glacial deposits (e.g. Birkeland and Janda, 1971; Hall and Heiny, in press).

4.3 Sedimentary Component

Sedimentary rocks comprise a significant component of the petrology of the glacial deposits of many locations in the Rocky Mountains. Of particular interest are calcareous shales and siltstones and limestones whose weathering may influence the development of $CaCO_3$-rich horizons in the post-glacial soil profiles.

In the Wind River Range, Bull Lake and Pinedale moraines at Bull Lake along the northeastern flank contain a significant sedimentary component which is believed to have influenced the development of Bca and Cca horizons that may contain as much as 40 percent and 20 percent equivalent $CaCO_3$, respectively (Table 7). In contrast, moraines of the same size near Boulder Lake along the southwestern flank contain only silicic igneous and metamorphic components. In this area post-Bull Lake soils include a maximum of only 12 percent equivalent $CaCO_3$, and post-Pinedale soils totally lack pedogenic $CaCO_3$. Because topographic, climatic, and vegetational factors appear to be very similar in the Bull Lake and Boulder Lake areas, the more extensive accumulation of pedogenic $CaCO_3$ in the Bull Lake area moraines is probably due to the weathering of calcareous materials that are lacking in the Boulder Lake area deposits.

Table 7. Maximum Percentage of Equivalent $CaCO_3$ in Profiles with and without a Calcareous Sedimentary Component

	AREA	AGE	
		PINEDALE	BULL LAKE
NORTHWEST WYOMING	BULL LAKE TYPE AREA	28	40
	BOULDER LAKE AREA	0	12
SOUTHWEST MONTANA	BEAR GULCH	14	32
	SOUTH BOULDER VALLEY	0	0

Similarly, in the Tobacco Root Range (Fig. 2), Bull Lake and Pinedale moraines in Bear Gulch include about 15 percent calcareous sedimentary rocks. The post-glacial soils include Bca and Cca horizons with 32 percent and 14 percent equivalent $CaCO_3$, respectively (Table 7). Moraines of the same ages in the South Boulder River valley contain only igneous and metamorphic rocks, and the post-glacial soils totally lack pedogenic $CaCO_3$.

The presence of a significant sedimentary component has also resulted in a finer-textured till at both Bull Lake, Wyoming, and Bear Gulch, Montana, compared to the till at sites in the same ranges derived from silicic igneous and metamorphic sources (Table 8). The finer texture should have favored the development of $CaCO_3$-rich horizons because the increased surface area favors more leaching of calcium in the upper horizons (Birkeland, 1974). Clay formation and the development of argillic B horizons are also favored, and such horizons (with up to 24 percent clay) are present, even in deposits of Pinedale age, at both Bull Lake, Wyoming, and Bear Gulch, Montana.

Table 8. Comparison of Particle-size Distribution of Unweathered Tills with and without a Significant Sedimentary Component

	AREA	SAND	SILT	CLAY
NORTHWEST WYOMING	BULL LAKE TYPE AREA	48	36	16
	BOULDER LAKE AREA	75	19	6
SOUTHWEST MONTANA	BEAR GULCH	53	30	17
	SOUTH BOULDER VALLEY	83	14	3

5 INFLUENCE OF LOESS

The presence of eolian material, or loess, on the surface of glacial and periglacial deposits in mountains has been reported in numerous studies (e.g., Bouma and others, 1969; Marchand, 1970; Retzer, 1974; Birkeland, 1974; King and Brewster, 1976; Hall and Heiny, in press). In a study of soil contamination in the White Mountains in eastern California, Marchand (1970) found that eolian materials constitute 0.7 to 34 percent of the total soil and up to 50 percent of the silt fraction. Birkeland (1974) recognizes loess in the upper horizons of soil profiles in the Colorado Rocky Mountains on the basis of its fine texture and absence of coarse quartz and feldspar grains. Retzer (1974) suggests that such fine-grained materials in the central Rocky Mountains originate as dust from desert areas to the west and southwest and are carried eastward by prevailing winds where they accumulate at a rate averaging 0.005 mm/yr.

It is clear that the presence of loess may substantially affect weathering and soil formation. Bouma and others (1969) report that alpine podzols in the Alps have A horizons enriched in silt, and King and Brewster (1976) have studied podzols in Banff National Park, Alberta, in which the entire solum is developed in eolian materials, including volcanic ash. Loess deposition during soil formation results in a cumulative soil profile (Birkeland, 1974) in which the A horizon undergoes vertical accretion, and the former A horizon can become a B horizon.

In the Tobacco Root Range in Montana, deposition of loess on the surface of moraines and periglacial deposits and mixing of the loess down into the profiles may often explain more of the variation in clay content than different degrees of weathering due to relative age. The high clay content may result from loess deposition in at least two ways: 1) in till and periglacial materials to which loess has been added more parent material is available for the formation of clay as a weathering product; and 2) some clay may be physically transported onto a site as a component of the loess and then mixed down into the profile. Most of the extra clay is probably pedogenic.

That loess is present in the upper part of many soil profiles is shown by the greater abundance of fine silt, and in places coarse silt, near the surface (Fig. 9). However, in the field a boundary between the loess and the underlying till or glacial deposits is often difficult to determine. Rather, the loess appears to have sifted down into the underlying materials, and perhaps the two materials have been further mixed by bioturbation. In the glaciated basins of the Tobacco Root Range the amount of fine silt in the solum of profiles does not vary systematically with the amount in the underlying C horizon (Fig. 10), which is further evidence that much of it was introduced later as loess. That the presence of loess has influenced weathering and abundance of pedogenic clay seems to be demonstrated by the covariance of clay enrichment (% clay in solum - % clay in C horizon) and % fine silt in the solum for these same deposits (Fig. 11).

Although the data on other weathering characteristics are too sparse to be definitive, the presence of abundant loess in a profile may also lead to increases in free iron enrichment, cation exchange capacity, and the abundance of expandable clay minerals. The presence of loess is also often accompanied by an increase in the 7 Å clay minerals, dominantly kaolinite, probably as a detrital component. Throughout the Tobacco Root Range study area, wherever the loess influence is greatest, there is considerably more weathering, and the factor of time since deposition becomes secondary and obscured.

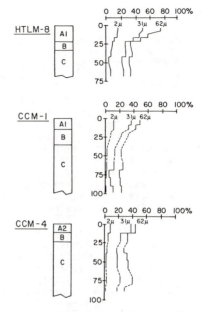

Figure 9. Vertical distribution of grain sizes in selected profiles in moraines of the Tobacco Root Range.

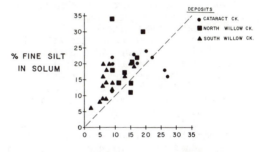

Figure 10. Comparison of fine silt content of solum and C horizon in glacial and periglacial deposits of the Tobacco Root Range.

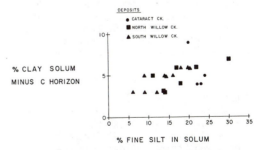

Figure 11. Covariance of clay enrichment and fine silt in solum in glacial and periglacial deposits of the Tobacco Root Range.

6 SUMMARY AND CONCLUSIONS

The weathering of moraines in southwestern Montana and northwestern Wyoming is to an important extent influenced by variations in the petrology of the till. In and on a given deposit it can be expected that coarse-grained igneous and metamorphic clasts will be more weathered than fine-grained clasts of the same composition. In deposits that contain more than one rock type the degree of clast weathering is influenced more by grain size controls on permeability than by clast composition. Thus, in most cases, silicic igneous clasts are more weathered than gneissic clasts which are, in turn, more weathered than mafic clasts.

Variations in till petrology also affect the weathering of the deposit as a whole. Deposits in which the silicic igneous component is considerably more abundant than the gneissic component are likely to have soil profiles with more compact and defined horizons. The solum in such profiles is likely to be more enriched in clay and free iron and more depleted of hornblende. Deposits derived from mixed silicic igneous and gneissic sources are likely to be weathered to a lesser degree. Among the clay mineral weathering products, kaolin minerals and illite are often associated with tills rich in silicic igneous materials, but expandable clay minerals, particularly montmorillonite, are more often associated with tills in which gneissic and mafic materials are also abundant. A significant percentage of calcareous sedimentary materials in a till favors development of soils containing Bca and Cca horizons.

Loess added to and mixed down into glacial deposits promotes more weathering, especially the pedogenic formation of clay. Thus, in studies designed to establish glacial chronology by relative age dating, the parent material must be as lithologically homogeneous as possible. If not, differences in weathering due to variations in petrology may be more significant than those attributed to the time since deposition.

7 REFERENCES

Barnhisel, R.I. & C.I. Rich, 1967, Clay mineral formation in different rock types of a weathering boulder conglomerate, Soil Sci.Soc.Amer.Proc., 31:627-631.

Birkeland, P.W. 1964, Pleistocene glaciation of the northern Sierra Nevada, north of Lake Tahoe, California, J.Geol., 72:910-825.

Birkeland, P.W. 1974, Pedology, Weathering, and Geomorphological Research, 285 pp. Oxford Univ. Press, London, Toronto, New York.

Birkeland, P.W. & R.J. Janda, 1971, Clay mineralogy of soils developed from Quaternary deposits of the eastern Sierra Nevada, California, Geol.Soc.Amer.Bull., 82:2495-2514.

Blackwelder, E. 1915, Post-Cretaceous history of the mountains of central Wyoming, J.Geol., 23:97-117,193-217,307-340.

Blackwelder, E. 1931, Pleistocene glaciation in the Sierra Nevada and Basin Ranges, Geol.Soc.Amer.Bull., 42:865-922.

Bouma, J., J. Hoeks, L. Van der Plas & B. Van Scherrenburg, 1969, Genesis and morphology of some alpine podzol profiles, J.Soil Sci., 20:384-398.

Burke, R.M. & P.W. Birkeland, 1979, Re-evaluation of multiparameter relative dating techniques and their application to the glacial sequence along the eastern escarpment of the Sierra Nevada, California, Quaternary Res., 11:21-51.

Currey, D.R. 1974, Probable pre-Neoglacial age of the type Temple Lake moraine, Wyoming, Arctic and Alpine Res., 6:293-300.

Hall, R.D. & J.S. Heiny, 1981, Glacial and post-glacial physical stratigraphy and chronology, North Willow Creek and Cataract Creek drainage basins, Tobacco Root Range, southwestern Montana. In press.

Hall, R.D. & W.R. Roy, 1979, Weathering characteristics and soil development of glacial deposits at Bull Lake, Wyoming, Geol.Soc.Amer.Abstr., 11:274.

Holmes, G.W. & J.H. Moss, 1955, Pleistocene geology of the southwestern Wind River Mountains, Wyoming, Geol.Soc.Amer.Bull., 66:629-653.

Isherwood, D. & A. Street, 1976, Biotite-induced grussification of the Boulder Creek granodiorite, Boulder County, Colorado, Geol.Soc.Amer.Bull., 87:366-370.

Jenny, H. 1941, Factors of Soil Formation, A System of Quantitative Pedology. 281pp. McGraw-Hill, New York.

King, R.H. & G.R. Brewster, 1976, Characteristics and genesis of some subalpine podzols (spodosols), Banff National Park, Alberta, Arctic and Alpine Res., 8:91-104.

Mahaney, W.C. 1978, Late-Quaternary stratigraphy and soils in the Wind River Mountains, western Wyoming. In W.C. Mahaney (ed.), Quaternary Soils Symposium, 223-264, York University, Toronto.

Marchand, D.E. 1970, Soil contamination in the White Mountains, eastern California, Geol.Soc.Amer.Bull., 81:2497-2506.

Miller, C.D. & P.W. Birkeland, 1974, Probable pre-Neoglacial age of the type Temple Lake moraine, Wyoming: discussion and additional relative-age data, Arctic and Alpine Res., 6:301-306.

Murphy, J.F. & G.M. Richmond, 1965, Geologic map of the Bull Lake West Quadrangle, Fremont County, Wyoming, U.S.Geol.Sur. Quad. Map, GQ-432.

Retzer, J.L. 1974, Alpine soils. In J.D. Ives and R.G. Barry (eds.), Arctic and Alpine Environments, 771-802, Methuen, London, New York.

Richmond, G.M. 1948, Modification of Blackwelder's sequence of Pleistocene glaciation in the Wind River Mountains, Wyoming, Geol.Soc.Amer.Bull., 59:1400-1401.

Richmond, G.M. 1962, Three pre-Bull Lake tills in the Wind River Mountains, Wyoming, U.S.Geol.Sur.Prof. Paper, 450-D, 132-136.

Richmond, G.M. 1964, Three pre-Bull Lake tills in the Wind River Mountains, Wyoming-a reinterpretation, U.S.Geol.Sur. Prof. Paper, 501-D, 104-109.

Richmond, G.M. 1976, Pleistocene stratigraphy and chronology in the mountains of western Wyoming. In W.C. Mahaney (ed.), Quaternary Stratigraphy of North America, 353-379, Dowden, Hutchinson, and Ross, Stroudsburg.

Richmond, G.M. & J.F. Murphy, 1965, Geologic map of the Bull Lake East Quadrangle, Fremont County, Wyoming, U.S.Geol.Sur. Quad. Map, GQ-431.

Roy, W.R. 1980, Glacial chronology of the South Boulder Valley, Tobacco Root Range, Montana. Master's Thesis, 163pp., Indiana University, Bloomington.

Roy, W.R. & R.D. Hall, 1981, Glacial geology of the South Boulder Valley, Tobacco Root Range, Montana, Montana Geol.Soc. Guidebook. In press.

Smith, J.L. 1970, Petrology, mineralogy, and chemistry of the Tobacco Root Batholith, Madison County, Montana. Ph.D. dissertation, 164pp., Indiana University, Bloomington.

Vitaliano, C.J. & W.S. Cordua, 1979, Geologic map of southern Tobacco Root Mountains, Madison County, Montana, Geol. Soc.Amer. Map, MC-31.

The enrichment of quartz in tills

SYLVI HALDORSEN
University of Ås, Norway

During a glacial phase the sedimentological prosesses result in a mechanical
comminution, and the rate of comminution
depends on the mechanical resistance of
the rocks and their constituent minerals
and on the types of comminution processes
involved (see for instance Dreimanis &
Vagners 1971, Haldorsen 1977, 1978, 1981,
in press and Slatt & Eyles 1981). The
purpose of the present work is
to study if this comminution may result in
a mineralogical fractionation and in combination with other glacigenic processes may
produce glacial sediments with a bulk
petrographical composition differing significantly from that of the source rocks.

1 METHODS

Major elemental analyses were performed by
atomic absorption spectrometry. The mineralogical analyses were carried out by microscopy of thin sections. The fractionation
in different particle-size intervals was
performed in setting tubes and the grain-
size analysis was carried out by sieving
and hydrometer method.

2 SOURCE ROCKS

The investigation was carried out in
Åstadalen in southeastern Norway (Fig.1).
This area was chosen because the bedrock
is homogeneous and detailed information
about its composition is available. The
bedrock consists of Late Precambrian sedimentary rocks of the Brøttum Formation
(Fig.1). In the northern part there are
sandstones alternating with some layers of
silty shale, and in the southern part there
is a more uniform sandstone. Samples taken
from 24 outcrops in the south yielded the

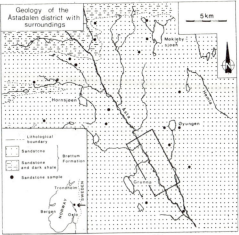

Fig.1 Keymap showing the bedrock geology
in Åstadalen. Framed: the studied
area.

average composition shown in Fig.2 and
Table 1. The glacial sediments discussed
here are from this southern sandstone
area (Figs.1 & 3).

3 SUBGLACIAL COMMINUTION PROCESSES

The transition from bedrock to glacial till
and subsequently to other glacigenic
sediments involves crushing caused by
percussion, compression and frost cracking.
In addition there may be significant abrasive grinding. These processes are closely
connected during glacial transport.
Separate processes of crushing and abrasion of sandstone samples from Åstadalen
were simulated by means of several mill
experiments (Haldorsen 1981 and in press).
The following are the main conclu-

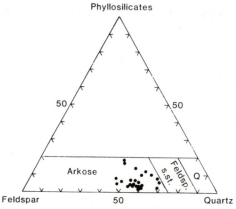

Fig.2 Modal composition of the homogeneous
sandstone area in Åstadalen (24
samples see Fig.1). (J.O.Englund,
pers.comm. 1980).
Q = quartzite.

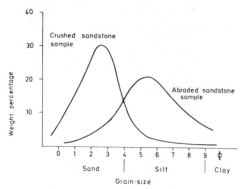

Fig.4 Grain-size distribution of one artificially crushed and one abraded
sample of the Brøttum sandstone. The
sample was first crushed in a ball
mill and was afterwards fractionated
by sieving and settling in a sedimentation cylinder. The abraded material
was formed by attrition of gravel
grains of size 16-32mm. Further explanation about the experiments are
given in the text and by Haldorsen
(1981).

sions from the experiments:

1. With moderate crushing in a ball mill
the grains were crushed until the material
was dominated by individual minerals.
Grain-size analyses of this material
(Fig.4) were compared with the primary
size distribution of minerals in the
bedrock (Haldorsen 1981) and a good
correspondence was found. It was concluded
that the crushing mainly produced cracks

between mineral grains and that each mineral retained its size distribution well
during this crushing. The mineral composition and the geochemical composition
within 1Φ - interval of one crushed sample
are shown in Figs. 5 & 6. These are, in
accordance with the conclusion from
previous papers (Haldorsen 1981 and in
press.) believed to reflect the primary
composition of the bedrock sample.

2. In another mill experiment some angular gravel grains from the same bedrock
sample, were tightly packed to prevent
crushing occurring; the comminution process
now was an abrasive grinding. The experiment was finished when the gravel grains
had reached a subangular to subrounded
stage. The abrasion created cracks across
minerals and produced a rock flour consisting of coarse to medium silt (Fig.4).
An increase in just the coarse and medium
silt in tills from Åstadalen corresponds
well with an increase in glacial abrasion
(Haldorsen 1981). The artificial abrasion
experiment seemed to be illustrative for
the process of glacial abrasion.

Both the crushed and the abraded material
show the same trend; there is a relative
decrease in the content of quartz with
decreasing grain-size from coarse silt to
clay (Fig.5). This is reflected in an
increase of the Al_2O_3 relative to SiO_2 and
in K_2O (Fig.6). The main difference is
that the abrasion gave an enrichment of

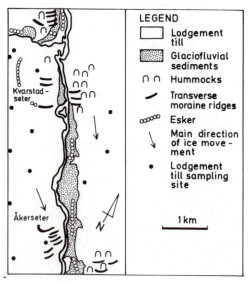

Fig.3   Quaternary geology of the
investigated area.

Table 1 Chemical composition of bedrock and the fraction finer than 2 mm of glacigenic sediments from Åstadalen

	Bedrock N=24		Lodgement till N=12		Melt-out till 1 N=8		Glaciofluvial N=25		Melt-out till 2 N=10	
	$\bar{X}$	SD	$\bar{X}$	SD	$\bar{X}$	SD	$\bar{X}$	SD	$\bar{X}$	SD
SiO_2	81.5	4.9	80.9	2.7	*82.2	2.9	**84.9	2.9	**83.9	3.1
Al_2O_3	8.4	2.0	8.6	1.2	8.4	1.6	** 7.0	1.1	* 7.8	1.7
Fe_2O_3	2.1	1.1	2.3	1.8.	2.1	1.4	* 1.5	1.6	1.9	1.9
MgO	0.6	0.4	0.8	0.3	0.9	0.4	* 0.6	0.5	* 0.6	0.4
Na_2O	1.8	0.4	2.0	2.0	** 1.7	0.2	** 1.4	0.4	** 1.5	0.4
K_2O	3.2	0.5	2.9	0.2	** 2.5	0.3	** 2.3	0.3	** 2.4	0.3
CaO	0.5	0.4	0.6	0.2	0.6	0.2	0.5	0.2	0.6	0.2
	98.2		98.1		98.5		98.2		98.7	

** Significant difference from lodgement till at a 5 % level of significance
* Significant difference from lodgement till at a 10 % level of significance

mica and feldspar in the silt (Fig.5) because these minerals were more easily comminuted than the quartz. The quartz grains partly retained their original size and were thereby enriched in the sand (Fig.5). This is reflected in the geochemistry of the sand by the 'excess' of SiO_2 and 'deficiency' of Al_2O_3, Na_2O and K_2O (Fig.6).

4. GLACIGENIC SEDIMENTS

The Åstadalen area represents a typical inland area where the glacial deposits mainly were formed after the Weichselian maximum. An active glacial phase was followed by a phase with rather stagnant ice (Haldorsen 1982).

The formation of different glacial sediments is schematically shown in Fig.7.

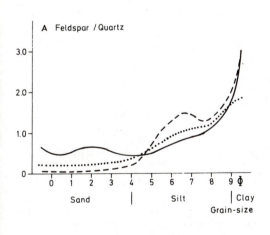

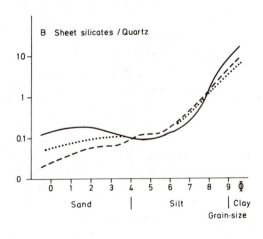

——— Crushed sandstone   — — — Abraded sandstone
·········· Lodgement till

Fig.5 Mineralogical composition of different particle size intervals of the crushed and abraded sandstone sample (for further explanation, see Fig.4) and one lodgement till sample.

143

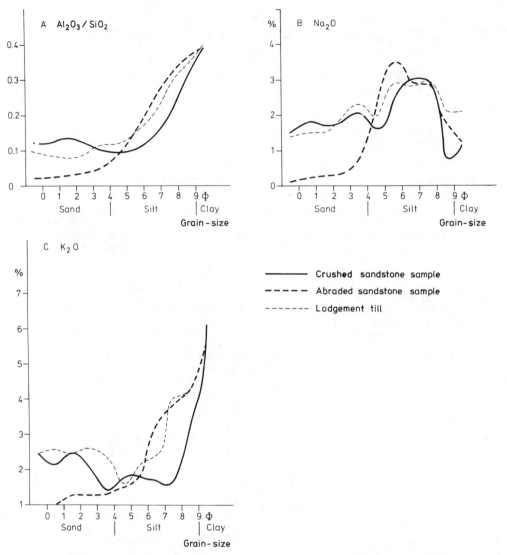

Fig.6 Distribution of some geochemical components in different size grades of crushed and abraded sandstone sample (further explanation, see Fig.4) and one lodgement till sample.

4.1 Lodgement till

In the Weichselian the sandstone bedrock was glacially eroded and comminuted. During the active glacial phase the debris was deposited as lodgement till (Fig.7, point A). That means till deposited from an active glacier, by lodging of clasts one by one, and a plastering on of finer grained material (see Boulton 1976). In Åstadalen lodgement till lies directly on the bedrock, and no underlying Quaternary sediments were observed. The lodgement till is thus regarded as the oldest glacial sediment in the area.

A detailed geochemical study of the silt + clay fractions of lodgement till including data for 125 till samples from the southern sandstone area west of Åsta (Fig.1) has been presented (Haldorsen in press). The samples were collected from C-horizons. South of the influence from the shale horizons in the north the till is homogeneous both in vertical sections and

Till forming processes and till types

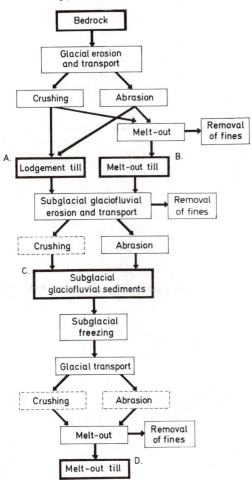

Fig.7 Postulated formation of different
glacigenic sediments in Åstadalen.

regionally. From this till 13 samples within the present studied area (Fig.3) were
further analysed. The bulk geochemistry
of the material finer than 2 mm is given in
Table 1. This also represents the parent
bedrock composition, as more than 90 % of
the sandstone minerals are found in the
sand and silt (Fig.4). The geochemistry of
the lodgement till accords well with the
geochemistry of the bedrock. There are no
indications of primary quartz enrichment in
this till type.

The lodgement till is rather rich in silt
(Fig.8, curve A) because much of its
material was derived by abrasion (Haldorsen
1981). As the formation of the lodgement
till involved both crushing and abrasion

(Fig.7) its mineralogy and geochemistry
resembles those of the artificially crushed
and artificially abraded materials (Figs.5
& 6).

4.2 Subglacial melt-out till, type 1

A subglacial melt-out till is formed by a
passive melt-out process of stagnant
debris-rich basal ice (Fig.7, point B). It
is recognized by its low degree of compaction and in many cases by the lenses of
sorted sediments which indicate the presence of running melt-water (Dreimanis,
1976). Such a subglacial melt-out till is
found in the transverse moraine ridges in
Astadalen (Fig.3). These were formed by
the stacking of debris-rich basal ice
(Haldorsen & Shaw, 1982) (Fig.9). The
ridges are in some cases found to rest on
a compact lodgement till. At least parts
of the melt-out till thus were deposited
after the lodgement till.

In the lateral end of the ridges, i.e.
away from the centre of the valley, the
coarse material of the melt-out till is
similar to the coarse material of the
lodgement till, which is reflected both by
the roundness of gravels (Fig.10A & B) and
the mineralogy of the sand (Fig.11). These
characteristics indicate that the melt-out
till was formed from the same type of
debris as the lodgement till.

The silt content is lower in this melt-
-out till than in the lodgement till
(Fig.8) mainly due to the removal of fines
by water during the melt-out process (Fig.
7).

The removed fine-grained material was
clearly richer in feldspar and sheet silicates than the remaining sand (see Fig.5).
The latter therefore became enriched in
quartz relative to the original lodgement
till and has a higher content of SiO_2 and
a lower content of Al_2O_3, Fe_2O_3, MgO,
Na_2O and K_2O. (Table 1). In Fig.12 this
quartz enrichment is shown schematically
and listed as a quartz enrichment of type 1.

4.3 Glaciofluvial sediments

Subglacially formed glaciofluvial sediments
are found in hummocks, eskers and terraces
in Åstadalen (Fig.3). They were deposited
from the time when the glacier became
sufficiently thin for the melt-water to
penetrate down to the base. This deposition
continued until the deglaciation was
completed. Glaciofluvial erosion of tills
is several places witnessed by the many
meltwater channels which penetrate the

145

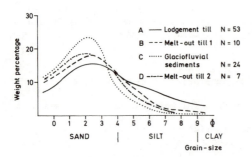

Fig.8 Grain-size distribution of glacigenic sediments in the investigated
part of Åstadalen.

surface of the lodgement till. The glaciofluvial sediments are sometimes found to
rest on the till. The surface of the
underlying till sheet has then features
reflecting the influence of meltwater
activity. Based on these observations it
is concluded that the dominant part of the
glaciofluvial sediments are formed either
by erosion of tills (Fig.7, points A-C) or
from basal ice debris.

Bulk samples were taken of most of the
glaciofluvial sediments in the area, from
road cuttings and gravel pits in hummocks,
eskers and terraces.

The glaciofluvial material can be
divided into two groups.

1. Small eskers and kames. The source is

probably local. The gravel of these deposits is significantly more rounded than
the gravel in the lodgement till and the
melt-out till, type 1. (Fig.10 A-C). This
shows that considerable abrasion occurred
even during a short glaciofluvial transport.

2. Main eskers and terraces along the
center of the valley (Fig.3) The gravel
material has a relatively high degree of
roundness (Fig.10 D).

The glaciofluvial material has a sand
which is significantly richer in quartz than
the two tills described above (Fig.11).
As the glaciofluvial material in this case
mainly originates from till or basal drift,
incorporation of weathered material seems
unlikely. Thus, the enrichment of quartz
in the sand is most likely the result of
grinding processes occurring during the
glaciofluvial transport. By applying the
results from the artificial grinding this
may be explained in the following way:
The glacial transport involved both an
abrasion and a crushing. During the abrasion a rock flour, mainly silt, was formed.
The silt was enriched by feldspar and sheet
silicates and the remaining sand by quartz
(Fig.5). The crushing produced mainly sand
rich in both feldspar and quartz (Fig.5).
However, a greater amount of sand was
produced by the crushing than by the
abrasion (Haldorsen 1981: 96), and any
great enrichment of quartz was therefore
not formed in the sand during the glacial
transport. During the glaciofluvial trans-

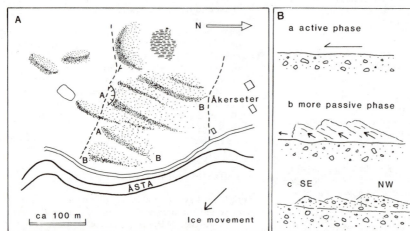

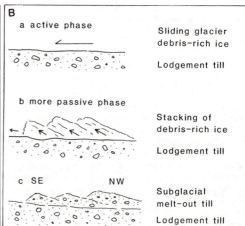

Fig.9 A. Transverse moraine ridges at Åkerseter (see Fig.3). A: subglacial melt-out
till, type 1, B: subglacial melt-out till, type 2.
B. Postulated formation of the ridges, a: active glacial phase with deposition
of lodgement till, b: more passive glacial phase with folding/stacking of
basal ice, c: final morainic morphology and till sequence.

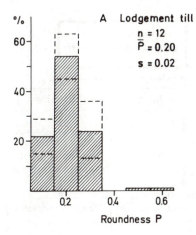
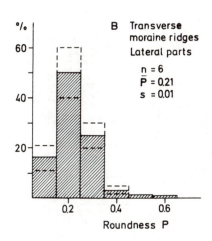
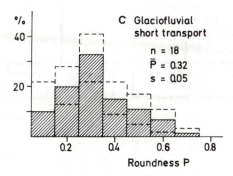
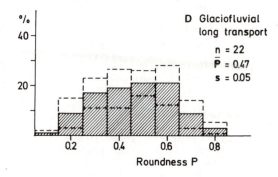
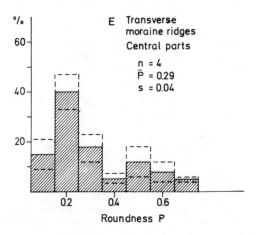

Fig.10 Degree of roundness of gravel 16-32 mm for different sediments in Åstadalen. Roundness index after Krumbein (1941).

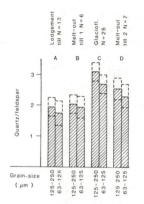
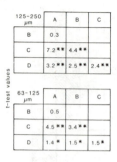

Fig.11 Quartz/feldspar ratio for the fine
sand grades of sediments from
Åstadalen. One standard deviation
is stipled. Right part of the
figure gives statistical significance of the difference between
the sediments.
**Significant difference at a 5 %
level of significance.
*Significant difference at a 10 %
level of significance.

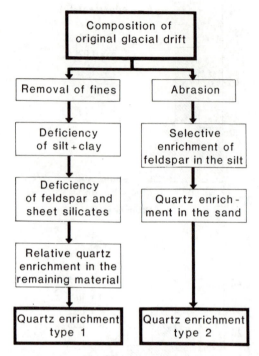

Fig.12 Two alternative ways of quartz
enrichments in sediments from
Åstadalen.

port, on the other hand, the main grinding process was an abrasion. Cracking of
clasts resulted in fragments which were
afterwards exposed to abrasion. Crushing
down to separate minerals did certainly
occur, but was probably quite subordinate
compared with the crushing during the
glacial transport. In this way a relatively important component of sand rich
in quartz was formed during the glaciofluvial transport. When this was mixed
with sand from the eroded till, a sand-
sized material with a significant enrichment of quartz was formed.

The enrichment of quartz in the sand,
caused by selective comminution of feldspar and sheet silicates is listed in
Fig. 12 as 'quartz enrihment of type 2'.

No extensive silt deposits are found in
Åstadalen today, and the glaciofluvial
sediments are poor in silt (Fig.8). Most
of the silt produced by the glaciofluvial
abrasion, and silt eroded from till must
have been removed from the area by meltwater (Fig.7). Consequently the glaciofluvial fraction finer than 2 mm is richer
in quartz, and SiO_2, than both the lodgement till matrix and the melt-out till,
type 1 (Table 1). This is an enrichment of
quartz of type 1 (Fig.12). The conclusion
is that the glaciofluvial sediments have
enrichment of quartz of both type 1 and
type 2.

4.4 Subglacial melt-out till, type 2

In the centre of the valley the till of
the transverse moraine ridges (Fig.3)
occasionally contains a considerable
amount of rounded gravel material (Fig.10
E). This indicates incorporation of a
water transported sediments. Such till is
located in areas where glaciofluvial
sediments are abundant and in some places
is found to rest directly on glaciofluvial
sediments. Such till has been interpreted
as a subglacial melt-out till with a
depositional history like that of the type
1 subglacial melt-out till.

The incorporation of glaciofluvial material may have occurred in places where
subglacial melt-water streams changed
course and glaciofluvial material locally
froze on to the glacier sole. The till contains rather angular gravel material too
(Fig.10 E), indicating that the incorporated material was mixed with ordinary basal
debris during the glacial transport.

Samples were taken from the three
sections where such till is exposed. The
content of silt is on average lower than
in the type 1 melt-out till (Fig.8), partly
as a result of the incorporation of

148

glaciofluvial sand. The sand grades of this till are significantly richer in quartz than the same grades of the original tills (Fig.11). From the discussion above it is concluded that this is the result of incorporated glaciofluvial sediments.

The SiO_2 content of the bulk matrix is higher in type 2 melt-out till than in type 1 (Table 1), due to the enrichment of quartz in the sand and to the lower content of fine-grained material (Figs.8, 11 & 12). Type 2 melt-out till thus has both a quartz enrichment of type 1 and a quartz enrichment of type 2.

5 DISCUSSION AND CONCLUSION

An enrichment of quartz was formed in tills from Åstadalen purely by subglacial processes and without influence from intervening or preceeding ice-free periods. One general conclusion is that the abrasion resulted in another size distribution of minerals than the crushing. The abrasion process alone was responsible for the enrichment of quartz in the sand while the combination of mechanical comminution and removal of fines by meltwater gave an enrichment of quartz in the bulk sediment.

The study reflects the general principle that a quartz grain, because of its great mechanical resistance, in many cases ends up in coarser fractions than for instance the feldspars. If the bedrock is variable, with coarse-grained types alternating with more fine-grained ones, the process of abrasion combined with a general winnowing of fines by water may result in an enrichment of quartz in both the sand and the silt. This effect would be even stronger with repeated cycles of till formation and fluvial activity.

The model is only valid in an 'open system' with a removal of fine-grained material. In a 'closed system' where the removed component was put back into the budget, there would have been a balance between the bulk sediment composition and the bedrock. Such a 'closed system' in the present study is represented by the relation between the bedrock and the lodgement till.

However, an enrichment of quartz in the till compared with the source rocks is a characteristic of many Scandinavian tills (Korbøl & Jørgensen 1973, Collini (in Lindén) 1975, Rosenqvist 1975a, 1975b, Perttunen 1977, Roaldset 1978). The great enrichment of quartz found in tills from Numedal, southern Norway has been related to an incorporation of preglacially (Tertiary) weathering components (Rosen-

qvist 1963, 1975a, 1975b). The processes described in the present paper can probably not serve as an alternative explanation of the quartz enrichment in tills like those in Numedalen. The present study may, on the other hand, be an alternative or additional explanation of a certain enrichment of quartz found in many other Scandinavian tills. This is also supported by the conclusion that much of the Scandinavian till material is of a multi-cyclic origin (Gillberg 1977). A quartz enrichment should then be a property related to many tills also outside Scandinavia, and such tills may even be more frequent than tills with a composition quite in accordance with the related bedrocks.

6 REFERENCES

Boulton, G.S. 1976, A genetic classification of tills and criteria for distinguishing tills of different origin. In: Stankowski, W. (ed.): Till, its genesis and diagenesis. Univ.Mickiewicza W. Poznaniu. Ser.Geografia 12: 65-80.

Dreimanis, A. 1976, Tills: their origin and properties. In: Legget, R.F. (ed.) Glacial till. An inter-diciplinary study, p.11-49. Royal Soc.Can.Spec.Publ.12.

Dreimanis, A. & Vagners, U. 1971, Bimodal distribution of rock and mineral fragments in basal tills. In:Goldthwait, R.P. (ed.): Till/A symposium, p.27-37. Ohio State Univ. Press.

Englund, J.-O. 1972, Sedimentological and structural investigations of the Hedmark Group in the Tretten - Øyer - Fåberg district. Norg.geol.Unders.276: 59 pp.

Gillberg, G. 1977, Redeposition: a process in till formation. Geol.Fören.Stockh. Förh.99: 246-253.

Haldorsen, S. 1977, The petrography of tills - a study from Ringsaker, southeastern Norway. Norg.geol.Unders.336: 36 pp.

Haldorsen, S. 1978, Glacial comminution of mineral grains. Norsk geol.Tidsskr.58: 241-243.

Haldorsen, S. 1981, The grain-size distribution of subglacial till and its relation to glacial crushing and abrasion. Boreas 10: 91-105.

Haldorsen, S. 1982, The genesis of tills
from Åstadalen, southeastern Norway.
Norsk geol.Tidsskr.62: 17-38.

Haldorsen, S. (in press): Mineralogy and
geochemistry of basal till and their
relation to till forming processes.
Norsk geol.Tidsskr.62.

Haldorsen, S. & Shaw, J. 1982, Melt-out
till and the problems of recognising
genetic varieties of till. Boreas 11,
(in press).

Korbøl, B. & Jørgensen, P. 1973, Faktorer
som er bestemmende for kvartære jordart-
ers innhold av kvarts. Frost i jord 11,
1973: 31-35.

Krumbein, W.C. 1941, Measurement and
geological significance of shape and
roundness of sedimentary particles.
J.Sed.Petr.11: 64-72.

Lindén, A. 1975: Till petrographical studies in an Archaean bedrock area in southern central Sweden. Striae 1: 57 pp.

Perttunen, M. 1977, The lithologic relation
between till and bedrock in the region of
Hämeenlinna, southern Finland.
Bull.Geol.Surv.Finl.291: 68 pp.

Roaldset, E. 1978, Mineralogical and chemical changes during weathering, transport
and sedimentation in different environments, with particular references to the
distribution of yttrium and lanthanoide
elements. Dr.Phil.thesis. Univ.of Oslo.

Rosenqvist, I.Th, What is the origin of the
hydrous micas of the Fennoscandia. Bull.
Geol.Inst.Univ.Upps.40: 265-268.

Rosenqvist, I.Th. 1957a, Chemical investigations of tills in the Numedal.
Geol.Fören.Stockh.Förh.97: 284-286.

Rosenqvist, I.Th. 1975b, Origin and mineralogy of glacial and interglacial clays
of southern Norway. Clays and Clay Min.
23: 153-159.

Slatt, R.M. & Eyles, N. 1981, Petrology of
glacial sand: implications for the origin
and mechanical durability of lithic
fragments. Sed.28: 171-183.

INQUA Symposia on the Genesis and Lithology of Quaternary Deposits / USA 1981 / Argentina 1982

Chronology and style of glaciation in the Wildhorse Canyon area: Idaho

KEITH B.BRUGGER & EDWARD B.EVENSON
Lehigh University, Bethlehem, PA, USA

ROBERT A.STEWART
Iowa State University, Ames, IA, USA

1. ABSTRACT

The Wildhorse Canyon system in south-central Idaho consists of a large trunk valley, Wildhorse Creek proper, and a tributary valley system to the east, the Fall Creek drainage. The entire drainage was glaciated at least twice during the Pleistocene. The glaciations are tentatively correlated to the Bull Lake (older) and Pinedale (younger) Glaciations in the Wind River Mountains of Wyoming (See Editors Note below).

The upper reaches of Wildhorse Creek and Fall Creek are underlain by distinctly different bedrock types. Thus, each catchment area produced tills with unique pebble and heavy mineral suites. The distribution of these suites, preserved in terminal moraines, demonstrates unmixed ice stream flow and allows inferences can be made concerning the flow paths and dynamics of the individual ice streams constituting the composite glacier. These inferences include: (1) dominance of the Wildhorse Creek glacier during glacial maximum; (2) waning contribution of ice from Fall Creek during deglaciation; and (3) separation of the constituent ice streams prior to Late Pinedale time. During latest Pinedale, the Wildhorse Creek and Fall Creek glaciers retreated continuously, while the retreat of smaller tributary glaciers was punctuated by minor stillstands of ice. The differences in the activities of the glaciers may be a consequence of a rising ELA and/or a difference in their response times to a climatic cooling.

[Editors Note: Although Cotter & Evenson (this volume) and Evenson et al, (in press) have proposed new stratigraphic terms for the glacial deposits of central Idaho, the terms "Pinedale Glaciation" and "Bull Lake Glaciation" have been retained in this paper due to time and editorial constraints. In Wildhorse Canyon, the terms "Pinedale" and "Bull Lake" have been replaced by the local names "Wildhorse Canyon Advance", and "Devil's Bedstead Advance", respectively. On a regional scale the terms "Pinedale" and "Bull Lake" have been replaced by "Potholes Glaciation" and "Copper Basin Glaciation", respectively.]

2. INTRODUCTION

Recent studies (Clague, 1975; Evenson, et al, 1979) have shown the value of till provenance investigations in the interpretation of the glacial history of deglaciated mountainous terrains. The purpose of this paper is to present an understanding of the history and style of glaciation in the Wildhorse Canyon area, which has evolved through traditional mapping techniques (Stewart, 1977) and detailed provenance investigations (Brugger, 1983). A second objective is to suggest that the dynamics of individual ice streams in a composite glacier are reflected in the till provenance.

2.1 LOCATION AND GEOMORPHIC ENVIRONMENT

Wildhorse Canyon lies along the central crest of the Pioneer Mountains, which divide the drainage of the Big Lost and Big Wood Rivers in south-central Idaho (Figure 1). The Wildhorse Creek system drains a catchment area of some 200 sq. km. before merging with the East Fork of the Big Lost River, which drains the adjacent Copper Basin. Ultimately, the East Fork joins the North Fork, Summit Creek, and Kane Creek to constitute headwaters of the Big Lost River (Figure 2).

The Wildhorse Canyon drainage system includes the trunk valley of the Wildhorse Creek and the major tributary valleys of Fall Creek and the East Fork of Fall Creek; minor tributaries are from Moose and Surprise Valleys, and the Boulder Creek Valley. The canyon floor averages

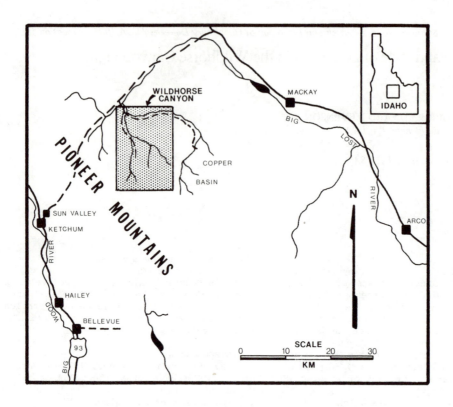

Figure 1. Location of Wildhorse Canyon study area.

approximately 2250m in elevation and is surrounded by steep mountainous ridges with several peaks reaching over 3300m. Hyndman Peak, at 3681m, is the highest mountain in the area. This extreme relief may be the result of late Tertiary to early Quaternary block faulting (Dover 1966; 1969). Subsequent modification of the composite valley by at least two Quaternary glaciations has created a spectacular alpine landscape. High in the headwaters of trunk and tributary valleys, vigorous glacial erosion has left multiple cirque basins separated by aretes and horns. Downvalley, the present underfit streams are incised into till deposits and terraced outwash gravels.

2.2 BEDROCK GEOLOGY

Four lithologic groups can be distinguished in the bedrock of the Wildhorse Canyon area: Precambrian metamorphic rocks, Paleozoic carbonate and clastic rocks, Cretaceous granitic intrusives, and Tertiary volcanic rocks (Figure 3). Each group is, to a degree

reflecting its areal distribution, a source of sediment in the tills. The distribution and the distinctiveness of the lithologies enhanced the success of this study.

The headwaters of Wildhorse Creek flow northward through Precambrian rocks of the Wildhorse Canyon Dome, composed predominantly of the Wildhorse Canyon Migmatitic Gneiss Complex (Dover, 1966; 1969) and lesser amounts of schists and quartites of the Hyndman Formation, restricted to cirque headwalls flanking the summit of Hyndman Peak.

The upper reaches of the Fall Creek drainage is underlain by Cretaceous/Tertiary granitic intrusives (Dover, 1966; 1969; 1981). These gneissose quartz diorites and quartz monzonites occur as an extensive intrusive sheet which separates the migmatites from the overlying Paleozoic sedimentary rocks. In addition, relatively large outliers of this granitic body are exposed in cirques along the western and southern boundaries of the upper Wildhorse Creek catchment basin.

The Wildhorse thrust (Dover, 1981) marks the contact of the early to middle

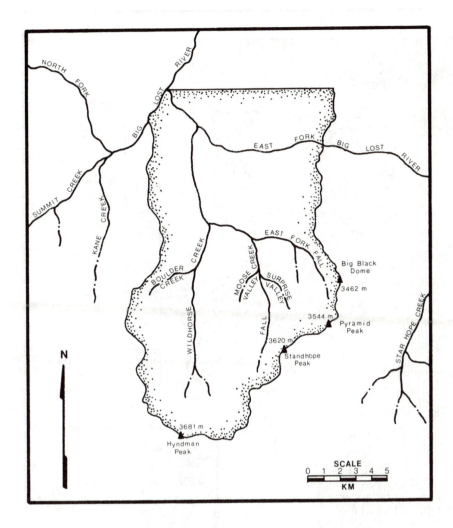

Figure 2.   Major glaciated tributary valleys in the study area.

Paleozoic clastic and carbonate rocks of the Copper Basin Formation (Nelson and Ross, 1969; Paull et al., 1972; Dover, 1981) toward the north, with the intrusive complex of the Pioneer Mountains to the south.   The Copper Basin Formation consists of a thick flysch sequence (Dover, 1981) consisting dominantly of argillites and quartzites, with minor conglomerate and carbonate units. Autochthonous calcareous rocks of Ordovician through Devonian age are exposed in the Wildhorse Window, near the junction of Fall Creek and Wildhorse Creek.   Isolated remnants of Challis Volcanics lie on the boundaries of the Wildhorse Canyon drainage.

These represent the remaining deposits of a once vast accumulation of andesitic to latitic flows and pyroclastics extruded during a period of Tertiary volcanism.

3. GLACIAL GEOLOGY OF WILDHORSE CANYON AND
   ADJACENT AREAS

Glacial deposits in the Pioneer Mountains were first noted by those studying the bedrock geology (Umpleby et al., 1930; Dover, 1966; 1969; 1981; Nelson and Ross, 1969).   Nelson and Ross (1969) originally mapped deposits of two ages in the Copper Basin, eight kilometers to the east of

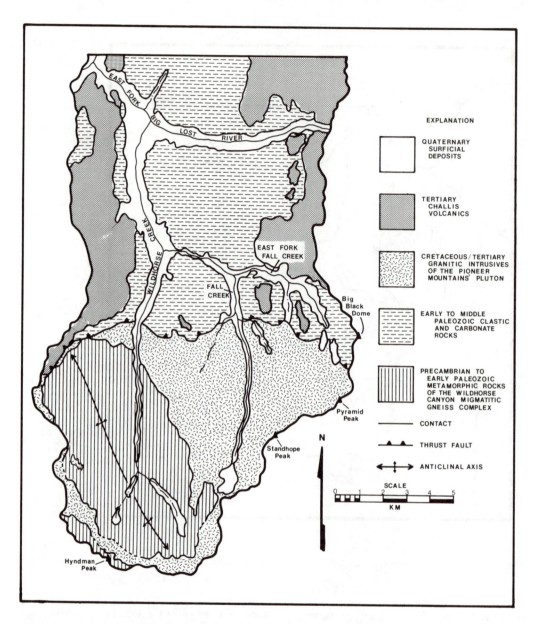

Figure 3. Generalized bedrock map of Wildhorse Canyon showing major
lithologic groups (modified after Dover, et al., 1976).

Wildhorse Canyon (Figure 1). These deposits
were correlated on the basis of similarities
in morphological and weathering
characteristics with those of the Bull Lake
(older) and Pinedale (younger) Glaciation in
the Wind River Mountains, Wyoming
(Blackwelder, 1915; Richmond, 1965; 1976).

    Subsequent studies concerned
specifically with the glacial geology of the
area began with Wigley (1976) and Pasquini

(1976). From his detailed mapping, Wigley
(1976) recognized deposits of three late
Quaternary glaciations in the Copper Basin,
which he correlated tentatively with the
Bull Lake (oldest), Pinedale (middle), and
Neoglacial (youngest) Glaciations of the
Rocky Mountain Glacial Model (Mears, 1974;
and references therein). The Bull Lake
Glaciation in the Copper Basin may have been
a single event; however poorly preserved

outwash remnants suggest the possibility of two stades of Bull Lake ice (Wigley, 1976). Wigley locally subdivided the Pinedale Glaciation into four stades based on recessional moraines. Several intervals of Neoglacial activity are implied by the cross-cutting and overriding relationships of rock glaciers and protalus ramparts in high, protected cirques.

A similar glacial chronology was developed by Cotter (1980) working in the drainage of the North Fork of the Big Lost River. Cotter (1980), however, developed a local stratigraphic nomenclature to "avoid both the implications of a long-distance correlation and the ambiguities of the terms, 'Bull Lake' and 'Pinedale'" (p. 84). Thus, he distinguished deposits of an older Kane advance, and of four "stadials" of a younger North Fork Advance. A full description of the glacial history of the North Fork drainage is given by Cotter and Evenson (this volume).

In Wildhorse Canyon, Stewart (1977) differentiated tills of Bull Lake and Pinedale age by their stratigraphic position and morphologic expression. Bull Lake till occurs as two large lateral moraines on the east and west valley walls at the canyon mouth (Figure 4). The Bull Lake moraines are situated topographically above the oldest Pinedale (I) moraines to a maximum elevation of 2560m (8400 ft). The Bull Lake moraines exhibit a subdued surface morphology typical of those throughout the Rocky Mountains, (Richmond, 1965; Mears, 1974) including the Pioneer Mountains (Wigley et al., 1978; Evenson et al., 1979). Surficial boulders are scarce and well weathered. The moraines are extensively dissected by small tributary streams. Bull Lake moraine remnants document a single advance of ice; however, as in the Copper Basin, two sets of outwash gravels 20-30m above the present drainage of the East Fork of the Big Lost River imply that the Bull Lake Glaciation consisted of two stades.

Pinedale deposits are the most widespread and best preserved in the study area. Pinedale till exists as a series of well developed terminal moraine complexes (Figure 4), each sourcing distinct outwash terraces. The morphology of these moraines is in marked contrast to that of the Bull Lake moraines. Pinedale moraine forms are steep sloped and sharp crested, and surface morphology is rugged, with kettles and other ice disintegration features still apparent. Surface bounders are relatively abundant; and dissection by streams is not as pronounced as in Bull Lake deposits. Moraines attributed to the Pinedale Glaciation (Pinedale I thru Pinedale IV) cannot be subdivided using morphologic

criteria, but can be differentiated on the basis of downvalley extent and stratigraphic association with outwash terraces. On this basis, four stades (I-IV) of the Pinedale Glaciation are evident. Neoglacial deposits in Wildhorse Canyon consist of undifferentiated rock glaciers and pro-talus ramparts.

The glacial chronology of Wildhorse Canyon, established by relative dating techniques, is consistent with others developed within the Pioneer Mountains (Wigley, 1976; Cotter, 1980). Events recorded in the Pioneer Mountains include two stades of the older Bull Lake glaciation, four stades of the Pinedale glaciation, and several episodes of Neoglacial periglacial activity. Evidence of a still older glaciation (pre-Bull Lake) exists as highly weathered, elevated terrace gravels, exposed in the Copper Basin (Wigley, 1976) and along the East Fork of the Big Lost River (Stewart, 1977). The chronologies developed locally are similar to those used throughout the Rocky Mountains, but differ in the subdivision of the younger ("Pinedale") glaciation. Regionally, the Rocky Mountain Glacial Model (Mears, 1974) includes of three stades of the Pinedale Glaciation, although deposits of a Pinedale IV stade were reported in Wyoming and Montana by Graf (1971) and in Colorado by Kiver (1972). It has been suggested (Mears, 1974; Curry, 1974) that these deposits represent equivalents to the pre-altithermal Temple Lake Moraine in the Wind River Mountains (Moss, 1951; Miller and Birkeland, 1974; Curry, 1974) and those of the Satanta Peak Advance in the Colorado Front Range (Benedict, 1973). Richmond (pers. comm., 1981) now recognizes four stades of the Pinedale Glaciation in the Wind River Mountains and includes the Temple Lake Moraine in this fourth stade.

Despite the lack of direct stratigraphic or absolute age correlation, glacial events of Wildhorse Canyon are tentatively correlated with the Bull Lake and Pinedale Glaciations of the Rocky Mountain Glacial Model. These terms are used here pending designation of local type sections (eg. Evenson et al., in press) and do not imply demonstrated age equivalence with the type Bull Lake and Pinedale sections in the Wind River Mountains, Wyoming. Instead, the nomenclature is maintained to provide a regional context for local events.

4. METHODS

Detailed study of the till provenance involved two phases of fieldwork. Initially a reconnaissance geologic survey of Wildhorse Canyon was undertaken to become

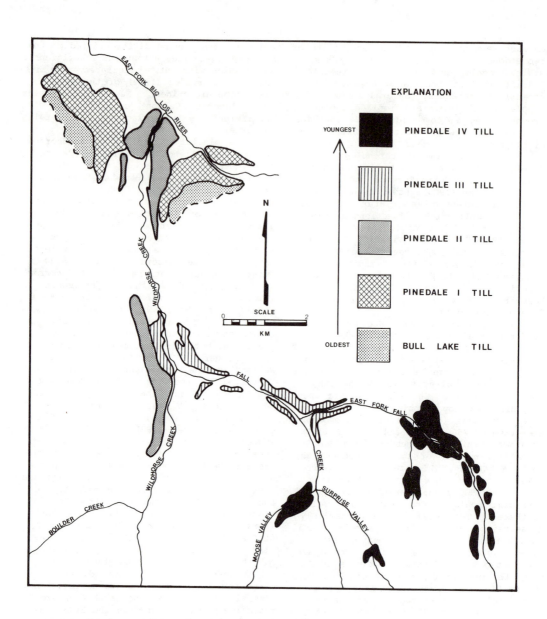

Figure 4. Generalized map showing major till deposits in Wildhorse Canyon
(modified after Stewart, 1977).

familiar with the distribution of the
bedrock types, and their distinguishing
gross lithologic features. This facilitated
the identification of clasts in the tills,
and hence their sources. The second phase
consisted of extensive sampling of the
tills. Sampling sites were located by
superimposing a grid of variable dimensions
on the area enclosed by a deposit on aerial
photographs. The dimensions of the sampling

grid for a particular deposit were an
arbitrary function of its location, and to a
lesser degree its size (eg. a large moraine
near the confluence of major tributary
valleys was given a heavier sampling density
than a small occurrence of till confined to
a cirque basin). At each of 82 sampling
sites thus determined (Figure 5A-C), 100
pebbles were collected from a depth of 45-90
cm and classified on the basis of lithology.

Additionally, a sample of the till matrix (fine fraction) was collected for later laboratory analysis of heavy mineral suites.

In the laboratory each matrix sample was seived into its constituent 1, 2, 3, and 4 phi size fractions. The 3 phi size was selected for heavy mineral analysis as it may represent the terminal grade of most heavy minerals in glacial transport (Dreimanis and Vagners, 1971). Separation of the heavy minerals was accomplished with tetrabromoethane adjusted to a specific gravity of 2.75 with dimethylformanide. The heavy residues were mounted on petrographic slides with thermoplastic quartz cement. Using a petrographic microscope with mechanical stage, a point count of 400 grains was performed on 46 selected samples (Figure 5). Grains were identified by standard petrographic techniques including: color, birefringence, form, cleavage, relief, optical character, extinction and dispersion.

The pebble and heavy mineral data were analyzed using the statistical methods of factor and cluster analysis. Factor analysis is designed to extract hypothetical "factors" among variables (R-mode) or samples (Q-mode) containing all the essential information inherent in the larger set of data. In essence, this is done by the construction of n-number of mutually orthogonal reference vectors, or factors, composed of linear combinations of the variables (samples), each accounting for a percentage of the total sample variance. Thus the composition of an individual sample can be expressed in terms of these factor axes by a factor score. The factor axes can be oriented to make interpretations easier by a technique referred to as "variamax rotation" (Klovan, 1975). While the results of R-mode factor analysis are reported in this study, Q-mode analysis produced similar results.

Factor scores of the samples were subjected to Q-mode cluster analysis in order to group samples of similar composition, and presumably provenance. The analysis here (modified after Parks, 1970) uses a distance coefficient as a measure of similarity and provides a heirarchical dendrogram on which groups can be chosen at any desired level of similarity. For a detailed discussion of these methods the reader is referred to Parks (1966) and Klovan (1975).

5. RESULTS AND DATA ANALYSIS

Reconnaisance of the bedrock geology in Wildhorse Canyon confirmed the distribution of lithologic types shown in Figure 3. Outcrops of Precambrian gneisses and metasedimentary rocks are unique to the Wildhorse Creek and Boulder Creek catchment areas. Cretaceous/Tertiary granitic intrusives, while not being unique to the Fall Creek drainage, exclusively underlie the headwaters of Upper Fall Creek and its tributaries (Moose and Surprise Valleys). Minor amounts of these intrusive rocks also occur in the headwater areas of the East Fork of Fall Creek, and they crop out within the catchment of Wildhorse Creek. The lower reaches of Wildhorse, Fall and East Fork Creeks are cut into Paleozoic clastic and carbonate rocks with Tertiary volcanic rocks exposed high on valley walls and interfluves (Fig. 3).

In outcrop and hand specimen, these major rock groups (Precambrian metamorphics, granitic intrusives, Paleozoics and volcanics) can easily be distinguished by their gross lithologic features. The Precambrian metamorphic suite consists largely of clear, light-colored quartzitic gneisses to gneissose quartzites and, to a lesser extent, biotite and pelitic schists, phyllites, and amphibolites. The granites, quartz monzonites, and quartz diorites are coarse-grained to porphyritic in texture with distinctive phenocrysts of pink feldspar. The Paleozoic lithologies are of obvious sedimentary origin and included dark quartzites and argillites, limestones, and conglomerates. The volcanic rocks are typically andesitic in composition, displaying phenocrysts of plagioclase feldspar set in a dark-colored, fine-grained matrix.

The Precambrian metamorphic and Cretaceous granitic pebbles are the most useful indicators of till provenance because of the distribution of these rock-types in the tributary canyons (Fig. 3). Clasts of the Paleozoic sedimentary units and Tertiary volcanics are not as diagnostic of a particular source area, as all major ice streams flow through areas dominated by this bedrock type (Fig. 3). Moreover, significant numbers of Tertiary volcanic pebbles appear in so few samples that they are excluded from further discussion.

5.1 PEBBLE LITHOLOGY DATA

Pebble composition (percent Precambrian metamorphic vs. Cretaceous granitic) of the tills in the Wildhorse Canyon system are shown in Figure 5A-C. Till pebble assemblages of the Pinedale tills deposited at the mouth of the canyon (Figure 5A) are dominated (88% average) by Paleozoic carbonate and clastic lithologies, but the major differences in till composition and therefore provenance are reflected by the distribution and percentages of metamorphic

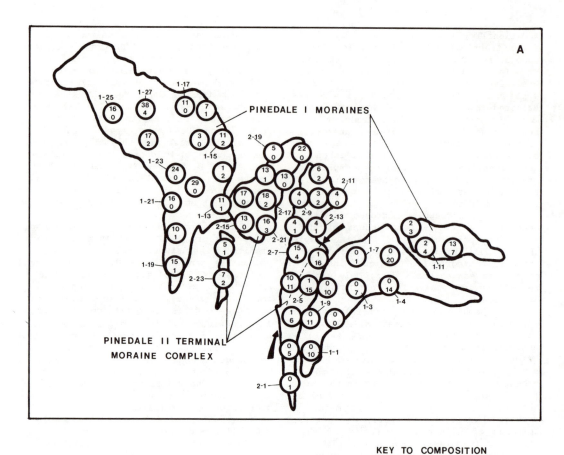

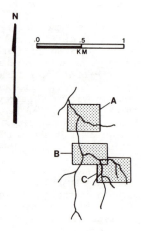

Figure 5. Sample locations and pebble compositions of the tills in Wildhorse Canyon. Heavy mineral compositions are listed in Table 1. Distinct break in pebble composition in (A) is noted by arrows.

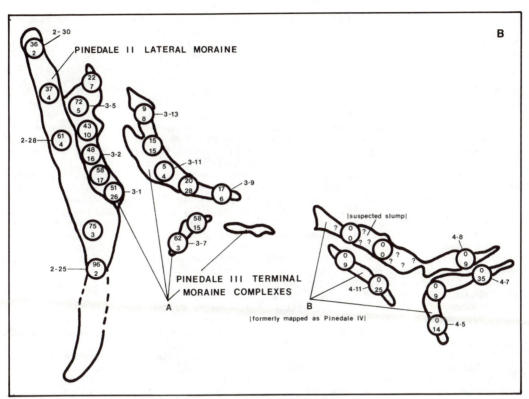

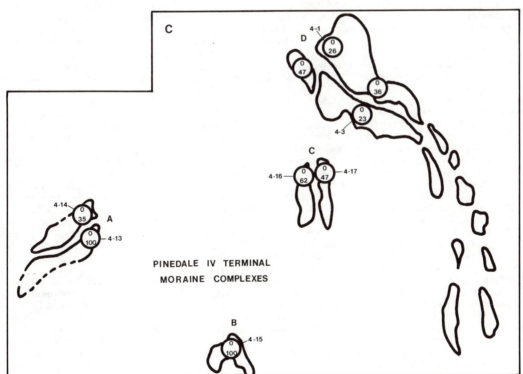

and granitic lithologies. For example, tills from the larger Pinedale I moraine segment on the east side of the valley (Fig. 5A) contain only the granitic indicator lithology while both indicator lithologies (Precambrian metamorphics and Cretaceous granitics) occur with nearly equal frequency in the smaller Pinedale I moraine segment on the far east side of the valley (Fig. 5A). The Pinedale moraines on the west side of the valley are characterized by an average till pebble assemblage of 15% metamorphic and 1% granitic lithologies.

Figure 5A clearly shows a distinct compositional break in the tills of the Pinedale II terminal moraine complex. With respect to the indicator lithologies, till to the east of this break has substantially higher proportions of granitic clasts than of metamorphic clasts. In fact no sample contained more than 1% metamorphic pebbles whereas the samples average 9% granitic pebbles. Till to the west of the break has, in contrast, an average of 9% metamorphic and 2% granitic lithologies in its pebble size fraction.

Pebble assemblages in the tills deposited near the confluence of Wildhorse Creek and Fall Creek are shown in Figure 5B. These tills, which include the large Pinedale II lateral moraine on the west and a Pinedale III terminal moraine (complex "A"), are composed of clastic and carbonate, metamorphic, and granitic lithologies. The percentages of clastic and carbonate pebbles are variable, but are much lower than those in tills deposited at the canyon mouth. With the exception of the Pinedale III moraine remnant northeast of the confluence, this decrease is accompanied by a relatively large increase in the percentages of metamorphic clasts, and only a slight increase in the percentage of granitic pebbles. Overall, the composition of the tills deposited here can be described as having a mean composition of 46% metamorphic and 10% granitic lithologies.

Pinedale III terminal moraine complex "B" is also shown in Figure 5B. This complex, previously mapped as Pinedale IV by Stewart (1977) is reassigned to the Pinedale III advance, on the basis of provenance analysis. Uncertainty exists as to the downvalley extent of the moraine segment on the north side of the junction of Fall Creek and the East Fork of Fall Creek. Till which forms this moraine is contiguous with a diamicton having a hummocky morphologic expression, but composed solely of Paleozoic clastics and carbonates; granitic erratics, which are common in the adjoining till valley, are conspicuously absent in pebble, cobble, and boulder size-fractions. The

morphology and composition of the diamicton are thus diagnostic of a slump from the Paleozoic bedrock outcropping on the valley wall above. Neglecting the two samples obtained in this area of suspected slump, pebble assemblages are dominated by clastic and carbonate lithologies with subordinate amounts of granitic clasts, but contain no metamorphic clasts.

Pebble assemblages in Pinedale IV moraines in the valley of the East Fork of Fall Creek (Figure 5C, complex "C" and "D") consist of clastic, carbonate, and subordinate granitic lithologies, which account for an average of 40% of the till composition. Correlative deposits high in Surprise Valley (Figure 5C, complex "B") and on the south side of the mouth of Moose Valley (complex "A") are composed entirely of the local granitic bedrock. No Precambrian metamorphics occur in the tills of this area.

5.2 HEAVY MINERAL DATA

Petrographic study of the heavy minerals in the tills revealed seven species which appear in a sufficient number of samples to warrant their use as indicators of provenance. These are: hornblende, biotite, clinopyroxene + diopside (henceforth called "CPX"), calcite, sphene, monazite, and garnet. These seven minerals account for an average of 82% (340) of the grains counted for each of the 46 samples obtained at the locations shown in Figure 5 A-C. The remaining heavy minerals consist primarily of: grains altered beyond recognition, undifferentiated opaque minerals, and species which are present in low numbers in very few samples.

The heavy mineral compositions of the till samples are listed in Table I. Subtle relationships among the data are obviously difficult to discern "by eye", but some general trends are evident. These trends can be outlined as follows:
(1) Hornblende and biotite are ubiquitous in relatively high proportions throughout the sample population and therefore are not useful in determining source areas. The other five minerals are consistently present in varying amounts in distinct sub-populations, which can be interpreted to reveal the source area of the till.
(2) CPX is absent in till forming the larger Pinedale I moraine on the east side of the mouth of Wildhorse Canyon (Figure 5A and Table 1). On the west side, till of equivalent age averages 30% CPX in its heavy mineral assemblage. In the Pinedale II till deposited at the canyon mouth, CPX is virtually absent east of the compositional

TABLE 1.   Compositions of the Heavy Mineral Samples (Locations Shown in Figure 5)

Sample	Hbl	Biotite	CPX	Calcite	Sphene	Monazite	Garnet
1-1	56	24	0	17	3	<1	0
1-3	42	39	0	16	2	1	0
1-4	34	22	0	42	2	1	0
1-7	33	53	0	12	2	1	0
1-9	44	19	0	35	2	2	0
1-11	71	8	6	8	6	1	0
1-13	25	19	54	0	1	0	0
1-15	54	25	20	<1	1	1	1
1-17	67	19	10	2	1	0	1
1-19	50	22	27	0	0	1	<1
1-21	44	17	38	1	0	<1	1
1-23	34	18	46	3	0	0	1
1-25	37	33	29	0	1	0	1.
1-27	39	48	13	0	<1	<1	<1
2-1	28	24	1	45	3	0	0
2-5	31	13	<1	54	1	1	0
2-7	40	14	8	34	2	<1	0
2-9	36	28	9	25	1	0	1
2-11	45	22	8	22	1	1	0
2-13	51	11	10	27	2	0	0
2-15	51	28	11	9	1	<1	<1
2-17	52	14	23	8	2	1	1
2-19	44	37	7	9	2	1	<1
2-21	37	37	14	12	1	1	1
2-23	27	30	31	11	1	1	0
2-25	39	21	39	0	1	0	<1
2-28	40	25	35	0	<1	0	1
2-30	41	30	28	0	1	4	<1
3-1	49	17	31	0	0	0	2
3-3	50	12	35	1	1	1	1
3-5	57	6	36	1	0	0	1
3-7	44	52	2	0	1	<1	0
3-9	56	35	3	2	2	2	0
3-11	55	34	1	8	2	1	0
3-13	50	25	2	20	2	1	0
4-1	52	38	0	<1	7	2	0
4-3	65	10	1	0	21	5	0
4-5	62	32	0	0	6	1	0
4-7	34	60	0	0	6	1	0
4-8	58	36	0	0	4	2	0
4-11	51	44	1	0	4	1	0
4-13	35	63	<1	0	2	<1	0
4-14	57	39	<1	0	3	1	0
4-15	49	45	0	0	5	2	0
4-16	64	32	0	<1	2	1	0
4-17	62	32	0	0	5	1	0

break noted in Figure 5A, while being present in moderate quantities (13%) east of this break. CPX comprises 34% of the heavy mineral fraction in both the large Pinedale II lateral moraine and the Pinedale III moraine segment west of Wildhorse Creek (Figure 5B), However, CPX is a minor component in Pinedale III till deposited immediately east of the confluence of Wildhorse and Fall Creeks, typically making

up 2% of the heavy mineral assemblage. Furthermore, CPX becomes insignificant in the tills lying upvalley in the Fall Creek and East Fork drainages, i.e. the Pinedale III terminal complex B, and Pinedale IV terminal complexes A, B, C, and D.

The presence of garnet seems to be associated with high percentages of CPX. The distribution of both CPX and garnet suggests that these two minerals are related

161

to the metamorphic source area in the upper reaches of Wildhorse Canyon.

(3) Sphene and monazite are present in tills on the eastern side of the trunk valley, and in tills deposited in the Fall Creek drainages. There is an inverse relationship between the abundance of sphene + monazite and CPX + garnet. This suggests that sphene and monazite are associated with the granitic intrusives in the Fall Creek drainage.

(4) Calcite is present in appreciable quantities only in the tills at the mouth of the canyon, especially on the east side of the canyon. Calcite occurrence is related to the abundance of Paleozoic rocks, which occur only in the lower reaches of the canyons.

5.3 STATISTICAL ANALYSIS

Because of the multivariate nature of the data, pebble and heavy mineral data were subjected to R-mode factor analysis in order to determine the interrelationships between variables. Resulting factor scores of the data were then analyzed by Q-mode cluster analysis. Combined, these two techniques proved extremely useful in grouping samples of similar composition, particularly with the heavy mineral data.

The Precambrian, Paleozoic, and Cretaceous lithologies were used as variables for the factor analysis of the pebble data. This resulted in two factors accounting for 49% and 39% of the total sample variance respectively (Figure 6A). Factor 1 suggests a negative correlation between Paleozoic lithologies and Cretaceous granitics. Similarly, Factor 2 implies a negative correlation between Precambrian metamorphic pebbles and the Paleozoic lithologies.

On the basis of the factor scores, cluster analysis defined four compositional groups at a distance coefficient of about .15 (Figure 6B). The actual compositions of these groups are shown in Figure 6c. Samples in both groups 1 and 2 are composed predominantly of clastic and carbonate pebbles. However, group 1 is skewed toward a metamorphic composition while group 2 is more strongly skewed toward the granitic end-member. Group 3 includes samples whose assemblages are moderate to high in metamorphic clasts. The two samples (4-13, 4-15) which comprise group 4 consist of 100% granitic pebbles.

The cluster groups associated with the individual till samples are illustrated as maps in Figure 7A-C. At the mouth of the canyon, group 1 samples (sedimentary units + metamorphics) are located on the west and

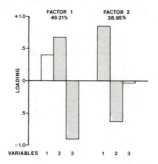

Figure 6. (A) Factor loadings on pebble variables: 1-metamorphic; 2-clastic and carbonate; 3-granitic.

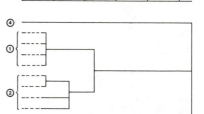

(B) Dendrogram showing clustering of pebble groups (dashed where simplified).

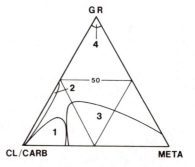

(C) Ternary plot of pebble group compositions: meta-metamorphic; cl/carb-clastic/carbonate; gr-granitic.

central portions of the canyon, while group 2 samples (sedimentary units + granites) are deposited on the east side of the canyon. The compositional break between these two statistically defined groups is identical to the break defined in Figure 5A, from the raw data. At the confluence of Wildhorse Creek and Fall Creek (Figure 7B) tills of Pinedale II lateral moraine and Pinedale III moraine complex A are statistically classified into group 1 (sedimentary + metamorphic) or group 3 (metamorphic + sedimentary), while tills in the Fall Creek drainage are classified into group 2 (sedimentary + granite) and group 4 (granite).

R-mode factor analysis using the seven indicator minerals as variables extracted 4 factors accounting for 80% of the total variance in the heavy mineral assemblages (Figure 8A). Factor 1 explains 42% of the variance as an inverse relationship between CPX + garnet and sphene + monazite; that is the presence of CPX and garnet is associated with the absence of sphene and monazite, as noted before. Factor 2, explaining 23%, indicates a similar relationship between hornblende (and possibly biotite) and calcite. Factor 3 (14%) suggests a weak negative correlation may exist between biotite and hornblende in some samples. Factor 4 accounts for only 9% of the sample variance and reveals no significant relationships among the variables.

Cluster analysis resolved the heavy mineral assemblages of tills in Wildhorse Canyon into seven compositional groups at a level of .15 (Figure 8B). It is difficult to graphically describe these groups in terms of their actual composition because such a description would involve seven-dimensional space. Instead, the average composition of all samples belonging to each group is shown below in Table 2.

This statistical representation of mineral assemblages in the tills is shown in Figure 7A-C. The distribution of the cluster groups is most striking at the mouth of the canyon (Figure 7A). Here the compositional break discussed previously also marks an abrupt change in the heavy mineral composition of the tills. Mineral assemblages to the east of the break are classified as group A, or in one case, group C. Assemblages in the tills west of the break are represented by group B in the Pinedale II moraine, and groups D and E in the Pinedale I moraine. Upvalley, the Pinedale II lateral moraine and Pinedale III moraine segment deposited on the west side of Wildhorse Creek (Figure 7B) are composed of mineral assemblages best described by groups D and E. The mineralogic composition of the remaining segments of Pinedale III terminal complex A is diverse, including groups A, C, and G. Mineral suites of the tills deposited within the Fall Creek drainage appear relatively homogeneous, represented by group G with one exception (Figure 7C).

6. DISCUSSION
6.1 TILL PROVENANCE

Pebble compositions of the tills in Wildhorse Canyon clearly indicate the source areas from which the tills were derived, and therefore reveal the flow path of the ice stream responsible for their deposition. Thus, till containing high proportions of

Table 2. Average Composition (%) of Heavy Mineral Groups.

GROUP	Hornblende	Biotite	CPX	Calcite	Sphene	Monazite	Garnet
A	42	29	.3	26	2	.6	0
B	42	25	14	17	2	.6	.3
C	65	24	4	1	4	2	0
D	44	27	30	0	.5	.3	.5
E	49	14	34	1	.6	.3	.6
F	70	11	.6	0	14	4	0
G	11	.6	.3	0	4	.9	0

From Table 2, the significant mineral assemblage present in each of the seven groups are, in decreasing abundance, as follows:

     Group A--calcite, sphene
     Group B--calcite, CPX, sphene, garnet
     Group C--hornblende, CPX equaling sphene, monazite
     Group D--CPX, garnet
     Group E--CPX, garnet
     Group F--hronblende, sphene, monazite
     Group G--biotite, sphene

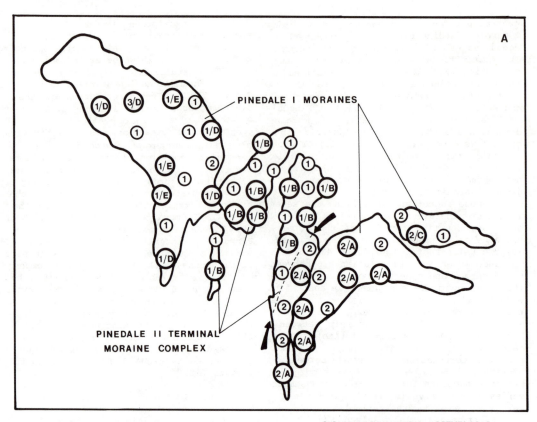

PINEDALE I MORAINES

PINEDALE II TERMINAL
MORAINE COMPLEX

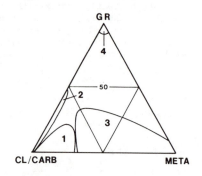

SIGNIFICANT MINERAL ASSEMBLAGES

A   CALCITE, SPHENE
B   CALCITE, CPX, SPHENE
C   HBL, CPX=SPHENE, MONAZITE
D   CPX
E   CPX
F   HBL, SPHENE, MONAZITE
G   BIOTITE, SPHENE

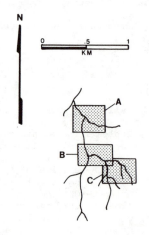

Figure 7.   Distribution of statistically
defined compositional groups. Keys
to composition, and locations
shown in insets. Break in till
composition in (A) noted by arrows.

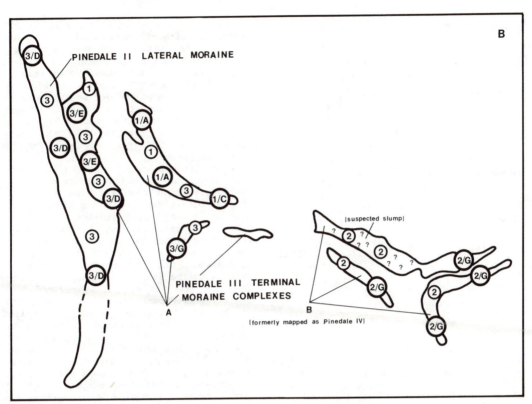

PINEDALE II LATERAL MORAINE

PINEDALE III TERMINAL
MORAINE COMPLEXES

[suspected slump]

[formerly mapped as Pinedale IV]

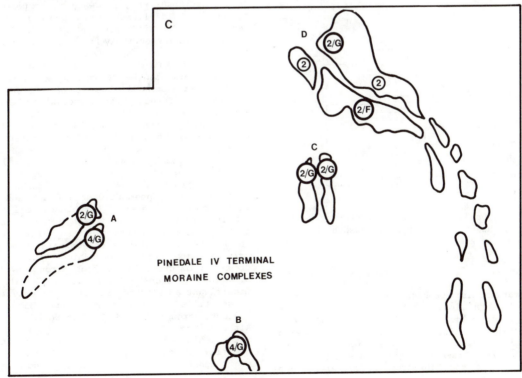

PINEDALE IV TERMINAL
MORAINE COMPLEXES

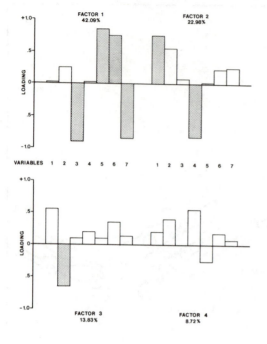

Figure 8. (A) Factor loadings on heavy
mineral variables: 1-hornblende;
2-biotite; 3-CPS; 4-calcite;
5-sphene; 6-monazite; 7-garnet.

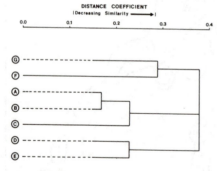

(B) Dendrogram showing clustering
of heavy mineral groups (dashed
where simplified).

metamorphic pebbles were deposited by
glaciers flowing out of Wildhorse Canyon
proper, including the Boulder Creek
drainage. Similarly, those tills having
high proportions of granitic clasts and
lacking metamorphic clasts are attributed to
ice from the Fall Creek source areas. The

few granitic pebbles which occur in the till
associated with the metamorphic source are
derived from outcrops of the intrusive
within the Wildhorse Catchment. Near their
respective sources, these indicator
lithologies assume relatively high
percentages in the pebble assemblages of the
tills. Downvalley, their numbers decrease,
reflecting the dilution caused by the
incorporation of clastic and carbonate
debris as the glaciers traversed the
Paleozoic terrane.

Statistical analysis of the provenance
data provides an objective means to
distinguish subtle differences in till
compositions. Moreover, it affords a
sorting of samples of similar composition
into geologically meaningful groups. Among
the compositional groups defined above for
pebble data, two types of suites are
recognized. The presence of metamorphic
pebbles in the samples of group 1 indicates
that the till was ultimately derived from
the Wildhorse source area. The composition
of the till represented by group 2 implies
that these tills originated in the granitic
bedrock of the Fall Creek drainage.
However, pebble assemblages of both groups
are dominanted by clastic and carbonate
lithologies, and these groups are therefore
interpreted as "diluted source suites". In
contrast, groups 3 and 4 are interpreted as
"source suites"; that is, each is composed
predominantly of lithologies exposed in the
bedrock of the Wildhorse and Fall Creek
source areas respectively, with little or no
dilution.

"Source suites" and "diluted source
suites" could also be recognized among the
heavy mineral groups, but it was first
necessary to identify the source(s) of each
mineral species. This was greatly
facilitated by the petrographic studies of
Dover (1966, 1969, 1981), and to a lesser
extent, the mineral's association with
particular pebble groups. Hornblende and
biotite occur as both characterizing and
minor accessory minerals in the rocks
underlying the upper Wildhorse Creek and
Fall Creek drainages. Hornblende and
biotite are thus found in high proportions
in tills from both source areas (Table 1).
The source of CPX (diopside) appears to be a
calc-silicate marble bed which "forms
prominent outcrops in the high cirques west
of Wildhorse Creek" (Dover, 1981, p. 24).
Significant amounts of CPX are found only in
those tills derived from metamorphic
bedrock, groups 1 and 3 (Figure 7).
Although calcite is present in the
metamorphic bedrock, it is not abundant in
the tills deposited upvalley. Calcite is a
major constituent of the mineral assemblages

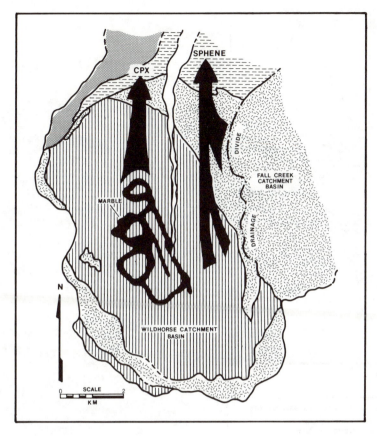

Figure 9. Idealized dispersal paths of heavy mineral in the Wildhorse Catchment (see text for discussion; bedrock symbols explained in Figure 3).

in the diluted tills at the mouth of the canyon (Table 2, Figure 5), which suggests that the calcite grains, like the carbonate clasts, were incorporated as dilutants from the Paleozoic rocks. Sphene and monazite are common in the granitic rocks (Dover, 1969) and this is reflected in their association with tills rich in granitic clasts (groups 2 and 4; Figure 7). Garnet occurs locally in Dover's (1969, 1981) basal unit of the migmatic gneiss complex. This unit is exposed in cirque basins on the western side of Wildhorse Creek where it is surrounded by the CPX-bearing marble (Dover, 1981, Plate 2). These exposures, poor in terms of providing supraglacial debris, may well explain the scarcity of garnet in the tills, as well as its close association with CPX (Table 2).

The fact that heavy mineral sources are often unevenly distributed in areas of otherwise grossly homogeneous bedrock gives rise to asymmetrical dispersal patterns for

heavy minerals in the Wildhorse Catchment are illustrated in Figure 9. CPX (and garnet) was preferentially carried by ice emanating from the cirques in the west. The Precambrian bedrock underlying eastern Wildhorse provided fewer distinctive minerals.

The distribution of carbonate rocks also leads to an asymmetric dispersal of calcite. Within the study area, these rocks are limited to a single outcrop directly northeast of the junction of the Wildhorse and Fall Creek valleys (Dover, 1981). Consequently, the presence of calcite, while being a dilutant, serves to indicate the proximity of the former ice streams to the eastern margin of the trunk valley. Paleozoic clastic rocks are thought to have contributed little to the heavy mineral compositions of the tills.

Of the seven heavy minerals selected as possible indicators, five were useful in determining the provenance of the tills.

167

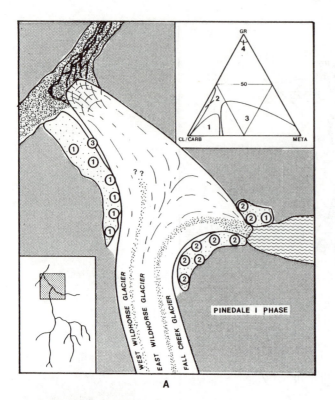

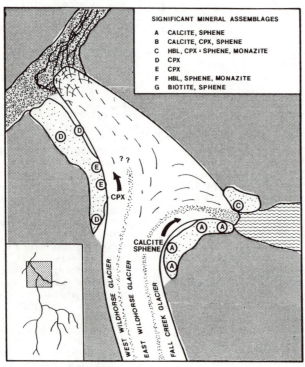

Figure 10. Reconstructions of ice stream geometries, and medial positions during the Pinedale I phase of glaciation based on (A) pebble provenance, and (B) heavy mineral provenance. (Note the presence of the ice-dammed lake to the east of the terminus.)

These are CPX, calcite, sphene, monazite, and garnet. Based on this conclusion and the considerations discussed above, the heavy mineral groups are interpreted as follows:

1. Group A is a dilute suite, characterized by calcite and sphene, which is spatially associated with pebble group 2 (sedimentary + granite) near the mouth of the canyon, and with pebble group 1 (sedimentary + metamorphic) near the confluence of Fall Creek and Wildhorse Creek. Thus, no unique source can be matched to group A.

2. Group B, composed of calcite, CPX, sphene and garnet is a dilute suite derived from Wildhorse Canyon, and is associated with pebble group 1 (sedimentary + metamorphic). The lack of abundant CPX and the presence of sphene suggest a source from the east side of Wildhorse Canyon, where the calc-silicate marble unit is poorly exposed, and the granite unit is prominent.

3. Group C is interpreted as a "mixed" source suite, containing both CPX, from a metamorphic source, and sphene from a granitic source. No genetic interpretation is made of this group, as only two samples were classified into this group.

4. Groups D and E are characterized by CPX and garnet, and are interpreted as a source suite from the metamorphic source. Samples which are classified into these groups are located in the Pinedale I terminal moraine, on the west side of the canyon mouth (Figure 7a), the Pinedale II lateral moraine, and the western portions of the Pinedale III terminal moraine complex A. The undiluted nature of these groups support the conclusion that Paleozoic clastic units supplied few components to the heavy mineral suite.

5. Groups F and G are characterized by sphene and monazite, and are interpreted as source suites from the granitic source in Fall Creek.

Table 3 summarizes the provenance of the tills represented by each of the compositional groups, and also presents correlations among corresponding pebble and heavy mineral groups.

6.2 GLACIAL HISTORY OF WILDHORSE CANYON

The distribution of provenance suites preserved in terminal moraines shown in Figure 7 demonstrates unmixed ice stream flow. After coalescing in the trunk valley, ice streams from Wildhorse Canyon and Fall Creek may have become thinned and attenuated in response to differing flow regimes, but did not mix as fluvial systems do. Each glacier maintained its identity, transporting debris characteristic of its respective source, and ultimately depositing this debris at the terminus as a moraine. The source area and flow paths of the ice streams are thus delineated by the distribution of unique pebble and heavy mineral suites. Furthermore, the medial position separating the adjacent ice streams is marked by a lateral change in till composition across the moraine. Study of the till provenance then, allows reconstruction of lobal geometries from which inferences can be made concerning the dynamics of the individual glaciers. These influences provide insights into the glacier history of Wildhorse Canyon which could not have been gleaned solely on the basis of moraine geometry.

During the Pinedale I phase of glaciation in Wildhorse Canyon, glaciers from Wildhorse Creek (East and west Wildhorse Glaciers) and Fall Creek merged to form a large composite glacier which terminated at the valley mouth as an "expanded foot" glacier (Figure 10). The composition of the large lateral moraine on the western side of the valley is fairly homogeneous and indicates that the till was deposited by ice from the western side of Wildhorse Canyon proper, referred to here as the West Wildhorse glacier. To the east a subtle change in composition occurs between the larger lateral segment on the west side of the canyon and the smaller one north of

Table 3. Interpretations and correlations of till compositional groups.

Provenance	Representative Composition Groups			
	Pebble Suites		Heavy Mineral Suites	
	Source	Diluted	Source	Diluted
Western Wildhorse	3	1	D,E	---
Eastern Wildhorse	3	1	---	B(A?)
Fall Creek	4	2	F,G	A
Mixed (?)	---	---	C(?)	---

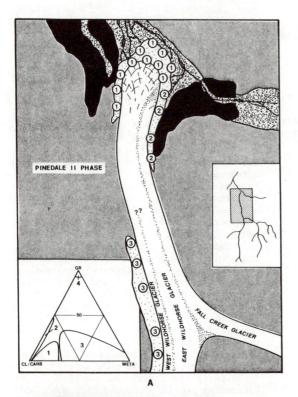

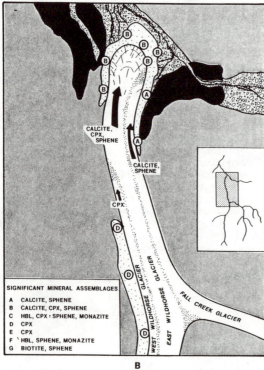

Figure 11. Reconstructions of ice stream geometries, and medial positions during the Pinedale II phase of glaciation based on (A) pebble provenance, and (B) heavy mineral provenance (shaded areas represent older till deposits).

the East Fork of the Big Lost River. Clearly, both the pebble (Group 2) and heavy mineral (Assemblage A) composition of the larger lateral moraine segment establishes that it was deposited by the Fall Creek glacier (Fig. 10). Although one sample in the smaller deposit classifies within group 1, the two other samples contain more metamorphic clast than does any other sample within group 2 (Figure 5A). At a lower level of similarity, cluster analysis placed these two samples in a subgroup of group 2 which Brugger (unpublished data) has interpreted as a mixed source suite representing the medial moraine. The heavy mineral assemblage in the till, group C, may also indicate a mixed suite. Therefore, the inferred position of the medial moraine separating the Wildhorse glacier(s) and the Fall Creek glacier is placed east of this deposit as shown in Figure 10. This conclusion is supported by the presence of ice rafted Precambrian boulders from Upper Wildhorse Canyon in the lake that occupied the East Fork valley during the Pinedale and earlier glacial maxim (Stewart, 1977; Evenson, et al., 1979). Along the East Fork valley, Precambrian gneissic erratics are found concentrated along a paleoshoreline of Glacial Lake East Fork at an elevation of about 2200 m (7200 ft.) for some 7 km up the valley. The Wildhorse Canyon Dome is the only local source of gneiss as no Precambrian gneiss underlies the Copper Basin. The only reasonable explanation for the boulders is that they were carried by the Wildhorse glacier to the edge of Glacial Lake East Fork, calved off with large icebergs and were ice rafted to their present positions, where they were freed by melting. Evenson, et al. (1979) have presented evidence (flood rafted gneissic boulders 10 km downstream of the terminus of the Wildhorse Canyon moraine) for subsequent catastrophic drainage of the lake. These erratics require a source of gneissic debris at the calving terminus. Thus the (East) Wildhorse glacier must constitute at least part of this terminus, supporting the reconstructed geometry shown in Figure 10.

The inferred position of the medial moraine suggests that the Wildhorse glaciers dominated the composite glacier, with ice from Fall Creek assuming a more lateral position (Figure 10), suggeting that a greater amount of ice flowed out from the Wildhorse Catchment area than from the Fall Creek drainage. Glaciological reconstructions of the Pinedale glaciers and ice flux calculations (Brugger, 1983) indicate that at the cross-sections just upstream of their confluence the proportion of ice from the Wildhorse Catchment to that from Fall Creek

is approximately 2.5:1.

The well-preserved Pinedale II terminal moraine (Figure 11) allows accurate reconstruction of the ice margin geometry at this phase of glaciation. The medial (between Wildhorse and East Creek ice) position along the terminal moraine is clearly defined by the abrupt break in both pebble and heavy mineral composition (Fig. 11A and B). This position again suggests a continued dominance of the Wildhorse Glacier. The distinction between the tills deposited by the East and West Wildhorse Glacier and the Fall Creek Glacier can be made on the basis of heavy mineral assemblages (Figure 11B). The West Wildhorse Glacier is responsible for deposition of the large lateral moraine (Fig. 11, group D) but apparently was never manifested at the snout. The west and south portions of the terminal moraine are represented by group B and reflect a diluted East Wildhorse source. However, in the absence of a continuous deposit of till along the western margin, the position of the medial moraine, between east and west Wildhorse ice streams, shown in Figure 11B is tentative, and therefore any dynamic implication would be speculative.

By Pinedale III time, the waning contribution of ice from Fall Creek resulted in a separation of the Wildhorse and Fall Creek glaciers (Figure 12A). This separation is clearly demonstrated by the provenance of pebbles and heavy minerals (Fig. 12A and C). This ice deployment contrasts greatly with the original interpretation of Stewart (1977) who postulated that the glacier retreated from its Pinedale II position to separate positions within the two valleys, and then readvanced, coalescing to form the Pinedale III terminal moraine (referred to in this work as complex A) (Figure 12B). The basis for the new interpretation is the pebble composition of the till constituting terminal complex A, particularly along the northeastern side of the moraine. If the two glaciers coalesced, as previously assumed, pebble compositions here would not have the provenance of the Wildhorse Canyon metamorphic bedrock, but instead would have a Fall Creek provenance. Therefore, all of moraine complex A is attributed to the Wildhorse glacier. The heavy mineral data (Figure 12C) neither supports or contradicts this conclusion. The revised interpretation requires that the Pinedale IV moraine mapped by Stewart (1977) immediately upvalley becomes the corresponding terminus for the Pinedale III ice of the Fall Creek glacier; hence this moraine was reassigned a Pinedale III age and is referred to as complex B.

171

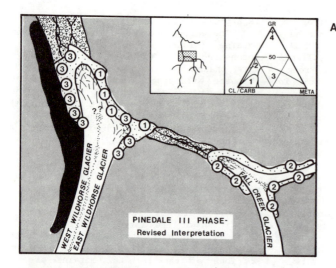

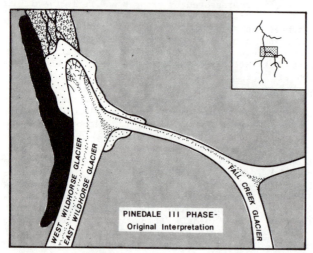

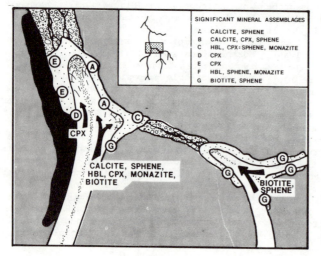

Figure 12.

(A) Reconstruction of ice stream geometries during the Pinedale III phase of glaciation based on pebble provenance, (B) original interpretation of ice lobe geometry prior to till provenance investigation. (C) Reconstruction based on heavy mineral provenance (see text for discussion).

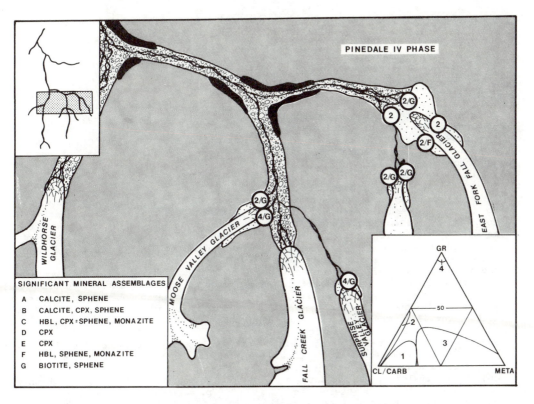

Figure 13. Reconstruction of glacier positions during the Pinedale IV phase of glaciation. Note lack of deposits fronting the Wildhorse and Fall Creek Glaciers, implying continuous retreat during this time (see text for discussion).

The Wildhorse and Fall Creek glaciers apparently receded rapidly from their respective Pinedale III positions with no stillstands. This continuous retreat is suggested by the lack of deposits upvalley from the Pinedale III termini. However, the tributary glaciers sourced from the drainages of the East Fork of Fall Creek, and Moose and Surprise Valleys were occasionally active, as shown by small Pinedale IV moraines deposited here (Figure 13). This may have been a consequence of a rising equilibrium line altitude (ELA) and the area-altitude distribution of the glaciers. The Wildhorse and Fall Creek glaciers, with most of their area at relatively lower elevations, were "starved" of snow accumulation when the ELA rose. In contrast, the catchment basins of the East Fork of Fall Creek, Moose and Surprise Valleys are quite high in elevation, so that these glaciers were protected from the rise in ELA and remained active somewhat longer.

An alternative explanation for the difference in the dynamic behavior of these glaciers may be due to the differences in their response times to climatic cooling. The response of a glacier to climatic change is essentially a response to changes in mass balance (Meier, 1965; Nye, 1960), and is a function of the longitudinal strain rate (du/dx, the change in the surface velocity with distance); the greater the strain rate, the quicker the glacier can respond to changes is mass balance (typically 2.5 to 25 years for valley glaciers, Paterson, 1981, p. 252). Preliminary calculations of longitudinal strain rates for the reconstructed Pinedale IV glaciers (Brugger, unpublished data) indicate that significant differences in response times may exist. Accordingly, if the climatic cooling during Pinedale IV time was of sufficiently short duration, the larger Wildhorse and Fall Creek glaciers, with longer response times, might have reacted sluggishly (or not at all) while the smaller glaciers paused in their retreat.

7. CONCLUSIONS

Based on this study of Wildhorse Canyon, it is concluded that:

1. Till provenance delineates glacier flow patterns, and thus aids which aid in the reconstruction of ice stream geometries.

2. Changes in till provenance across moraines indicate the medial positions separating adjacent ice streams; this provides a measure of the relative magnitudes of the individual ice stream in a composite glacier and hence an indication of their dynamics.

3. In this study, pebble provenance data is easier to analyze, but is sensitive to only the largest differences in till composition, while heavy mineral data is more difficult to analyze, but is sensitive to minor compositional differences. While most aspects of deglaciation were reflected in both data sets, some aspects would not be noticed by the use of only one methodology, hence, a combined methodology is a much more powerful tool. Multivariate statistical analysis is important in distinguishing the subtlest differences in till composition, and is a powerful tool in objectively analyzing complex data.

4. Provenance analysis is useful in conjunction with field mapping in two ways. It provides additional supporting evidence in resolving problems discovered during field mapping (for example, the existence of the Glacial Lake East Fork). In addition, provenance studies may reveal some aspects of glacial history that are completely undetectable by field mapping, for example, the separation of the Wildhorse Canyon and Fall Creek glaciers during the Pinedale III phase.

8. ACKNOWLEDGEMENTS

We thank George Thomas for assistance in the field. We also extend our deepest thanks to our friend Vic Johnson who provided logistical support for many years. Discussions with Jim Cotter, Dr. Bobb Carson, Jim Zigmont, Jack Ridge, Jim Bloomfield, Dr. Jim Parks, and Dr. J. Donald Ryan were helpful and are gratefully acknowledged. Discussions with Dr. Roger Hooke stimulated some ideas on glacier dynamics, but the correctness of our interpretations rests with us. Patti Burns and Laurie Cambiotti aided in preparation of the manuscript. Lehigh University provided partial funding for this work. Finally, one of us (K.A. Brugger) wishes to thank Charles W. and Jean A. Brugger for their continued support and encouragement.

9. REFERENCES

Benedict, J. B., 1973, Chronology of cirque glaciation, Colorado Front Range, Quat. Research, V. 3, p. 584-599.

Blackwelder, E., 1915, Post-Cretaceous history of the mountains of central western Wyoming, Jour. Geology, V. 23, p. 307-340.

Brugger, K. A., 1983, A till provenance investigation and its implication for the style of glaciation of Wildhorse Canyon, Custer County, Idaho, Unpublished M.S. Thesis, Lehigh Univ.

Clague, J. J., 1975, Glacial flow patterns and the origin of late Wisconsinan till in the southern Rocky Mountain trench, British Columbia, G.S.A. Bull., V. 86, p. 721-731.

Cotter, J. F. P., 1980, The glacial geology of the North Fork of the Big Lost River, Custer County, Idaho, Unpublished M.S. Thesis, Lehigh Univ., 102p.

Cotter, J. F. P. and Evenson, E. B. (this volume) Glacial history and stratigraphy of the North Fork of the Big Lost River, Pioneer Mountains, Idaho.

Currey, D. R., 1974, Probable Pre-Neoglacial age of the Temple Lake Moraine, Wyoming, Arctic and Alpine Research, V. 6, p. 293-300.

Dover, J. H., 1966, Bedrock geology of the Pioneer Mountains, Blaine and Custer Counties, Idaho, Ph.D. Thesis, Univ. of Washington, 138p.

Dover, J. H., 1969, Bedrock geology of the Pioneer Mountains, Blaine and Custer Counties, central Idaho, Idaho Bureau of Mines and Geology Pamphlet No. 142, 66p.

Dover, J. H., 1981, Geology of the Boulder-Pioneer wilderness study area, Blaine and Custer Counties, Idaho, U.S.G.S. Bull., 1475A.

Dover, J. H., Hall, W. E., Hobbs, S. W., Tschanz, C. M., Batchelder, J. M., and Simons, F. S., 1976, Geologic map of the Pioneer Mountains Region, Blaine and Custer Counties, Idaho, U.S. Geol. Survey Open File Report 76-75.

Dreimanis, A. and Vagners, U. J., 1971, Bimodal distribution of rock and mineral fragments in basal till, in Goldthwaite, R. P., ed., Till: A Symposium, Ohio State Univ. Press, p. 237-250.

Evenson, E. B., Pasquini, T. A., Stewart, R. A., and Stephens, G. C., 1979, Systematic provenance investigations in areas of alpine glaciation: applications to glacial geology and mineral exploration: in Schluchter, C. ed., Moraines and Varves, A. A. Balkema, Rotterdam, p. 25-42.

Evenson, E. B., Cotter, J. F. P., and Clinch, J. M., Glaciation of the Pioneer Mountains: a proposed model for Idaho: in Cenozoic of Idaho, R. Breckenridge, ed., Idaho Bureau of Mines (in press).

Graf, W. L., 1971, Quantitative analysis of Pinedale landforms, Beartooth Mountains, Montana and Wyoming, Arctic and Alpine Research, V. 3, p. 253-261.

Kiver, E. P., 1972, Two late Pinedale advances in the southern Medicine Bow Mountains, Colorado, Contribution to Geology of the Univ. of Wyoming, V. 11, no. 1, p. 1-8.

Klovan, J. E., 1975, R- and Q-mode factor analysis in McCammon, R. B., ed., Concepts in Geostatistics, New York, Springer-Verlag, 230p.

Mears, B., 1974, The evolution of the Rocky Mountain glacial model, in Coates, D. R., ed., Glacial Geomorphology, New York State Univ. Publisher, p. 11-40.

Meier, M. F., 1965, Glaciers and climate, in Wright, H. E. and Frey, D. G., eds., The Quaternary of the United States, Princeton Univ. Press, N.J.

Miller, C. D. and Birkeland, P. W., 1974, Probable Pre-Neoglacial age of the type Temple Lake Moraine, Wyoming: discussion and additional relative-age data, Arctic and Alpine Research, V. 6, p. 301-306.

Moss, J. H., 1974, The relation of river terrace formation to glaciation in the Shoshone River Basin, western Wyoming, in Coates, D. R., ed., Glacial Geomorphology, New York State Univ. Publisher, p. 293-315.

Nelson, W. H. and Ross, C. P., 1969, Geology of the Mackay 30-minute quadrangle, Idaho, U.S.G.S. Open File Report, 215p.

Nye, J. F., 1960, The response of glaciers and ice sheets to seasonal and climatic changes, Proc. Roy. Soc. A, V. 256, no. 1287, p. 559-584.

Parks, J. M., 1966, Cluster analysis aplied to multivariate geologic problems, Jour. Geology, V. 74, p. 703-715.

Parks, J. M., 1970, Fortran IV program for Q-mode cluster analysis on distance function with printed dendogram, Kansas Geol. Survey Computer Contr. 46, 32p.

Paterson, W. S. B., 1981, The Physics of Glaciers, Pergamon Press, New York, 380p.

Pasquini, T. A., 1976, Provenance investigation of the glacial geology of the Copper Basin, Idaho, Unpbl. M.S. Thesis, Lehigh Univ., 136p.

Paull, R. A., Wolbrink, M. A., Volkmann, R. G., and Grover, R. L., 1972, Stratigraphy of the Copper Basin group, Pioneer Mountains, south-central Idaho, A.A.P.G. Bull., V. 56, p. 1370-1401.

Richmond, G. M., 1965, Glaciation of the Rocky Mountains, in Wright, H. E. and Frey, D. G., eds., The Quaternary of the United States, Princeton Univ. Press, Princeton, N.J., p. 217-230.

Richmond, G. M., 1976, Pleistocene stratigraphy and chronology in the mountains of western Wyoming, in Mahaney, W. C., ed., Quaternary Stratigraphy of North America, John Wiley and Sons, Inc., New York, p. 353-380.

Stewart, R. A., 1977, The glacial geology of Wildhorse Canyon, Custer County, Idaho, Unpbl. M.S. Thesis, Lehigh Univ., 103p.

Umpleby, S. G., Westgate, L. G., and Ross, C. P., 1930, Geology and ore deposits of the Wood River Basin, Idaho, U.S.G.S. Bull., V. 814, 250p.

Wigley, W. C., 1976, The glacial geology of the Copper Basin, Custer County, Idaho: A morphologic and pedologic approach, Unpbl. M.S. Thesis, Lehigh Univ., 104p.

Wigley, W. C., Pasquini, T. A., and Evenson, E. B., 1978, Glacial history of the Copper Basin, Idaho: A pedologic, provenance and morphologic approach, in Mahaney, W. C., ed., Quaternary Soils, Geo. Abstracts Publ., Norwich, England, p. 265-307.

Applied glacial geology

Evaluation of past-producing gold mine properties by drift prospecting: An example from Matachewan, Ontario, Canada

ROBERT A.STEWART*
University of Western Ontario, London, Canada

EDMOND H.VAN HEES
Pamour Porcupine Mines Ltd., Timmins, Ontario, Canada

1 INTRODUCTION

1.1 Location and physiography

The Matachewan mining camp is located in Powell Township, northern Ontario (Fig. 1), about 80 km southeast of Timmins and 50 km west of Kirkland Lake, two other famous gold mining camps. The study area is located at longitude 80° 40' 30", latitude 47° 56' 30" (N.T.S. sheet 41 P/15). The topography is subdued, with relief of roughly 183 m. Elevations vary between 305 m and 488 m above sea level. Rounded, subdued, ice-sculpted, streamlined landforms characterize the landscape. The landscape of the mining camp is heavily forested, and in the study area only about 5% of the land surface is outcrop (van Wiechen, 1981).

1.2 Scope and purpose of the study

The analysis of glacial dispersal of rocks and minerals has long been a popular exploration tool in Fennoscandia (Kauranne, 1976). In recent years, the method has also gained widespread use in Canada. Some studies have concentrated on the distribution of mineralized clasts, such as the sphalerite-bearing boulders at George Lake, Saskatchewan (Karup-Moller and Brummer, 1970). Others have emphasized geochemical dispersal trains, such as those associated with the Kamkotia massive sulfide deposit near Timmins, Ontario (Skinner, 1972). Both geochemical and boulder dispersal trains are common to certain mineral occurrences, as discussed by DiLabio (1981), for the Icon-Sullivan copper deposit near Lac

* Present Address: Department of Earth Sciences, Iowa State University, Ames, Iowa USA

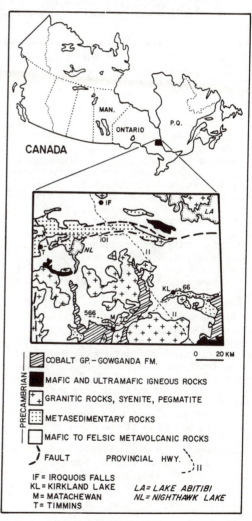

Fig. 1. Location of Matachewan and regional geology (after Ont. Geol. Surv., 1980).

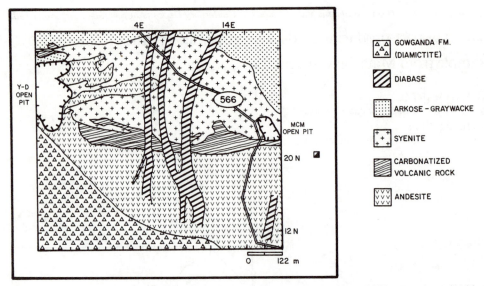

Fig. 2. Bedrock geology of the study area (after van Wiechen, 1981; North and Allen, 1948). Open-pit workings are of the Young-Davidson (Y-D) and Matachewan Consolidated Mine (MCM) properties. Shaft east of line 20N, 20E is the principal access shaft to the MCM underground workings.

Mistassini, Québec, and Szabo et al. (1975) for the Mo-W-Bi-Cu-Sn-Zn mineralization at Mt. Pleasant, New Brunswick.

Most published studies concerning glacial dispersal of ore-bearing bedrock fragments have dealt with drift prospecting techniques in either of two ways. The first is the application of glacial geology as a primary exploration method as one of the first steps in an exploration program (DiLabio, 1981; Karup-Moller and Brummer, 1970). Secondly, drift prospecting has been utilized after the fact of discovery to decipher specific mechanisms of debris entrainment and transport by glaciers (Shilts, 1976; Dreimanis, 1956). In this study, we have used till prospecting methods to re-evaluate glaciated terrain encompassing two former producing gold mine properties, the Matachewan Consolidated (MCM) and Young-Davidson (Y-D) Mines. Our aims were two-fold: first and foremost, a rapid appraisal of this terrain was necessary to ascertain the presence or absence of additional gold mineralization; secondly, an examination of past glacier processes in the mining camp area was undertaken to determine the systematics of debris dispersal as a guide for future exploration in the region.

1.3 History of the mining camp

Gold was first discovered in the Matachewan area in 1916. Actual production began in 1934 at two mines, the Young-Davidson and Matachewan Consolidated. The Young-Davidson ore body was developed primarily as an open pit, which produced 6,128,272 tons of ore yielding 585,690 ounces of gold and 131,989 ounces of silver. The Matachewan Consolidated property began as an underground mine with the development of ore bodies in mafic volcanic rocks. Later in the life of the mine, ore was also extracted from mineralized syenitic rock by open pit methods. Production totaled 3,535,200 tons of ore from which were extracted 370,427 ounces of gold and 133,170 ounces of silver (Lovell, 1967). Both mines closed by 1956 due to escalating costs and a fixed price of gold.

The price of gold had risen significantly by 1979, and production from the two properties was revived when they were acquired by Pamour Porcupine Mines Ltd. in the period June, 1979, to May, 1980. Development has continued since then on an intermittent basis, mainly from small open pits on the Matachewan Consolidated property.

A rapid, effective exploration program became necessary in the event that provincial highway 566 (Fig. 2), overlying a potential ore zone, had to be moved in 1981. To this end, sampling of the surficial sediments was undertaken in conjunction with detailed bedrock mapping of the properties. This paper will concentrate on the results of the former program: positive indications of gold reflected favorably mineralized bedrock, whereas negative evidences of gold were also useful for future considerations in the possible re-positioning of the highway.

2 REGIONAL GEOLOGIC FRAMEWORK

2.1 Bedrock geology

The mine properties are within the world-famous Abitibi greenstone belt of northern Ontario and Québec (Fig. 1). This belt of Archean rocks is in the Superior structural province; the Matachewan district lies in the southwestern portion of the greenstone belt. The reader is referred to the work of Hutchinson et al. (1971) for a discussion of the regional bedrock geology and its relationship to economic mineralization (including gold).

The Matachewan area is the southwestern extension of the Kirkland Lake-Larder Lake "break." The "break" actually comprises south-dipping sedimentary rocks of the Timiskaming series, which overlie ultra-basic to acidic volcanic rocks of the Keewatin series (Thomson, 1948). Sedimentary rocks are mainly graywackes and arkoses, with interbedded conglomerates. Ultra-basic to basic volcanic rocks include talc-chlorite schists, serpentinites, basalts and andesites, with the latter two rocks dominating. Associated with this basic suite of rocks are felsic extrusive rocks, including trachytes, tuffs, stratified tuffs, breccias and agglomerates. Rocks of the two series are commonly intermixed at the eastern end of the belt (Larder Lake area). In the Matachewan district, a clearer relationship is seen: sedimentary strata (graywacke and arkose) underlie volcanic rocks, and all strata dip to the south. The entire volcano-sedimentary belt is complexly faulted (Thomson, 1948, Fig. 1). On a regional basis, most significant ore deposits are located along or to the north of the main Kirkland Lake fault, or break. Flat-lying rocks of the Proterozoic Cobalt Group (mainly diamictite of the Gowganda Formation) unconformably overlie the Archean basement (Lovell, 1967).

2.2 Quaternary geology

The Quaternary geology of the Matachewan area has not been studied before, although numerous reports of the bedrock geology give cursory descriptions of the surficial deposits, which include till, glaciofluvial sediments and post-glacial bogs, swamps and muskeg. Matachewan is to the south of the maximum extent of early Holocene glacial pulses, which include the Cochrane and Rupert advances related to Hudson Bay ice (Prest, 1970; Hillaire-Marcel et al., 1981), and the Hurricana interlobate moraine, which resulted from the combined effects of the Hudson Bay and Nouveau Québec ice domes (Hardy, 1977).

The Matachewan region was within the limits of glacial lakes Barlow and Ojibway (Vincent and Hardy, 1979); however, the characteristic varved sediments deposited in these lakes are not present in the area. Glacier movement in this part of Ontario was generally due north-south (Prest et al., 1968). Where there are no glaciolacustrine sediments, glacial deposits are limited to a single thin till sheet.

3 GEOLOGY OF THE MATACHEWAN MINING CAMP

3.1 Bedrock geology

The following description of lithology and structure of the rocks in the study area is summarized from van Wiechen (1981). Volcanic flows and fragmental rocks are areally most abundant (Fig. 2). Their compositions are chiefly andesitic and basaltic, with subordinate amounts of dacitic and rhyolitic rocks. Occurring as interbeds in these volcanic rocks are layers of "carbonatized volcanics," a chromium-rich carbonate rock containing abundant emerald to apple-green muscovite and darker green chlorite. Carbonate minerals comprise calcite, dolomite, siderite and ankerite. This volcanic assemblage is overlain by sedimentary rocks of the Timiskaming Series, which includes argillite, lithic arkose and graywacke. Both volcanic and sedimentary rocks contain bodies of syenitic, porphyritic rock.

Flat-lying beds of the Proterozoic (Huronian) Gowganda Formation (diamictite) unconformably overlie the Archean rocks. Archean and Proterozoic rocks in the study area are cut by diabase dikes of multiple ages.

Tectonism has deformed and metamorphosed all rocks in the study area. Metamorphism attained greenschist facies; folding is commonly isoclinal with near-vertical axial planes. Beds dip steeply to the south. Small faults and shear zones are common in rocks of the mining area.

3.2 Mineral deposits

In the Young-Davidson Mine, gold mineralization occurs in cherry- and brick-red

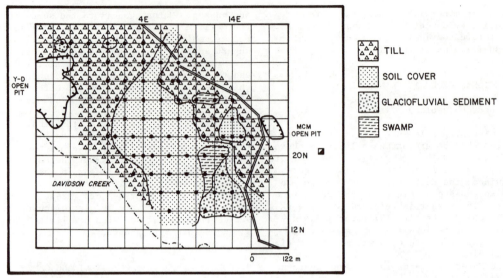

Fig. 3. Surficial geology of the study area, also showing locations of surficial sediment samples (black circles).

phases of syenitic porphyry. The ore body was a tapering cone, which extended downward approximately 305 m (1000 ft). The reddish phases of the syenite contained about 2% disseminated pyrite; 30-50% of the gold occurred as fine-grained inclusions in the pyrite (North and Allen, 1948). Superimposed on the syenitic ore rock are a series of east-west trending hair-like fractures which contain auriferous pyrite and free gold in a quartz matrix. Chalcopyrite is commonly associated with the pyrite throughout the ore, and minor amounts of tourmaline are also associated with the gold.

Two types of ore occur in the Matachewan Consolidated Mine property: irregular ore bodies of limited extent in mafic volcanic rocks and as a large mass in syenitic porphyry at the west end of the property (Derry et al., 1948). The latter rock is similar to the syenite on the Young-Davidson property. Gold occurs in disseminated pyrite and in small fractures as free gold in quartz and also in pyrite. This syenitic rock was interpreted as a felsic crystal tuff by van Wiechen (1981) and a product of submarine volcanism.

The second type of ore is restricted to pyritic mafic volcanic rocks, including the "carbonatized volcanic" rock discussed above. Van Wiechen (1981) interpreted this rock as a waterlain crystal tuff deposited in association with the syenite prophyry. Large fragments of this distinctive rock form the only boulder train

in the study area (see below). Gold occurs in the mafic volcanic rocks mainly in disseminated pyrite and less commonly as coarse, free gold in lenticular quartz veins which may also contain auriferous pyrite and minor amounts of scheelite.

3.3 Quaternary geology

The study area is covered by a thin till sheet which usually ranges in thickness from 1 to 3 m, although a few deep pockets in excess of 5 m were also encountered. Podzolic soils have developed on the till. Where till is thin or absent around bedrock knobs, podzolic soils alone (derived from till) are present (Fig. 3). Glaciofluvial sand and gravel occur in two locations; paleoflow was to the south. Swamps developed in local hollows.

The detailed nature of this prospecting program required that the direction of local ice movement be determined as accurately as possible in the study area. To this end, till clast fabrics and striation orientations were measured wherever practicable (Fig. 4). The till is typical of "Canadian Shield tills" (Scott, 1976), in that it is coarse-grained and sandy, with very low clay contents (Fig. 5).

Till fabrics (azimuth and plunge of elongated clasts) were analyzed numerically by the eigenvalue method of Mark (1974), which computes a "preferred" azimuth and plunge (V_1) for each set of data. The

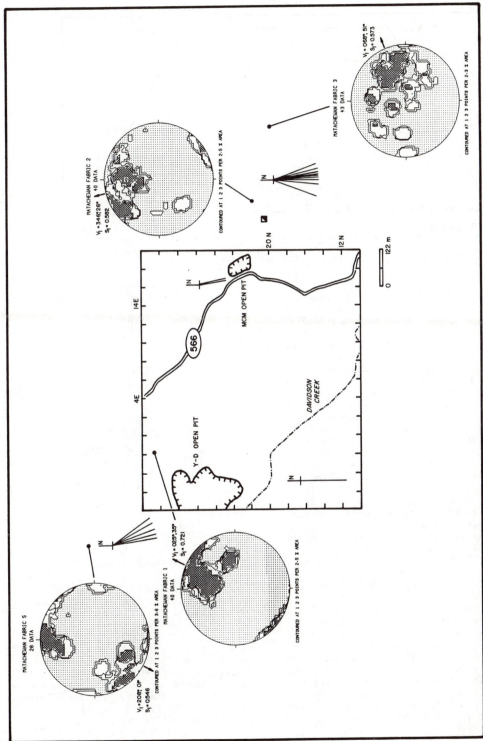

Fig. 4. Plan grid of the study area, showing locations of striation measurements and till clast fabric analyses. Lower hemisphere, equal-area Lambert stereograms contoured after the method of Starkey (1977). Arrows show directions of principal eigenvectors (V1).

183

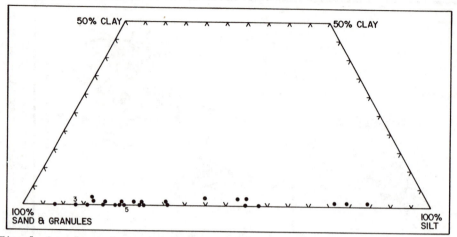

Fig. 5. Granulometric composition of tills from the study area. Sand + granules =
-4 mm, + 0.063 mm; silt = -0.063 mm, + 0.002 mm; clay = -0.002 mm.

clustering of the clasts is moderately
strong to diffuse (Fig. 4). The most
scattered clast fabrics were measured on
the up-glacier side of a bedrock knob,
where transverse and parallel modes were
present (Fig. 4, till fabric 3). Other
fabrics were oriented parallel to ice movement and were measured in till deposited
on level bedrock surfaces. Elongated
clasts oriented parallel to glacier movement commonly showed a northward (up-
glacier) imbrication. This effect has
also been observed in tills forming by
subglacial lodgement (Boulton, 1971).

On a local scale, the orientations of
striations agree with the clustering of
elongate clasts in the till (Fig. 4).
Glacier movement in the mine area, as
indicated by striations, varied between
north-northwest and north-northeast. This
is consistent with striation measurements
obtained from nearby townships (including
Powell Twp.) which show a prominent north-
south maximum (Fig. 6). Only one age of
striations was noted at the various sites
observed.

Few structures were apparent in the till
and were limited to uncommon lenses of
fine sand and silt. The generally strong
parallel alignment and north-imbrication
of clasts in the till, in conjunction with
its very local clast provenance (Fig. 4,
Table 1), suggests that the till derived
primarily from the basal zone of the
glacier. Strong fabric-forming processes
and abundant local debris are to be expected in this environment (Boulton, 1971;
Dreimanis, 1976), from which lodgement,
melt-out, or other varieties of subglacial
till may form. We are confident that the

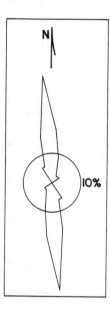

Fig. 6. Rose diagram of regional striation
orientations, including those from the
study area.

data obtained during this investigation
support a subglacial origin for the till,
possibly by lodgement. The lack of definitive structures in the exposures examined
precludes more than this tentative interpretation of a specific genetic process.

Table 1. Pebble lithology analyses at till fabric sites 1 and 2 (see Fig. 4 for locations). Local components derive from within 1-3 km of study area; distal components from distances greater than 3 km.

Local lithologies (range-%)	Local to distal lithologies (range-%)	Distal lithologies (range-%)
Red syenitic rock (12-32)	Diabase (13-18)	Granite (3-5)
Mafic volcanic rock (15-32)		Gneiss (2)
Arkose-graywacke (31-35)		

4 METHODS

A grid was surveyed on the area between the Young-Davidson and Matachewan consolidated properties (Fig. 3), with a sample spacing of 61 m (200 ft), which was adjusted according to local geomorphic impediments, mainly excessively deep swamps. The sampling area is heavily forested by conifers (black spruce, balsam fir, jackpine and cedar) and lesser amounts of deciduous trees (poplar and birch), with abundant thick stands of alder.

Samples of 2-5 kg were obtained by backhoe or with a 2-person gasoline engine auger. At each sample site, an effort was made to penetrate to unweathered till (if present) beneath the modern podzolic soil profile. In some cases where fresh till was not present or could not be sampled, the deepest material encountered was sampled and described sedimentologically. The soil developed on till showed a typical podzolic horizon sequence (described according to the Canadian Soil Survey Committee, 1978): LFH (5 to 10 cm); A (10 cm), Ae (about 10 cm), Bf (30 to 40 cm), Cox or Cg (to bedrock). Sampling depth varied between 30 cm and 1.5 m for holes drilled with the auger and 1.2 to 3 m for pits excavated by backhoe.

From each sample, a 250 g split was taken for heavy mineral analysis of the -2 mm, + 0.250 mm size fraction. A sample of the silt-plus-clay fraction (-0.063 mm) was also recovered for separate analysis. Two successive heavy liquid separations were performed. The first utilized bromoform (S.G. = 2.89) to obtain the total heavy fraction, after which magnetic minerals and rock fragments were removed. This was followed by a methylene iodide (S.G. = 3.32) separation to procure an intermediate fraction (between 3.32 and 2.89) and a "superheavy" fraction (>3.32). The former

was found to contain quartz fragments with free gold by Thomson and Guindon (1979) in their study of tills down-glacier from Kirkland Lake district gold mines, and it was hoped that this method would also prove fruitful in this study.

Gold analysis was performed by fire assay (Pamour Porcupine Mines Ltd. assay laboratory) with subsequent neutron activation analysis (X-Ray Assay Laboratories Ltd., 1980) on fresh till and other surficial sediment samples (-4ϕ (16 mm) and finer) and also on the silt-plus-clay fraction alone ($+4\phi$ (0.063 mm) and finer) of selected till samples. In this way, the gold content of "whole till" samples could also be evaluated with respect to the fine fraction (silt-plus-clay).

5 RESULTS OF FIELD AND LABORATORY INVESTIGATION OF DISPERSAL TRAINS IN THE STUDY AREA

Detailed mapping and subsequent heavy mineral and geochemical analysis revealed the presence of a boulder train with associated mineralogical and geochemical dispersal trains. The extent of each is consistent with the local direction of ice movement. Smaller scale variations in the latter two dispersal fans result in part from postglacial colluvial effects.

5.1 "Carbonatized volcanic" boulder train

A single boulder train is present in the study area (Fig. 7A), comprising large (0.5 to 2 m), angular boulders of the "carbonatized volcanic" rock common in the Matachewan Consolidated Mine workings. The boulders contain abundant apple-green chromium-rich muscovite, calcite and ankerite; the latter weathers to a distinctive

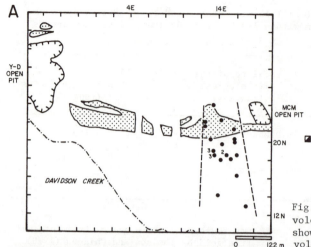

Fig. 7A. Dispersal train of carbonatized
volcanic rock fragments. Stippled area
shows inferred subcrop of carbonatized
volcanic rock.

Fig. 7B. Boulder of carbonatized
volcanic rock. Gray streaks
near lens cap are rusty-colored
weathered ankerite. Lens cap is
6 cm in diameter.

Fig. 7C. View north along line 16E;
axis of boulder dispersal train
roughly parallels the tree line to
the right of the transmission towers.
The maximum extent of the boulder
dispersal train is near the viewer.
Carbonatized volcanic rock subcrops
on the ridge in the distance.

rusty color (Fig. 7B). The boulder train is ribbon-shaped, 122-183 m wide, and approximately 366 m long (Fig. 7A,C). Chip samples were taken from each boulder and assayed. The boulders contain very little gold, with an average of 0.01 troy ounces per ton (opt) for seven samples and trace amounts (<0.01 opt) in the other 15 samples.

The boulders were quite rich in silver, with values varying between 0.02 and 0.32 opt ($\overline{x}$ = 0.07) for 22 samples. In ore mined on the Y-D and MCM properties, this relationship was reversed: gold dominated over silver by a 3:1 ratio (Lovell, 1967). The results of the boulder sampling program were nonetheless successful in that high silver values in a rock type favorable to gold mineralization nearby serve as a good pathfinder for additional gold mineralization (cf. Rose et al., 1979, Chapter 5).

The apex of the boulder train corresponds well to the inferred subcrop extent of the "carbonatized volcanic" (cf. Fig. 2). As only about 5% of the study area is outcrop (van Wiechen, 1981), much of the bedrock geology is inferred and/or projected upward from the known subsurface mine geology. The thin drift cover in the area of the apex (1-2 m) suggests that the source of the boulder train should not be far removed from the apex. The apex region, along with the terrain up-glacier, was scrutinized carefully but without success for outcroppings of source bedrock. It is inferred that the boulders derived from bedrock which is at present drift-covered.

5.2 Mineralogical dispersal train

The only heavy mineral found of significance to the exploration program was pyrite, in the "intermediate" heavy fraction (3.32 > S.G. > 2.89). No free visible gold (in fraction >3.32) or gold in quartz (3.32 > S.G. > 2.89) was observed in any sample. A few grains of chalcopyrite were seen in the intermediate fraction of several of the glaciofluvial sediment samples. The pyrite grains were commonly euhedral and altered to hematite on weathered surfaces. Pyrite grains were found for the most part in arkosic rock fragments, but occasionally some pyritic syenite was observed.

The pyrite follows three trends, of which two are directly related to glacial dispersal (Fig. 8). The first of these is an irregular linear belt, about 244 m long, which is oriented northwest-southeast on the southwest side of the grid. The second trend occurs on the northeast side of the sample grid and has the shape of a short (153 m) blunt lobe. The amount of pyrite increases to the northeast. Samples comprising these two trends include only till or soil derived from till. A third arcuate belt of pyrite is present in the southeast corner of the grid; pyrite content increases to the east. This dispersal train probably includes sediment reworked from pyritic till by glaciofluvial action (cf. Fig. 3).

When the mineralogical (pyritic) dispersal trends (Fig. 8) are compared with local

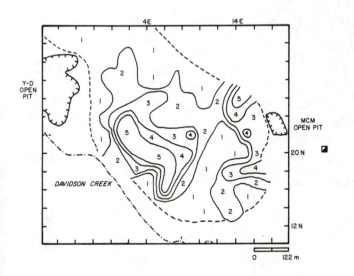

EXPLANATION

Contour	% Pyritic rock fragments
1	0-4
2	5-9
3	10-14
4	15-19
5	20+

Fig. 8. Contour diagram of the percentage pyritic rock fragments in the intermediate heavy fraction (3.32 > S.G. > 2.89) of surficial sediment samples in the study area.

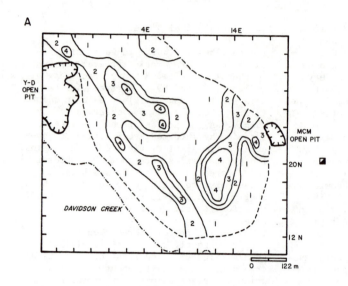

Fig. 9A. Contour diagram of gold in parts per billion (PPB) in surficial sediment samples in the study area.

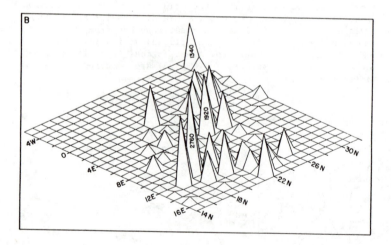

Fig. 9B. Perspective diagram of the gold dispersal trains. East dispersal train is in foreground; selected peak heights are in parts per billion.

directions of ice movement (Fig. 4), a source area near the north end of the study area is suggested. To test this hypothesis, a test pit was excavated over the inferred arkose-syenite contact in the northeast corner of the grid. Due to unusually thick till (5+ m) beyond the capabilities of the backhoe, bedrock was not reached. The till contained numerous clasts of pyritic arkose and graywacke. Closer field reconnaissance of a swamp abutting the test pit disclosed several small outcrops of pyritic arkose. Similarly mineralized rocks are also exposed to the east of the MCM property (van Wiechen, 1981; Fig. 2). The extent of the arkosic sedimentary rocks beneath the glacial cover is presumably extensive, judging by the widespread distribution of such rock fragments in the till.

Pyrite-bearing syenite rock particles were prevalent in the light (S.G. < 2.89) fraction of most till samples but were not counted separately. The presence of the syenite in the till was still of immediate interest, as it confirmed the continuity of this rock type between the two mine properties, even though bedrock was totally obscured in this area by glacial sediments. Pyrite apparently had little effect on the specific gravity of most syenitic rock fragments, probably owing to its thorough dissemination and relative scarcity (2%) in that particular rock.

5.3 Dispersal of gold in the surficial sediments

The results of gold analyses on surficial sediments are shown in Fig. 9. Two gold dispersal trains are present. "Anomalous" gold values were defined as those in excess of 200 parts per billion (ppb). This particular threshold was chosen as it roughly corresponds to 0.01 troy ounces gold per short ton, and gold contents below this level were reported as "trace" to "nil" by the company assay laboratory. Gold and silver analysis by fire assay alone is amenable to very rapid, mass-processing of large batches of samples. The additional accuracy gained by neutron activation analysis may often be an unaffordable luxury in terms of time and money for an exploration program. Therefore, for prospecting purposes in areas of known or suspected gold-silver mineralization such as Matachewan, values of "trace" and "nil" have little prospecting significance.

The west dispersal train is ribbon-shaped, 793 m long, and trends northwest to southeast (Fig. 9A,B). Gold values vary from 210 ppb to 1920 ppb (0.056 opt). The latter is significant in that this level was close to typical cut-off grades for local mining operations at the time of writing. Peak gold values in this dispersal train were obtained from till and till derived soil overlying and down-glacier from syenite and carbonatized volcanic rock.

The east dispersal train is tongue-shaped and trends to the south 354 m from an area adjacent to the MCM open pit. Gold contents vary between 210 ppb and 2760 ppb (0.083 opt). The latter value was "ore grade" at the time of writing. High gold values are found in glacial drift covering syenite and carbonatized volcanic rock. Peak values were found in organic-rich swamp sediment just to the south of the carbonatized volcanics (cf. Fig. 3). These rocks subcrop at the crest of a hill (cf. Fig. 2 and 7A); moreover, the swamp occurs down-glacier from this ridge at the base of the hill. Apparently auriferous drift, deriving from carbonatized volcanic and syenitic rocks, was initially deposited on the hill slope and base. Post-glacial swamp development and groundwater flow seem to have enriched the gold content of the paludal sediments in this small basin. Reconcentration of metals, especially in the presence of organic matter as in this study, is common and effective in this environment (Rose et al., 1979, Chapters 8 and 14).

The modes in which gold was dispersed compare favorably to the distribution of pyritic rock fragments as well (Fig. 8) in terms of the directions of glacial transport and sense of elongation of the mineralogical and geochemical dispersal trains. Two source areas for the gold are suggested, on the east and west sides of the northern part of the sample grid. These results confirmed the continuity of gold mineralization in the drift-covered outcrops of syenitic and carbonatized volcanic rocks between the two mine sites, as hoped in the initial planning of this study. Pyritic rock fragments derived principally from the sedimentary assemblage (arkose and graywacke), which is essentially barren of gold, with only minor contributions from the two volcanic host rocks.

5.4 Distribution of gold according to granulometry of the till

Previous considerations of grain size and pebble provenance of the till demonstrated that it was dominated by coarsely comminuted, local bedrock. In an effort to determine any effect that gross grain size

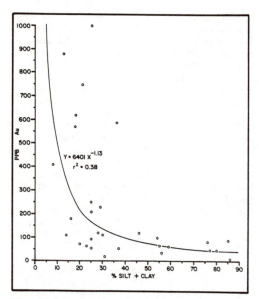

Fig. 10. Variation of gold concentrations (PPB) with respect to silt-plus-clay contents of fresh till samples.

characteristics may have exerted on the distribution of gold in till, the silt-plus-clay (-0.063 mm) fraction of 30 samples of fresh till (from the C horizon) was analyzed by the same method as the "bulk" samples. The results are presented in Fig. 10.

The gold content of fresh till has a general tendency to increase sharply with decreasing amounts of silt-plus-clay. A power curve has been fitted to these data to quantify the trend, although the low correlation coefficient ($r^2 = 0.38$) indicates that this trend is not particularly strong.

The adsorptive capacity of the fine fraction is probably low, due to the lack of scavenging agents in the clay (-0.002 mm) fraction, such as organic matter and clay minerals. Most gold in this size range is likely found in finely comminuted mineralized grains in the silt fraction (cf. Fig. 5). Individual auriferous rock and mineral fragments in the sand and granule range account for the balance of the gold in the samples analyzed.

6 ECONOMIC IMPLICATIONS OF THE BOULDER, MINERALOGICAL AND GEOCHEMICAL DISPERSAL TRAINS

Peak gold values in the east and west dispersal trains (Fig. 9A,B) were, respec-

tively, about 9X and 13X the assigned (albeit somewhat arbitrary) background. Shilts (1976) suggested that similar ratios, obtained from dispersal trains elsewhere in glaciated parts of Canada, indicated close proximity to the source, usually within a few hundreds of metres. Based on these data, the nature of the carbonatized-volcanic boulder train, and the ubiquity of syenite fragments in the light fraction of the surficial sediments, it was decided to drill-test the inferred subcrop of the syenite and carbonatized volcanic rock in the northern portion of the grid.

A percussion and diamond-drilling program was implemented to intercept the syenite at surface to shallow depths, whereas the carbonatized volcanic rock was penetrated only on the subsurface. To date, ore was proved to a depth of about 31 m in the cherry- and brick-red phase of the syenite between the two mine properties. Even more encouraging were assays as high as 0.2 opt Au obtained from the carbonatized volcanic rock at depth. The hypothesis that the high silver contents of chip samples from the boulder train were potential indicators of gold mineralization elsewhere in that rock unit would appear to be valid.

An important facet of this study is that any gold mineralization in subcropping carbonatized volcanic rock was reflected mainly as a "shadow anomaly" in the east and west geochemical dispersal trains, on the down-glacier side of similarly mineralized syenite. The boulder train indicated no significant gold mineralization in the near-surface outcrop by the east dispersal train, although trace amounts of gold in the carbonatized volcanic rocks may have contributed to both east and west dispersal trains. Clearly, where multiple sources for dispersal trains are thought to exist, each should be tested, as definitive dispersion by glaciers from each discrete source may not have occurred.

7 GLACIAL HISTORY OF THE YOUNG-DAVIDSON AND MATACHEWAN CONSOLIDATED MINES AREA

The dispersal trains discussed in this article are short (168–793 m) and narrow (61–152 m); furthermore, their size and genesis can be related to specific processes at work during the glaciation of the area. Our hypothetical model of glacial erosion and deposition in the study area is depicted in Fig. 11. There is a clear analogy, relative to form and process, which can be made between the geomorphology of this landscape and a rôche moutonée (Figs. 7 and 11). The stoss side of the

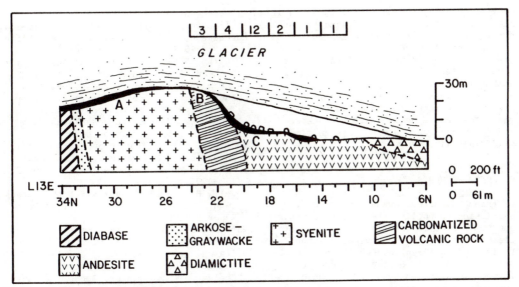

Fig. 11. Schematic longitudinal cross-section through the study area along line 13E (north to left), during active glaciation. Lodgement till is deposited on the stoss side of the hill at A, quarrying occurs at B, deposition of lee-side till in subglacial cavity at C. Upper scale shows number of boulders carbonatized volcanic rock per 61 m east-west along line 13E (compare with Fig. 7A).

hill was eroded and striated by the glacier, which deposited a thin layer of till from its basal debris zone on the bedrock surface. This interpretation of the till genesis is based on the observed strong clast alignments, imbricated to the north and parallel to striation orientations, and the abundant, locally-derived rock fragments (Figs. 4, 8; Table 1).

We suggest that the boulder, mineralogical and geochemical dispersal trains originated subglacially, with respective apexes forming in subglacially deposited (lodgement?) till at the north end of the study area (Fig. 11, A). The boulder train of carbonatized volcanic rock developed in response to quarrying at the ridgecrest (Fig. 11, B); boulders were transported only a short distance in the glacier sole or tumbled down the slope into the subglacial cavity. The latter process seems to have dominated, as most boulders occur at the foot of the hillslope (Fig. 11). The bodies and tails of the mineralogical and geochemical dispersal trains formed in lee-side till (Fig. 11, C) as debris was deposited from beneath the glacier sole and slumped into the subglacial cavity (cf. Boulton, 1971, Fig. 12; Sugden and John, 1976, Fig. 11.3 and discussion).

Debris slumping into a subglacial cavity on the lee-side of a bedrock knob moves only a short distance, as observed by

Boulton (1971). A similarly short transport distance is exhibited by the three dispersal trains in our study. The down-glacier limits are restricted to the lee-side of the hill, with all three dispersal trains terminating within 366 m of the ridgecrest.

8 CONCLUSIONS

1. Detailed geomorphic analyses, such as in the Matachewan area, can be instrumental in predicting the source and extent of dispersal trains.
2. Based on geomorphological considerations and the nature of three distinct dispersal trains, a favorably mineralized source area was confirmed between the two mine properties.
3. Where multiple rock units in stratigraphic contact are economically mineralized, those in the down-glacier direction from the first such unit will contribute material to geochemical dispersal trains as "shadow anomalies," if the type of mineralization (Au, Ag, etc.) is the same. If this phenomenon is not anticipated, potential targets may be overlooked even on a detailed scale.
4. The geochemical dispersion of gold in the surficial sediments depended on the disseminated auriferous pyrite and gold contained in the syenitic rock fragments.

191

Heavy liquid separation recovered an insignificant fraction of these fragments, as pyrite and gold were apparently too sparse to substantially increase the specific gravity of the syenite to values greater than 2.89 (bromoform). For this reason, bulk till samples were more useful for gold determinations. The source rocks for most pyritic rock fragments recovered in the intermediate heavy fraction (arkose and graywacke) were barren of significant gold mineralization.

9 ACKNOWLEDGMENTS

We thank Pamour Porcupine Mines Ltd. for permission to publish the results of this study. The style and clarity of the article benefited from the reviews of John Lemish and Kathleen Galloway (ISU). Granulometric analysis of the till was performed by Karin Lootsma (UWO); financial support for this aspect of the study was provided by NSERC grant A-4215 to A. Dreimanis (UWO).

10 REFERENCES

Boulton, G. S., 1971, Till Genesis and Fabric in Svalbard, Spitsbergen, in: Till-A Symposium, R. P. Goldthwait, ed., Ohio State Univ. Press, 41-72, Columbus, Ohio.

Canada Soil Survey Committee, 1978, The Canadian System of Soil Classification: Agriculture Can. Publication 1646, Ottawa.

Derry, D. R., Hopper, C. H. and McGowan, H. S., 1948, Matachewan Consolidated Mine, in: Structural Geology of Canadian Ore Deposits, Can. Inst. Mining and Metallurgy, 638-643.

DiLabio, R. N. W., 1981, Glacial dispersal of rocks and minerals at the south end of Lac Mistassini, Quebec, with special reference to the Icon dispersal train: Geol. Surv. Can. Bull. 323, 46 p.

Dreimanis, A., 1956, Steep Rock iron ore boulder train: Geological Assoc. of Can., Proceedings, Part I, 27-70.

Dreimanis, A., 1976, Tills-their origin and properties, in: Glacial Till, R. F. Legget, ed., Royal Soc. Can. Spec. Pub. 12, 11-49, Ottawa.

Hardy, L., 1977, La déglaciation et les épisodes lacustres et marin sur le versant québécois des basses terres de la baie de James: Géographie physique et quaternaire, v. 31, 261-273.

Hillaire-Marcel, C., Occhietti, S. and Vincent, J-S., 1981, Sakami moraine,

Quebec: A 500-km-long moraine without climatic control: Geology, v. 9, 210-214.

Hutchinson, R. W., Ridler, R. H. and Suffel, G. G., 1971, Metallogenic relationships in the Abitibi Belt, Canada: A model for Archean metallogeny: Canadian Inst. Mining and Metallurgy Bull., v. 64, 48-57.

Karup-Moller, S., and Brummer, J. J., 1970, The George Lake Zinc Deposit, Wollaston Lake Area, Northeastern Saskatchewan: Economic Geology, v. 65, 862-874.

Kauranne, L. K., 1976, Conceptual models in exploration geochemistry, Norden, 1975: Jour. Geochem. Exploration, v. 5, 173-420.

Lovell, H. L., 1967, Geology of the Matachewan Area: Ontario Dept. of Mines Geol. Report 51, 61 p.

Mark, D. M., 1974, On the interpretation of till fabrics: Geology, v. 2, 101-104.

North, H. H., and Allen, C. C., 1948, Young-Davidson Mine, in: Structural Geology of Canadian Ore Deposits, Canadian Inst. Mining and Metallurgy, 633-637.

Ontario Geological Survey, 1980, Geological highway map, Northern Ontario: Ontario Geological Survey, Map 2440.

Prest, V. K., 1970, Quaternary Geology of Canada, in: Geology and Economic Minerals of Canada, R. J. W. Douglas, ed., Geol. Survey Can. Econ. Geol. Report 1, 676-764.

Prest, V. K., et al., 1968, Glacial map of Canada, scale 1:5 million: Geol. Survey Can. Map 1253A.

Rose, A. W., Hawkes, H. E. and Webb, J. S., 1979, Geochemistry in Mineral Exploration 2nd edition: Academic Press, New York, 657 p.

Scott, J. S., 1976, Geology of Canadian Tills, in: Glacial Till, R. F. Legget, ed., Royal Soc. Can. Spec. Pub. 12, 50-66, Ottawa.

Shilts, W. W., 1976, Glacial Till and Mineral Exploration, in: Glacial Till, R. F. Legget, ed., Royal Soc. Can. Spec. Pub. 12, 205-224, Ottawa.

Skinner, R. G., 1972, Drift prospecting in the Abitibi Clay Belt: Geol. Survey Can. Open File Report 116, 27 p.

Starkey, J., 1977, The contouring of orientation data represented in spherical projection: Can. Jour. Earth Sci., v. 14, 268-277.

Sugden, D. E., and John, B. S., 1976, Glaciers and Landscape: Edward Arnold Ltd., London, 376 p.

Szabo, N. L., Govett, G. J. and Lajtai, E. Z., 1975, Dispersion trends of elements and indicator pebbles in glacial till around Mt. Pleasant, New Brunswick, Canada: Can. Jour. Earth Sci., v. 12, 1534-1557.

Thomson, J. E., 1948, Regional structure
of the Kirkland Lake-Larder Lake Area,
in: Structural Geology of Canadian Ore
Deposits, Can. Inst. Mining and Metallurgy, 627-632.

Thomson, I., and Guindon, D., 1979, A geo-
chemical reconnaissance basal till survey
and related research in the Kirkland
Lake area, District of Timiskaming:
Ontario Geol. Survey Current Res., 1979,
Special Project S14, 171-174.

van Wiechen, A., 1981, Felsic porphyritic
rocks and their role in the formation of
gold deposits at the Pamour Matachewan
No. 4 open pit mine, Matachewan, Ontario:
Unpub. B.Sc. Thesis, Dept. of Geology,
University of Western Ontario.

Vincent, J-S., and Hardy, L., 1979, The
evolution of glacial lakes Barlow and
Ojibway in Quebec and Ontario: Geol.
Survey Can. Bull. 316, 18 p.

X-Ray Assay Laboratories Ltd., 1980, Neutron Activation Analysis: Unpub. manuscript, XRAL Ltd., 1885 Leslie St., Don
Mills, Ontario, Canada.

Active alpine glaciers as a tool for bedrock mapping and mineral exploration: A case study from Trident Glacier, Alaska

GEORGE C.STEPHENS
George Washington University, Washington, DC, USA

EDWARD B.EVENSON
Lehigh University, Bethlehem, PA, USA

RICHARD B.TRIPP & DAVID DETRA
US Geological Survey, Golden, CO, USA

1 INTRODUCTION

Although there have been many advances in geochemical exploration using ground moraine from continental ice sheets (for example, Bradshaw, 1975; Shilts, 1976; Kujansuu, 1976; and Levinson, 1980), little has been written on geochemical exploration in alpine-glaciated regions. Conventional geochemical exploration in rugged actively-glaciated terrains such as the Canadian Coast Range, the Chugach Mountains, the Alaska Range, and the high Andes is largely impossible. Lack of streams and soil development in high alpine regions prevent the collection of adequate stream sediment or soil samples necessary in standard geochemical exploration programs. Bedrock geochemistry is difficult, expensive, and time-consuming because of rugged, often inaccessible terrain coupled with widespread snow and ice cover at higher elevations. This paper describes a new geochemical sampling method for such regions and examines the feasibility of using medial moraine systems to broadly characterize the bedrock geology of an alpine glacier catchment area.

We have previously proposed using medial moraines in complex alpine glacier systems as geochemical sampling sites, and have demonstrated the usefulness of alpine moraines in determining bedrock-source regions for morainal debris (Evenson and Stephens, 1978; Evenson, Pasquini, Stewart, and Stephens, 1979). Because the debris in medial moraines is derived either subglacially or supraglacially from valley walls adjacent to the ice, it should accurately reflect the geochemical and lithologic composition of the bedrock from which it is derived. Further, because glacier movement is by laminar flow (Figure 1), no mixing or dilution occurs between adjacent moraines

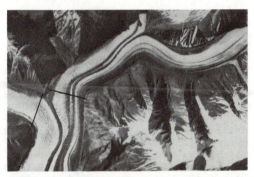

Figure 1. The central portion of the Trident Glacier, showing the well-developed medial moraine system and location of traverse.

except by melting or ablation of the ice surface with resultant flowage of surface materials. Therefore each stripe has its own unique lithologic and geochemical composition. The bedrock source of each moraine can be determined by aerial photo analysis. Systematic collection and analysis of samples from these moraines can be used to characterize the lithology and geochemistry of the glacier catchment area.

In addition, field and laboratory examination of the pebble, cobble, and boulder fractions in each medial moraine yields information about lithologic distribution, minor structural features, visible mineralization and associated hydrothermal alteration, and ore genesis within the area. This information, coupled with conventional aerial photo interpretation techniques, allows the compilation of a generalized, moraine-based geologic map of the area.

This paper presents the results of the first field test of our method on the Trident Glacier in the central Alaska Range, Alaska (Figure 2). The Trident Glac-

Figure 2. Index map of Alaska, showing the location of the Alaska Range and the Trident Glacier study area.

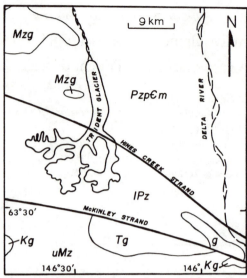

Figure 3. Geologic map of the Trident Glacier area. PzpЄm = pre-Cambrian/Lower Paleozoic metamorphic rocks, lPz = Lower Paleozoic metamorphic rocks, uMz = Upper Mesozoic sedimentary rocks, Mzg, Kg, Tg, and g = granitic intrusive rocks (after Beikman, 1974).

ier is located approximately 350 km northeast of Anchorage and 25 km west of the Delta River on the northern side of the Alaska Range. The Trident Glacier is a large complex system consisting of three major branches which coalesce to form one major alpine glacier. The three branches of the Trident are, in turn, fed by thirty smaller tributary ice streams. As these individual ice streams merge, medial moraines are developed between adjacent ice masses.

Little has been published about the bedrock geology in the vicinity of the Trident Glacier: the following summary is based largely on maps by H.M. Beikman (1974, 1980). Two branches of the Denali Fault, the Hines Creek Strand and the McKinley Strand, occur in the vicinity of Trident Glacier. The McKinley Strand is marked by a topographically low, linear, zone which is occupied by the Black Rapids and Susitna Glaciers directly to the south of the Trident Glacier. The McKinley Strand delineates the boundary between Lower Paleozoic rocks to the north and Upper Mesozoic sedimentary rocks to the south. The catchment of the Trident Glacier is composed chiefly of regionally metamorphosed sedimentary rocks of Lower Paleozoic age (principally the Jarvis Creek Schist), intruded by a series of Mesozoic to Tertiary igneous rocks (Figure 3).

2 SAMPLE COLLECTION AND ANALYTICAL METHODS

A single traverse was made across the glacier to collect geochemical samples and to examine boulders and cobbles in the moraines for lithologic and structural information. The traverse location is constrained by a number of factors:

1. the traverse must be below the junction of the lowest ice tributary,

2. it must also be below the snow line so as to provide maximum exposure of the moraines,

3. the moraines should be sampled as far up-ice as possible to minimize mixing due to melting and ablation, and

4. the traverse location should provide easy, rapid, and safe access to the moraines in an area relatively free of crevasses, snow bridges, and icefalls. The traverse on the Trident Glacier was located as nearly as possible to meet these constraints (Figure 4).

Each of the thirty medial moraines was examined and classified as single or compound; that is, either the moraine was uniform throughout with respect to lithologic variation, or was laterally variable due to material from two fundamentally different bedrock sources (Figure 5). The compound moraines were divided into right- and left-hand sample sites, with the "handedness" based on viewing the moraine in the up-ice direction.

The following types of samples and data were collected at each sample locality.

1. Fifty boulders (25 cm or larger in diameter) were examined; the rock types present, their relative percentages, hydrothermal alteration effects, visible mineralization, and structural features (small folds and faults, veins, cross-cutting relationships, etc.) were recorded.

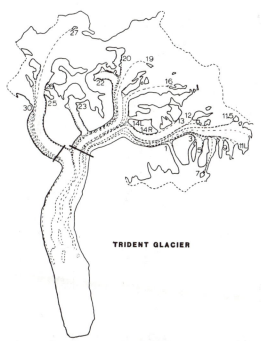

TRIDENT GLACIER

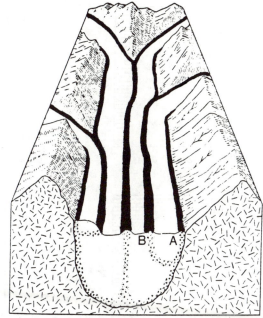

Figure 4. Map of the Trident Glacier, showing location of traverse (heavy lines) and numbers of selected moraines.

Figure 5. Cross-section sketch of a compound alpine glacier. A = simple medial moraine, B = compound medial moraine (after Sharp, 1960).

2. Fifty cobbles (10-25 cm in diameter) were also examined as described above.

3. Fifty pebbles (1-4 cm in diameter) were collected and bagged for archival purposes and laboratory study.

4. Two fine-grained sand, silt, and clay samples were collected, where possible, from each site; one sample of supraglacial debris collected from a surface stream on the ice, and a second sample of englacial melt-out debris (Figure 6). The heavy mineral (specific gravity greater than 2.85) 2.5 to 4.0 phi size fractions of these samples were concentrated by bromoform separation. A "non-magnetic" fraction of the heavy mineral concentrates was separated using a Frantz isodynamic magnetic separator at a setting of 1.0 ampere. Multi-element analyses by semi-quantitative emission spectroscopy were performed on the non-magnetic heavy mineral concentrates by the U.S. Geological Survey, Branch of Exploration Research.

Eleven elements (Ag, As, Cr, Cu, Mo, Ni, Pb, Sb, Sn, W, and Zn) representative of ore mineral assemblages were selected for study. Because the total number of samples is small, conventional statistical methods for determining anomalous geochemical values were not used. The concentrations of each element (in parts per million) for each

Figure 6. Medial moraine showing accumulation of englacial melt-out debris on top of ice.

moraine were plotted in increasing order and a curve was fitted to each data set. Anomalously high values were defined as those lying above the abrupt break in slope

197

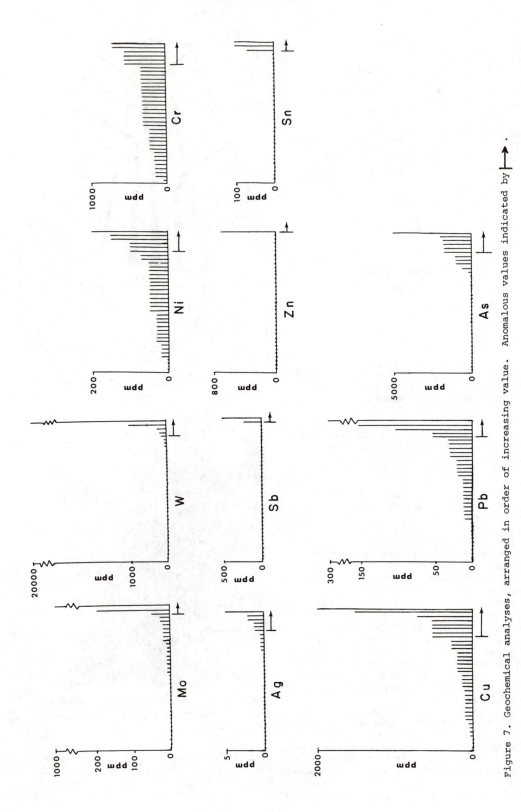

Figure 7. Geochemical analyses, arranged in order of increasing value. Anomalous values indicated by ↑.

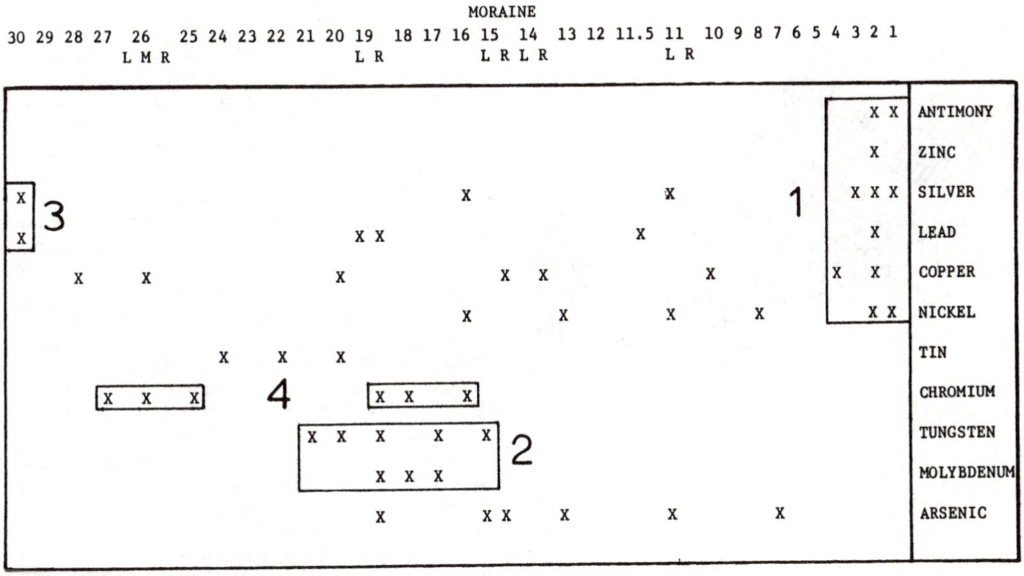

Figure 8. Geochemical anomalies of the Trident Glacier catchment: 1 = multi-element anomaly, 2 = tungsten-molybdenum anomaly, 3 = silver-lead anomaly, and 4 = chromium anomaly. L = left side, M = middle, R = right side.

in the curve (Figure 7). Geochemical results from the surface stream sediment samples are similar in terms of anomaly definition to those from the englacial melt-out samples. For this reason, only the results for the englacial melt-out samples are discussed in this paper. The geochemical anomalies are summarized in Figure 8.

Each of the sampled moraines was traced to its bedrock source using coventional black and white aerial photos. Because several of the moraines disappear beneath new snow or firn at high elevations, a one-to-one correlation between moraines and source areas is not always possible; however enough moraines can be traced directly to their sources to bracket and limit the extent of uncertainty of the other correlations.

3 GEOLOGICAL AND GEOCHEMICAL RESULTS

Based on the cobble and boulder data, several maps were produced. Because these maps are based on relative percentages of various rock types within the moraines, the units shown are not conventional geologic formations, but rather are "terrane units". Figure 9 illustrates the distribution of major rock types and terrane units within the area. No sedimentary rocks were found in any of the moraines. Morainal debris is composed of regional and contact metamorphic rocks and hypabyssal to plutonic igne-

ous rocks. The terrane units shown on Figure 9 are based on the igneous/metamorphic ratio in the moraines. From these data we infer the presence of a large intrusive body at the western edge of the area, and another at the southeastern edge. The remainder of the area is dominated by metamorphic rock units with lesser contributions from small, isolated, igneous bodies.

The relative percentages of rock types presented in Figures 9 through 13 were determined by combining the sample counts from both the boulder and cobble fractions. For most moraines, the compositions of the boulder and the cobble fractions differ significantly from one another. Because some rock types, such as phyllite, schist, and marble, disintegrate easily they are more abundant in the smaller size cobbles than in the boulders, whereas more-resistant rock types such as gabbro, granodiorite, and granitic gneiss are more abundant in the boulder fraction. Some rock types, such as vein quartz and aplite (dikes) tend to be sparsely represented in the boulder fraction because of the thin tabular shape of the original bodies from which they were derived.

3.1 Lithologies

The catchment area is dominated by thinly layered, foliated, light to dark gray, phyllitic schist and tan metaquartzite.

199

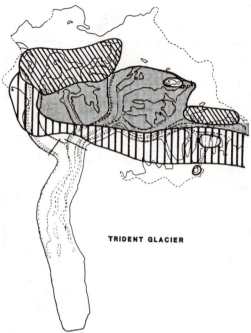

TRIDENT GLACIER

Figure 9. Terrane units of the Trident
Glacier catchment area. Diagonally lined
area = >90% intrusive, stippled area =
mixed igneous and metamorphic terrane
(>60% intrusive), vertically-ruled area =
mixed igneous and metamorphic terrane
(<40% intrusive), unpatterned area = >90%
metamorphic.

Lesser amounts of marble, calc-silicate,
and metasiltstone are also present. Minor
amounts of amphibolite are found in moraines 19R, 19L, 24, and 25. Moraines 6 to
9 and 16 are composed entirely of metamorphic rocks, whereas moraines 10 to 11L and
30 are of mixed metamorphic and igneous
origin (metamorphic rocks >60%). Moraine
30 contains a distinctly different suite
of metamorphic rocks than the other moraines. It consists of light greenish gray
quartzite and mica schist, metaconglomerate,
and dark gray limestone.

The igneous rocks of the area comprise
two major plutonic centers plus several
smaller plutonic bodies, as well as lenses
and dikes of hypabyssal rocks. The westernmost plutonic body is centered on moraines
11.5 to 13, which contain over 90% igneous
cobbles and boulders. The plutonic rocks at
this locality are massive to foliated, medium grained, equigranular, biotite-bearing
monzonites and granodiorites, some containing minor amounts of hornblende. The feldspars and mafic minerals in these felsic
rocks are unaltered and display no evidence

of hydrothermal alteration.

The southeastern-most plutonic complex
includes moraines 20 through 22 and 26
through 29. This is a mixed igneous terrane
consisting of medium grained, foliated
monzonites (5-30%), fine grained, equigranular dioritic rocks (20-50%), and massive,
medium to coarse grained, equigranular
gabbro (5-25%). Quartz-feldspar pegmatites
are also most abundant within this area
(3-11%).

Moraines 14R to 15L, 17 to 19L, and 23
through 25 are derived from composite igneous and metamorphic sources with greater
than 60% igneous contribution. Minor aplite
(Figure 10), quartz-feldspar pegmatite
(Figure 11), and fine grained, biotite-
bearing quartz diorite dikes(?) occur
throughout this area. Quartz veins (Figure
12) cross-cut metamorphic fabrics in boulders and were presumably derived from, and
emplaced near, the edges of the major plutonic bodies. No evidence of hydrothermal
alteration was found in any of the moraines.

3.2 Structural Information

In addition to lithologic differences between the cobble and boulder size fractions,
structural information is more commonly
preserved in the larger blocks of material.
Minor folds and planar features such as
joints, veins, small faults, and dikes are
commonly seen in boulders. The fold style
as well as the types and relative abundance
of planar features within the catchment can
be inferred from the boulders. In addition,
cross-cutting relationships (for example,
veins and dikes) or the nature of igneous
contacts (concordant or discordant) can
commonly be observed and yield valuable
information about the stratigraphic and
structural history of the area. For example, within the Trident Glacier area,
numerous, tight, isoclinal folds are found
in the boulder fraction of the moraines.
Quartz veins are sparse, and aplite dikes
are generally narrow (2-6 cm). All contacts
observed in the boulders between the foliated metamorphic rocks and the plutonic
rocks were sharp and discordant.

This structural information was integrated
with data from conventional aerial photo
interpretation. Erosional characteristics
of the metamorphic and plutonic rocks aided
in determining the locations of major lithologic contacts (Figures 14 and 15). Aerial
photo analysis also aided in determining
the strike direction of the regional foliation and fold orientations in the metamorphic terranes.

Using these techniques, we have determined that the Trident Glacier catchment

200

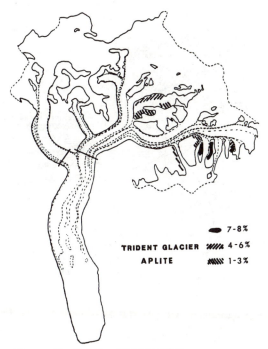

Figure 10. Aplite distribution.

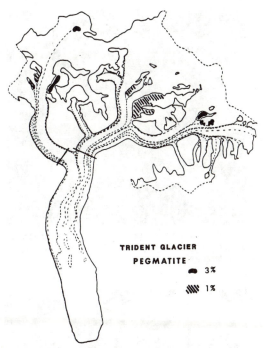

Figure 11. Quartz-feldspar pegmatite distribution.

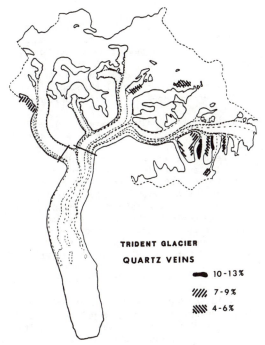

Figure 12. Quartz vein distribution.

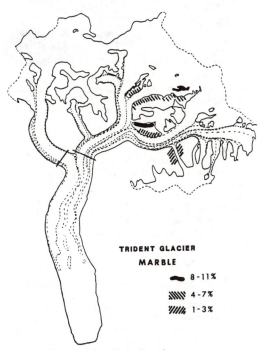

Figure 13. Marble distribution.

Figure 14. Aerial photo showing characteristic craggy topography developed on igneous terranes.

Figure 15. Aerial photo showing control of
erosional features by foliation in metasedimentary terranes.

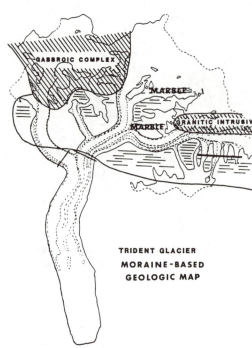

TRIDENT GLACIER
MORAINE-BASED
GEOLOGIC MAP

Figure 16. Moraine-based geologic map.
Diagonally-ruled area = plutonic rocks.
Foliation in metamorphic rocks shown by
short east-west lines. Trace of east-west
trending fault shown by heavy line.

area is composed of regionally metamorphosed schists and phyllites with minor
interlayered quartzite and marble, and dis-
cordantly intruded by two major igneous
bodies (Figure 16). The metamorphic rocks
are well-foliated and isoclinally folded
around east-west trending, gently-plunging
fold axes. Axial planes of the folds, as
well as the major foliation, dip steeply to
the north or south.

3.3 Geochemistry and Visible Mineralization

Development of efficient geochemical exploration methods for metallic mineral resources is one of the major goals of this
study. After determining anomalous geochemical values for the eleven selected elements (Figure 8), these values were plotted

on a map of the glacier catchment (Figure
17). Four major types of anomalies are
present:
     1. a multi-element (Cu, Ag, Pb, Zn, Sb,
Ni) anomaly in moraines 1 through 4;
     2. a strong Mo-W anomaly in moraines 15L
through 21;
     3. a Ag-Pb anomaly in moraine 30; and
     4. a pair of Cr anomalies in moraines
16 through 19R and 25 through 27.
Of these four anomaly groups only the
chromium anomaly is believed to reflect
lithologic variation within the catchment
in that the chromium anomaly is spatially
related to a large gabbroic intrusive and
probably reflects the relatively high
chromium content of these rocks. The other
three anomaly groups represent valid exploration targets, although the Ag-Pb
anomaly in moraine 30 is of low priority
because it is based on values from only
one moraine.
     Field and laboratory examination of visi-
ble sulfides was used to supplement the
geochemical results. In the field, visible
mineralization was classified as: vein-type,
disseminated, or stratiform, leading to a
genetic model for mineralization within the

202

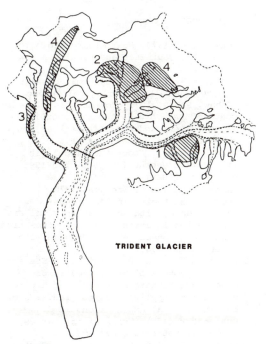

Figure 17. Location of geochemical anomalies shown in Figure 8.

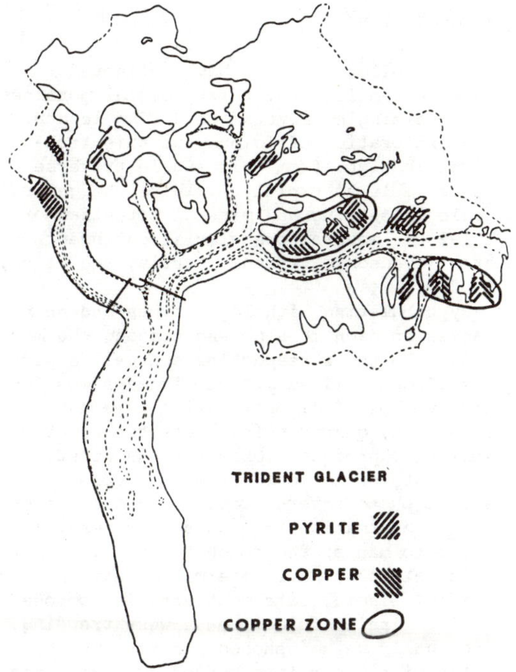

Figure 18. Location of visible pyrite and
copper-bearing minerals.

area. Figure 18 presents the data from this
phase of our study.

No visible tungsten or molybdenum mineralization was seen in the field although
scheelite and powellite have subsequently
been identified by microscopic and ultraviolet examination of the non-magnetic,
heavy-mineral fraction. Visible pyrite is
ubiquitous in small amounts within the area.
Most of the pyrite is disseminated or occurs
as thin (0.5-1.0 mm) veinlets within dark
gray to black phyllite. Much of this pyrite
is directly related to metamorphism of the
host rocks and is not related to potential
ore-bearing systems.

In moraines 8L, 11L, 13, and 14R, significant amounts of copper mineralization
were encountered. The copper occurs as
chalcopyrite, bornite, azurite, or mala-
chite. Chalcopyrite is finely disseminated
within the metasedimentary rocks or occurs
in veinlets, commonly associated with pyrite and quartz. One large, eighteen kilo-
gram, massive-sulfide cobble (15x20 cm) was
found on moraine 11L and is composed chiefly of pyrite with lesser amounts of chalcopyrite, bornite, arsenopyrite(?), and less
than five percent interstitial quartz.

By combining the geochemical data with
the visible-sulfide data, two exploration
targets are outlined. One is the multi-
element copper-lead-zinc-silver area in

the westernmost major tributary (moraines
1-4), and the other is the tungsten-molybdenum anomaly in moraines 15L-21.

3.4 Mineral Genesis

To construct genetic models for possible
mineral deposits, the location of the geochemical anomalies and the visible-sulfide
occurences were combined with information
from the moraine-based geologic map (Figure 16). The copper mineralization can best
be explained as a product of a hydrothermal
vein system which is genetically related to
the westernmost monzonite-granodiorite intrusive complex. Based on composition,
texture, and lack of widespread hydrothermal alteration effects in the plutonic
rocks, there seems to be little likelihood
of a well-developed porphyry copper system.
Likewise, we do not believe that the massive-sulfide cobble from moraine 11L represents a conventional volcanogenic massive-
sulfide deposit, but rather is a fragment
of a sulfide-rich, gangue-free vein located
near the periphery of the plutonic complex.

The tungsten-molybdenum anomaly is probably a skarn deposit associated spatially
with the southeasternmost plutonic complex
of gabbro, diorite, and monzonite.

4 CONCLUSIONS

In this pilot study we have collected moraine samples from thirty medial moraines along a single traverse of 3.2 km length. The exploration efficiency of this technique is high: from this single traverse we are able to rapidly evaluate the general geologic setting and mineral potential of a 192 km^2 mountainous glacial catchment area (Evenson, Stephens, Curtin, and Tripp, 1982).

By tabulating lithologic types and abundances in each moraine and tracing the moraines to their respective sources, a series of generalized lithologic maps was derived. Maps of mineralization type and abundance, quartz veins, pegmatites, etc. were also produced. Observations on structural styles in boulders were coupled with aerial photo interpretation and the lithologic information to produce a generalized geologic map of the catchment area. Using this method we have determined that the Trident Glacier catchment area is composed of isoclinally folded, east-west trending, regionally metamorphosed schists and phyllites with minor interbedded quartzite and marble, intruded by two major igneous bodies - one granitic and one gabbroic. Hydrothermal veins associated with the granitic intrusive contain high concentrations of Cu, Pb, Zn, and Ag. Skarns, associated spatially with the gabbroic intrusive, have anomalously high W and Mo concentrations.

We believe that our technique will be of interest to explorationists who need a rapid, accurate method for evaluating the mineral potential of high mountainous glaciated terrains. We believe that although our moraine-based geologic map is not a substitute for a conventional bedrock geologic map, it does yield much valuable reconnaissance-level information about the bedrock types and structures of an area.

5 ACKNOWLEDGEMENTS

We are grateful for field and laboratory support from the U.S. Geological Survey, Branch of Exploration Research. Gary Curtin of the USGS provided valuable support, suggestions, and encouragement. The manuscript benefitted from careful reviews by Harley King and Gary Winkler of the USGS. The George Washington University Committee on Research provided travel funds for one of us (GCS) and partial logistical support for the project.

6 REFERENCES

Beikman, H.M. 1974, Preliminary geologic map of the southeast quadrant of Alaska, USGS Map MF 612.

Beikman, H.M. 1980, Geologic map of Alaska (1:2,500,000), U.S. Geological Survey.

Bradshaw, P.M.D. 1975, Common glacial sediments of the Shield, their properties, distribution, and possible uses as geochemical sampling media. In P.M.D. Bradshaw (ed.), Conceptual models in exploration geochemistry, p. 189-199. New York, Elsevier.

Evenson, E.B. & G.C. Stephens 1978, Till provenance studies in the Pioneer Mountains, Idaho, with application to mineral exploration in areas of alpine glaciation (abs), GSA Annual Meeting Program with Abstracts 10:398.

Evenson, E.B., T.A. Pasquini, R.A. Stewart, & G.C. Stephens 1979, Systematic provenance investigations in areas of alpine glaciation: applications to glacial geology and mineral exploration. In C. Schluchter (ed.), Moraines and varves, p. 25-42. Rotterdam, Balkema.

Evenson, E.B., G.C. Stephens, G.C. Curtin, & R.B. Tripp 1982, Geochemical exploration using englacial debris, USGS Circular 844, 108-110.

Kujansuu, R. 1976, Glaciogeological surveys for ore-prospecting purposes in northern Finland. In R.F. Legget (ed.), Glacial till. p. 225-239. Ottawa, The Royal Society of Canada Special Pub. 12.

Levinson, A.A. 1980, Introduction to exploration geochemistry (2nd ed.). Wilmette, Illinois, Applied Pub. Ltd.

Sharp, R.P. 1960, Glaciers (Condon Lectures). Eugene, Oregon, Univ. of Oregon Press.

Shilts, W.W. 1976, Glacial till and mineral exploration. In R.F. Legget (ed.), Glacial till, p. 205-224. Ottawa, The Royal Society of Canada Special Pub. 12.

Geotechnical studies at Anfiteatro Moraine, Rio Limay valley, Argentina

ELOY DEPIANTE
Agua y Energía Eléctrica de la Nación, Neuquén, Argentina

ABSTRACT

Anfiteatro Moraine will be a natural earth-fill dam when Segunda Angostura Arc Dam is finished. This natural granular-material body has a northwest-southeast orientation, it is 700 m wide at the base, 500 m wide at the maximum level of the proposed reservoir and it is lying on the same basaltic and andesitic volcanics which the Arc Dam will be built on.

As a proposed natural dam, it is necessary to know about some aspects of its stability and its hydraulic performance when the reservoir is filled and a hydraulic gradient of 30 m is created.

This landform was studied from the geological, hydrogeological and geotechnical points of view, considering the evidence of its glacigenic origin and the expected favourable conditions to act as a natural dam. Groundwater contained in this landform was considered as water-table, which is influenced by the level of the Río Limay. The bottom of the pre-existing valley was studied by means of seismic-refraction procedures. The 'bed-rock' is composed of volcanic rocks, with velocities of 3000 m per sec, and it appears as a wide valley, regularly extended from Rincón Grande to Anfiteatro in a northwest-southeast cross-section. In order to know the water-table position, 24 water-table gauges were drilled in both localities. The water-table iso-levels show a tilt from Rincón Grande to Anfiteatro. Two holes were drilled down to bedrock. The first hole is located at Rincón Grande, with a thickness of 37 m, including clay, silt, sand, gravel and erratic blocks. Similar boulders were also found when drilling the water-table gauges. Geotechnical studies include mechanical analysis of core samples, with a large variety of grain sizes, semi-quantitative tests of permeability ("open-end") and pumping tests.

The analysis of field evidence and test data suggests that an important part of the moraine is composed of till and other relatively impervious glacigenic materials. This has been proved by drilling and the water-table gauges observation, which shows that natural recharge coming from the western slope of the valley has an isopiezic surface of high gradient when it reaches the moraine.

Thus, the moraine is expected to behave as a relatively impervious material when the reservoir is filled, as it did during glacial retreat and a temporary glacial lake was filled behind it.

INTRODUCTION

Río Limay is the outlet of Lake Nahuel Huapi, a glacial lake in the Northern Patagonian Andes. In the head area, this stream has eroded glacigenic deposits of the Nahuel-Huapi Drift (Flint & Fidalgo 1964) which correspond to the last great ice age (Wisconsin) in this area.

The upper reach of this river flows through a 1 Km-long narrow valley named "Primera Angostura" (=First Gorge) which has been cut into the Bariloche Moraine, Nahuel-Huapi Drift. From here downstream, the Limay River follows a wider valley of glacial origin, excavated during one or more pre-Nahuel Huapi glaciations. Sixteen kilometres further downstream, the river has partially eroded an older morainic belt formed during the El Cóndor Glaciation or, perhaps, during both El Cóndor and Pichileufú Glaciations (Flint & Fidalgo 1964; Rabassa 1982).

Impeded by this morainic arch, the Limay cut a very narrow gorge through the volcanic bedrock (Ventana Formation, Eocene; González Bonorino 1978, Rabassa 1974, Dessanti 1972). This gorge is locally

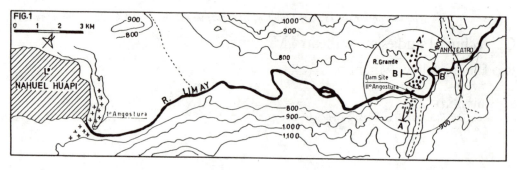

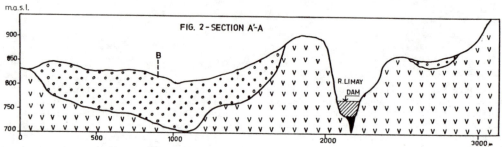

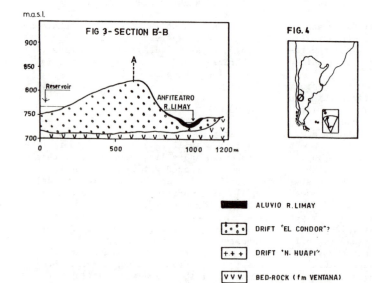

ALUVIO R.LIMAY

DRIFT "EL CONDOR"?

DRIFT "N. HUAPI"

BED-ROCK (fm VENTANA)

Figure 1. General map of the upper reach of Río Limay Valley. A-A' and B-B' correspond to cross-sections of Figures 2 and 3.

Figure 2. Cross-section A-A'. B shows position of Cross-section B-B'.

Figure 3. Cross-section B-B'. A shows position of Cross-section A-A'.
References are the same for the three figures.

Figure 4. General location map of the area in Argentina and the Province of Neuquén.

known as "Segunda Angostura" (=Second Gorge) (Figure 1).

The building of a hydroelectric dam and its impounding reservoir in "Segunda Angostura" has been proposed and investigated by Agua y Energía Eléctrica de la Nación (Agua y Energía Eléctrica de la Nación 1979, 1980; Böhm 1954; Depiante & Di Salvo 1981), to provide energy supply to the region and deliver the energy surplus to the national energy network.

MAIN FEATURES OF THE PROJECT

Between its source and Segunda Angostura, Río Limay descends 25 m in elevation; an arc dam will be constructed to take advantage of this drop, with the creation of a 1200 Ha, artificial reservoir. This lake will have the same level as Lake Nahuel Huapi, and thus it will not modify the maximum registered flood level of the latter. This project includes the building of a gravity arc-dam with a hydroelectrical power-house incorporated to the structure. Its main characteristics are:

Installed capacity .... 120,000 Kw
Maximum dam height above
  foundation level .... 67 m
Dam height above stream bed 35 m
Crest length .... 153 m
Río Limay mean annual dis-
  charge .... 224 m^3/sec

The study area is located at lat. 40°57'S and long. 71°03'W. San Carlos de Bariloche is the nearest city, with a population of 45,000 and 30 Km to the southwest (Figure 4).

PURPOSE OF THE STUDY

Río Limay changed its channel position after the ice retreat during El Cóndor Glaciation and it superimposed its bed over the volcanic bedrock. Thus, the sediments of the Anfiteatro Moraine were preserved from stream erosion. This moraine is expected to provide a lateral closure for the reservoir. This shutting shall fulfil the usual required conditions for earth-dams. It must be steady under the normal operation conditions and it should not be affected by internal erosion due to fines washing, piping or seeping, which could lead towards the failing or destruction of the landform, taken as a natural dam. The slopes must be also stable once that water saturation conditions are achieved after the filling of the reservoir.

METHODOLOGY OF INVESTIGATION

The problem of using a terminal morainic belt as a natural earth-dam (Wahlstrom 1974) is a quite non-typical situation for geotechnical studies in Argentina. Thus, the methodology herein presented is rather new for local standards and it certainly needs further discussion.

The study level up to the present is such as required for Economic-Technical Project Feasibility. The subject is not exhausted and the problem is considered still open because the Executive Project will certainly require a more detailed and quantitative approach of the physical and hydrophysical parameters.

It was considered of high significance at this stage of the project to determine the necessary protection of the soil against seepage. This protection, if necessary at a large scale, could eventually affect the economic feasibility of the works.

The studies were oriented mainly to establish the internal and external geometry of the moraine and to obtain reliable representative permeability values of the landform soils.

The dimensions and limits of the sedimentary body give evidence of its glacial origin. It is an elongated, rather high belt of significant local relief, tranverse to the valley axis and its sides are uneven, with a gentler slope upstream.

Bedrock topography was surveyed by means of seismic refraction techniques and drilling. Preliminary analysis of samples obtained from the holes and many field observations in natural or artificial outcrops suggest an internal geometry of high complexity, with large variations in sediment grain-size and sedimentary structures in very short distances.

The hydrogeological parameters were investigated by means of pumping tests, to provide significant average values for the entire sedimentary body of the moraine, which should be considered as a whole entity.

SEISMIC REFRACTION STUDIES

The seismic refraction method provided excellent results thanks to the large difference between wave velocity in the glacial sediments and the volcanic bedrock. The results obtained are also consistent with the depths observed in the drillings.

Six seismic lines were carried out with a 10 m-interval between receivers and a total recording length of 3850 m. They have yielded enough information about the bedrock surface so as to enable us to draw schematic isolevel contour-lines on it. The cross-sections prepared (see Figures 2 and 3) are representative of the basal geometry of the geomorphic unit.

The transversal(to the valley) section A-A' (Figure 2) shows a much wider buried valley compared with the present Limay Valley. The longitudinal section B-B' (Figure 3) exhibits a bedrock high downstream (northeastwards) which could be interpreted perhaps as the terminus of the glacier advance. In both sections, the vertical scale has been exaggerated.

DRILLINGS

Among the drillings performed, two of them reached the volcanic bedrock, traversing the entire sedimentary thickness. The wells were drilled following the technique of percussion drilling with casing and discontinuous spoon sampling. The existence of large boulders, interpreted as glacially-derived erratics, made the penetration slow and difficult. These boulders had to be cored by means of rotary diamond bit. Afterwards, the boulders were blasted with explosives to continue the advance later on by the technique described before. When it was assumed that bedrock was finally reached, drilling was continued by the rotary method and core sampling, to avoid mistakes and to check the weathering condition of the rock. The first well was drilled at the southern slope of the moraine (Rincón Grande side) where 52 m of glaciofluvial sediments were measured to the bedrock.

The second hole was placed at the northern slope of the glacial landform (Anfiteatro side) where 37 m of gravels and boulders with sandy matrix were drilled to the bedrock. The altitude of the volcanic bedrock surface is 7 m higher in this hole than in the first one.

In both holes bedrock is composed of basalts with columnar jointing, the same rock which also outcrops precisely at the gorge where the arc dam will be built.

Twenty-four additional observation wells were drilled on both sides of the moraine to the water-table level. These holes were drilled in the same way as those described above (percussion method) but not to the bedrock surface. They were stopped instead when reaching water-table level and then they were provided with PVC casing, screened to record water-table oscillations.

HYDROGEOLOGICAL STUDIES

The phreatometric surface was defined using the observation wells. The shape of this surface suggests a general West to East, gentle gradient thanks to rain and snowmelt recharge in the higher portions of the valley.

The axis of the morainic belt is also coincident with a ground-water, phreatic divide towards Anfiteatro and Rincón Grande, respectively. Hydraulic gradients are larger in both, northwards and southwards, longitudinal directions.

While the drillings were being completed, several infiltration (open-end) tests were performed in the saturated zones. These tests provided semi-quantitative permeability values.

Permeability was found to be, in general, rather low, especially along the southern slope of the moraine. This was considered highly favourable for the project because this upstream slope will be supporting the hydraulic pressure of the reservoir (Depiante 1981; Depiante & Di Salvo 1981).

Two pumping tests were also performed using as main pumping wells the two deep-holes drilled down to the bedrock surface. The phreatometric observation wells were taken as observation holes for the water-table depression. The methodology used for these tests is, although rather complex, a standard procedure which will not be discussed further in this paper (Zappi 1980).

The permeability values obtained are considered acceptable in terms of the site and the test characteristics, over a large areal extent. These values will be sufficient for this natural earth-dam (the Moraine) which is expected to provide an impervious lateral shutting to the planned reservoir.

CONCLUSIONS

From the geological and geomorphological analysis of the site and the projected reservoir area, it looks quite possible that the Anfiteatro Moraine had already acted as a "natural earth-dam" for a pre-existing glacial lake when ice receded from its maximum extent during the El Cóndor Glaciation and before the deepening of the present stream-bed of Río Limay. Thus, the former Lake Nahuel-Huapi would have probably extended till the Anfiteatro Moraine during that glacial epoch or, perhaps, a pro-glacial lake evolved in between the ice-front and the moraine. The age, extent and permanence of this ancient glacial lake in the landscape is still uncertain, but glacial-lake sediments and gravel lake beaches provide a good evidence of its existence.

This paleogeographic scheme has been suggested before by several authors; this interpretation is highly significant from the geotechnical point of view because it indicates that the morainic belt would have been already "tested" as a "natural earth-dam" during the geological past. However, it should be taken into consideration that conditions may have been different then than those today, mainly because of the continued landscape modification during the Wisconsin and the Holocene, which has lowered the bottom of the valley and oversteepened the valley slopes. In fact, it seems highly probable that the height of the water column in the aforementioned Pleistocene glacial lake was much smaller than the depth of the proposed reservoir. Presently, the morainic belt has sufficient width (even at its narrowest cross-section) so as to keep the hydraulic gradient low enough between the reservoir (Rincón Grande slope) and the Río Limay (Anfiteatro slope). Besides, no preferred ground-water flow lines are foreseen within the moraine, mainly due to the high complexity of its inner structure and the lack of continuity of potentially-active aquifers.

To increase the security factor of the dam, several improvement works have been recommended, like the water-proofing of the southern slope with clay blankets and a rip-rap protection against waves. This previously mentioned slope is the side of the moraine which will support the reservoir head due to the weight and height of the water mass.

On the other slope or free-face of the "natural earth-dam", a collecting pipe or buried drain will be excavated along the foot of the moraine. This pipe is expected to provide an easier drainage towards the river by means of filters of sand and gravel following the usual filters' regulations, to catch and control the water seepage which could eventually get through the moraine.

ACKNOWLEDGEMENTS

The author is greatly indebted to Agua y Energía Eléctrica de la Nación and its authorities for the permission to publish this report.

REFERENCES

Agua y Energía Eléctrica de la Nación (1980). Aprovechamiento hidroeléctrico "Segunda Angostura", Río Limay. Factibilidad. Unpublished tech.report, Buenos Aires.

Agua y Energía Eléctrica de la Nación (1979). "Segunda Angostura". Jefatura Estudios y Proyectos Región IV, Rev.Unica Apuaye, 5(13): 26-33, Buenos Aires.

Böhn, K.E. (1954). Informe sobre los resultados obtenidos en el estudio geológico de la región de Segunda Angostura en el valle del Río Limay. Agua y Energía Eléctrica de la Nación, unpublished tech.report, Buenos Aires.

Depiante, E. (1981). Determinazione in situ dell'impermeabilità delle rocce per il projetto delle Dighe. Unpubl.Thesis, 16° Corso Internazionale di Idrologia, Universitá degli Studi di Padova, Istituto di Idraulica, Padova, Italy.

Depiante, E. & Di Salvo, C. (1981). Geología y clasificación geotécnica del macizo rocoso del Aprovechamiento Hidroeléctrico Segunda Angostura, Río Limay, provincias de Río Negro y Neuquén. Actas VIII Congr. Geol.Arg., San Luis, 2: 81-101.

Dessanti, R.N. (1972). Andes Patagónicos Septentrionales. in: A.L.Leanza, editor, "Geología Regional Argentina", Acad.Nac. Cienc., p.439-507, Córdoba.

Flint, R.F. & Fidalgo, F. (1964). Glacial geology of the east flank of the Argentine Andes between latitude 39°10'S and latitude 41°20'S. Geol.Soc.Amer.Bull., 75: 335-352.

González Bonorino, F. (1978). Geología de la región de San Carlos de Bariloche: un estudio de las formaciones terciarias del Grupo Nahuel Huapi. Asoc.Geol.Arg. Rev., 33(3): 175-210.

Rabassa, J. (1974). Geología de la región de Pilcaniyeu-Comallo, Provincia de Río Negro. Fundación Bariloche, Public. N°17, Depart.Rec.Nat.Energía, San Carlos de Bariloche.

Rabassa, J. (1982). Excursions' guidebook. INQUA Commission on Lithology and Genesis of Quaternary Deposits, Southamerican Regional Meeting, Neuquén, Argentina, March-April 1982, Departamento de Geografía, Universidad del Comahue, 150 pp.

Wahlstrom, E.E. (1974). Dams, dam foundations and reservoir sites. Elsevier Sci. Publ.Co., Amsterdam.

Zappi, A. (1980). Ensayos de bombeo, Segunda Angostura. Agua y Energía Eléctrica de la Nación, unpubl.tech.report, Buenos Aires.

Glaciofluvial and glaciolacustrine deposits

The sedimentology of glacial, fluvial and lacustrine deposits from northwestern Santa Cruz Province, Argentina

LUIS SPALLETTI
Universidad Nacional de La Plata, Argentina

ABSTRACT

Sediments from San Lorenzo valley-glacier, Río del Oro and Lago Pueyrredón, located in Southern Patagonian Andes, have been studied. It has been established that, in ablation tills, texture and composition of pebbles are influenced by provenance and englacial-subglacial movement, whereas in aqueous environments they are strongly affected by provenance, selective transportation and abrasion. Petrofabric analysis of glacial and fluvioglacial pebbles has yielded significant evidence of preferred orientation. Till deposits show A axes arranged parallel and transversal to the local axis of active moraine, whereas fluvial gravels reveal preferred orientation of A axes parallel to the direction of transportation and upcurrent dipping imbrication. Textural parameters, based on graphic measurements of grain-size distribution from fine gravels to clays, show that textural recognition of glacial, fluvial and lacustrine sediments is only possible when the processes of transportation-deposition are the result of very different flow conditions. Size distributions are related to several mechanisms of transportation, such as surficial reptation, saltation and suspension.

INTRODUCTION

The purpose of this work has been to contribute to the knowledge of the sedimentary properties of modern alpine glacial, fluvial and lacustrine (coastal) deposits, and to increase our understanding of the processes of transportation and deposition by ice, running water and wave action in such environments where the dynamic conditions respond rapidly to climatic seasonal changes.

The sediments formed at the northern end of San Lorenzo Glacier, Río del Oro braided valley-train and Lago Pueyrredón coast were studied. This area is located in the Southern Patagonian Andes, northwestern Santa Cruz Province, Argentina (lat.47°25'-47°41'S; long.71°58'-72°22'W)(Figure 1).

This is one of the coldest regions of continental Argentina. During January (summer) the average temperature ranges from 5°C to 10°C, while in July (winter) it decreases to -10°C. In the Andes, the mean annual precipitation (rain + snow) is over 3000 mm; in the piedmont area it diminishes to 800 mm, and further to the east in the pre-Andean region, it is only 116 mm.

Local bedrock is shown in Fig.1. It primarily consists of Paleozoic phyllites and quartzites of the Río Lácteo Formation (Leanza 1972). Rhyolites, andesites and breccias belonging to the Cretaceous "Vulcanitas del Cerro San Lorenzo" (Riggi 1958) are also frequently exposed. The main body of the Cerro San Lorenzo is composed of Cretaceous granites and tonalites, and an alkali-feldspathic Miocene granite (Ramos & Palma 1981). In the eastern side of the study area, there is a fringe of Jurassic rhyolites and tuffs (Quemado Fm.), marine wackes and shales (Cretaceous, Kachaike Fm.) and continental siliciclastic sedimentites and tuffs (Cretaceous, Chubut Group). Landforms and Quaternary deposits of Cerro San Lorenzo, Río del Oro and the western coast of Lago Pueyrredón were described and mapped by Spalletti (1975).

Cerro San Lorenzo (3360 m a.s.l.) is partially covered by an outlet valley-glacier coming from the Northern Patagonian Ice-Cap.

As a consequence of Pleistocene and Holocene glaciations, erosional features such as arêtes, horns, monuments, cirques and quarried surfaces are frequent. Main glacial

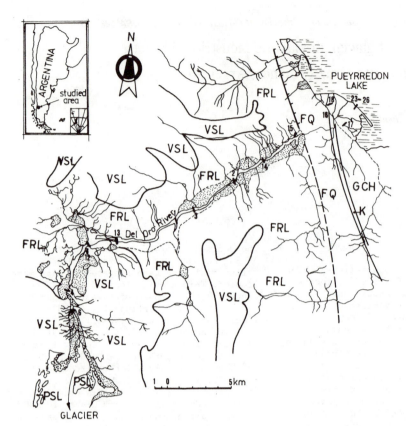

Figure 1. Location
and geological map
of Río del Oro area

Numbers indicate
sampling localities

FRL: Río Lácteo Fm.
(Early Paleozoic)
FQ: Quemado Fm.
(Jurassic)
K: Kachaike Fm. (K)

GCH: Chubut Group (K)

VSL: Vulcanitas
del Cerro San
Lorenzo (K)

PSL: Plutonitas
del Cerro
San Lorenzo
(Cretaceous-
Tertiary)

Locality 6 is at
the shore of Lago
Muñoz, a small, pro-
glacial lake

depositional forms are inactive and active
end moraines, and kame terraces, all formed
in the proximal region by recent glacial
activity.

Three groups of moraines are placed between
the edge of the glacier and the head canyon
of Río del Oro, along 3600 m. The outermost
is the oldest and is represented by a set
of hills with an elevation of 45-50 m above
the valley floor; this end moraine is characterized by intense soil genesis. The
second is an inactive end moraine (16-20 m
above the valley floor) that occupies the
whole section of the valley around Lago
Muñoz (Figure 1). The younger glacial deposits correspond to an active ice-contact
end moraine. All moraines are composed of
unstratified drift with a large proportion
of psephitic material; they are also characterized by sharp and irregular knob-and-
kettle topography.

Kame terraces appear as two thin string-
shaped crests 3000 m-long and 20 m-wide,
located in both margins of the proximal
valley. These deposits are formed by unstratified psephitic sediments and are drained
by meltwater streams. Laterally, they grade

into the youngest moraines and at their
distal stretches they show evidence of
degradation by debris-slides.

Río del Oro valley-train heads at the
northern terminus (1000 m a.s.l.) of San
Lorenzo Glacier, emerging from beneath the
ice as a swift and extremely turbulent
meltwater stream. It flows 33 Km as a
braided river within a valley which has been
cut a few hundred metres below the hilly
uplands of the Southern Patagonian Andes
and it forms a typical arch-like delta at
Lago Pueyrredón (111 m a.s.l.).

The valley walls are characterized by steep
slopes, some of them over 45°, composed of
bedrock (especially the higher slope) and
gravitational (mid and foot slope)-alluvial
(foot slope) sediments. Six terrace levels,
alluvial fans at the mouth of tributaries,
talus, several types of hillslopes and three
bedrock constrictions which concentrate
the flow in one channel with rectilineous
pattern, have been recognized within the
valley.

Main geomorphological characteristics of
Río del Oro are summarized in Table 1;
the origin of braiding has been discussed

214

elsewhere (Spalletti 1975).

TABLE 1

MAIN GEOMORPHOLOGICAL CHARACTERISTICS OF
RIO DEL ORO

Drainage area	270 Km^2
Drainage density (length of channels/drainage area)	1.037 Km^{-1}
Length of channels	280 Km

Stream ordering

Order number	Number of channels of each order	Bifurcation ratio
5	1	
		3.00
4	3	
		5.33
3	16	
		3.50
2	56	
		3.33
1	187	
Mean		3.79

Stretch	Sinuosity
Head waters	1.05
Upper-middle	1.03
Middle-lower	1.20
Mouth	1.33
Delta channels	1.12
Mean	1.098

Table 2 shows some hydraulic parameters of
the channels at different stretches of Río
del Oro whose flow is unsteady, non-uniform
and turbulent, of rapid (supercritical)
régime at head waters and tranquil (subcri-
tical) at the distal stretches.

Bed-forming materials are cobbles, pebbles
and boulders, with subordinate amounts of
sand and silt at the top of emerged bars.
The streambed is characterized by transverse barchanoid bars, antidunes and plane
beds. Islands dividing channels are lens-
shaped longitudinal (braid) bars.
The subaerial deltaic platform of Río del
Oro has a longitude of 4000 m and amplitude
of 1600 m parallel to the river flow (Fig.
1). Four sections were recognized in the
delta area: active channels, flood plain or
interdistributary area, spit, and coast of
Lago Pueyrredón.
The drainage pattern is fan-shaped with
three active areas, each of them having an
intricate network of braided channels. As
it is shown in Table 2, the dimensions,
velocity and discharge of these channels
are lower than those of the channels forming the braided valley-train. The bed material in the delta platform creeks is fine
to medium gravel, sand and silt, and the
observed sedimentary structures are plane
beds, transverse bars and ripples.
The deltaic floodplain is characterized by
a wide flat platform covered by fine sand
and silt, and it is cut by abandoned
braided channels with a fine to medium
gravel bed. There is no evidence of levées,
crevasses and crevasse-splay deposits in
the interdistributary area.
The spit appears at the southern edge of
the subaerial deltaic platform and extend
into the Lago Pueyrredón over the subaqueous platform. It is 1300 m-long, 100 m-wide
and 15 m-high, being covered by a chain

TABLE 2.   RIO DEL ORO HYDRAULIC PARAMETERS

	w	d	R	A	v	D	Re	F
Subglacial channel mouth	2.0	0.60	0.375	1.20	3.000	3.600	11.393 x10^5	1.30
Outwash plain	3.0	0.30	0.250	0.90	2.000	1.800	3.797 x10^5	1.16
Upper-middle reach	8.0	0.60	0.521	4.80	1.538	7.382	5.840 x10^5	0.66
Upper-middle reach	18.0	0.80	0.734	14.40			5.415 x10^6	0.61
Upper-middle reach	15.0	0.60	0.555	9.00	1.426	12.834	0.528 x10^5	0.35
Upper-middle reach	3.5	0.18		0.63	0.465	0.293	1.841 x10^5	0.51
Upper-middle reach	10.0	0.32		3.20	0.909	2.909		
Middle-lower reach	10.0	1.40	1.093	14.00			3.961 x10^5	0.27
Río del Oro mouth	20.0	0.80	0.740	16.00	0.777	12.432	0.767 x10^5	0.43
Main channel deltaic platf.	2.0	0.20	0.166	0.40	0.606	0.242	0.711 x10^6	0.63
Main channel deltaic platf.	3.5	0.15	0.138	0.53	0.750	0.394		

References
w : width (m)
d : depth (m)
R : hydraulic radius
A : channel area (m^2)
v : velocity (m.seg^{-1})

D : discharge (m^3.seg^{-1})
Re : Reynolds number
F : Froude number

platf.: platform

of eolian sand dunes.
In the deltaic environment, coastal deposits of Lago Pueyrredón are mainly composed of shingle. Coarser materials concentrate at a beach ridge and at a subaqueous bar. Between them, at the beach, there is a 1.6 m-wide zone with the characteristics of a sand-run.

PROCEDURE

Twenty-five samples were analyzed; their characteristics have been discussed elsewhere (Spalletti & Gutiérrez 1976). Table 3 shows their distribution in terms of environment, while the location of samples along Río del Oro is shown in Fig.1. The position of sampling stations was established taking into account the environment of deposition, geomorphology of the area and -in the case of fluvial sediments- equidistance among stations. All the samples represent available surficial populations (Griffiths 1967) and were obtained from the youngest sedimentation unit of the environment.

Magnitude of A, B and C axes, roundness (by reference to the chart of images by Krumbein 1941) and lithological composition of 200 clasts larger than 16 mm in diameter, within an area of 2 m^2 per station, were determined for the textural study of tills and gravels. Fabric determinations have been based on the azimuth and dip of A axis of 100 clasts between 32 and 128 mm in diameter for each station. Materials smaller than 16 mm were collected (0.5 Kg) in all localities for textural analysis by sieve and pipette. Textural properties of the gravels, such as size, sphericity, flatness, F factor (Spalletti & Lluch 1972), geometricity (shape according to the Zingg classification) and C/B ratio were computed for each environment. The values of the geometricity ratio (Spalletti 1976) and correlation between pairs of properties were also analyzed. Petrofabric studies of glacial and fluvial deposits were based upon stereographic frequency representations, rose diagrams of azimuths and histograms of dip values.

TABLE 3. ENVIRONMENTS, SUB-ENVIRONMENTS AND DISTRIBUTION OF SAMPLES

Environments	Subenvironments	Sample	Observations
Glacial		7,9	Inactive lateral moraine
		5	Inactive end moraine
		11	Active end moraine
Fluvioglacial		7 bis	Kame-terrace
Lacustrine		8 bis	Lago Muñoz
Mixed fluvial-lacustrine	Tributary channel	8	
	Distributary channel	6	
Fluvial	Outwash plain (channel)	10	Río del Oro
	Braided valley-train	4,13,12, 3,2,1,15,16	Río del Oro
Deltaic	Distributary channels	20,18,17	
	Eolian dune	22	
	Swamp	19	
Lacustrine	Bay	21	Lago Pueyrredón
	Berm	23	
	Sand-run (Beach)	25	
	Littoral string	24	
	Bar	26	

The method proposed by Curray (1956) was used to establish mean circular vector orientation and magnitude.

Laboratory procedures for sieving and sedimentation grain-size analysis of <16 mm sediments were those recommended by Carver

(1971). Statistical treatment of data was based on graphic measures (mean, st.deviation, skewness, kurtosis; Folk & Ward 1957), CM diagrams (Passega 1957) and cumulative frequency distributions on probability paper (Visher 1965, 1969).

SEDIMENTOLOGY OF PSEPHITIC DEPOSITS

TILLS

Glacial sediments are poorly sorted and unstratified rudites, including scattered boulders up to 7 m in diameter. The lithology and texture of ablation tills are shown in Table 4. Clastic composition is mainly granitic, revealing provenance from the San Lorenzo Batholith. The higher proportion of grey granites in the coarser particles is the consequence of joint separation, distribution and hardness of different granitic varieties.

The mean grain-size of tills ranges from 62 to 70 mm and does not indicate any major temporal and areal change in the competence of the glacier movement. It is also evident that degradation processes on inactive moraines are negligible. Clast-size heterogeneity suggests subglacial and englacial movement of the debris (Flint 1955).
Roundness of till clasts is very low, with mean value of 0.33. It increases with size and is higher in white granitic clasts. This behaviour has been detected in other glacial deposits by Sharp (1949), Holmes (1960) and Slatt (1971). Slight differences in roundness among clasts of different lithologies and sizes reveal slow increase in rounding during subglacial transport of coarser clasts or partial mobilization by meltwater. However, low roundness values in the majority of till clasts suggest englacial transport of debris (Mills 1977a). The higher

TABLE 4. LITHOLOGY AND TEXTURE OF TILLS

Station	LITHOLOGY (%)									
	V	PV	M	PS	GG	WG	RG	GeG	A	Q
5	0.00	0.00	0.00	0.00	44.04	2.93	44.04	8.89	0.00	0.00
9	0.00	0.00	0.00	0.00	24.02	0.00	57.03	17.58	1.17	0.00
11	0.00	0.00	0.00	0.00	64.84	2.44	18.75	12.79	0.98	0.00
Mean	0.00	0.00	0.00	0.00	45.56	1.85	38.70	13.15	0.74	0.00

Station	TEXTURE										
	S	R	SDR	F	SDF	ϕ	SD ϕ	FF	SDFF	C/B	SD C/B
5	69.9	0.361	0.106	1.953	0.828	0.717	0.111	1.017	0.127	0.674	0.186
9	69.4	0.352	0.115	1.984	0.738	0.692	0.100	1.042	0.152	0.671	0.180
11	62.3	0.284	0.106	2.450	1.683	0.648	0.111	1.050	0.168	0.617	0.212
Mean	66.9	0.329	0.115	2.149	1.224	0.683	0.107	1.037	0.152	0.652	0.196

Station	%P	%L	%O	%E	n
5	19.04	8.89	37.50	34.47	168
9	25.88	7.03	35.25	31.74	170
11	23.73	14.36	39.55	22.27	202
Mean	22.96	10.37	37.59	29.07	540

References: Station 5: inactive end moraine; Station 9: inactive lateral moraine; Station 11: active moraine (ablation till).
V: vulcanites; PV: porphiric vulcanites; M: metamorphics; PS: psammites; GG: grey granites; WG: white granites; RG: red granites; GeG: green granites; A: applites; Q: quartz and quartzites.
S: mean size (mm); R: roundness; SDR: st.deviation of roundness; F: flatness; SDF: st.dev. of flatness; ϕ: sphericity; SDϕ: st.dev.of sphericity; FF: F factor; SDFF: st.dev. of F factor; C/B: ratio of C and B axes; SD C/B: st.dev. of C/B; E: ecuants; P: prolates; L: laminars; O: oblates; n: number of clasts.

roundness of inactive moraine tills may be the consequence of glacial and proglacial populations mixing - up by glacier readvance over outwash-plain deposits (Nichols & Miller 1951 ; Slatt 1971).
Geometricity of inactive moraines is OEPL (oblate, equant, prolate and laminar, in this order) while that of active moraine is OPEL. In the younger deposits there is a high proportion of laminar and oblate shapes (geometricity ratio: 0.853) in relation to the deposits of inactive moraines (geometricity ratio: 1.361-1.154). The other morphological properties show changes in concordance with geometricity variations. Active moraine materials have lower values of sphericity and C/B, and higher values of flatness than those from inactive moraines. Shape does not change with grain-size but it does so with lithology. The relative abundance of spherical and prolate clasts and the higher values of sphericity, C/B and geometricity ratio in inactive tills suggest "contamination" of glacial with proglacial or glaciofluvial pebbles and cobbles.
Petrofabric diagram of till shows marked anisotropy (Table 5; Fig.2), with two E-W

and N-S poles. There is also a poorly defined horizontal girdle. The rose distribution of azimuths shows a bimodal pattern (Fig.2) and, as in other glacial deposits (Andrews & Smithson 1966), the position of the principal mode is coincident with the major axis of the moraine. Statistical bidimensional analysis of azimuths reveals that the orientation of the resultant vector (v_o) deflects only 6° from the direction of glacial flow (Table 5). Its percentage magnitude is extremely low, denoting a high degree of dispersion of individual data. The values of dip of A axes are low; however, 10% to 20% of clasts dip more than 30°(cf. Holmes 1941; Andrews & Smithson 1966; Lindsay 1970). Mean dip values are comparable to other glacial deposits studied by Wright (1962), whereas its standard deviation is higher than that established by Andrews & Smithson (1966).
In our opinion, fabric of till is controlled by processes of transport and deposition. The bimodal pattern seems to be intermediate between the englacial and subglacial fabrics of Lindsay (1970). The principal mode has been formed by parallel orientation of englacial clasts transported by protracted

TABLE 5. FABRIC. STATISTICAL DATA

Sample	Location	$v_o^°$	$A_o^°$	$\Delta^°$	L%	MD	SD
11	Till (active moraine)	331	325	6	5.50	19	18.45
4	Río del Oro	3	34	31	17.30	13	11.49
1	Río del Oro	6	45	39	12.00	14	13.60
16	Río del Oro (mouth)	331	333	2	9.50	11	10.23

References: $v_o^°$: mean semicircular vector; $A_o^°$: local direction of flow; $\Delta^°$: $v_o^°-A_o^°$; L%: percentage magnitude of mean semicircular vector; MD: mean dip of A axes; SD: standard deviation of dip.

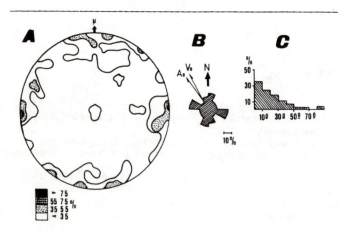

Figure 2. Fabric of till
A: petrofabric diagram
B: histogram of azimuths
A_o : transport direction
V_o : mean semicircular vector
C: histogram of dip of major axes

flow rotation. The secondary mode, transversal to the axis of the moraine, may be the consequence of glacier basal sliding (Holmes 1941). The action of vertical processes of deposition along radial and transverse crevasses or glacier ablation is suggested by the presence of some vertical clasts (Figure 2). Our petrofabric diagram resembles that found by Mills (1977a) in ablation till.

RIO DEL ORO GRAVELS

Fluvial gravels have been deposited in the main channels of the braided valley-train and constitute transverse and braid (longitudinal) bars along the whole longitudinal profile of the river. An important decrease in the amount of psephitic materials is evident towards the deltaic platform, where distributary channel deposits are mainly sands and silts with scarce gravel lenses. Gravel composition is progressively enriched downstream (Table 6). At the river head, as in glacial deposits, clast lithology is essentially granitic. Along the proximal stretches of the river, the changes in composition are due to the incorporation of volcanic and metamorphic clasts from "Vulcanitas del Cerro San Lorenzo" and Río Lácteo Formation, respectively, and by the increase of red granitic pebbles with correlative decrease in grey granites (Figure 3). Metamorphic clasts are strongly increased between stations 13 and 2, but farther downstream the content of these rocks is reduced, whereas the percentage of grey and pink granites rises. On the deltaic platform pebbles and cobbles of red acid volcanics and welded-tuffs from Quemado Formation, and sandstones and tuffs from Chubut Group and Kachaike Formation are incorporated to the streambed.

Grain-size reaches high values at the mouth of the subglacial channel. These deposits are agglomerates, with boulders of 800 mm in diameter, revealing the large competence of proximal sedimentary processes. Towards the outwash-plain, gravel size is over 60 mm (Table 6, Fig.3) and from this place down to station 12 it diminishes to a minimum of 37 mm. Afterwards, grain-size shows a slight increase and remains between 45 and 50 mm (Fig.3). On the deltaic platform, it suddenly decreases to 25-30 mm. It is evident that clast reduction with transportation distance is the consequence of selective transportation with minor influence of abrasion. Partial increments in gravel size are assigned to lateral contributions of coarse materials by tributaries.

Fluvial pebbles are rounded to well-rounded. Their mean roundness varies from 0.46 to 0.65 . Along the braided valley-train this property changes inversely in relation to size; thus, where gravel size is coarse, roundness is low and viceversa (Table 6, Fig.3). Metamorphic and sedimentary clasts are less mature, whereas granitic and volcanic clasts show higher values of roundness.

The mean geometricity of fluvial gravels is OEPL (Fig.3, Table 6) with a geometricity ratio of 0.891. There are no major changes in shape along the river, except for a slight decrease in sphericity and C/B and a correlative increase of flatness and in the proportion of oblate clasts with transportation distance. Spherical pebbles are granitic and a marked enrichment in oblate shapes in coarser fractions has also been detected.

Roundness values distribution reveals that abrasion is more effective in places where size selective transportation is intense and evident. Roundness increases in the braided proximal and distal stretches of Río del Oro. It remains stationary or decreases in the intermediate sector, where lateral contribution through tributaries supply immature pebbles and cobbles. Low-resistant clasts, such as those of metamorphic and psammitic composition show little degree of roundness; this apparently anomalous behaviour may be explained by their proclivity to suspension selective transport on account of their oblate geometricity.

Shape analysis shows the relationship of all the properties with composition. Shape-selective sorting processes have been detected in a number of braided streams (Unrug 1957; Teruggi and others 1971; Spalletti 1976); notwithstanding, the pebbles of Río del Oro do not show remarkable changes in shape, except of the relative increase in oblate clasts towards the coarser size classes. This character may be interpreted as the proclivity of such populations to suspension selective transport.

Petrofabric study of fluvial gravels reveals preferred orientation of A axes parallel to the direction of transportation, up-current imbrication and a poorly-defined girdle of horizontal or low-dipping clasts (Figure 4). The same pattern has been described by Krumbein (1940, 1942), Schlee (1957) and Unrug (1957), and reproduced in experimental currents by Johansson (1963). As in other fluvial deposits (Andrews 1965, Rust 1972), there is a high dispersion of azimuth data (see values of L%, Table 5) and a marked deviation of the mean semi-circular vector (v_o) orientation away from the flow line (A_o) (Table 5, Fig.4). Mean dip values ($11°-14°$) are lower than those showed by Cailleux (1945), Unrug (1957)

TABLE 6.   LITHOLOGY AND STRUCTURE OF FLUVIAL GRAVELS (RIO DEL ORO)

LITHOLOGY

Station	V	PV	M	PS	GG	WG	RG	GeG	A	Q
6	0.00	0.00	0.00	0.00	47.66	8.98	21.58	20.70	0.88	0.00
4	9.57	0.00	1.46	0.00	33.30	24.22	26.17	1.46	0.00	3.52
12	3.13	0.00	0.00	0.00	23.54	16.70	49.71	6.25	0.49	0.00
13	3.22	0.00	1.86	0.00	18.16	12.60	60.25	3.71	0.00	0.00
1	0.39	0.00	30.37	0.00	7.71	3.32	41.50	13.48	0.00	2.83
3	17.29	0.00	41.80	0.00	26.37	0.39	13.38	0.39	0.00	0.00
2	0.00	0.00	91.80	0.00	1.37	0.00	1.86	0.00	0.00	4.69
14	3.13	0.00	68.26	0.00	9.47	0.00	11.23	1.76	0.00	5.86
15	2.44	0.00	42.77	0.00	11.33	0.00	33.98	8.79	0.00	0.49
16	3.32	13.77	36.13	9.96	12.79	0.39	19.43	1.86	0.39	1.37
17	5.18	5.18	42.09	7.81	13.09	0.00	15.72	2.54	2.54	5.18
18	4.20	4.20	46.58	1.37	5.66	1.86	19.92	5.18	0.00	9.96
Mean	4.46	1.85	35.00	1.22	16.44	5.86	27.07	5.09	0.18	2.84

TEXTURE

St.	S	R	SDR	F	SDF	SD	FF	SDFF	C/B	SDC/B	%E	%P	%L	%O	n
6	63.4	0.46	0.08	1.72	0.55	0.766	0.10	0.993	0.11	0.708	0.17	52.3	9.0	4.5	34.2	111
4	55.3	0.47	0.10	1.85	0.56	0.734	0.10	0.999	0.13	0.673	0.18	38.4	14.1	6.1	41.4	198
12	36.7	0.59	0.09	1.72	0.46	0.759	0.09	0.994	0.10	0.703	0.17	47.6	11.4	1.6	39.2	191
13	37.8	0.61	0.09	1.82	0.66	0.748	0.10	0.988	0.10	0.681	0.17	45.3	10.3	6.1	38.3	214
1	43.9	0.59	0.13	2.14	0.83	0.702	0.11	0.982	0.11	0.607	0.19	29.3	7.2	10.1	53.1	207
3	46.6	0.53	0.11	2.03	0.90	0.704	0.10	1.011	0.12	0.653	0.18	38.9	12.9	9.1	38.9	208
2	49.8	0.46	0.19	1.99	0.61	0.699	0.11	1.020	0.14	0.646	0.17	29.6	15.7	13.9	40.6	209
14	45.2	0.54	0.13	2.09	0.72	0.697	0.10	0.995	0.11	0.617	0.17	27.2	11.2	13.5	47.9	221
15	44.8	0.65	0.11	2.13	0.66	0.702	0.10	0.972	0.11	0.594	0.17	27.1	7.3	8.3	57.1	203
16	48.8	0.55	0.13	2.03	0.77	0.709	0.11	1.001	0.12	0.643	0.18	31.4	13.8	10.9	43.8	210
17	25.8	0.54	0.13	1.79	0.45	0.728	0.10	1.026	0.14	0.701	0.18	31.5	26.3	5.2	36.8	38
18	29.9	0.61	0.11	2.02	0.67	0.700	0.10	1.010	0.13	0.643	0.19	26.7	16.6	10.0	46.6	210
Mean	44.5	0.56	0.13	1.97	0.70	0.718	0.10	0.998	0.12	0.649	0.18	34.9	12.2	8.8	44.1	2220

References as in Table 4

and Potter & Pettijohn (1963). Standard deviation of dip (Table 5) and dip histograms (Fig. 4) reveal that the main range of inclinations is between 0° and 28°. The sub-horizontal population parallel to flow direction in fluvial fabrics is the consequence of transport phenomena under upper-flow régime conditions (Rust 1972). Such conditions inhibit the mobilization by rolling and then the formation of a pole perpendicular to the current direction. On the other hand, the high concentration of clasts on the streambed favoured the formation of imbricate structures (Cailleux 1945) which are characteristic in these gravels (Fig. 4).

Furthermore, the presence of a sub-horizontal transverse population in the petrofabric diagram, corresponding to the mouth of Río del Oro is assigned to the deposition of coarse cobbles and boulders of local provenance (Quemado Formation), which are too coarse to be reoriented parallel to the transport line by rapid flows.

COASTAL GRAVELS

The gravels of the coast of Lago Pueyrredón have a varied composition, being metamorphic and granitic clasts the most abundant. Metamorphic clasts are common in the coastal string, whereas there is a strong increase in red granitic pebbles towards the subaqueous bar (Figure 5, Table 7). The mean size of these gravels is different in each sub-environment. At the storm deposit (berm) it is 20 mm, at the coastal string rises to 26 mm, afterwards it decreases to 16 mm in the sand-run and finally it increases up to 37 mm in the subaqueous (breaking) bar (Fig. 5, Table 7). The finer components are those of psammites

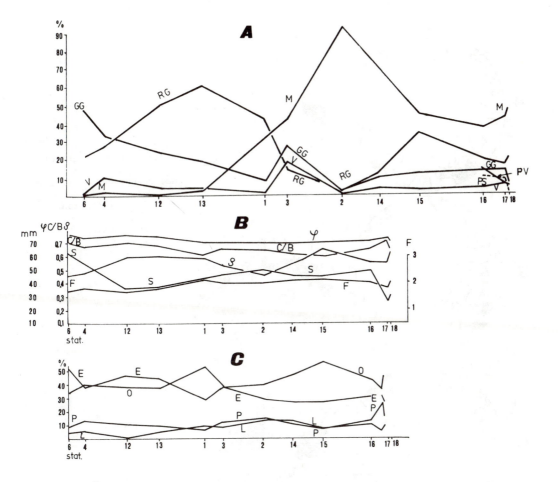

Figure 3. Variations in lithology and texture of fluvial gravels with the distance of transportation.

A. Lithology. RG: red granites; GG: grey granites; V: vulcanites; PS: psammites; PV: porphiric vulcanites (Quemado Formation); M: metamorphics.

B. φ: sphericity; C/B: axial ratio; ρ: roundness; F: flatness; S: mean size.

C. Geometricity. E: ecuant; O: oblate; P: prolate; L: laminar.

and metamorphics, being the coarser those derived from the volcanic Quemado Formation.

The mean lacustrine gravel roundness is higher than that of the river deposits. The sediments of the coastal string show a slightly decrease in rounding (Table 7). Roundness increases with the gravel size and it is higher in granitic and psammitic pebbles.

Shape variations among sub-environments are significative. Equidimensional clasts prevail in the subaqueous bar and oblate pebbles are important in the coastal string (Fig. 5), whereas the content of these

opposite geometricities is intermediate in the berm and sand-run. The geometricity ratio reflects also this behaviour and shows higher values at the bar and the lowest at the coastal string. Pebbles of granite and pyroclastic volcanics (from Quemado Formation) are richer in equant geometricity, as well as those of psammites, metamorphics and volcanics (from "Vulcanitas del Cerro San Lorenzo") have propensity to oblate geometricity. Sphericity and C/B ratio show lower values in the coastal beach ridge and increase markedly towards the subaqueous bar. Platity (flatness) behaves in an opposite way. Sphericity and

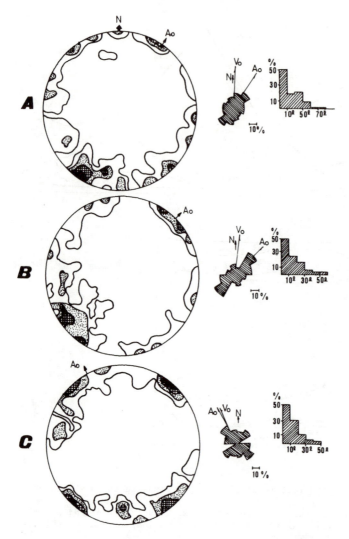

Figure 4.

Fabric of fluvial gravels.

Same references as in Figure 2.

A: head-water station

B: braided valley-train

C: delta apex

C/B also increase towards the coarser gra-vel-size fractions and are higher in grani-tic and volcanic pebbles.

As in other coastal deposits (cf. Spalletti & Lluch 1973), the quantitative analysis of textural properties reveals significa-tive variations among sub-environments in the Lago Pueyrredón shingle coast. They are the consequence of the hydrodynamic action of waves and associated currents which cause textural modifications in the deposits, through both abrasion and selec-tive transportation.

Gravel-size distribution indicates that the sub-lacustrine bar has been deposited under high flow-régime conditions. The gravels are also coarse in the coastal beach ridge (Fig. 5). On the other hand, the smallest grain size has been detected in the beach (sand-run) zone; however, these finer depo-sits are not the product of low-energy conditions, since they represent an area of transport of pebbles and cobbles by swash and back-wash currents.

Shape (and composition) distribution reveals the intensity of selective transportation processes (Bluck 1967; Allen 1970; Spalletti & Lluch 1973). The tendency of sub-lacus-trine bar clasts to equant and spherical geometricities may be interpreted as an evidence of tractive current deposition by back-wash and/or rip currents which origi-nate deposits enriched in shapes of high sedimentation velocity. The deposition of the low sedimentation velocity shapes, such as oblate and laminar geometricities, take place at the coastal string, where the pebbles are selectively carried by surf

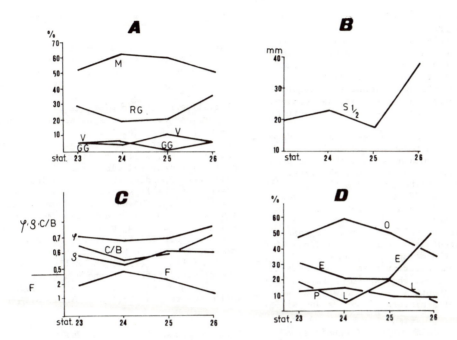

Figure 5. Variations in lithology and texture of coastal lacustrine gravels.
Same references as in Figure 3.
A: Lithology; B: Mean size; C: Textural properties; D: Geometricity.

TABLE 7. LITHOLOGY AND TEXTURE OF COASTAL GRAVELS

Station	V	PV	M	PS	GG	WG	RG	GeG	A	Q
23 Berm	6.45	0.00	52.15	0.00	4.30	0.00	28.22	6.45	0.00	2.15
24 String	3.61	1.76	62.89	1.76	5.47	0.00	18.46	0.00	1.76	3.61
25 Sand-run	9.96	0.00	59.96	0.00	0.00	0.00	19.92	0.00	0.00	9.96
26 Bar	4.69	2.73	50.00	1.86	4.69	0.00	34.86	0.88	0.00	0.00
Mean	5.09	1.85	54.17	1.39	4.63	0.00	28.70	1.85	0.46	1.85

Station	S	R	SDR	F	SDF	ϕ	SDϕ	FF	SDFF	C/B	SD C/B
23	19.9	0.587	0.12	1.968	0.62	0.709	0.10	1.006	0.12	0.650	0.177
24	23.0	0.522	0.12	2.413	0.95	0.680	0.09	0.958	0.13	0.555	0.209
25	17.0	0.610	0.09	2.147	0.64	0.694	0.10	0.983	0.13	0.590	0.165
26	38.8	0.599	0.13	1.683	0.48	0.771	0.10	0.993	0.10	0.713	0.158
Mean	29.9	0.578	0.12	1.948	0.73	0.731	0.10	0.986	0.12	0.654	0.188

Station	%E	%P	%L	%O	n
23	30.37	12.99	8.69	47.75	46
24	20.31	14.75	5.47	59.18	54
25	19.92	9.96	19.92	50.00	10
26	50.88	7.52	5.57	35.84	106
Mean	37.50	10.64	6.94	44.91	216

References as in Table 5

and/or swash suspensive currents arising after the break of the waves. On account of the close relation between shape and composition, the tractive-suspensive process of transportation tends also to modify the lithology of the deposits, being those of the coastal string enriched in metamorphic clasts whereas the gravels of the subaqueous bar have a higher percentage of granitic varieties.

The effectiveness of abrasion may be deduced through the increase of low-resistance lithologies (such as metamorphic and psammitic) towards the finer gravel sizes. Roundness distribution shows that abrasion processes are effective in those places where the sediments are deposited by tractive currents (bar). Therefore, there is significant vinculation between abrasion and the agents of transport and deposition acting in the coastal environment.

Textural characteristics of berm deposits (Table 7) reveal that storm waves are fair selective agents and originate a sediment composed of a mixture of shapes and sizes.

GRAIN-SIZE DISTRIBUTION OF SEDIMENTS FINER THAN 16 MM

Histograms of grain-size distribution of the San Lorenzo area deposits are shown in Figure 6 and the values of the statistical parameters (Folk & Ward 1957) are listed in Table 8.

Glacial deposits are mixtites with variable proportions of gravel, sand and mud, having bi- or polymodal distribution. In two samples the higher frequencies are in the gravel fractions, while the other is mainly psammitic. Textural classification (Folk 1954) corresponding to these samples is gravel for the first two and gravelly sand for the third. The mean is medium- to coarse sand and the standard deviation shows values around 2.0, denoting a poor- to very poor sorting (Folk 1966); however, our data are smaller than those published by Slatt (1971), Landim & Frakes (1968) and Mills (1977b). The glacial drift also shows a tendency to positive skewness (see also Mills 1977a) and meso- to platykurtosis (Table 8).

TABLE 8. VALUES OF FOLK & WARD's (1957) GRAIN-SIZE GRAPHIC MEASURES

	L.Muñoz	Fluv + Lacus			Fluvial (Río del Oro)						
Sample	8bis	8	6	4	13	10	12	1	3	2	15
Median(Md)	3.30	−0.10	0.65	1.10	−0.50	2.40	0.95	2.20	2.20	2.00	1.35
Mean (Mz)	3.28	0.08	2.22	1.13	0.08	2.48	0.96	2.24	2.28	2.07	1.35
St.Dev.	0.707	1.614	3.570	2.327	2.077	1.049	0.656	0.606	0.714	0.636	0.530
Skewness	0.076	0.344	0.574	0.146	0.571	0.132	0.020	0.059	−0.069	0.171	0.041
Kurtosis	1.254	1.355	0.601	1.241	1.205	1.213	1.001	1.120	0.638	1.356	1.083

	Fluv	Delta			Delta	Delta	Lago Pueyrredón			Kame
		Fluv	Fluv	Fluv	Dune	Swamp	Bay	Coast	Coast	terrace
Sample	16	17	18	22	19	21	23	25		7bis
Median(Md)	2.70	1.90	1.80	−0.60	0.90	8.85	5.90	0.70	0.20	−1.40
Mean (Mz)	2.97	1.77	1.80	−0.16	0.91	7.91	6.28	0.71	−0.20	−1.17
St.Dev.	1.508	1.159	0.560	1.538	0.362	2.536	2.014	0.664	0.952	0.767
Skewness	0.357	−0.221	0.005	0.503	0.014	−0.458	0.371	−0.194	−0.471	0.728
Kurtosis	1.288	1.621	1.027	0.938	0.984	0.662	1.188	1.581	0.701	2.049

	Glacial			
	Inactive			Active
Sample	9	7	5	11
Median(Md)	1.60	0.00	2.25	0.35
Mean (Mz)	1.54	0.45	1.98	0.70
St.Dev.	2.267	1.949	1.640	2.005
Skewness	0.075	0.356	0.150	0.315
Kurtosis	1.059	0.685	0.900	0.841

The similitude between active and inactive moraine deposits reveals the absence of postdepositional changes in the older deposits, and no major variations in the glacial flow in both episodes. Notwithstanding, the little amount of pelitic material in San Lorenzo tills may be the consequence of partial washout of the finer fractions by meltwater (cf. Spalletti 1975).

As it has been shown elsewhere (Spalletti 1972), the grain-size distribution depends on the nature of the debris and the charac-

224

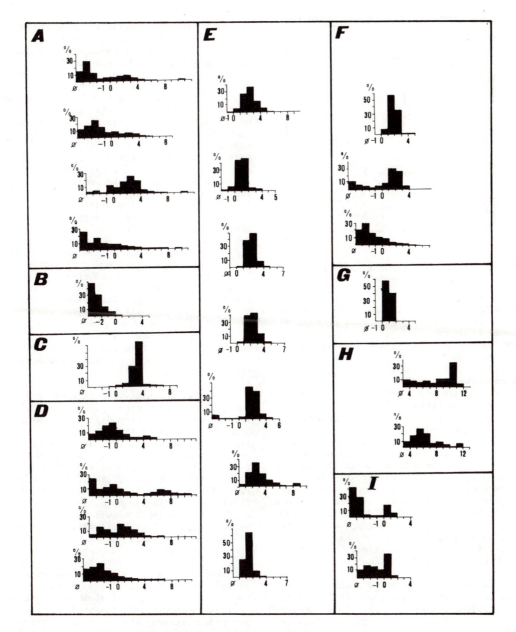

Figure 6. Histograms of sediments and samples finer than 16 mm.
A: tills; B: kame terrace; C: Lago Muñoz; D: outwash plain; E: braided valley-train;
F: delta channels; G: dune; H: swamp and bay; I: coast of Lago Pueyrredón.

teristics of the depositional agents. These samples, composed of complex distributions, were deposited by a single transport agent distinguished by its high viscosity, as it is shown by standard deviation values. Moreover, the tendency to strong positive skewness may be used as an indicator of high competence of the glacier (see Spalletti 1972).

CM distributions in the glacial sediments (Figure 7) is closely related to the pattern of glacial deposits studied by Landim & Frakes (1968) and some mass-flow deposits (Bull 1962; Spalletti 1972). It is characterized by coarse values of C and variable values of M.
Cumulative frequency distributions resemble those assigned to slumps and density currents

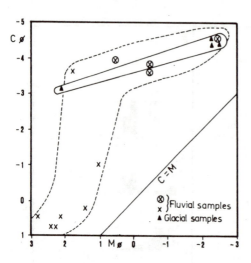

Figure 7. CM pattern of fluvial and
glacial samples

in shallow water environments (Visher 1969).
They show a marked development of populations transported by suspension (Figure 8).
The matrix of kame-terrace deposits has a
coarse unimodal distribution. The mean and
median fall in granule size, standard deviation reveals moderate sorting, skewness is
very positive and kurtosis is high (Table 8).
All these characteristics suggest that the
deposit was formed by a fluid, selective
and competent current.
Lago Muñoz, located near the terminus of
the San Lorenzo Glacier, was formed as a
frontal standing body by meltwater streams
dammed by an arc of inactive marginal
moraines. Lake deposits show a marked mode
in very-fine sand (Fig. 6) containing more
than 60 % of the whole distribution. Cumulative representation is composed of three
sections of different slope, being the
finer (suspension population; Moss 1962)
the best sorted. Statistical parameters
show very-fine sand mean grain-size, moderately good sorting, symmetrical skewness
and leptokurtosis. The absence of coarser
populations composed of granules, pebbles
and cobbles suggests negligible influence
of ice-rafting or, at least, that the glacier was not in contact with the body of
standing water.
The deposits of the upstream reaches of
Río del Oro resemble those studied by
Dyer (1970) and Eynon & Walker (1974). They
have a coarse (sandy) mean size (from 0 ϕ
to 2.2 ϕ) and poor to very poor sorting
(Table 8), being it comparable to the values found by Smith (1970) in proximal
transverse bar deposits of braided rivers.
Positive to very positive skewness and

leptokurtosis are similar to the figures
shown by Williams & Rust (1969) for the
same sedimentary environment. The proximal
deposits of Río del Oro are complex populations grading from gravel to clay, and the
samples analyzed here represent only a portion of the deposit ("disconform sample" of
Spalletti & Gutiérrez 1976). They were formed
under conditions of upper flow régime and
competence, marked fluctuations of velocity
and low fluidality. It is also possible
that some of these sediments would be originated by mixtures of fluvial and lacustrine (Lago Muñoz) populations.
The remaining deposits of Río del Oro, that
is, those composing the middle and downstream reaches, are sands deposited by
weak currents as a matrix of previously
deposited open gravels. The modal classes
appear between 1 ϕ and 3ϕ, and the distributions lack gravel and muddy materials.
These general features have also been described by Williams & Rust (1969) and Rust
(1972) in braided deposits. Mean is fine
sand (>2ϕ) and, in all but two samples,
standard deviation shows moderately well-
sorted distributions (Table 8), being finer
and better sorted than the deposits described by Smith (1970, 1974) and Mazzoni (1977).
The outwash-plain sample closely resembles
those of same environment studied by Landim
& Frakes (1968). Likewise, skewness and
kurtosis data are comparable to Mazzoni's
values, which come from deposits of semi-
arid braided rivers from Argentina.
Textural features of these deposits are the
result of both types of particle transportation (Moss 1962; Spencer 1963; Klovan 1966;
Visher 1969) and the nature of the source
rocks (Mazzoni 1977). The bulk volume of
fluvial materials has been transported in
saltation by selective currents of moderate
kynetic energy. The infill of the open gravels with these sands occurs shortly after
the vertical aggradation of braided channels
by gravel materials (Fahnestock 1963), that
is, when the longitudinal bars are emerging
as islands in the braided valley-train.
The CM diagrams (Fig. 7) show that upstream
fluvial sediments were transported by roll-
ing-graded suspensions (Passega & Byranjee
1969) under alternating transitional and
lower flow régimes (Williams & Rust 1969).
On the other hand, fluvial sands acting as
matrix of open gravels have been formed by
graded suspensions (Passega 1964) or by
saltation mechanisms under tranquil flow.
Characteristics of grain-size distributions
reveal that these deposits are the result
of tranquil and turbulent, or sub-critical
currents (Spalletti & Gutiérrez 1976).
Stream deposits of Río del Oro delta are
extremely variable in textural properties

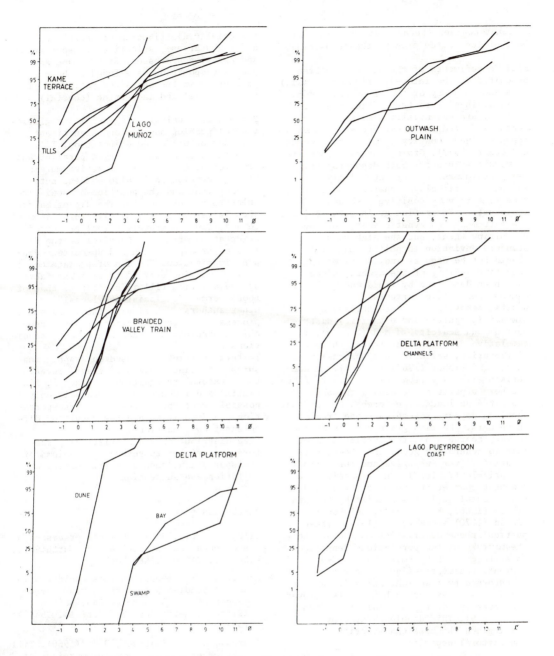

Figure 8. Cumulative distributions of sediments and samples finer than 16 mm.

(Figures 6 and 8; Table 8) denoting systematic changes in hydrodynamic characteristics of the streams draining the deltaic platform. More than 97% of the eolian sands deposited in the deltaic spit, are composed of two size classes (between 0 φ and 2 φ, Figure 6); size features resemble those analized by Mazzoni (1977) and are characterized by the absence of psephitic materials, unimodality, presence of an important saltation segment, symmetrical and mesokurtic distributions (Table 8). However, in comparison with other eolian sediments (Di Paola 1967; Solohub & Klowan 1970) this sand is

227

somewhat coarser, because of the neighbour-hood of the source area to the depositing place.

Pelitic sediments of the delta have been deposited in the inter-distributary (swamp) area and in a bay of the Lago Pueyrredón. Their grain-size features (Fig. 6 and 8; Table 8) are essentially similar to other deposits formed in lacustrine environments (Spencer 1963; Coakley & Rust 1968; Thomas and others 1972). Grain-size data show dis-criminant values for both deposits; the swamp sediments are very negatively skewed and platykurtic clays, whereas the lacust-rine silt is very positively skewed and leptokurtic. Their fine grain-size reflects the extremely low kynetic energy of the depositing agent, but the high values of standard deviation in pelitic materials should not be used as indicators of the selective ability of the agents, since they have been deposited by fluid and weak aqueous suspensive currents.

Coastal deposits are characterized by bimodal (psephitic and psammitic) distri-butions and scarcity of mud (Fig. 6). Their cumulative distributions reveal the presence of reptation, saltation and suspension popu-lations of Visher (1969) (Fig. 8). In com-parison with data from Fox and others (1966), the berm deposit correlates with sands deposited on lacustrine forebeaches and the sand-run deposit resembles those from off-shore bars. The mean grain-size is medium to very coarse sand, the sorting is mode-rate to moderately good, the skewness is negative to very negative and the kurtosis is variable (Table 8). These samples are somewhat coarser and more poorly sorted than coastal sediments studied by Fox and others (1966), Beal (1970), Solohub & Klovan (1970) and Mazzoni (1977). These particular characteristics may be due to the mixture of two populations deposited in a steep beach by small waves. Swash currents sedimented fine gravels and sands transported by traction, saltation and sus-pension, whereas backwash currents removed the finer (pelitic) materials (Friedman 1967, 1979). This double process printed sorting and skewness characteristics to the littoral deposits.

CONCLUSIONS

The analysis of psephitic materials has proved that texture and composition of till pebbles are influenced by provenance and englacial and subglacial movement within the valley glacier. For aqueous environ-ments, textural properties and clast litho-logy are strongly affected by provenance (especially distribution of source rocks and tributaries), selective transportation and abrasion processes. Size, shape and gravel composition were found to be useful parameters in the characterization of gla-cial, fluvial and lacustrine (coastal) depo-sits.

Petrofabric analysis of glacial and glacio-fluvial pebbles has yielded significant evidence of preferred orientation. Till deposits show A axes arranged parallel and transversal to the axis of active moraine. Strong correlation is also evident between ice-movement and the position of semi-cir-cular resultant vectors. The dip of major axes is variable, but mainly ranges from 0° to 37°. Fluvial gravels reveal preferred orientation of A axes parallel to the direc-tion of transportation and up-current dip-ping imbrication. The dip of prolate clasts is low, varying from 0° to 28°. The fabric of the ablation till is similar to that of those termed "englacial-subglacial" by other authors. The fluvial fabric is due to processes of transportation of dense pse-phitic populations under high energy condi-tions.

Textural parameters, based on graphic mea-sures of fine gravels to clays, reveal that textural recognition of glacial, flu-viatile and lacustrine sediments is only possible when the processes of transporta-tion and deposition are the result of dif-ferent flow conditions.

Size distribution and statistical parame-ters are related to several mechanisms of transportation, such as surficial reptation, saltation and suspension.

REFERENCES

Allen, J.R.L. (1970). Physical processes of sedimentation. Amer.Elsevier, Earth.Sci., Ser. 1, 248 pp., New York.

Andrews, J.T. (1965). Surface boulder orien-tation studies around the northwestern margin of the Barnes Ice Cap, Baffin Island, Canada. Jour.Sed.Petrol., 35 (3): 753-757.

Andrews, J.T. & Smithson, B.B. (1966). Till fabrics of the cross-valley moraines of north-central Baffin Island, Northwest Territories, Canada. Geol.Soc.Amer.Bull., 77(3): 271-290.

Beall, A.D. (1970). Textural differentia-tion within the fine sand grade. J.Geol. 78 (1): 77-94.

Bluck, B.J. (1967). Sedimentation of beach gravels: examples from South Wales. Jour.Sed.Petrol., 37: 128-156.

Bull, W. (1962). Relation of textural (CM) patterns to depositional environment of alluvial-fan deposits. Jour.Sed.Petrol., 32(2): 211-216.

Cailleux, A. (1945). Distinction des galets marines et fluviatiles. Bull.Geol.Soc. France, (5), 15: 375-404.

Carver, R. (1971). Procedures in sedimentary rocks. Wiley Interscience, 653 pp., New York.

Coakley, J.P. & Rust, B.R. (1968). Sedimentation in an Arctic lake. Jour.Sed.Petrol. 38 (4): 1290-1300.

Curray, J.R. (1956). The analysis of two-dimensional data. Jour.Geol., 64 (2): 117-131.

Di Paola, E.C. (1967). Contribución al estudio de sedimentos eólicos en los alrededores de Tunuyán, Provincia de Mendoza. Asoc.Geol.Arg.Rev., 22 (4): 281-290.

Dyer, K. (1970). Grain-size parameters for sandy gravels. Jour.Sed.Petrol., 40 (2): 629-641.

Eynon, G. & Walker, R.G. (1974). Facies relationships in Pleistocene outwash gravels, Southern Ontario: a model for bar growth in braided rivers. Sedimentology, 21 (1): 43-70.

Fahnestock, R.K. (1963). Morphology and hydrology of a glacial stream, White River, Mount Rainier, Washington. U.S. Geol.Survey, Prof.Paper 422-A: 1-70.

Folk, R.L. (1954). The distinction between grain size and mineral composition in sedimentary rock nomenclature. Jour.Geol., 62: 344-359.

Folk, R.L. (1966). A review of grain-size parameters. Sedimentology, 6: 73-93.

Folk, R.L. & Ward, W. (1957). Brazos River Bar: a study in the significance of grain-size parameters. Jour.Sed.Petrol., 27 (1): 3-27.

Fox, W.T.; Ladd, J.W. & Martin, M.K. (1966). A profile of the fou moment measures perpendicular to a shore line, South Haven, Michigan. Jour.Sed.Petrol., 36 (4): 1126-1130.

Friedman, G.M. (1967). Dynamic processes and statistical parameters compared for size frequency distribution of beach and river sands. Jour.Sed.Petrol., 37: 327-354.

Friedman, G.M. (1979). Address of the retiring president of the International Association of Sedimentologists: Differences in size distributions of populations of particles among sands of various origins. Sedimentology, 26: 3-32.

Griffiths, J.C. (1967). Scientific method in analysis of sediments. McGraw-Hill, Earth & Planetary Sci., 508 pp., New York.

Holmes, C.D. (1941). Till fabric. Geol.Soc. Amer.Bull., 52 (9): 1299-1354.

Holmes, C.D. (1960). Evolution of till-stone shapes, central New York. Geol.Soc.Amer.

Bull., 71 (11): 1645-1660.

Johansson, C.E. (1963). Orientation of pebbles in running water. A laboratory study. Geogr.Annaler, 45 (2-3): 85-112.

Klovan, J.E. (1966). The use of factor analysis in determining depositional environment from grain-size distributions. Jour. Sed.Petrol., 36 (1): 115-125.

Krumbein, W.C. (1940). Flood gravels of San Gabriel Canyon, California. Geol.Soc.Amer.Bull., 51 (5): 639-676.

Krumbein, W.C. (1941). Measurement and geological significance of shape and roundness of sedimentary particles. Jour.Sed. Petrol., 11: 64-72.

Krumbein, W.C. (1942). Flood deposits of Arroyo Seco, Los Angeles County, California. Geol.Soc.Amer.Bull., 53: 1355-1402.

Landim, P.M.B. & Frakes, L.A. (1968). Distinction between tills and other diamictons. Jour.Sed.Petrol., 38 (4): 1213-1223.

Leanza, A.F. (1972). Andes Patagónicos Australes. in: Geología Regional Argentina, Acad.Nac.Cienc.Córdoba, p. 689-706.

Lindsay, J.F. (1970). Clast fabric of till and its development. Jour.Sed.Petrol., 40 (2): 629-641.

Mazzoni, M.M. (1977). El uso de medidas estadísticas texturales en el estudio ambiental de arenas. Obra Centenario Museo de La Plata, t.IV, Geol., p. 179-223, La Plata.

Moss, A.J. (1962). The physical nature of common sandy and pebbly deposits. Part 1. Amer.J.Sci., 260: 337-373.

Mills, H. (1977a). Differentiation of glacier environments by sediment characteristics: Athabasca Glacier, Alberta, Canada. Jour.Sed.Petrol., 47 (2): 728-737.

Mills, H. (1977b). Textural characteristics of drift from some representative Cordilleran glaciers. Geol.Soc.Amer.Bull., 88: 1135-1143.

Nichols, R.L. & Miller, M.M. (1951). Glacial geology of Ameghino Valley, Lago Argentino, Patagonia. Geogr.Rev., 41 (2): 274-294.

Passega, R. (1957). Texture as characteristic of clastic deposition. Amer. Assoc. Petrol.Geol.Bull., 41: 1952-1984.

Passega, R. (1964). Grain-size representation by CM patterns as a geological tool. Jour.Sed.Petrol., 34: 830-847.

Passega, R. & Biranjee, R. (1969). Grain-size image of clastic deposits. Sedimentology, 13: 830-847.

Potter, P.E. & Pettijohn, F.J. (1963). Paleocurrents and basin analysis. Springer-Verlag, 296 pp., Berlin.

Ramos, V. & Palma, M. (1981). El batolito granítico del Monte San Lorenzo, Cordillera Patagónica, Prov. de Santa Cruz. Actas VIII Congr.Geol.Arg., 3: 257-280.

Riggi, J.C. (1958). Resumen geológico de la zona de los lagos Pueyrredón y Posadas, provincia de Santa Cruz. Rev.Asoc.Geol.

Arg., 12 (2): 65-97.

Rust, B.R. (1972). Pebble orientation in fluvial sediments. Jour.Sed.Petrol.,42 (2): 384-388.

Schlee, J. (1957). Fluvial gravel fabric. Jour.Sed.Petrol., 27: 162-176.

Sharp, R.P. (1949). Studies of superglacial debris on valley glaciers. Amer. J. Sci., 247 (5): 289-315.

Slatt, R.M. (1971). Texture of ice-cored deposits from ten Alaskan valley glaciers. Jour.Sed.Petrol., 41 (3): 828-834.

Smith, N.D. (1970). The braided stream depositional environment: comparison of the Platte River with some Silurian rocks, North Central Appalachians. Geol.Soc.Am. Bull., 81: 2993-3014.

Smith, N.D. (1974). Sedimentology and bar formation in the Upper Kicking Horse River, a braided outwash stream. Jour. Geol., 82 (2): 205-224.

Solohub, J.D. & Klovan, J.E. (1970). Evaluation of grain-size parameters in lacustrine environments. Jour.Sed.Petrol., 40 (1): 81-101.

Spalletti, L.A. (1972). Sedimentología de los cenoglomerados de Volcán, provincia de Jujuy. Rev.Mus.La Plata, N.S., Secc. Geol., n° 66, 8: 137-225.

Spalletti, L.A. (1975). Estudio del glaciar septentrional del Monte San Lorenzo y del Río del Oro (provincia de Santa Cruz). I. Aspectos generales: Geomorfología. Asoc.Geol.Arg.Rev., 30 (1): 17-43.

Spalletti, L.A. (1976). Sedimentología de gravas glaciales, fluviales y lacustres de la región del Cerro San Lorenzo (prov. de Santa Cruz). Asoc.Geol.Arg.Rev., 31 (4): 241-259.

Spalletti, L.A. & Gutiérrez, R. (1976). Estudio granulométrico de sedimentos glaciales, fluviales y lacustres de la región del Monte San Lorenzo, Provincia de Santa Cruz. Asoc.Geol.Arg.Rev., 31 (2): 92-117.

Spalletti, L.A. & Lluch, J.J. (1972). Contribución al estudio morfométrico de clastos. Rev.Museo de La Plata, N.S., Secc.Geol., 8: 117-147.

Spalletti, L.A. & Lluch, J.J. (1973). Estudio sedimentológico comparativo de los Rodados Patagónicos y del cordón psefítico del Lago Pellegrini (Provincia de Río Negro). Asoc.Arg.Min.Petrol.Sed.Rev. (AMPS), 4 (4): 105-141.

Spencer, D.W. (1963). The interpretation of grain-size distribution curves of clastic sediments. Jour.Sed.Petrol., 33 (1): 180-190.

Teruggi, M.E.; Mazzoni, M.M. & Spalletti, L.A. (1971). Sedimentología de las gravas del Río Sarmiento (Provincia de La Rioja). Rev.Mus.La Plata, N.S., Secc. Geol., 7: 77-146.

Thomas, R.L.; Kemp, A.L.W. & Lewis, C.F. (1972). Distribution, composition and characteristics of the surficial sediments of Lake Ontario. Jour.Sed.Petrol., 42 (1): 66-84.

Unrug, R. (1957). Recent transport and sedimentation of gravels in the Dunajec Valley, Western Carpathians. Acta Geol.Polonica, 7 (2): 252-257.

Visher, G.S. (1965). Fluvial process as interpreted from ancient and recent fluvial deposits. in: Middleton, G.V., editor, Primary sedimentary structures and their hydrodynamic interpretation. Soc.Econ. Pal.Min., Spec.Public., 12: 116-132.

Visher, G.S. (1969). Grain-size distributions and depositional processes. Jour.Sed. Petrol., 39 (3): 1074-1106.

Williams, P.F. & Rust, B.R. (1969). The sedimentology of a braided river. Jour.Sed. Petrol., 39 (2): 649-679.

Wright, H.E. (1962). Role of the Wadena lobe in the Wisconsin glaciation of Minnesota. Geol.Soc.Amer.Bull., 73 (1): 73-100.

Some depositional models in glaciolacustrine environments (southern Pyrenees)

DAVID SERRAT & JOAN M.VILAPLANA
Universitat de Barcelona, Spain

CARLES E.MARTÍ
Instituto de Estudios Pirenaicos, Jaca, Spain

1 INTRODUCTION

Clearly orientated in an east-westerly direction between 42º and 43º N latitude, the Pyrenees is a mountain range that extends from the Mediterranean Sea to the Bay of Biscay, serving as isthmus between the Iberian Peninsula and the European continent. Through more the half of its 400 km in length there are summitting peaks, from the Puigmal Massif (2912 m.a.s.l.) in the east to the Balaitous Massif (3144 m.a.s.l.) in the west, above or very close to 3000 m in height. The highest peak is the Aneto (Maladeta Massif) 3414 m.a.s.l. and in the centre of the range.

Geologically we are dealing with an Alpine folded mountain range, formed at the line of junction between the Iberian and European Plate, that has an axial zone of Paleozoic materials, with Hercynian granites, and a Prepyrenean zone on both slopes, though wider in the south where it meets the Ebro basin than where it meets the Aquitaine basin in the north, with Secondary and Tertiary materials showing structures pertaining to tangential tectonics.

The east-west setting of the Pyrenees cuts obliquely across the incoming Atlantic storms from the north and north-west, creating a marked climatic contrast over the western sector of the mountain range, such that the southern slopes are dry and the northern wet. On the otherhand, the eastern sector is influenced by the Mediterranean lows that distribute the rainfall equally over both slopes, from the Puigmal to the Mediterranean Sea.

2 THE QUATERNARY GLACIATIONS IN THE PYRENEES

These mountainous massifs have been affected by the Quaternary glaciations, just as the very Albrecht PENCK had already demonstrated in 1883 by distinguishing the four alpine glaciations in the Pyrenees. In certain cases his interpretations were based on the existing river terraces in the Prepyrenean zone.

Later and more detailed investigations have tried to find sediments belonging to the glaciations of the Lower Quaternary, but as yet the interpretations, that are normally based on the existence of isolated erratic blocks, do not give a sufficiently clear idea of the palaeo-environment or the implicit importance of these probable glaciations.

However, the remains of glacial sediments belonging to the recent Quaternary, are much more common and can be found in nearly all of the central Pyrenean valleys. In spite of this, the studies that have been carried out mainly on the specific landforms and weathering of glacial materials, have not been able up till now to find out how long the phases lasted or how many there were, given that the only evidence available is very localised in time and space, within certain priviliged valleys whose great width has preserved the deposits from the rivers and torrents. It is very difficult to compare different

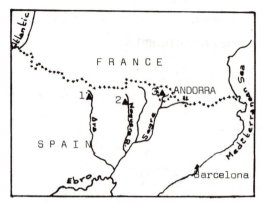

Fig. 1 Situation of the studied examples in the Pyrenees. 1.- Linás de Broto; 2.- Llestui; 3.- La Massana (Andorra)

Pyrenaen valleys and even more so if we want to compare them with the Alps.

Our investigations are centred on defining the most complete and continuous local stratigraphic and lithostratigraphic units for each valley, according to the guide-lines marked out in the Dijon meeting (1978) of the Sub-commission on European Quaternary Stratigraphy (SEQS), so as to be able to stablish similarities between adjacent valleys doing without the terminology on alpine glaciations.

In order to do this, we needed much more continuous registers and deposits than could be provided by the ridges of the terminal moraines studied up till now, so we began a regional detailed study that allowed us to distinguish the preserved accumulations of glacial origin in the middle parts of the valleys and generally associated with blockages in the tributary valleys by the main glacier, and the consequent lacustrine and deltaic deposits and tills than can normally be correlated. The examples dealt with in this paper belong to the Ara River Valley (obstruction of the Linás de Broto), the Noguera Ribagorçana – Llauset Valleys (obstruction of the Llestui) and the Valira River Valley (obstruction of the La Massana – Andorra)(Fig. 1).

We have adopted the provisional nomenclature of the Commission on Genesis and Lithology of Quaternary Deposits of the INQUA, in the sedimentological studies

that we have carried out.

3 GLACIOLACUSTRINE DEPOSITS OF THE SOUTHERN PYRENEES

Obstructions in the tributary valleys by the glacier that flowed through the main valley, are responsible for the present-day glaciolacustrine deposits on the southern slopes of the Pyrenees. The glacial tongue that caused the obstructions had a thickness of about 300–400 m and slightly entered into the tributary valleys, where it deposited moraine that held back its own lacustrine environment.

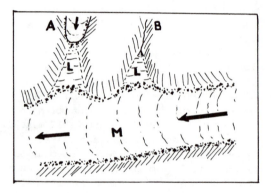

Fig. 2 Glaciolacustrine environments (L) obstructed by a main glacier (M). A and B, tributary valleys with glacier (A) or without (B)

The blocked-up tributary valleys could or could not have a glacier (Fig. 2), conditioning the lacustrian environment that received the sediments and water sent down from the tributary valley : when there was a small glacier the environment would be proglacial and in its absence, fluvial.

3.1 Linás de Broto glaciolacustrine formation

The first example dealt with in this paper, refers to the deposits that dammed up the lake obstructed by the glacier of

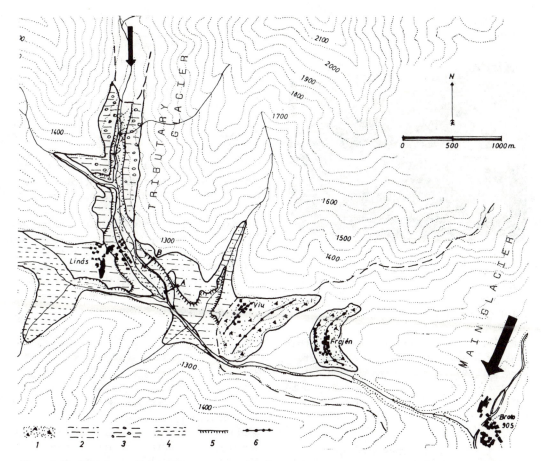

Fig. 3 Sedimentary formations in the Linás de Broto zone. 1.- Morainic deposits; 2.- Glaciolacustrine deposits; 3.- Glaciofluvial deposits in correlation with 2; 4.- Terraces and torrential deposits, posterior to 1, 2 and 3; 5.- Scarped border; 6.- Morainic ridge. A and B are the vertical sequences of the figure 5.

the Ara River valley - which had a length of up to 35 km - where it met its tributary, the Sorrosal Barranco. Where they converged, the tongue of the main glacier was 400 m thick in ice. There was another glacier tongue in the tributary valley, but it never reached the main valley. The deposits that dammed up this lake are well-known since PENCK (1883), though no detailed study has yet been done of them.

It is the intention of this paper at this outcrop, to study the glaciolacustrine and fluvial deposits that are visibly up to 60 m thick, while trying to deduce its relationships with the main glacier tongue, and test if the repetitions

observed in the cycles correspond only to the local characteristics, or whether they can be extrapolated to the whole of the mountain range.

This outcrop lies in the Alto Aragón, which is an area of the Pyrenees, climatically transitional in nature between the influence of the atlantic climate, from the north and west, and the mediterranean climate, that comes in specially from the south along the bottom of the valleys. As a result, the heads of the glacial valleys are normally subjected to atlantic-type rainfall (about 2000 $1/m^2$), while the lower parts, where the glaciers used to terminate, clearly feel the mediterranean-type rain and temperatures.

Fig. 4 View of sedimentary formation of Linás de Broto.

3.1.1 Description of the deposits

There are three types of formations that clearly stand out from amongst the studied sediments, and which succeed each other in a rather regular fashion. We interpret this succession as more or less complex repetitive cycles (Fig. 5).

3.1.1.1 Glaciolacustrine complex

Here we are dealing with clayey silt and striated pebbles and blocks. They start off being massive without any visible bedding, and later lead on to a rhythmic varve-type deposits to the order of a millimetre, with sandy intercalations somewhat thicker, but as a whole, lacking in sedimentary structures.

3.1.1.2 Low energy fluvial and deltaic deposits

Predominantly sandy levels (medium to fine sands) with occasional appearances of clay layers with flame structures; the sands are most frequently arranged into climbing ripples, commonly in-phase and more rarely in drift. Normal current ripples as well as sandy levels with parallel lamination are scarce. All of these

wave forms can be found made to stand out by the underlying clay drapes.

Some of the layers show inclined foresets, where the gravels progressively wedge out into and alternate with sandy beds containing climbing ripples.

3.1.1.3 Fluvial gravels

Medium-size rounded pebbles represent the typically fluvial deposits (centil 5-10 cm). Occasionally there is parallel lamination clearly visible in the levels containing smaller-sized gravels. Dispersedly in other occasions, these episodes present sand lenses with cross-bedding, which make one think of a braided river deposition. As a whole, the bottom limit to the gravels is an erosive surface, fitting in to poorly-marked channels (0.2 - 1 m).

3.1.2 Genetic interpretation

Let us place ourselves within the context of a glacially obstructed zone with a small, more or less permanent, lake. From there we can appreciate the uniformity of the massive deposits of column A (Fig.5), lacking in bedding surfaces and formed by clayey silts and striated blocks, previously described in the section on glaciolacustrine deposits. However it is difficult to see whether we are really dealing with ice-rafted pebbles. The grain-size of the sand, silt and clay fraction ($<$ 2 mm) (Fig. 6) is practically the same in all the glaciolacustrine levels but otherwise perceptibly different to the percentages of clay in the lateral tills in the principal, as well as in the tributary valleys. This criterion does not clarify things excessively, as there are probably grain-size differences between the basal tills and the tills from the lateral moraines (although we have endeavoured to sample the lodgement till in the latter). Presumably, the former is more similar to vertical section A. Nevertheless, there are not appreciable grain-

234

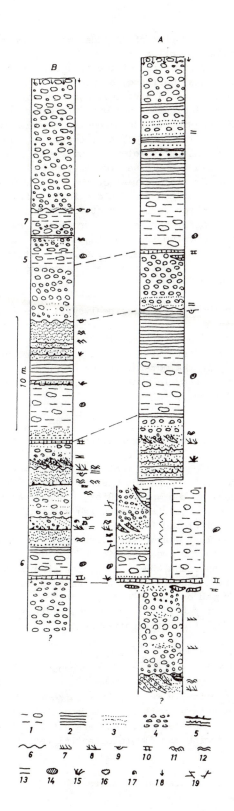

Fig. 5 Vertical sequences of Linás de Broto. 1.-Lacustrine clays with striated pebbles; 2.-Layered clays and rhythmites; 3.-Sands; 4.-Fluvial pebbles; 5.-Sands with clay layers; 6.-Erosive contact; 7.-Cross-bedding; 8.-Foreset beds; 9.-Scour and fill; 10.-Carbonate cemented beds; 11.-Cross lamination; 12.-Climbing ripples 13.-Parallel lamination; 14.-Striated pebbles; 15.-Flame structures; 16.-Clay balls; 17.-Convoluted beds; 18.-Soil weathering; 19.-Sinsedimentary glacitectonic structures (inverse and normal faults).

The numbers 5, 6, 7 and 9 placed on the grafic, correspond with the samples analysed in the figure 6.

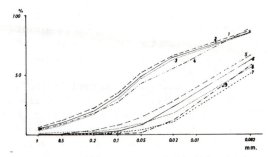

Fig. 6 Granulometric analysis (cumulative curves) of the Linás de Broto sediments.

1 and 2.- matrix of lateral moraine of the main glacier, Viu area; 3 and 4.- matrix of lateral moraines of the tributary glacier; 5, 6 and 7.- lacustrine deposits (columnar section B, Fig. 5); 8.- massive deposit under lodgement till; 9.- stratified clays (columnar section A, Fig. 5).

size differences between the supposed waterlain till and the associated typical bedded lacustrine deposits (Fig. 5, A and B) which Dreimanis considers (1979) as being an important criterion to distinguish between the glaciomarine and glaciolacustrine sediments, although this same author admits that the glaciolacustrine clayey silts could be as poorly sorted as the tills.

It is obvious what relation exists between the glacier tongue and those massive

235

Fig. 7 Sinsedimentary glacitectonic struc
tures in the glaciolacustrine deposits of
Linás de Broto.

deposits, as there are in a series of
inverse faults, that can only be explained by glaciotectonics. The geometry of
the formation tells us that it comes from
the main valley.

On top of these last deposits there are
rhythmic deposits with alternating clay
to silt and fine sands. Since the grain-
size distribution is basically the same
as in the forementioned deposits, this a-
rrangement seems to indicate accumulation
in a calm lake that would have been later
replaced by other sediments deposited in
other low energy current environments,
with large amounts of silts and sands nor
mally giving climbing ripples, as in the
case of medium to fine sands, and parallel lamination when the type of material
indicated higher energies (coarse sand
and gravel). Sporadically appearing in
the profile, the wedged foresets are always under the river gravels and erosively in contact with them.

Frequently the upper part of the river
gravels are irregularly cemented with car
bonates. Locally these fill in channels,
which we take to be the end of the cycle
as the materials that are directly above
them are again the glaciolacustrine silts
and clays.

3.1.3 Local conclusions

The succession of episodes may be summa-
rised as follows :

a. Obstruction of the tributary valley
(Sorrosal Barranco) by ice from the glacier of the Ara Valley, and formation of
a small glacial lake with its corresponding sediments (occasionally associated
with compressive glacial tectonics) that
were initially waterlain till and later
clearly of a lacustrine origin.

b. Partial retreat of the ice and depo
sition of sands and clays in deltaic and
low-energy water current formations.

c. A further and more pronounced retreat of the ice, causing the lake to
empty into the main valley. Deposition
of gravels with carbonate cementing on
the top.

In the field, this ideal cycle-type which
can be more complex especially in the mo
re distant parts of the main valley, who
lly repeats itself three times. The bottom of the outcrop is the upper part of
another cycle whose earlier phases are
unknown.

3.2 Llestui glaciolacustrine formation

The second model that we have studied is
to be found in the Llauset Valley, tribu
tary of the Noguera Ribagorçana River Va
lley. A glacial tongue ten kilometres
long, flowed through this valley during
the last Quaternary Glaciation, to join
on to the great glacier of the Noguera
Ribagorçana, 30 km long. The glaciolacus
trine deposits that we have studied, was
formed on the left margin of the Llauset
glacier (Fig. 8) by the obstruction of a
small fluviotorrential basin.

3.2.1 Description of the deposits

Genetically we can distinguish three sedimentary units in this deposit intimate
ly inter-related. They are glacial, lacustrine and fluviotorrential facies.

3.2.1.1 Glacial deposits

236

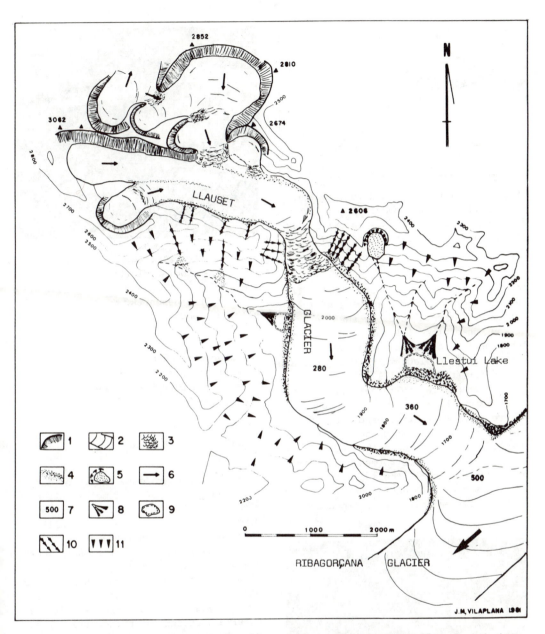

Fig. 8 Interpretation of the glacial environment in the Llauset Valley area during the last Quaternary Glaciation. 1.- glacial cirque; 2.- glacier; 3.- crevasses; 4.- supraglacial debris; 5.- nivation cirque; 6.- ice flow; 7.- thickness of the glacial tongue; 8.- alluvial fan; 9.- glacial lake; 10.- avalanching channels; 11.- slopes with generalized periglacial processes.

These are formed by till deposits of the Llauset glacier, and are essentially res ponsible for the blockage of the lake. The deposit, about 80 m thick can be di-vided into a lower part, subglacial till, and an upper part, supraglacial till.

The subglacial till is formed of stria ted boulders, pebbles and gravels contai

237

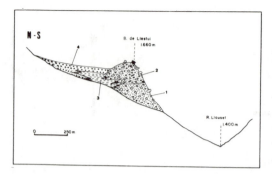

Fig. 9 Profile of the sedimentary forma
tion of Llestui. 1.- subglacial till;
2.- supraglacial till; 3.- lacustrine de
posits; 4.- alluvial fan deposits.

Fig. 11 General view of Llestui formation. 1.- subglacial till; 2.- supraglacial till; 3.- lacustrine deposits; 4.-
alluvial fan deposits. We can see the
alluvial landform.

ned in a grey clayey compact matrix. In
certain sections, the particles of the
matrix are orientated along shear planes.
At the top of the till and in the northern part, the glacial deposits included fragments of lacustrine sediments are
also deformed by the till.

The supraglacial till rests directly on
top of the subglacial till, although the
re are remains of structured and deformed
lacustrine clay and sands on a rather i-
rregular contact between the two. On the
whole, the deposit is formed by a great
abundance of blocks, the majority of
which are granitic; and by schist and gra
nite pebbles and gravels within a beige-

coloured silty sand-matrix. There is a
general tendency for the clasts to dip
towards the interior of the lake basin
(NW) (Fig. 9).

3.2.1.2 Lacustrine deposits

These deposits are essentially of clay
and sand rhythms of centimetre thickness
with parallel lamination. One can well
appreciate the sedimentary instability of
this depositional environment in the
south-eastern margin of the basin where
the glacial deposits lie upon lacustrine
deposits. At the opposite side of the ba
sin, there is an observable transition
zone between the lacustrine and fluvial
material, belonging to the alluvial fan
that functioned as a fan delta. There are
abundant load and current structures such
as: clay injections, flame structures and
ripples in the sandy layers. Other intraformational structures, such as small
slumps, faulted beds, and ice-rafted pebbles or boulders (Fig. 10) stand out
amongst these sediments and characterize
the instability of the glaciolacustrine
environment. Although the lake deposit is
not visible and the centre of the basin

Fig. 10 Small conjugated fractures in
the lacustrine rhytmites of Llestui formation made up of sand layers (dark colour) and clay layers (light colour).

is now very eroded, we estimate the total thickness to be 20 m in the central part of the lacustrine basin of Llestui.

3.2.1.3 Alluvial fan deposits

Here, we deal with two coalescent alluvial fans, that were once functioning as a delta. They are mainly formed of levels of poorly rounded pebbles, dipping about 5º towards the centre of the basin. At the apex the pebbles dip somewhat greater, but distally tend to become horizontal. There, we are able to see the lateral change from alluvial to lacustrine facies, alread commented upon previously. Boulders and layers of sand as well as lenses of sand appear only sporadically. This alluvial fan formation is estimated to be 40 m thick, covering the top of the lacustrine formation and even leaning on the lateral moraine.

3.2.2 Discussion on the depositional model

Fortunately we are dealing with a well-preserved sedimentary formation, having good outcrops and being of relatively small size. All these things ease the difficulty of establishing a depositional glaciolacustrine model.

We are therefore dealing with the creation of a juxtaposed glaciolacustrine basin, blocked by the Llauset glacier. The sedimentary formation depends on the modest amount of materials washed down on both sides along the two torrents. Laterally these pass on to the centre-of-basin sediments, resting on lateral moraine that also progrades onto the lacustrine deposits. Also noteworthy are the upper levels of the cones that cover the lacustrine deposits and even lean on the lateral moraine. From this we gather that the basin was filled in completely.

From the large quantities of sedimentary material with respect to the limited size of the obstructed, makes us think that the blockage lasted quite a long time. This would indicate a period of stability for the glacier. When the glacier retreated completely and the obstruction ceased to exit, the base level of the outlet of the lake changed abruptly and created a deep incision in the previously deposited sediments

3.3 La Massana glaciolacustrine formation (Andorran Pyrenees)

The third example that we are presenting can be found at the confluence of the Valira d'Ordino river valley and the Arinsal river valley, both of which are tributaries of the Valira d'Orient. These rivers are responsible for the main part of the drainage network of the Principat d'Andorra.

In this case the ice dam was provoked by the great glacier of the Valira d'Orient valley, that held back the proglacial discharge from the Ordino and Arinsal glaciers. Because of the important post-glacial incisive fluvial activity we are left with only isolated remains, sometimes very difficult to correlate between themselves.

3.3.1 Description of the deposits

As with the previous models and following the same line of treatment we are also able to group the different types of deposits into three sedimentary units, genetically different.

3.3.1.1 Glacial deposits

In the most northern sector we find clayey till with an abundant clayey silt matrix, grey in colour, compact, and containing glacial clasts of varying sizes. Clear structures are not commonly found, and there is only the occasional shear plane. As we do not have any definite criteria to state otherwise, we believe that we are dealing with a basal lodgement till.

In the most southern zone of the basin,

Fig. 12 Ice-rafted-pebbles included in
the bottomset sediments of the La Massana lake.

the till has very distinctive traits. Ins
tead of clayey, its matrix is sandy, with
clear structures of water circulation within the sand lenses as well as several
wash out structures, belonging to a basal
melt-out till.

Fig. 13 Foreset sediments of the deltaic
unit in the La Massana formation.

3.3.1.2 Glaciofluvial deposits

At several points of the basin the previous unit is covered by fluvial facies
sediments, containing levels of rounded
pebbles, sand and silt beds with climbing ripples and occasional clay. The top
of this formation corresponds to filled
in level, remains of an outwash plain,
about 70 m above the present level of the
river.

3.3.1.3 Glaciolacustrine deposits

Near La Massana and its environs, there
are several very important remains of se
diment lake accumulations. We distinguish
the following two types:

   A) Deltaic deposits. This is the best
preserved deltaic formation in the glaciolacustrine complex of La Massana, from
which we believe we have been able to
distinguish three typical deltaic units:

   a) Bottomset sediments. When close to
the delta front, these sediments can be
best represented by centimetre levels of
clay and silt with parallel lamination
containing ice-rafted-pebbles (Fig. 12)
alternating with sandy levels. In the
centre of the basin the sediments are
clearly varved and are described in the
following paragraph - B.

   b) Foreset sediments. We find these covering the most proximal facies of the
bottomset sediments. They are formed of
well-bedded gravels and sands dipping
about 25º towards the centre of the basin (S) (Fig. 13).

   c) Topset sediments. These materials
are resting erosively on top of the fore
set beds. At the base there is a bed of
poorly classified materials, which is fo
llowed by an interval of sand with current megaripples underlying horizontally structured sands and gravels. The
poorly graded materials reappear at the
top of the unit as small blocks, gravels
and sand lenses, all within a silty matrix. The surface of the sedimentary unit
belongs to the level of an ancient deltaic plain, 50 m above the present river
level (Fig. 14).

Fig. 14 Erosive contact (dotted line) be
tween the foreset sediments and the topset sediments. The sedimentary structures show the current directions from
right to left.

B) Lake-bottom deposits. Lacustrine se
diments isolated by post-glacial fluvial
activity are to be found in the south and
to the west of the deltaic formation
(Fig. 16). They are centimetre rhythms
of fine sand or light silt in the lower
part, and dark finely-laminated clays in
the upper part. Sometimes the parallel
lamination is distorted by ice-rafted-pe
bbles. Although the tranquility of this
sedimentary environment expels the existence of sedimentary flow structures,
there are noteworthy cases of intraformational crumplings (Fig. 15) affecting
a small number of cycles, and probably
caused by, as COHEN (1979) points out, a
slope depositional environment. In some
sectors, these rhythmites are slightly
tilted and affected by small normal
faults. Laterally they lean on the clayey
till.

3.3.2 Discussion about the depositional
      environment

A small phase of recession and a later

stabilising phase in the dynamics of the
glaciation arrived after the glaciation
had reached a maximum in its phase, during which the Ordino and Arinsal glaciers had joined the Valira d'Orien , to
whom the basal till of La Massana belongs. While the confluent glacial tongues separated and the small Ordino and
Arinsal glaciers receded up-valley, the
principal valley glacier became stabili-
sed, obstructing and slightly entering
into the affluent valley. The blockage
held back the glacial deposits that had
already been deposited previously and on
top of which a small out-wash plain later built-up (Fig. 16). During the phase
of stability there developed a proglacial lake in contact with the fronts of
two small glaciers to the north, and with
the clayey till from the south.

Fig. 15 Intraformational crumplings in
rhythmite sediments of La Massana formation. To see two normal faults of low an
gle

From the palaeocurrent measurements and
the special situation of the delta forma
tion, we have interpreted the delta as
being the result   of the accumulations
of a juxtaglacial river from the Ordino
glacier (Fig. 16). In spite of not having any proof, we have not rejected the
possibility of subglacial discharges in
those areas where the front directly came into contact with the lake. What has
definitely been demonstrated is that the
moraine was released from the front of
the glacier in icebergs. This sedimentary formation, important in volume acts

241

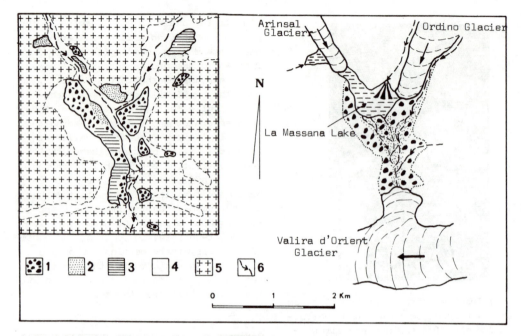

Figure 16 La Massana area (Andorra). On the left: Map showing the Quaternary deposits. 1.- till deposits; 2.- lacustrine deposits; 3.- glaciofluvial terrace; 4.-Alluvial and colluvial deposits; 5.- Paleozoic bedrock; 6.- stream. On the right: Interpretation of the Quaternary glaciolacustrine environment.

as a register of a stable period that maintained the obstruction during an important lapse of time, though this type of sedimentation is relatively quick according to Gustavson and al. (1975).

Within this glaciolacustrine environment we have clearly defined a deltaic facies of coarse materials and current structures that indicate a high sedimentation rate; a fine sedimentary facies, on the bottom of the bassin, definitely deposited in seasonal succession from sediments in suspension (detailed studies of the pollen are being carried out). We should however underline the special depositional conditions at the foot of the distal prodeltal slope, where small avalanches, turbidity currents and certain sin sedimentary deformations within the rhythmites could have taken place. As a final word, let us say that the normal faults of small slip, that locally disrupt the rhythmites, were probably caused

by internal movements as the sediments settled or even by small collapses inside the sedimentary mass.

4 CONCLUSIONS

This study has allowed us to see the possibilities that are offered by applying sedimentological techniques to the field of geomorphology for reconstructing the glacial history of the southern slope of the Pyrenees.

Following the criteria adopted at the INQUA Commission on Genesis and Lithology of Quaternary Deposits, we have distinguished genetically different tills within the glacial sedimentary formations. The relation between these deposits and the associated glaciolacustrine and glaciofluvial deposits has permitted us to interpret the depositional models of Liñás de Broto, Llestui and La Massana (Andorra).

Fig. 17 Artistical picture showing the reconstruction of the glaciolacustrine environment of La Massana (Andorra) during the last Quaternary glaciation. The La Massana Lake is in contact with the Ordino glacier (on the right) and the Arinsal glacier.

Though this is not a stratigraphical paper, we think it is interesting to make a brief reference to this aspect. As was already mentioned in the introduction these studied deposits correspond to the last Quaternary Glaciation. As we also pointed out in Figure 18, the southern slope of the Pyrenees underwent a maximum in the glaciation during this period, and to this pertains the deposition of the La Massana till. In the phase that followed, the glacier became stabilized and, during a rather prolonged period of time, the glacial obturation of the tributary valleys produced the in-filling of the glaciolacustrine environments that they themselves had caused. This phase was registered in the lacustrine sediments that we have studied and which reveals us three models. Subsequently, as distinguished by some small pulsations,

the glaciers retreated completely. Lastly, one final advance brought down a cold and dry climate and gave way to a phase

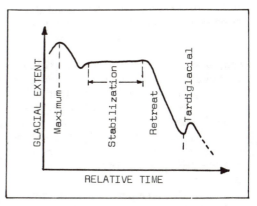

Fig. 18 Relationship between relative ti̲me and glacial extent during the last Quaternary Glaciation (Southern Pyrenees).

243

of rock glaciers and small cirque glaciers, called Tardiglacial by Serrat (1979).

We think that the absolute datings presently under way and deeper probing of current studies as aided by the introduction of new techniques, will lead to the establishment of a Quaternary stratigraphy for the Pyrenees as well as a better understanding of the evolution of the Quaternary

5 ACKNOWLEDGEMENTS

We wish to express our sincere gratitude to all the members of the INQUA Commission on Genesis and Lithology of Quaternary Deposits, particularly to Christian Schlüchter of the Federal Institute of Technology (ETH-Hönggerberg), Zürich, for the advice and help received in preparing this paper.

We should also like to thank to the Department of Geomorphology and Tectonics of the Faculty of Geology, University of Barcelona, under whose auspices we have carried out this paper.

6 BIBLIOGRAPHY

Alimen, E. 1964, Le Quaternaire des Pyrénées de la Bigorre, Mém. Carte Géol. France. Paris.

Alimen, H., Solé Sabarís, L., Virgili, C. 1957, Comparaison des formations glaciaires des versants N et S des Pyrénées, Res.Com. V Con.Int.INQUA, 9-10 Madrid-Barcelona.

Cohen, J.M. 1979, Deltaic sedimentation in glacial Lake Blessington, County Wicklow, Ireland. In Ch.Schlüchter (ed) Moraines and Varves, p. 357-367. Rotterdam, Balkema.

Fontboté, J.M., Solé Sabarís, L., Alimen, H. 1957, Livret guide de l'excursion N.- PYRENEES, V Con.Int.INQUA, Madrid Barcelona.

Gustavson, T.C., Ashley, G.M., Boothroyd J.C., 1975, Depositional sequences in glaciolacustrine deltas. In A.V.Jopling B.C.McDonald (ed.), Glaciofluvial and glaciolacustrine sedimentation, p. 264 280, S.E.P.M., Spec.Publ., 23.

Martí Bono, C.E., Serrat, D., González, M.C. 1978, Los fenómenos glaciares en la vertiente meridional de los Pirineos, V Col.Geog.Granada-1977, p. 67-73 Granada.

Miskovsky, J.C. 1974, Le Quaternaire du Midi Méditerranéen, Stratigraphie et paléoclimatologie, Etud.Quaternaire, 3 Paris.

Nussbaum, F. 1946, Orographische und morphologische Untersuchungen in den östlichen Pyrenäen, Jahr.Geog.Gess.Bern, XXXV-XXXVI: 245 p. Bern.

Penck, A., 1883, Die Eiszeit in den Pyrenäen, Mitt.Ver.Erdk. In french: Bull. Soc.Hist.Nat. 19: 105-200, Toulouse.

Reineck - Singh 1973, Depositional Sedimentary Environments. New York, Springer-Verlag.

Serrat, D. 1979, Rock glacier morainic deposits in the eastern Pyrenees. In Ch.Sclüchter (ed.), Moraines and Varves, p. 93-100. Rotterdam, Balkema.

Taillefer, F. 1967, Extent of Pleistocene Glaciation in the Pyrenees. In H.E. Wright - W.H.Osburn (ed.), Arctic and Alpine Environments, p. 255-266. India na, Univ.Press.

Taillefer, F. 1969, Les glaciations des Pyrénées, Et.Fran.sur le Quat., INQUA Paris, p. 19-32

Vilaplana, J.M., Serrat, D. 1982, Els diposits d'origen glacial de la cubeta de La Massana - Ordino (Andorra): llur significació paleo-geogràfica, Acta Geológica Hispánica 14: 433-440

Deposition in a thermokarst sinkhole on a valley glacier, Mt. Tronador, Argentina

S.RUBULIS
Hidronor S.A., San Carlos de Bariloche, Argentina

ABSTRACT

The specific features of a sediment, such as grain size, shape, orientation and packing of grains and the nature of the sedimentary structures are determined mainly by small-scale mechanisms operating during transport, deposition and early compaction and deformation of the deposited materials (Blatt et al. 1972). Distinctive units of a deposit can be a clue to the interpretation of the genesis and the environment at the time of deposition. Complex deposits are produced in the terminal area of a Mt. Tronador valley glacier due to thermokarst and restriction of ice movement by an important endmoraine. Observations of deposition in a thermokarst sinkhole penetrating to the glacier bed indicate the source of the material from subglacial and intraglacial tunnels for some of the stratified and nonstratified deposits. Subsequent deformation of these deposits by movement and collapse of ice is frequent. The deposits, as studied in this stagnant sector of the glacier are characterized by different units of sorted, stratified and nonstratified material of fairly uniform size within the unit, but highly variable from one unit to the other, even when the units are in contact. These units are often covered or underlain by ablation till.

1. INTRODUCTION

Mt. Tronador at 71° 53' W and 41° 10' S is a composite volcanic cone superimposed on granitic and metamorphic relief. Probably Miocene to Pleistocene age can be given for the volcanics (Greco 1974, Gonzalez Bonorino 1973). Proportion of pyrolastic material is high. R. Manso Glacier or Ventis-

quero Negro, the black glacier, is a valley glacier flowing out of the Mt. Tronador ice cap and facing S and SE. The highest elevation of this mountain and glacier is 3554 m a.s.l. At about 2000 m a.s.l. an ice fall separates the upper 5 km-long clean part from debris-covered, regenerated glacier some 700 m below (Rabassa et al. 1978). The regenerated part of the glacier 3 km downvalley abuts against an end moraine. This moraine is broken by the present river outlet and on the SW side by a probable neoglacial event (Lawrence and Lawrence 1959). The glacier front is controlled by restricted ice movement because of an end moraine and has more or less pronounced thermokarst features according to the prevalence of positive or negative mass balance of this part of the glacier.

The ice thickness at the ice fall has been estimated at over 100 m; measured 500 m from the end moraine, more than 50 m. Due to downwasting in the period from 1977 to 1981, the difference in elevation of the highest point of the glacier to the glacier bed, was reduced from 66 m to 48 m at 200 m from the front. The ablation area below the ice fall is 1.8 km^2 with some 10 km^2 of accumulation area above; an additional 5 km^2 is producing meltwater, incorporated in the glacier from the valley sides, but with no avalanching of snow or ice from this source to the glacier surface.

2. CLIMATE AND HYDROLOGY

Climate for Pampa Linda meteorological station, 4 kms away and 105 m below in elevation to the R. Manso glacier corresponds to: Perhumid, microthermic with little or no deficiency of water, Thornthwaite classification AC'$_2$ra' (Gallopin 1978).

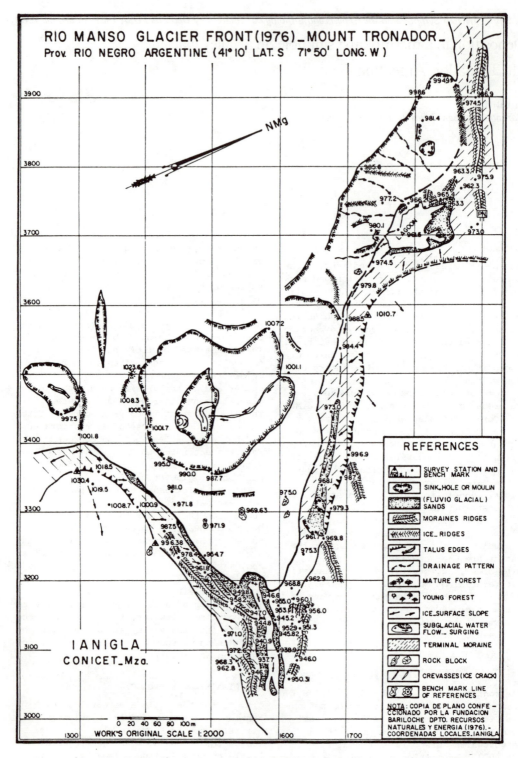

Figure 1 : R. Manso Glacier, generalized map of frontal parts (mapped 1976)

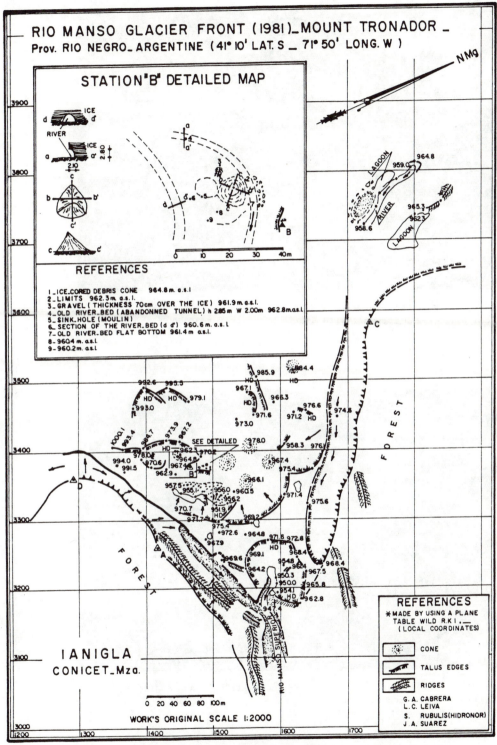

Figure 2 : R. Manso Glacier, detailed map with morphological and
glaciological features of the frontal parts (mapped 1981)

247

Figure 3 : Manso Glacier with avalanche cone in right upper corner.
Conical hillock and ice-wall with debris bands in foreground

Figure 4 : Conical hillock (sample RM 1), kettlehole and supraglacial pond connected to subglacial stream

The yearly precipitation is close to 3000 mm, 50% of this amount falling in 3 months: May, June and July. The average yearly temperature is below 6.0^o and the monthly average of the two coldest months, June and July, is close to 0^oC.

The stream discharge is highly variable and produces frequent flooding when fresh snow is washed downslope by intense warm rainfall. Measured at a gauging station of Upper R. Manso, fed by Mt. Tronador glaciers, on 7th to 8th of June, 1981, an increase from 21.2 m^2/s to 100 m^2/s was observed in 20 hours and a decrease in 8 hours from 108 m^2/s to 47.4 m^2/s. The daily discharge of 7.8 m^2/s on May 1st, increased to 35.3 m^2/s on May 2nd, 1981

The subglacial and intraglacial tunnels will tend to fluctuate about a steady state appropriate to average water supply of the preceding weeks (Shreve 1972). The variability of the meltwater discharge from the glacier is greater than from the Manso river, occasionally increased even more by opening and closure of new and old passages and by events, such as collapse of ice or opening of a crevasse. Thus conditions for alternating high and low energy transport of the sediments are frequent.

3. DEPOSITIONAL ENVIRONMENT AND SEDIMENTOLOGY

Thermokarst features (Clayton 1964; Embleton and King 1975) develop on the lowermost 1000 m of the regenerated part of R. Manso glacier. Debris cover increases over this part of the glacier from 20 to 30 cm to about 50 cm at the snout of the glacier. The surface is extremely uneven, steep ice walls of 10 to 20 m are common. Supraglacial pools and sinkholes develop adjacent to these walls. Some of the sinkholes are preserved even in positive mass balance conditions for this sector (Figs. 1, 2 and 3).

One of the sinkholes observed on 1970 aerial photographs was positioned in 1976 over a curve of a subglacial stream, where the flow inverted from downglacier to upglacier in a horizontal distance of 80 m. The probable control for this stream was bedrock topography, a 4 m drop in water level was mapped. The "travelling" sinkhole, apparently, on reaching an appropriate position had penetrated to the glacier bed and established a more permanent residence. This sinkhole has been increasing

in extension up to now, partially by capturing nearby sinkholes and at present is in contact with the end moraine, a few remnants of ice still attached to the proximal slope.

The subglacial stream has maintained this position until 1980. From 1976 on the dividing crest inside the curve of the stream superimposed units of moderately sorted, washed sediments were deposited similar to Eisrandschwemmkegel (German et al. 1979), but with fines washed out. At the present time the deposits are left as conical hillocks, the subglacial stream is obstructed and has found a new tunnel under the ice.

A neighbouring sinkhole mapped in 1976 has been closed by glacier movement before 1980 and the deposits deformed and pushed 13 m up an ice slope in contact with the end moraine, similar to Press-Schuppe, described by Gripp (1979). On the southwest side of the enlarged sinkhole, downwasting has exposed an englacial tunnel, that can be related to sediments mapped in 1981. On the inside of a meander a conical ice-cored hillock partially eroded on one side by a temporary pool, is covered by moderately sorted, very weakly stratified sand (sample RM1 Table and Figs. 4 and 5). The stratification has a concentric peel structure corresponding to the outer shape, according to Keller 1952 this is typical for kames (Reineck and Singh 1980). The size distribution has only one mode in 0 to 2 phi sizes (72% of the sample) corresponding to mixed suspension and bed load transport, the median and mean at 1.25 phi. On one side of this hillock a terrace is superimposed and has close to 50% of pebble-sized deposits. The sediment is polymodal, very poorly sorted with the principal mode in -6 to -5 phi size, the median at -2 phi (sample RM2 Table 1).

The surrounding ice wall contains a remnant of the subglacial tunnel that had washed out these deposits. This tunnel has a curved roof, similar to most ice and snow caves and a flat floor, covered with sand and granule-sized material similar to sample RM1. The dimensions are given in a detailed map (Fig. 2) and are similar to the dimensions of a subglacial stream that has replaced the tunnel. The continuation of the tunnel floor is preserved on one side of the conical hillock. On the other side a temporary pool with a bottom connection to the present subglacial stream was related in March 1981 to the pulsating flow of this stream as could be seen from surging and subsiding of a fountain above the connection

TABLE 1 : Size distribution

Size Ø units	Weight %			Cumulative %		
	RM1	RM2	RM3	RM1	RM2	RM3
-6...-5		35,0	2,0		25,0	2,0
-5...-4		12,2	3,7		37,2	5,7
-4...-3	0,1	7,3	6,7	0,1	44,5	12,4
-3...-2	0,3	4,7	12,7	0,4	49,2	25,1
-2...-1	1,9	9,5	28,1	2,3	58,7	53,2
-1... 0	7,5	7,9	27,7	9,8	66,6	80,9
0... 1	34,6	15,2	18,4	44,4	81,8	99,3
1... 2	37,4	13,5	0,6	81,8	95,3	99,9
2... 3	15,1	3,7	0,05	96,9	99,0	99,95
3+++++	3,1	1,0	0,05	100	100	100

TABLE 2 : Roundness %

Sample	RM1		RM2			RM3		
Ø units	-X...-2	-2...-1	-X...-1	-3...-2	-2...-1	-X...-3	-3...-2	-2...-1
Total n	44	131	104	131	148	161	248	113
Rounded								1,8
Subrounded	15,9	10,7	12,5	0,8	0	0,6	0	0,9
Subangular	13,6	35,1	23,1	40,5	35,8	29,8	27,0	40,7
Angular	59,1	48,1	59,6	55,7	62,8	65,9	71,4	54,0
Very Ang.	11,4	6,1	4,8	3,0	1,4	3,7	1,6	2,6

Figure 5 : Conical hillock and adjacent terrace (samples RM 1 and RM 2)

Figure 6 : Supraglacial stream connected with subglacial stream, as seen below roof of remnant tunnel

to the subglacial stream. In May 1981 this pool had drained and contained stratified lake sediments.

On the downglacier side of the hillock a deposit of subangular to rounded cobbles up to 30 cm in diameter was observed. Between the englacial tunnel and the temporary lake a 70 cm sequence of poorly sorted stratified deposits (sample RM3 Table 1) formed a small terrace with the dominant mode in -2 to 0 phi sizes (55.8% of granules and very coarse sand) with 99.3% of the sediments coarser than coarse sand, the median and mean at -1 phi. Each of the 3 samples taken has different characteristics, but all have less than 3% of very fine sand and silt and contain no clasts of more than -6 phi. The degree of roundness was estimated visually using the scale of Powers (1953) and gave results similar to Lawson (1979) for basal dispersed pebbles (Table 2). The granulometric analysis on the 3 dried samples was performed over an ASTM sieve series at 1 phi intervals for granule and sand sizes, pebbles sizes were sorted visually (Blatt et al, 1972; King 1978; Reineck and Sing 1980; Folk and Ward 1957). The rest of the sinkhole contained similar hillocks and lenses of stratified material, some of the deposits were spalled by downwasting of subjacent ice. Alluvial and solifluction fans were observed below inflexions of the surrounding ice walls, but most of the material was ablation till, somewhat sorted by the irregular surface and differential melting of ice and washed by rain and meltwater.

4. CLASSIFICATION

One of the difficulties met in the studies of natural systems and processes is that all the variables are dependent to a higher or lower degree on each other and, on trying to define too closely, we are left with nothing but the definition. The deposition from, on or under a valley glacier as in this study of deposition in a thermokarst sinkhole on Mt. Tronador valley glacier is fairly representative of this problem.

The deposition is mainly by fluvioglacial processes, but direct deposition is very important as well as deformation by glacier movement and solifluction. The closest approximation found for a classification of these deposits is moulin kame, as defined by Reid (1967): "Conical hills of outwash sands and gravels deposited in depressions in a glacier or as alluvial cone against a steep ice face at the terminus of a glacier". Most of the outwash sands and gravels in this sinkhole have their source in subglacial and englacial tunnels (Fig. 6); the process of their genesis is very similar to the eskers as described by K. Gripp (1978), accumulation in the tunnels and washout when the hydrostatic pressure has increased.

As Reid (1967) has very well stated there is some question as to whether all or even a majority of ice sinkhole lakes (and sinkholes) originate through the enlargement of moulins. Sinkhole kame could be a more appropriate term. This term could include all deposits in sinkholes, such as: Lake deposits, delta deposits, outwash fans, temporary stream channel deposits, ablation till, and flow till from the surrounding and subjacent ice and deformation till produced by the movement of the glacier.

5. CONCLUSIONS

Deposition by a glacier takes place during melting and even pure glacial deposits have been affected by running water. The quantity of water present during and immediately after deposition gives a distinct character to the deposits (Reineck and Singh 1980; Sugden and John 1976).

To produce the deposits found in a sinkhole of the valley glacier of Mt. Tronador, the following mechanism is proposed: Meltwater and precipitation wash the less coarse material downslope the uneven surface of the glacier surface, most of all from exposed ice walls. These materials accumulate in depressions in ice and in subglacial and englacial tunnels and cavities. As the competence for sediment transport decreases on reaching wider or more ample sections the coarse fraction is skewed towards the coarser sizes, but as these sizes are dropped the distribution tends to become symmetrical (Blatt et al. 1972).

On increase of meltwater supply the increase of hydrostatic pressure produces upwelling and pulsating turbulent flow at the outlet of englacial or subglacial tunnels in a similar way to that described by Gripp (1978) and will discontinuously transport available sediments and deposit this material very often as separate units close to the outlet on reaching the sinkhole. This material can be deformed by glacier movement or melting of subjacent ice or be affected by fluvioglacial processes in the sinkhole.

6. ACKNOWLEDGEMENTS

To the "Instituto Argentino de Nivologia y Glaciologia" and J.C. Leiva, J.A. Suarez and G. Cabrera we transmit sincere thanks for field work in 1981 and for the drawing of maps, to Ing. A. Marcolin, "Instituto de Tecnologia Agropecuaria", for sieve analysis. We thank Mrs. M. Böndel for the revision of the English text, Mrs. S. Fremery and Mrs. Th. Frei for the typing of the manuscript.

7. REFERENCES

Blatt, H., Middleton G. and Murray, R. 1972, Origin of sedimentary rocks, Prentice Hall, N.J.

Clayton, L. 1964, Karst topography on stagnant glaciers, Journ. of Glac. vol. 5: 107-112.

Embleton, C. and King, C.A.M. 1975, Glacial Geomorphology, E. Arnold, London.

Folk, R.L. and Ward, W.C. 1957, Brazos River bar: a study in the significance of grain size parameters, Journ. of Sed. Petrol., V. 27: 3-26.

Gallopin, G.C. 1978, Estudio ecologico integrado de la cuenca del R. Manso sup. R. Negro, Argentina, Anales de Parques Nacionales, Vol. XIV: 179-188.

German, R., Mader, M., Kilger, B. 1979, Glacigenic and glaciofluvial sediments, typification and sediment parameters, Moraines and Varves (Ch. Schlüchter ed.), Balkema A.A., Rotterdam.

Gonzalez Bonorino, F. 1973, Geologia del area entre S. C. de Bariloche y Llao-Llao, Fundacion Bariloche, Dep. Rec. Nat. y Energ., No. 16.

Greco, R. 1974, Descripcion geologica de la Hoja 40a Monte Tronador, Serv. Nac. Geologico. (Unpublished)

Gripp, K. 1978, Die Entstehung von Geröllosern (Eskers) Eiszeitalter und Gegenwart, Vol. 28: 92-108.

Gripp, K. 1978, Glazigene Press-Schuppen, frontal und lateral, Moraines and Varves (Ch. Schlüchter ed.), Balkema A.A., Rotterdam.

Keller, G. 1952, Beitrat zur Frage Oser und Kames, Eiszeitalter und Gegenwart, Vol. 2: 127-132.

King, C.A.M. 1978, Techniques in geomorphology, E. Arnold, London.

Lawrence, D.B. and Lawrence, E.B. 1959, Recent glacial variations in Southern South America, Off. of Nav. Res. Contr. No. 641 (04), American Geograph. Soc., N.Y.

Lawson, D.E. 1979, Sedimentological Analysis of the western terminus region of Matanuska Glacier, Alaska, CRREL report, 79-9.

Powers, M.C. 1953, A roundness scale for sedimentary particles, Journ. Sed. Petrol, Vol. 23: 117-119.

Rabassa, J., Rubulis S., Suarez, J. 1978, Los Glaciares del Monte Tronador (R. Negro, Argentina), Anales de Parques Nacionales, Vol. XIV: 264-316.

Reid, J.R. 1967, Glaciers "Living and Dead", Proc. of North Dakota Acad. of Sc., Vol. XXI: 42-56.

Reineck, H.E., Singh, I.B. 1980, Depositional Sedimentary Environments, Springer Verlag, Berlin.

Schreve, R.L. 1972, Movement of water in glaciers, Journ. of Glac., Vol. 11: 205-214.

Sugden, D.E., John, B.S. 1976, Glaciers and Landscape, E. Arnold, London.

Subaquatic mass flows in a high energy ice marginal deltaic environment and problems with the identification of flow tills

JONATHAN M.COHEN
Marathon Mining Ireland Ltd., Dublin, Ireland

ABSTRACT

The occurrence of find-grained sand/silt/clay rhythmites within steeply dipping coarse gravel foreset sequences provides suitable conditions for the production of diamictons and a wide range of high matrix content pebble and gravel units. Occasional gravel avalanches which occurred during periods of predominantly low energy conditions of rhythmite deposition, caused overloading of the underlying finer grained sediments, the production of excessive pore pressures and failure. The combination of gravel and fine grained rhythmites resulted in a matrix supported deposit which fits the description of diamicton. These diamictons are internally indistinguishable from known flow tills in the same delta. It is suggested that this may cause complications in other similar cases in the recognition of flow tills and in the identification of glacial deltaic and ice marginal environments where flow tills are used as criteria.

The sedimentology of the rhythmites is presented, together with criteria for the identification of the rhythmites as an ice marginal deltaic foreset facies. It is suggested that the recognition of this facies may overcome the complications posed by such diamictons in the recognition of glacial depositional environments.

1. INTRODUCTION

Diamictons and high matrix content pebble and gravel units occur within fine grained rhythmites (clay to sand sized) in the foreset gravel sequences of a high energy, ice marginal, multilobate Gilbert type delta complex. The delta complex is 5km wide and up to 170m thick and is confined within a preglacial bedrock valley. Deposition occurred into proglacial Lake Blessington at the margin of the ice sheet which covered the midlands of Ireland during the Weichselian glaciation (fig. 1).

The general sedimentology of the delta is treated by Cohen (1979a, 1979b). Briefly, the delta is a multilobate construction with a wide variation in paleoflow directions and rapid vertical facies changes between alternating foreset gravels and bottomset sands. This is a function of shifting sources of sediment supply which caused temporary and local delta abandonment. Sudden cessations in flow resulted in the deposition of fine grained (clay to sand sized) rhythmites within the steeply (20-30°) dipping foreset gravel sequences. The rhythmites are exposed at 5 different locations in the Blessington delta and range from approx. 0.30m - 4.60m in thickness. Similar rhythmites are also exposed in the nearby Pollaphuca delta (Fig. 1).

2. SEDIMENTOLOGY OF THE RHYTHMITES

The rhythmites generally consist of fine to coarse sand, silt and thin clay laminae with intercalated, laterally discontinuous, pebble, gravel and diamicton units (fig. 2). Sequences are laterally variable in thickness both in sections which are transverse and longitudinal to the foreset dip direction. The lateral thickness variability in the transverse sections is due to the concentration of mass flow deposits along discrete flow paths and channels within the rhythmites. The suspension and current deposited sediments maintain relatively constant lateral thickness (Fig. 3).

The majority of the sand and finer grained units are internally structureless. The most common

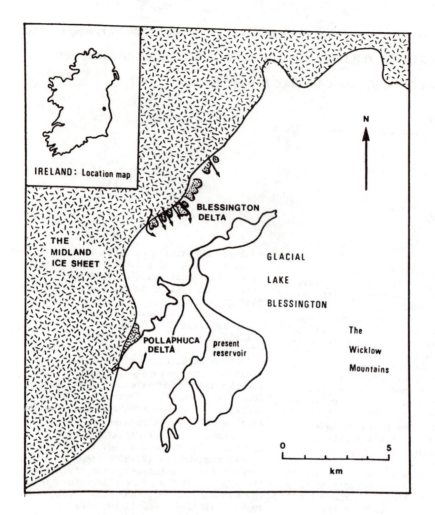

Figure 1. Location map showing the position of the Blessington and
Pollaphuca deltas marginal to the Midland Ice Sheet in the initial
phases of the final deglaciation. The lake was dammed to the east
by high ground of the Wicklow Mountains and by local glaciers.

sedimentary features are sole markings on
the bedding planes of individual laminae.
A wide variety of other sedimentary
features may be present, including:

a) Transverse erosional current
markings similar to those produced
artificially by Allen (1969).

b) Parting lineations which are
oriented parallel to current flow
and foreset dip direction.

c) Groove markings produced by
pebbles which avalanched down the
rhythmite covered foreset slope (Fig. 4).

d) Flame structures and floating
rip-up clasts produced by turbidity
current flow down the foreset slope.

e) Brittle deformation in the form
of minor faulting and shear planes, and
overfolding produced by grain flows.

f) Intraformational crumplings
produced by syndepositional downslope
mass movements.

g) An original structure, termed a
'squeeze-out' structure, has been
identified (Cohen 1979b) and attributed to
localized syndepositional compression and
lateral flowage of sand and silt between
confining clay layers.

Within the total rhythmite
successions, a spectrum of gravel and
pebble units exists, containing various
concentrations of fine grained matrices.

Figure 2. A typical rhythmite sequence within a foreset gravel
succession. Note the laterally discontinuous diamicton unit. Also
note that the rhythmites have been eroded at the base of the
diamicton.

NOTE : (a) flow into diagram,

(b) black lines represent clay laminae
and can be considered as time-lines,

(c) incorporation of first deposited sediments
within succeeding deposits,

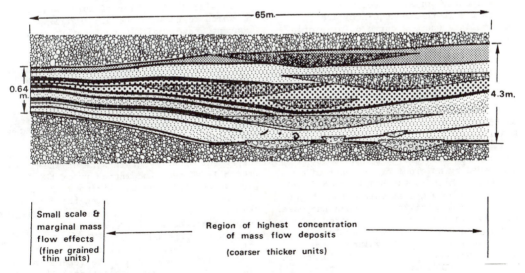

Figure 3. Generalized transverse rhythmite section showing the
relationship of the units and lateral thickness variability.

257

Figure 4.   A natural calcreted mould of groove markings produced by
stones which avalanched down the foreset slopes.

The grain size distributions of examples of these gravel units is shown in (Fig. 5). Also included is data for nearby known flow till units in the same delta. A spectrum of matrix grain distributions is apparent. Curve 8 (foreset gravels from beneath the rhythmite sequence) indicates the coarsest matrix with the lowest concentration of fines. Curves 7 and 4 (poorly sorted gravel units from within the rhythmite sequence) are intermediate whereas curve 5 (diamicton from within the rhythmite sequence) has the highest fine grain matrix content. The flow till samples represented by curves 2 and 3 have a slightly higher although similar matrix content to the diamictons within the rhythmites.

3. REMOBILIZATION AND THE FORMATION OF DIAMICTONS

The exposure illustrated in Figure 6 contains foreset gravel at the base succeeded by an erosional remnant of rhythmite which is further succeeded and surrounded by a gravel unit with a higher fine grained matrix content than the basal gravel. This succession indicates that the rhythmites have become partly reworked by a succeeding gravel avalanche and became incorporated into the avalanche to produce a higher fine grained matrix content. Essentially, the gravel avalanche transformed from a grain flow

where dispersive pressures were generated by grain to grain contact to a slurry flow where dispersive pressures were generated by the relatively viscous interstitital fluid containing the remobilized fine grained rhythmites. The result is a tendency towards the formation of a matrix supported structure. This increase in matrix content is illustrated by the finer grain size distribution of curve 4 (Fig. 5) which represents the high matrix content gravel unit in contrast to curve 8 which represents the basal gravel underlying the rhythmites.

It is proposed that the rapid loading of the unconsolidated rhythmites on the foreset slope by gravel avalanche is the mechanism of remobilization. This occurred by the formation of excessive pore pressures during the gravel avalanche (Morgenstern 1967). Conceivably, the grain size of the matrix is largely dependent upon the grain size of the remobilized bed, the intensity of the avalanche, dilution and mixing with the surrounding lake water during the flow of the remobilized mix. In some cases matrix supported pebble units exist where the matrix is composed of sand (lower part of Fig. 7). It appears that during the corresponding avalanches the foreset slope was covered by sand rather than silts and clays as in the case of the true diamictons. Where the avalanche was not

258

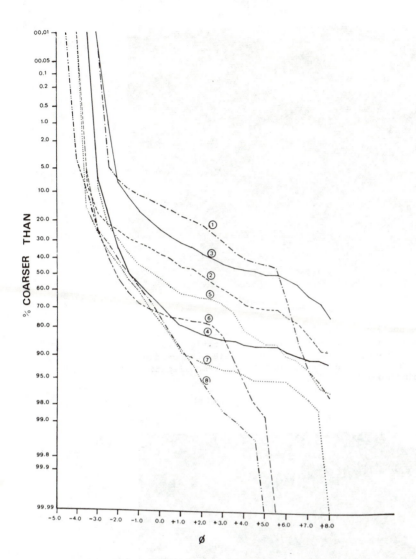

Figure 5. Cumulative grain size distribution of the matrix contents of
a variety of mass flow deposits associated with foreset rhythmites;
curve 1 - remobilized sand unit; curve 2 - flow till A; curve 3 -
flow till B; curve 4 - remobilized high matrix content gravel;
curve 5 - diamicton formed by remobilization; curve 6 - remobilized
sand matrix supported gravel A; curve 7 - remobilized sand matrix
supported gravel B; curve 8 - gravel with no apparent evidence of
remobilization from beneath the rhythmite sequence.

sufficiently intense, simple
stratification of pebble and gravel clasts
exist within relatively undisturbed sand
units (Fig. 7).

A similar mechanism was proposed by
Crowell (1957) to account for the origin
of pebbly mudstones (tilloids) in Jurassic
to Pliocene sediments in California. His
mechanisms involved the remobilization and
mixing of previously deposited strata of
alternating fine and coarse grained
sediments. It seems probable in this case
however, that the diamictons were produced
by the mixing of predeposited fine grained
sediments with subsequent avalanches of
gravel down the foreset slopes, rather

Figure 6. Transverse section through a foreset rhythmite sequence
   illustrating coarse gravel unmodified by remobilization (A)
   succeeded by an erosional remnant of rhythmites (B), succeeded by
   gravel with a high matrix content (C) due to the remobilization of
   the underlying rhythmites.

Figure 7. Sand and pebble units of the foreset rhythmite. Note the
   sand matrix supported structure of the pebble unit (lower part) and
   the isolated thin pebble bands.

than the mixing of alternating rhythmites and gravel beds. This is for two reasons: (a) the underlying rhythmites were eroded by the subsequent mass flow rather than incorporated by interstitial mixing, and (b) within the rhythmites, gravel containing the highest fine grained matrix contents (diamictons) are interbedded with the forest rhythmites. If Crowell's mechanism had operated in this case then relatively thick foreset gravel units should be found within the rhythmites representing the original interbedded fine and coarse sediments prior to failure and mixing. Only thin (less than 0.30m) gravel beds are found in these rhythmites and in most cases it is clear that the matrix content has been enhanced by remobilization.

There are three prerequisite conditions necessary for the production of diamictons by remobilization:-

a) The presence of a steep slope to help initiate and to perpetuate the gravitationally induced mass flow of the coarse clasts over the finer grained sediments.

b) The presence of a covering of fine grained sediment on the foreset slope.

c) The passage of coarse grained sediment induced by gravitational flow over the bed of finer grained sediment (current driven bedload is specifically excluded as the finer grained sediment on the bed would be winnowed away in most cases).

A high energy ice marginal glaciolacustrine delta provides a most suitable environment with these conditions. There is a supply of coarse gravel to form steep foreset slopes and fluctuations and sudden cessations in flow to allow fine grained sediments to be deposited on the steep slope. A low energy ice distal environment would not be suitable mainly because sudden cessations in sediment supply could not occur to cause local delta abandonment and foreset rhythmite formation. The sudden cessations in flow were caused by the occasional closure of separate meltwater tunnel feeders for the delta at the ice margin. In the case of an ice distal delta, the meltwater supply routes from various tunnel exists at the ice margin would become mixed over the intermediate subaerial outwash area before reaching the lake area (Cohen and Huddart, in press). Hence, the effect of the occasional closure of one tunnel mouth would not be effective at the distal delta providing that the total meltwater discharge was not greatly affected. Therefore alternating

sequences with marked contrasts in grain size such as the foreset rhythmite relationships are absent in ice distal deltas.

4. COMPARISONS BETWEEN SUBAQUATIC FLOW TILLS AND DIAMICTONS FORMED BY REMOBILIZATION

The deposition of flow tills in preglacial environments has been previously proposed by Carey and Ahmad (1961), Lavrushin (1969), Rukhima (1973), Drozdowski (1971) and Evenson et al (1977). The actual term 'subaquatic flow till' was applied by Evenson et al (1977) to describe sediments deposited by mass flow into a standing body of water from marginal ice. This term is preferred here as it emphasizes the mechanism of deposition by subaqaceous mass flow in contrast to the previously used term 'water-lain till' (Dreimanis 1969).

The restriction of foreset rhythmites and diamictons formed by remobilization to ice marginal environments is coincidental with the occurrence of flow tills. Confusion between flow tills and remobilization produced diamictons is likely because flow tills have been found to occur preferentially in association with foreset rhythmites in both the Blessington and Pollaphuca deltas. The reason is that foreset rhythmites represent a period of low meltwater input over part of the delta; therefore flow tills released from the adjacent ice are less likely to be reworked and washed by contemporaneous meltwater flow and tend to be preserved in association with the lower energy foreset rhythmite facies. True flow tills are recognized in the delta by the simple absence of erosional remnants of rhythmites or fine grained sediments either beneath or in lateral proximity. However, the possibility exists that flow tills may also occur in association with rhythmites that are partly eroded and contain diamictons formed by remobilization. In such cases distinctions would be difficult if not impossible to make.

Textural data is used to form comparisons between sediments which have been deposited as subaqautic flow tills and those sediments which have been busjbected to the proposed remobilization process. Both subaquatic flow tills and remobilized sediments which contain high percentages of fine grained matrices were deposited by similar subaqueous mass transport mechanisms, namely by 'slurry

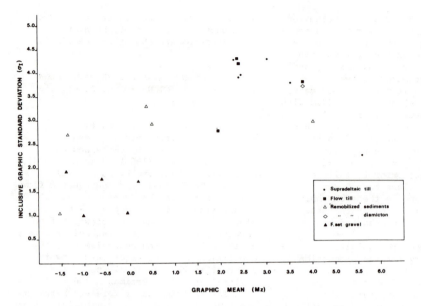

Figure 8. Mean grain size vs. standard deviation for the foreset
     sediments and supradeltaic till samples.

flow'. These sediments originated from
different sources. The subaquatic flow
tills were derived from unsorted sediments
supplied directly from the adjacent ice
margin, whereas the remobilized sediments
originated on the foreset slope by the
reworking of indigenous foreset sediments.
The distinction in origin can be made in
classifying the subaquatic flow tills as
allocthonous and the diamictons produced
by remobilization as autocthonous
sediments (i.e. formed within the
lacustrine system).

4.1 TEXTURAL ANALYSIS

Textural data for a range of foreset
sediments is presented which includes
samples of - (a) subaquatic flow tills,
(b) foreset gravels which have been
apparently unmodified by remobilization
and (c) foreset sediments and diamictons
which have been clearly modified and
produced by remobilization. The textural
data from supradeltaic till samples from
the same delta is included for comparative
purposes.

Grain size analysis has been carried
out using a combination of the sieving and
hydrometer techniques (British Standards
1967). Analysis was carried out at ½ Ø
intervals. The matrix content has been
recalculated from the full grain size
distributions and is presented here. The

grain size parameters are determined by
the method of Folk and Ward (1957). The
graphic size parameters include: mean
grain size, inclusive graphic standard
deviation, inclusive graphic skewness and
inclusive graphic kurtosis. The graphical
method was used to enable further
comparisons to be made with the results of
Landim and Frakes (1969) who worked on the
distinction between tills and other
diamictons.

The values for the various size
parameters are plotted against each other
(Figs 8, 9, 10, 11), to determine which
parameters are most effective in
discriminating between the samples.

Supradeltaic till and subaquatic flow
till samples contain the finest grained
and most poorly sorted matrix. The mean
grain size appears to be the most
discriminating parameter between the tills
generally and the remaining sediments.
There is a slight tendency for the flow
till samples to be more platykurtic (lower
positive values of kurtosis) and the
foreset gravels tend to be more
leptokurtic with the supradeltaic till
situated at intermediate positions.
Generally, kurtosis does not appear to
serve as a discriminating parameter for
these samples. The majority of the
samples are symmetrically skewed (around
zero), however there is a greater tendency
for the flow tills to be symmetrically

262

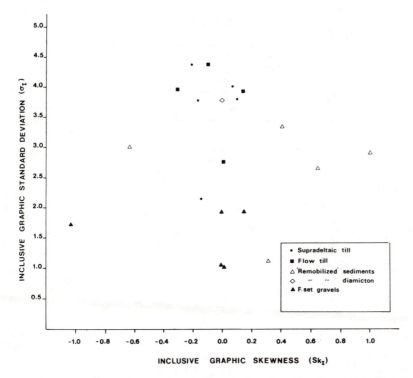

Figure 9. Skewness vs. standard deviation for the foreset sediments and supradeltaic till samples.

skewed than there is for the remobilized sediments.

4.2 GEOLOGICAL SIGNIFICANCE OF THE GRAIN SIZE DATA

The geological significance of the four parameters - mean, standard deviation, skewness and kurtosis is outlined by Landim and Frakes (1969) and Sahu (1964).

The supradeltaic till and the subaquatic flow tills contain the finest grained and the most poorly sorted matrix due to the lack of water sorting in contrast to the other foreset sediments. Landim and Frakes (1968) found similar results - "Till samples show the poorest sorting, have skewness symmetrically disposed around zero and also display relatively small values of mean grain size. Alluvial fan samples (which include subaerial mass flow deposits) are better sorted, positively skewed and have larger values for mean size. Outwash deposits show the best sorting, the highest values for kurtosis and negative skewness". They interpreted the till distributions as a

product of the lack of water sorting and the alluvial fan deposits, which are slightly better sorted due to the activity of some water sorting. They also stated that "subaqueous mudflow deposits may thus be expected to be even better sorted but have not yet been observed." The positive skewness values of subaerial mass flow deposits found by Landim and Frakes was attributed to transportation by mudflow and similar viscous types of mass movement agents. The diamictons produced by remobilization also tend to have higher positive values of skewness than subaquatic flow tills (Fig. 9). This suggests that the remobilization process is more effective in producing positive skewness values in sediments than the mass flow mechanisms which occurred during the deposition of flow tills directly from ice. The higher positive skewness values of the remobilized sediments indicates a relatively high excess of fine grained sediments. This may be due to the matrix content of the initial avalanche--during avalanche, the matrix content would have been generally reduced by mixing and washing with the surrounding fluid.

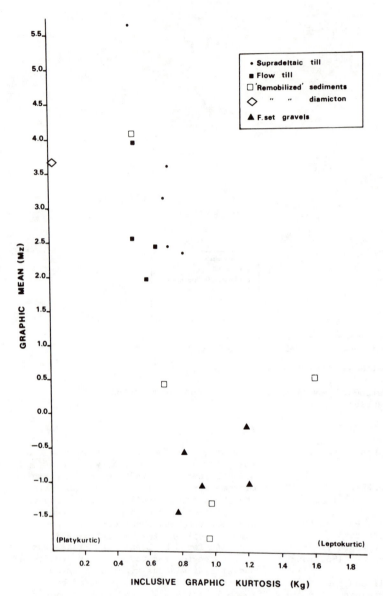

Figure 10. Kurtosis vs. mean grain size for the foreset sediments and supradeltaic till samples.

Following that, the avalanche remobilized and incorporated the predeposited fine grained sediments which covered the foreset slope to produce the final deposits which contained an excess of fine grained matrix in relation to the coarse grained matrix content.

It is suggested that the grain size distributions of remobilized sediments are a result of the combined initial grain size distributions of the first avalanche and that of the predeposited fine grained rhythmites which were subsequently remobilized. The greater the contrast in mean grain size of the two components, the greater are the values of positive skewness which may be expected in the final remobilized sediments.

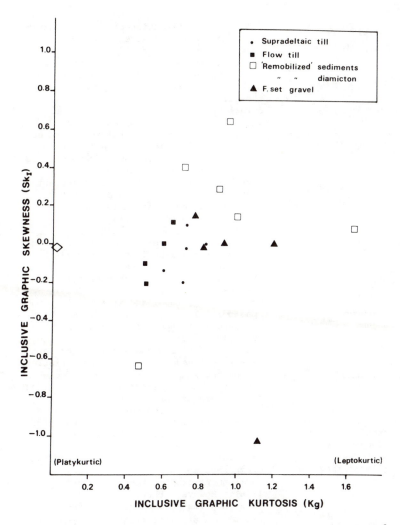

Figure 11. Kurtosis vs. skewness for the foreset sediments and supradeltaic till samples.

5. COMPARISONS BETWEEN RHYTHMITES WHICH OCCUR WITHIN FORESETS WITH DISTAL LAKE FLOOR RHYTHMITES

Exposure limitations may often not allow delta foreset dips and associated clear deltaic environment characteristics of rhythmites to be recognized. In such cases those rhythmites may be mistaken for distal lake floor rhythmites and associated diamictons may be interpreted as tills. The object here is to present criteria which may aid in the recognition and interpretation of rhythmites as a delta foreset facies and therefore support

the alternative that intercalated diamictons may not necessarily be tills. If on the other hand, the rhythmites prove to be distal lake floor deposits, then the remobilization hypothesis is less viable.

A summary of the sedimentary features found in foreset rhythmites is presented in Fig. 12. Additional features which are not illustrated include a variety of sole markings both of the gravitational tool formed and current formed types. It is suggested that the following two features may be used as primary criteria for these facies discrimination.

(a) Deposition by turbidity currents generated by the input of sediment laden

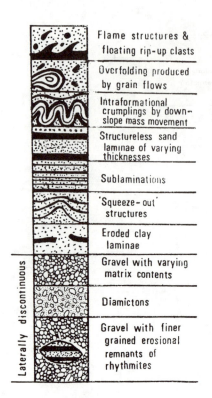

Flame structures &
floating rip-up clasts

Overfolding produced
by grain flows

Intraformational
crumplings by down-
slope mass movement

Structureless sand
laminae of varying
thicknesses

Sublaminations

'Squeeze-out'
structures

Eroded clay
laminae

Gravel with varying
matrix contents

Diamictons

Gravel with finer
grained erosional
remnants of
rhythmites

Laterally discontinuous

Figure 12. A summary of the sedimentary features found in foreset rhythmites which may be considered in the identification of the foreset rhythmite facies. Sole markings are not represented here.

A high correlation between the grain size and the thickness of laminae within foreset rhythmites should not exist. The coarsest laminae of these rhythmites may be the thinnest. This is because grain size and thickness are not proportional to the size of a mass flow, but a re determined by the composition of the initial slump. In contrast, a high correlation between the thickness and mean grain size of summer laminae (or the coarser laminae of rhythmite couplets) is characteristic of lake floor varves/rhythmites as the most competent turbidity currents should deposit the thickest and coarsest laminae.

(b) Regional controls (e.g. lake level fluctuations or changes in the position of the ice margin) should be most evident in the lake floor rhythmites in contrast to delta foreset rhythmites. In the latter case, the characteristics of the coarse laminae of each couplet is controlled by local and effectively random spasmodic slumping and mass flows, which are largely independent of regional controls. A random distribution of the coarse grained laminae should be present in vertical sequences and rapid variations in thickness and grain size would be expected in transverse/lateral directions, in the case of foreset rhythmites.

It is suggested that the following two features may be used as secondary or supportive criteria for the facies differentiation.

(c) The presence of intraformational crumplings would indicate a slope environment greater than 2°, (Shrock 1948). However, they do not necessarily infer a delta foreset slope. Other crumplings and intrastratal deformation features may be produced by mass flow. Such mass flows could occur on slopes as low as 3.5° (Shaw 1977) and are not necessarily restricted to foreset environments as they have been found within esker deposits (Saunderson 1975). It is concluded that there is a greater probability of finding these structures on the steep slope which can occur within a foreset environment, but they are not solely diagnostic of that environment.

(d) Sole markings (both tool and current formed) are common in the foreset environment, but can be found at the base of turbidity current deposits and are not necessarily characteristic of a steep slope environment. The tool markings may be used as criteria for a steep slope environment if it is found that the width of the tool markings represent clast sizes which are greater than those clasts which

meltwater or by generation from mass flows are characteristic of lake floor rhythmites (Shaw 1977). Deposition by mass flow is characteristic of the foreset slope depositional environment, together with deposition from suspension. The main criteria is that rapid lateral variations in thickness should occur within the foreset rhythmites in sections which are transverse to bed dips. This is because the mass flows and slump generated turbidity currents would produce localized deposition. There may also be a differential thickness variation in clay laminae due to the localized interruption of background clay suspension sedimentation in the areas of mass flow.

could be transported by turbidity currents in each case. The only mechanism remaining to form such large tool markings would be gravitational transport by rolling down a steep slope (Fig. 4).

6. SUMMARY

The high energy ice marginal glaciolacustrine environment provides suitable conditions for the formation of diamictons both by a process of remobilization and by the deposition of subaquatic flow tills. The final process of deposition of these diamictons is similar and the sediments may be difficult to distinguish internally. This may cause complications in the correct identification of till although fortunately, both types are typical of the same depositional environment.

The recognition of fine grained rhythmites associated with such laterally discontinuous diamicton units as a delta foreset facies supports the alternative remobilization process. This in turn supports the possibility that in such environments diamictons need not necessarily represent flow tills.

7. ACKNOWLEDGMENTS

The author would like to thank Marathon Mining Ireland Ltd. for financial support in presenting this paper.

8. REFERENCES

Allen, J. R. L., 1969; Erosional current marks of weakly cohesive mud beds; J. Sediment. Petrol. 39, 607-623.

British Standards, 1967; Methods of testing soils for civil engineering purposes; British Standards Institution B.S. 1377, 233p.

Carey, S. W. and Ahmad, N., 1961; Glacial marine sedimentation, In Rasch, G. O. (ed.); Geology of the Arctic, 2, 865-894, Univ. of Toronto Press.

Cohen, J. M., 1979a; Deltaic sedimentation in glacial Lake Blessington, Co. Wicklow, Ireland. In Schluchter, Ch. (ed.); Moraines and Varves, Proceedings of the Zurich Symposium, 1978, INQUA Commission on the Genesis and Lithology of Quaternary Sediments, 357-367, Balkema.

Cohen, J. M., 1979b; Aspects of Glacial Lake Sedimentation. Unpublished Ph.D. thesis, Trinity College, The University of Dublin, 344p.

Cohen, J. M. and Huddart, D., in press; Sedimentation in Tunsbergdalsvatnet, Jostedal, Norway; Geogr. Ann.

Crowell, J. C., 1957; Origin of pebbly mudstone; Bull. Geol. Soc. Am. 68, 993-1010.

Dreimanis, A., 1969; Selection of genetically significant parameters for investigation of tills; Zesk. Nauk. im. A. Mickiewicza W. Poznaniu Geografia 8, 15-29.

Drozdowski, E., 1974; Facial variation of tills in the profile of the second marine stratum at Sartowice (Lower Vistula Valley), (Polish with English summary); Zesk. Nauk. Univ. im. A. Mickiewicza W. Poznaniu, Geografia, 10, 121-136.

Evenson, E. B., Dreimanis, A., and Newsome, J. W., 1977; Subaquatic flow tills: a new interpretation for the genesis of some laminated till deposits; Boreas, 6, 115-133.

Folk, R. L. and Ward, W. C., 1957; Brazos river bar, a study in the significance of grain size parameters; J. Sediment. Petrol. 27, 3-26.

Landim, P. M. and Frakes, L. A., 1968; Distinction between till and other diamictons based on textural characteristics; J. Sediment. Petrol., 37, 1213-1223.

Lavrushin, Yu. A., 1968; Features of deposition and structure of the glaciomarine deposits under conditions of a fjord coast (based on examples from Spitsbergen) (translated from Russian). Litologiya in Poleznye Jskopaemye, 6, 63-79.

Morgenstern, N. R., 1967; Submarine slumping and the initiation of turbidity currents. In Richards, A. F. (ed.); Marine Geotechnique, 189-220, Univ. of Illinois Press.

Rukhina, E. V., 1973; Litologiya leonikouikh otlozhanii. Wedra Leingrad, 176p.

Sahu, B. K., 1964; Depositional mechanisms from the size analysis of clastic sediments; J. Sediment. Petrol. 34, 73-83.

Saunderson, H. C., 1975; Sedimentology of the Brampton esker and its associated deposits: an emperical test of theory. In Jopling, A. V. and McDonald, B. C. (eds.); Glaciofluvial and Glaciolacustrine Sedimentation, 155-176, Soc. of Econ. Paleont. Mineral. Spec. Publ. 23.

Shaw, J., 1977; Sedimentation in an Alpine lake during deglaciation, Okanagan Valley, B. C., Canada; Geogr. Ann. 59, 221-240.

Shrock, R. R., 1948; Sequences in layered rocks. 1st ed. N.Y. McGraw-Hill Book Co. Inc., 507p.

Pre-Pleistocene glaciations

Evidences for the neopaleozoic glaciation in Argentina*

CARLOS ROBERTO GONZÁLEZ
Universidad Nacional de Tucumán & CONICET, Argentina

ABSTRACT

From the Lower Carboniferous to the Lower Permian, glacial conditions were existing in parts of Argentina. An interglacial stage probably took place towards the end of the Carboniferous.
Some evidence of glaciation in two Carboniferous marine formations from Western Argentina and Patagonia is briefly discussed.

INTRODUCTION

Since the end of the Precambrian to the end of the Mesozoic, South America formed, together with Africa, Australia and Peninsular India, the great Gondwana Supercontinent. Remains of an ancient glacial age have been registered in Neopaleozoic sedimentary deposits in all these continents.
The southern part of South America was actively affected by the climatic changes that took place since the Carboniferous to the Lower Permian and which provided the conditions for the development of glaciers in the Gondwana Supercontinent.
The existence of sedimentary rocks of this age which are considered to be a direct or indirect product of glacial activity is known for the western mountainous belt of Argentina, especially in the Andean Precordillera of San Juan (Keidel 1922) and later on, in Extraandean Patagonia (Suero 1948). Although it is true that since then, the available literature is abundant and thematically varied, many doubts arose and generated extended discussions.
However, it is possible now to give additional proofs of Carboniferous glaciations

* This paper has been included in Project #42, IGCP, and in Program #73, Universidad Nacional de Tucumán, sponsored by the Fundación Miguel Lillo and CONICET

along the Precordillera of Western Argentina as well as in Central Chubut, though it is still necessary to continue the search for new data in these areas and to extend the investigations to others, with the aim of obtaining a better knowledge on the actual significance of this important paleoclimatic event.
For our purposes, only certain glacigenic features occurring in two lithostratigraphic units will be referred to in this paper. They are the Hoyada Verde Fm. of the Andean Precordillera of San Juan and the Las Salinas Fm. of Extraandean Patagonia, both having been deposited simulta-

Figure 1. Location map

neously with the onset of the Neopaleozoic Glaciation, i.e. in the Carboniferous (Figure 1)

There are no doubts that the study of glacigenic rocks presents many difficulties when they belong to ancient sedimentary formations. Their identification is still much more complex when they have been deposited below sea level, where the glacial processes are generally acting under such conditions which are markedly different to those of the continental environment and where some aspects of the sediment deposition are not directly related to the ice itself. But, to compensate this serious shortage, the sediments that have accumulated in subsident marine basins have larger probabilities of conservation in the stratigraphic record due to the more rapid deposition of younger beds on top.

SOME EVIDENCES OF THE CARBONIFEROUS GLACIATION

Both in the Hoyada Verde Fm. and the Las Salinas Fm., the existence of certain lithological, mineralogical, geomorphological and paleontological characteristics is fairly important for the interpretation of the sedimentary environment, because they show a genetic relationship with glacial processes.

(a) LITHOLOGICAL EVIDENCES
The diamictites (Flint and others 1960a, 1960b) are generally considered as the most conspicuous and diagnostic rocks of glacially-related sedimentary deposits, both in continental or marine environments. However, it is also true that the occurrence of this rock is not an indicator of glaciation by itself, because it appears too in other sedimentary environments generated by non-glacial phenomena (Crowell 1957). When dealing with ancient deposits, we prefer the terms *diamictite* or *mixtite* (Schermerhorn 1966) to others like *tillite* or *aquatillite* which have genetic connotations. Diamictites are essentially unsorted clastic mixtures with wide textural variation of their components, being generally classified as "pebbly shales" or "pebbly mudstones" (Folk 1954).

Both massive (without stratification or internal structures) and banded (various degrees of internal stratification) diamictites have been described. The lateral continuity of these diamictites is highly variable and some of them are filling ancient stream-beds with lens-like cross-section.

Frequently, the pelitic bodies contain pebbles of varied size and lithological composition which have interrupted or deformed the lamination at the site where they fell. The abundance of these pebbles (which are called "dropstones") suggests that they were carried by icebergs rather than plant-made rafts, moreover if we consider that the existence of the latter in ancient times may be even difficult to prove.

The rhythmic sequences with regularly-spaced sandy/shaly or psephitic/psammitic beds suggest a seasonal sedimentary régime. Generally, dropstones in only one of the rhythmite units would be indicative of iceberg formation periodicity with an epoch in which more intense pebble dropping occurred.

(b) GEOMORPHOLOGICAL EVIDENCES
Faceted, polished and striated clasts have been pointed as typical for glaciated areas. Similar characteristics may be formed by other processes in addition to ice (Fossa-Mancini 1943). However, in some cases, their occurrence can only be related to glacigenic processes. This is so when these striated and polished clasts appear concentrated, forming a boulder pavement beneath a diamictite. Thus, this surface should have been formed by glacial abrasion.

One of the most relevant examples, for its areal extent and other glacigenic evidences, is the striated-boulder pavement interbedded with diamictites in the lower portion of the Hoyada Verde Fm. (Figure 2). Paleochannels with diamictite filling would have been generated by subaqueous density currents, in some cases coming from the transition zone. Some of them would have been produced by gravitational sliding whereas others would have been formed by meltwater streams and slurries coming from beneath the glacier.

Large, semiconsolidated, sandy bodies would have been incorporated to the moving mass and dragged with the rest of the material (Frakes, Amos & Crowell 1969). The transport of this mixture would have been largely dependent on slope and it could have continued forward till distal zones where finally it would have stopped.

Soft-sediment deformation of various kinds have been frequently reported in association with these paleochannels and they have been interpreted as related with the mass-movement of the debris.

(c) MINERALOGICAL EVIDENCES
The coastal regions where glaciers advance into the sea may even form a floating ice-shelf may undergo several geochemical changes in ice-contact sea-water. It has been suggested (Carey & Ahmad 1961) that, under certain conditions, saline concentrates and cold brines are originated beneath the ice -shelf. These solutions accumulate at the bottom by differential

Figure 2. Partial view of the striated-boulder pavement of the Hoyada Verde Fm., east of Barreal (San Juan)

density and, if they find a protected site or closed basin they precipitate in crystals. The mineral thus formed would have probably been -originally- thenardite (Na_2SO_4) (Kemper & Schmitz 1975) and it would remain stable within a certain temperature range above which it would be metasomatically replaced by siderite or calcite. The final product is a pseudomorph known as *glendonite* (David and others 1905). In some cases, only the external mould of the crystal is preserved as hollows in the rocks or rather forming the void nucleii of concretions, occurring both as isolated individual crystals or as crystalline aggregates (González 1980b).

(d) BIOLOGICAL EVIDENCES

There are paleontological assemblages which for their occurrence and association may be used as environmental indicators. The benthonic organisms are especially sensitive to the environment variables which they depend upon.

As these variables are modified within certain threshold values, some of these biological assemblages may survive under extreme or even critical conditions; others being more sensitive or better adapted to a certain environment, will not tolerate these changes. Among the latter, those who are thermically-conditioned are the most significant as climatic indicators. *Eurydesma*, a bivalve genus found in coastal deposits of Gondwana Lower Permian and considered by Dickins (1957) as cold-water dweller, is one of them.

In the areas under study, the glacigenic sedimentary deposits (Las Salinas Fm. and Hoyada Verde Fm.) are intimately associated with marine beds carrying *Levipustula* zone invertebrate fossils. *Levipustula levis* is a brachiopod species which is relatively abundant in the Carboniferous marine beds of Western and Southern Argentina and it is used as fossil-guide in the biostratigraphy of this period (Amos & Rolleri 1956). As it has been pointed before (Roberts, Hunt & Thompson 1976), it is suggestive that the *Levipustula* zone paleogeographic distribution is very similar to that of the cold-faunas of the Permian, especially for the subpolar to polar biomas (Waterhouse & Bonham-Carter 1975). It is also important to mention that *Levipustula levis* has been found associated with several elements in the Las Salinas Fm., which integrated later on (Lower Permian) the *Eurydesma* faunal assemblage (González 1972b, 1977, 1978, 1981a).

THE CARBONIFEROUS GLACIGENIC DEPOSITS

Some general considerations on the Hoyada Verde Fm. and Las Salinas Fm. are presented below so as to provide the geological framework for the Carboniferous Glaciation evidence. More details may be found in the available literature (for example, Mésigos 1953; González 1972a).

THE HOYADA VERDE FM.

General aspects: east of Barreal, on the western margin of the Precordillera of San Juan, the Carboniferous sequence starts with the 335 m-thick Hoyada Verde Fm., diamictites, glacimarine sediments and shales (Mésigos 1953; Figure 3). The oldest beds of this Formation are almost entirely composed of glacigenic deposits, including massive diamictites.

Glacigenic features: an erosional feature (extensive striated-boulder pavement) has been identified within the psephitic masses of the lower portion of this unit (González 1981b). This pavement appears as an alignment of varied-size clasts, 200 m-long, through the diamictite beds, at the eastern flank of the Hoyada Verde Brachianticline. Clasts have been polished and striated by N-to-S flowing ice (González 1981b; Figure 2). The pavement stratigraphic location separates two distinctive diamictites in the complex, being the lower darker and richer in larger boulders. At the top of the lower diamictite (pavement level), re-transported marine shells and glendonite nodules are relatively abundant and are found *in situ* in the "Fossiliferous Shales" Member, a younger unit of this Formation. Undoubtedly, some of the most relevant evidences for glacier overriding are the polished and striated abrasion surfaces. The existence of a striated-boulder pavement in the Hoyada Verde Fm., where the glacigenic sedimentary rocks are mixed with beds containing marine shells, suggests a coastal environment origin. Although information is still fragmentary, two preliminary interpretations for the genesis of this striated-boulder pavement may be proposed:

(1) Periodical temperature changes affected the growth and recession of the glaciers during this ice-age, which may be correlated with sea-level changes on the rugged, western coast of the "Protoprecordillera" (Amos & Rolleri 1965; Rolleri & Baldis 1969). This would be partially suggested by the rhythmites with dropstones, which indicate a cyclic perturbation in the debris-transportation régime. Thus, during minimum teperature events, sea-level lowering may have exposed the previously-deposited diamicton beds (aquatill) which were later overriden by the advancing ice.

(2) The abrading effect of a continental glacier over its basement, either being it formed by bedrock or older drift, may still continue even below sea-level. The friction surface will continue as long as there is still ice-rock contact, that is, until the glacier forms its floating ice-tongue. Thus, the ice-bedrock contact may well be extended to a certain distance into the sea. In this case, the Hoyada Verde Fm. striated-boulder pavement would have been formed at the grounded-shelf zone, below sea-level, and it would not be

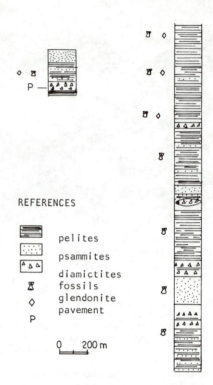

REFERENCES

	pelites
	psammites
	diamictites
𝕏	fossils
◇	glendonite
P	pavement

0   200 m

Figure 3. Comparative columns of the Las Salinas and Hoyada Verde Formations.

necessary to assume a sub-aereal exposure of the lower diamictite.

The channel-filling conglomerates, related to the glacial pavement, were originally attributed (González 1981b) to meltwater stream erosion (Lindsay 1970). Boulders and pebbles with Carboniferous fossils and glendonite nodules have been found within them.

The striated-boulder pavement is an erosional surface and so, a diastheme of unknown extent. There is also an apparent conformity between the diamictites underlying and overlying the pavement. Thus, whichever its origin had been, the time lapse between the lower diamictite deposition and the ice readvance may have been relatively short.

LAS SALINAS FM.

General aspects: the subsidence of the Central Patagonian Basin was initiated with the deposition of the Tepuel Group, during

274

lower Carboniferous. The oldest beds of this
sequence were accumulated in a continental
to littoral environment, changing afterwards into an entirely-marine facies till
the Lower Permian. The 2500 m-thick Las
Salinas Fm. (González 1972a) corresponds
to the lower portion of the Tepuel Group.
The evidences for Neopaleozoic Glaciation
in Extraandean Patagonia are found in this
unit. Like in the Hoyada Verde Fm., these
evidences appear within or, very close to,
the *Levipustula* zone levels, thus providing
a firm correlation between both formations.
(Figure 3).

Glacigenic features: some massive diamictites have been recognized within the Las
Salinas Fm. but, in general, most of them
are mainly banded, grading from very-coarse
stratification or clast orientation to true
banding. The banded diamictites have been
attributed (Frakes, Amos & Crowell 1969;
González 1972a) to subaqueous density-
current processes, coming from the proximal
zones where previously-deposited glacial
debris was already lying. These currents
have frequently produced soft-sediment
deformation over the overridden substratum.
Many small, lens-like diamictites form the
filling of paleochannels. Deformed sandstone bodies are occasionally found within
them after being dragged from their original, *in situ* location (Frakes, Amos &
Crowell 1969). The size of these bodies,
sometimes remarkable (González 1972a),
would be inversely related to the travelled
distance.

Dropped pebbles and boulders are frequent
within the fine-grained sedimentary rocks.
In many cases, the impact-induced deformation is clearly visible (Figure 4). Although
other interpretations have been suggested
to explain the origin of dropstones, the
abundance and large size of some of them
indicates that they were ice-rafted.

Void concretions and glendonite moulds are
found either scattered or concentrated in
several levels within the *Levipustula* zone,
especially at the upper half of the Formation (González 1980b). Rhythmite beds similar to those found in the Hoyada Verde Fm.
where mixtite beds and fine sandstones are
alternating, occur north of Arroyo Pescado
(Rolleri 1970). The mixtites are 20-30 cm
thick and have a finely-stratified wacky
matrix, abundant dropstones, and impact-
induced deformed lamination (Figure 5).
Sandstone beds are 3-5 cm thick, have no
matrix or dropstones, and show the predominance of tranquil deposition over
chaotic sedimentation. Laterally, the unit
is affected by subaqueous-sliding deformation (slumping). Probably, these deposits
were formed in a proximal, protected

Figure 4. A dropped pegmatite boulder, on
fine sediments of the Las Salinas
Fm., north of Arroyo Pescado
(Estancia Ap-Iwan, Chubut), showing impact deformation in the
stratification.

Figure 5. Partial view of the rhythmite
beds of Las Salinas Fm, north of
Arroyo Pescado, showing dropstones
in the laminated mixtite beds,
alternating with fine sandstones.

environment. Although no fossils have been found, the sequence is tentatively correlated with the middle to upper levels of the Las Salinas Fm., exclusively on a lithological basis.

SOME CONSIDERATIONS ON THE NEOPALEOZOIC GLACIATION IN ARGENTINA

After the Acadian movements, the western margin of South America was occupied by a mountainous belt during the Carboniferous. The "Protoprecordillera" extended at that time over a large portion of the present Precordillera of San Juan and Mendoza (Amos & Rolleri 1965; Rolleri & Baldis 1969). Similarly, in the Patagonian Andean region, the pre-Devonian and Carboniferous movements (Miller 1979) produced a more-or-less generalized up-lifting of the area. The Early Carboniferous physiography was showing a wide continental belt along the Pacific coast of South America, though areas of lower elevation were still permitting the marine transgression and embayments as the Central Patagonian Basin and perhaps also west of the "Protoprecordillera" came to existence.

Originally, a continental-type ice-sheet glaciation was assumed for large regions of Western and Southern Argentina. However, a lack of precise evidence sustained instead an Alpine-type glaciation on the higher terrains towards the end of the Devonian (Amos & Rolleri 1965; Frakes, Amos & Crowell 1969; González 1972a). This mountaineous belt was located, during the Carboniferous, in a position transversal to the present one and at more than lat. 60° S, on the basis of paleomagnetic data (Valencio 1973).

According to the available information, it seems evident that glaciers occupying these mountains were reaching the Pacific coast to the west. But, the almost-polar geographical position of South America, during the Neopaleozoic Ice-Age, was favouring the development of extended ice-sheets over the Argentine lowlands too, as it happened in the Paraná Basin. Then, these regions should have been under similar glacial conditions to those in other Gondwanic areas.

An interglacial epoch within the Upper Paleozoic climatic event, from the Lower Carboniferous to the Lower Permian, for Southernmost South America, has been discussed elsewhere (González 1981a). This interglacial epoch during the Upper Carboniferous (Stephanian) would be suggested by the absence of glacial evidence in the sedimentary formations of that age of Western and Southern Argentina,

as well as by the existence of Carboniferous to Permian, temperate-regions marine faunas along the Pacific coast (Douglass & Nestell 1976; González 1980a, 1981a).

REFERENCES

Amos, A.J. & Rolleri, E.O. (1965). El Carbónico marino en el valle de Calingasta-Uspallata (San Juan-Mendoza). Bol.Inf. Petrol., 368: 1-23, Buenos Aires.

Carey, S.W. & Ahmad, N. (1961). Glacial marine sedimentation. Publ.Dept.Geol., Univ.Tasmania, 87. Reprinted from Proc. 1st. Internat.Symp.Arctic Geol., 2:865-894

Crowell, J.C. (1957). Origin of pebbly mudstones. Bull.Geol.Soc.Amer., 68: 993-1010.

David, T.W.E.; Taylor, T.G.; Woolnough,W.G. & Foxall, H.G. (1905). Occurrence of the pseudomorph Glendonite in New South Wales. Rec.Geol.Survey, New South Wales, 8 (2): 161-179, Sydney.

Dickins, J.C.M. (1957). Lower Permian pelecypods and gastropods from the Carnarvon Basin, Western Australia. Bull.Bur.Min.Res. Geol. & Geophys., 41, 75 pp., Canberra.

Douglass, R.C. & Nestell, M.K. (1976). Late Paleozoic Foraminifera from Southern Chile. U.S.Geol.Survey, Prof.Paper 858, 49 pp.

Flint, R.F.; Sanders, J.E. & Rodgers, J. (1960). Symmictite: a name for nonsorted terrigenous sedimentary rocks that contain a wide range of particle sizes. Bull.Geol.Soc.Amer., 71: 507-510.

Flint, R.F.; Sanders, J.E. & Rodgers, J. (1960). Diamictite, a substitute term for Symmictite. Bull.Geol.Soc.Amer., 71: 1809.

Folk, R.L. (1954). The distinction between grain size and mineral composition in sedimentary-rock nomenclature. J.Geol., 62 (4): 344-359.

Fossa-Mancini, E. (1943). Supuestos vestigios de glaciaciones del Paleozoico en la Argentina. Rev.Mus.La Plata, N.S., 1, Geol., 10: 347-404.

Frakes, L.A.: Amos, A.J. & Crowell, J.C. (1969). Origin and stratigraphy of Late Paleozoic diamictites in Argentina and Bolivia. IUGS Symp. Gondwana Stratigr., Earth.Sci.UNESCO, 2: 821-843, Buenos Aires.

González, C.R. (1972a). La Formación Las Salinas, Paleozoico Superior de Chubut (República Argentina). Parte 1. Estratigrafía, facies y ambientes de sedimentación. Rev.Asoc.Geol.Arg., 27 (1): 95-115. Buenos Aires.

González, C.R. (1972b). La Formación Las Salinas, Paleozoico Superior de Chubut (República Argentina). Parte 2. Bivalvia:

Taxonomía y Paleoecología. Rev.Asoc.Geol. Arg., 27(2): 188-213. Buenos Aires.

González, C.R. (1977). Bivalvos del Carbónico Superior del Chubut, Argentina. Acta Geol.Lilloana, 14: 105-147. Tucumán.

González, C.R. (1978). *Orbiculopecten gen. nov.* (Aviculopectinidae, Bivalvia), from the Upper Carboniferous of Patagonia, Argentina. Jour.Paleontol., 52(5): 1086-1092. Kansas.

González, C.R. (1980a). Algunos Myalinidae (Bivalvia) del Paleozoico Superior de Chile. Actas IICong.Arg.Paleontol. y Bioestratig. y I Congr.Latinoamer.de Paleontol., 4: 23-29. Buenos Aires.

González, C.R. (1980b). Sobre la presencia de "glendonita" en el Paleozoico Superior de Patagonia. Rev.Asoc.Geol.Arg., 35(3): 417-420. Buenos Aires.

González, C.R. (1981a). El Paleozoico superior marino de la República Argentina, Bioestratigrafía y Paleoclimatología. Ameghiniana, 28 (1-2): 51-65. Buenos Aires.

González, C.R. (1981b). Pavimento glacial en el Carbónico de la Precordillera argentina. Rev.Asoc.Geol.Arg., 36 (3): 262-266. Buenos Aires.

Keidel, J. (1922). Sobre la distribución de los depósitos glaciarios del Pérmico en la Argentina. Bol.Acad.Nac.Cienc. Córdoba, 25: 239-367. Córdoba.

Kemper, E. & Schmitz, H.H. (1975). Stellate nodules from the Upper Deer Bay Formation (Valanginian) of Arctic Canada. Canada Geological Survey Papers, 75 (1C).

Lindsay, J.F. (1970). Depositional environment of Paleozoic glacial rocks in the Central Transantarctic Mountains. Bull. Geol.Soc.Amer., 81(4): 1149-1172.

Mésigos, M.G. (1953). El Paleozoico Superior de Barreal y su continuación austral, Sierra de Barreal, provincia de San Juan. Rev.Asoc.Geol.Arg., 8 (2): 65-109. Buenos Aires.

Miller, H. (1979). Unidades estratigráficas y estructurales del basamento andino en el archipiélago de Los Chonos, Aisén, Chile. II Congr.Geol.Chileno, 1979, pp. A103-A120, Arica.

Robert, J.; Hunt, J.W. & Thompson, D.M. (1976). Late Carboniferous marine invertebrate zones of eastern Australia. Alcheringa, 1: 197-225.

Rolleri, E.O. (1970). Discordancia en la base del Neopaleozoico al este de Esquel. Actas IV Jorn.Geol.Arg., 2: 273-319.

Rolleri, E.O. & Baldis, B. (1969). Paleogeography and distribution of Carboniferous deposits in the Argentine Precordillera. IUGS Symp.Gondwana Stratigraphy, Earth Sci.UNESCO, 2: 1005-1024. Buenos Aires.

Schermerhorn, L.J.G. (1966). Terminology of mixed coarse-fine sediments.

J.Sed.Petrol., 36: 831-835.

Suero, T. (1948). Descubrimiento del Paleozoico Superior en la zona extraandina del Chubut. Bol.Inf.Petrol., 287: 31-48. Buenos Aires.

Valencio, D.A. (1973). El significado estratigráfico y paleogeográfico de los estudios paleomagnéticos de formaciones del Paleozoico Superior y Mesozoico Inferior de América del Sur. Actas V Congr.Geol. Arg., 5: 71-79.

Waterhouse, J.B. & Bonham-Carter, G.F. (1975). Global distribution and character of Permian Biomas based on brachiopod assemblages. Canadian J. Earth Sci., 12 (7): 1085-1146.

Criteria for identifying old glacigenic deposits

O.LÓPEZ GAMUNDI
Universidad de Buenos Aires & CONICET, Argentina

A.J.AMOS
Universidad de Buenos Aires, Argentina

ABSTRACT

Evidence of ancient glacial episodes is concentrated in Precambrian, Ordovician-Silurian and Carboniferous-Permian times. Glacial continental and glacial marine sedimentary processes have been taken from studies of Pleistocene glaciations and glaciomarine deposits of Arctic and Antarctic areas. The characteristics of these glacial deposits serve to determine the origin of ancient deposits of doubtful origin.

Criteria for recognition of ancient glacigenic deposits have been subdivided, in order of importance, as follows:

(a) Erosional features: striated pavements and 'roche moutonnée' forms are the two main types of abraded rock surfaces. These surfaces contain different kinds of striations and other erosional features (i.e., crescentic features).

(b) Erosional-depositional features: include boulder pavements.

(c) Presence of diamictites with striated clasts.

(d) Associated lithologies: massive diamictites, esker-like sand bodies and glacilacustrine varvites with ice-rafted pebbles are the lithologies associated with glacial continental environment. Crudely to well stratified diamictites and pebbly shales with dropstones and marine fossils are indicative of glacimarine sedimentation.

(e) Textural criteria: differentiation, through textural parameters, of glacial diamictites and mass-flow and turbidity-current deposits has been relatively successful. Investigations are not so well developed to use them for recognition of ancient glacial deposits.

(f) Microtextures on quartz and garnet grains: chattermark trails on these grains are interpreted as caused by ice abrasion.

(g) Indirect criteria: include geochemical studies (i.e., Fe vs. Mn contents), cold-water marine fauna and paleomagnetic studies.

The combination of these criteria, and not the isolated utilization of one or other of these criteria, is the best way to study deposits of possible glacial origin.

INTRODUCTION

Evidence of ancient glacial deposits is concentrated in Precambrian, Ordovician-Silurian and Carboniferous-Permian times. The characteristics of glacial deposits and their associated erosional forms are used to determine the genesis of ancient deposits of doubtful origin. Those characteristics have been used frequently in order to obtain the criteria for identifying old glacigenic deposits (Schwarzbach 1964; Harland, Herod & Krinsley 1966; Schemerhorn 1966, 1974; Flint 1971; Hambrey & Harland 1979, 1981). The models of terrestrial-glacial and glacimarine sedimentation are taken from studies connected with Pleistocene glaciation and with glacimarine depositional processes that we can see acting nowadays in the Arctic and Antarctic zones. Terrestrial-glacial environments were studied mainly by Smith (1959), Flint (1971), Boulton (1972, 1978) and Dreimanis (1976), among others. The work of Carey & Ahmad (1961), Chriss & Frakes (1972), Holdsworth (1973), Dreimanis (1979), Kurtz & Anderson (1979) and Anderson, Kurtz, Domack & Balshaw (1980) is focused on glacimarine processes and their products. Most examples which are cited come from Late Paleozoic Gondwana glaciation, which is the best studied pre-Pleistocene

glacial episode. Potential of preservation
of the deposits must be remarked as an
important concept: glaciomarine sediments
are deposited in basins of continuing
sedimentation, and are likely to be rapidly
covered by other marine sediments. On the
contrary, terrestrial glacial deposits are
more exposed to penecontemporaneous erosion.
In addition to that, Late Paleozoic glacial
sequences of Brazil, India, South Africa
and western and central Australia lie in
stable, non-orogenic, cratonic areas. They
are not strongly deformed. In contrast,
those of eastern Australia, western Argentina and Bolivia lie within orogenic belts,
associated with rocks of unstable geosyn-
clines; then they are more likely to be
eroded and/or deformed (Dott, 1961).

CRITERIA

Criteria have been subdivided in order of
importance, as follows: (a) Erosional
features; (b) Erosional-depositional
features; (c) Presence of diamictites with
striated clasts; (d) Associated lithologies;
(e) Textural criteria; (f) Microtextures
on quartz and garnet grains and (g) Indirect
criteria.

(a) Erosional features: include striated
pavements and 'roche moutonnée' forms on
basement rocks. Generally these surfaces
are overlain by true tillites. On these
abraded bedrock surfaces many types of
striae and grooves can be observed. Two
sets of striae, generally intersecting at
a low angle, are present in many places.
Associated with these striae are the
chattermarks, crescentic fractures,
crescentic gouges and nailhead striations
(Gilbert 1905; Dreimanis 1956). Ice-flow
pattern can be deduced from those striations
and lineations. Studies were made on
abraded bedrock surfaces in Sahara (Late
Ordovician glacial episode, Fairbridge
1970) and in many pavements assigned to
Late Paleozoic glaciation in India (Smith
1963 a, 1963 b; Gosh & Mitra 1969; Ahmad
andothers 1981), Australia (Konecki and
others 1958; Bowen 1969; Grey and others
1977), Brazil (Rocha-Campos and others
1969; Rocha-Campos 1981; Rocha-Campos &
dos Santos 1981), and South-Africa (du Toit
1921; Haughton & Frommurze 1936; Stratten
1970; Heath 1972; Frakes & Crowell 1970;
Crowell & Frakes 1972; Schreuder & Genis
1974; Bond 1981 a, 1981 b; Martin 1981 a,
1981 b; von Brunn & Stratten 1981).

(b) Erosional-depositional features:
include striated boulder pavements which
have the same paleocurrent characteristics.

A subglacial environment is indicated by
the abraded surface of the boulders of the
pavement. This environment could be sub-
aereal or subaqueous because the formation
of this erosional feature on the boulders
depends only on glacial shear, which
causes the striations and other lineations.
Intertill and intratill varieties have
been recognized. Examples of boulder
pavements have been described from
Antarctica (Schmidt & Williams 1969;
Frakes, Matthews & Crowell 1971), Brazil
(Rocha-Campos and others 1976), Africa
(Stratten 1969, for the Karroo Basin, and
Rocha-Campos 1976, for Angola), India
(Gosh & Mitra 1969) and Argentina (Amos
& López Gamundi 1981 a, González 1981).

(c) Presence of diamictites with striated
clasts: since the studies of gravity-
driven flows and turbidity currents were
intensified, too much light has been shed
on the origin of tillite-like deposits,
considered traditionally as true tillites.
The possibility of a non-glacial, mass-
flow origin for some diamictites of
Precambrian (Winterer 1964, Dott 1961) and
Late Paleozoic ages (Frakes and others
1969; Frakes & Crowell 1969; Amos & López
Gamundi 1981 a) has been confirmed. The
differentiation between these two broad
fields, mass-flow or glacial origin, could
be possible due to studies which describe
characteristics of diamictites of those
two environments above mentioned (Crowell
1957; Dott 1963; Walker 1978; Mills 1977).
Slump folds (slump overfolds, Crowell 1957),
convolute bedding, slump balls (ten Haaf
1956), clastic dykes, distorted sandstone
masses, hook-like forms (Crowell 1957) and
other deformational structures are
considered to be connected with gravity
or current driven flows, though sometimes
they are also associated with flow tills
(Canuto & Rocha-Campos 1981). Other authors
think that some of these structures are
caused directly by ice-pushing (Martin
1964; Bell 1981).
Striae with parallel or subparallel pattern
(Wenworth 1936) suggest that the striae-
bearing stones were affected by glacial
abrasion and point towards a primary
glacial origin for the deposits which
contain the clasts. However, the presence
of striated clasts has been reported from
non-glacial diamictites (terrestrial
mudflows and some tectonic environments).
Winterer (1964), in his study of the Late
Precambrian pebbly mudstone of Normandy,
France, described a 'moulded' pattern
(Wenworth's classification) of striae,
but he adds that "...striae of stones of
Saint Germain d'Ectot are thus very much
more abundant than those on stones of

280

known glacial deposits, and possess a very different pattern". Finally, the same author concludes that "where muddy matrix includes even smaller pebbles and granules, the striae are continuous from one small clast to another...this is interpreted to indicate that striae were cut after the stones were buried in their matrix, by differential motions of the stones within the enclosing mudstone".

Some authors consider that abrasion between clasts is another cause of striations on them.   However, in diamictites with low clasts/matrix ratios, distance between particles is very large and then forbids much abrasion of clast on clast (Frakes, Amos & Crowell 1969). Thus, the presence of striated clasts with parallel or subparallel patterns is strongly in favour of a glacial origin for these diamictites with striated boulders and pebbles.

Mixed or dual origin seems to be a common characteristic in some diamictites; redeposition of glacial material into deeper basin areas, transported by different kinds of flows could be a reasonable mechanism for these deposits. Striated clasts in the diamictites may well have come from till, and a rapid deposition of till adjacent to a steep submarine slope could have provided ideal conditions for generation of gravity driven flows and slides (Frakes, Amos & Crowell 1969). Some Late Paleozoic diamictites from Central Patagonian Basin (Southern Argentina; González 1980, Amos & López Gamundi 1981 b) and Southern Hills of Buenos Aires Province (Central Argentina; Coates 1969) may have the above described origin.

(d) Associated lithologies
The presence of essentially non-stratified diamictites lying on abraded bedrock surfaces is indicative of a terrestrial glacial episode of sedimentation. These deposits are associated with thinly laminated shales and mudstones with isolated clasts (dropstones) and intrabasinal pebbles of till, and include also true varvites, probably deposited in ephemeral proglacial lakes (built by terminal moraine damming). The clasts deflect lamination of shales and varvites and are considered as ice-rafted stones (Crowell, 1964). Fining-upwards sequences composed of sandy or muddy diamictites (tillites), with veinules of sand, considered similar to pseudomorphs of ice-vein or veinules, and wedge-shape bodies of sandstone, that may correspond to pseudomorphs of ice-wedges, laminated mudstones and varvites, cyclically epeated, have been identified in Paraná

Basin (Rocha-Campos 1967). The lower part of this sequence is derived directly from ice-contact sedimentation. Shoestring sand bodies, sometimes cross-bedded, interpreted as eskers have been described from Upper Paleozoic rocks of Brazil (Frakes, Figueiredo Filho & Fulfaro 1968) and Falkland (Malvinas) Islands (Frakes & Crowell 1967). In glaciomarine sequences, diamictites appear as moderately to well-stratified deposits associated with pebbly shales with dropstones. This last kind of rock strongly suggests, as in the case of glacilacustrine environment, rafting processes. Negative evidence in favour of ice-rafting includes the absence of organic rafts preserved as fossils and the general paucity of carbonaceous matter. Proximal glaciomarine facies have the same characteristics of the terrestrial glacial environments. Subglacial channels of shoestring geometry (esker-like forms) within crudely stratified diamictites have been described from glaciomarine sequences of Western Argentina (López Gamundi 1982). Glaciomarine deposits are laterally connected with fine-grained marine layers.

(e) Textural criteria
The differentiation, through textural parameters, between true tillites and subaereal mudflows and, by other side, between glaciomarine deposits and subaqueous gravity mass-flow deposits and turbidites has resulted relatively successful (Landim & Berrios 1972; Landim & Frakes 1968; Frakes & Crowell 1975; Kurtz & Anderson 1979; Anderson and others 1980). Frakes & Crowell (1975) have applied studies of Landim & Frakes (1968) to Gondwana glaciers. These authors have differentiated, with two simple textural parameters, mean and inclusive geometric standard deviation (Folk & Ward 1957), different fields for terrestrial tills and glaciomarine deposits.

Anderson and others (1980) give textural characteristics for basal tills, compound glaciomarine sediments and residual glaciomarine sediments. Basal tills are massive, poorly sorted and homogeneous deposits. Compound glaciomarine sediments are crudely stratified and contain a current-derived fine mode (sorted mud fraction). Residual glaciomarine sediments have been winowed by currents, they are depleted in silt and clay and contain a moderately sorted sand fraction and are deposited from floating ice and reflect the action of marine currents.

(f) Microtextures on quartz and garnet grains
Chattermark trails, considered to be produced by glacial abrasion, have been

identified on quartz and garnet grains
(Krinsley & Donahue 1968; Folk 1975, 1977;
Bull 1977).
Gravenor & Gostin (1979) and Gravenor (1980)
have carried out studies in garnet grains
of the Late Paleozoic tillites of Paraná
Basin (Brazil), describing chattermark
trails.

(g) Indirect criteria
Indirect evidence which favours a glacial
origin is summarized in this item. These
criteria are: cold water marine fauna,
geochemical analysis and paleomagnetic data.
The presence of cold water marine faunas
has been identified in Late Paleozoic
sequences of Australia (Dickins, 1978) and
Argentina (González, 1972). A glacial origin
is better supported if glendonites, concretions originated by high concentrations of
marine salts when a glacier becomes bouyant
in contact with sea water, are present
(Carey & Ahmad, 1961).
Paleomagnetism has helped with paleogeographic reconstructions and determination of
paleolatitudes, which in many cases were an
important aid to solve paleoclimatic problems.
Geochemical analysis of Antarctic sediments
(Angino 1966), particularly Fe and Mn
contents, have been compared with those
from probable glacial marine deposits of
Late Paleozoic age (Frakes & Crowell 1975).

Finally, it is worthwhile to remark that
oceanographic studies demonstrate that there
is an overwhelming correlation between deep
sea turbidity currents and glaciation episodes in surrounding lands. Many authors
point out that frequency of turbidity
currents during glacial stages is much
higher than in postglacial times.
The ratio of turbidity currents in post-
glacial times to that of glacial times is
at least 1:10 in many areas.

The combination of these criteria and not
the isolated utilization of one or another,
is the best way to study deposits of
possible glacial origin.

REFERENCES

Ahmad, N. (1981). Late Paleozoic Talchir
tillites of Peninsular India. In:
Hambrey, M. & Harland, W., editors,
"Earth's Pre-Pleistocene glacial record",
Cambridge University Press, p.326-330,
Cambridge, U.K.
Ahmad, N.; Ghauri, K.K.; Abbas, S.M. &
Moakhar, C.R. (1975). Basal Talchir pave-
ments from Lower Hasdo Valley, M.P.,
India. Indian Geol.Ass.Bull., 9: 51-52.
Amos, A.J. & López Gamundi, O. (1981 a).
Las diamictitas del Paleozoico Superior

de la República Argentina: su edad e in-
terpretación. Actas VIII Congr.Geol.Arg.,
San Luis, t.3, p.41-58.
Amos, A.J. & López Gamundi, O. (1981 b).
Late Paleozoic diamictites of the Central
Patagonian Basin, Argentina. In: Hambrey,
M. & Harland, W., editors, "Earth's pre-
Pleistocene glacial record".Cambridge
University Press, Cambridge, U.K.
Anderson, J.B.; Kurtz, D.D.; Domack, E.W.
& Balshaw, K.M. (1980). Glacial and gla-
cial marine sediments of the Antarctic
Continental Shelf. Jour.Geology, 41.
Angino, E.E. (1966). Geochemistry of
Antarctic pelagic sediments. Geochim.
Cosmochim.Acta, 30: 939-961.
Bell, C. (1981). Soft sediment deformation
of sandstone related to Dwyka glaciation
in South Africa. Sedimentology, 28:321-329.
Bond, G. (1981 a). Late Paleozoic (Dwyka)
glaciation in the Middle Zambezi region.
In: Hambrey, M. & Harland, W., editors,
"Earth's pre-Pleistocene Glacial Record,
Cambridge University Press, Cambridge, U.K.
Bond, G. (1981 b). Late Paleozoic (Dwyka)
glaciation in the Sabi-Limpopo region,
Zimbawe. In: Hambrey, M. & Harland, W.,
editors, "Earth's pre-Pleistocene glacial
record". Cambridge University Press,
Cambridge, U.K.
Boulton, G.S. (1972). Modern Arctic glaciers
as depositional models for former ice-
sheets. Jour.Geol.Soc.London, 128(4):
361-393.
Boulton, G.S. (1978). Boulder shapes and
grain-size distributions of debris as
indicators of transport paths through a
glacier and till genesis. Sedimentology,
75: 773-799.
Bowen, R.L. (1969). Late Paleozoic Glaciation, Eastern Australia. In: Gondwana
Stratigraphy, IUGS Symposium, UNESCO,
2: 821-843, Paris.
Bull, P.A. (1977). Glacial deposits identified by chattermark trails in detrital
garnets. Comment.Geology, 5: 248.
Canuto, J. & Rocha-Campos, A.C. (1981).
Facies and sedimentary environments of
Late Paleozoic diamictites and associa-
ted rocks in Southern Paraná and Northern
Santa Catarina, Brazil. IUGS-UNESCO,
Project 42, Upper Paleozoic of South
America, Bulletin N° 4.
Carey, S.W. & Ahmad, N. (1961). Glacial
marine sedimentation. In: Raasch, G.,
editor, Proceed.First Inst.Symp.on
Arctic Geology, 2, Toronto, University
of Toronto Press, 865-894.
Chriss, T. & Frakes, L.A. (1972). Glacial
marine sedimentation in the Ross Sea.
In: Adie, R., editor, Antarctic Geology
and Geophysics, Oslo, Comm.Antarctic
Research, 747-762.

Coates, D.A. (1969). Stratigraphy and sedimentation of the Sauce Grande Formation, Sierra de la Ventana, Southern Buenos Aires Province, Argentina. In: Gondwana stratigraphy, IUGS Symposium, UNESCO, 2: 799-820, Paris.

Crowell, J. (1957). Origin of pebbly mudstones. Geol.Soc.Am.Bull., 68: 993-1010.

Crowell, J. (1964). Climatic significance of sedimentary deposits containing dispersal megaclasts. In: Nairn, A.E.M., editor, Problems in Paleoclimatology, Interscience, London.

Crowell, J. & Frakes, L.A. (1972). Late Paleozoic Glaciation: Part V, Karroo Basin, South Africa. Geol.Soc.Am.Bull., 83: 2887-2912.

Dickins, J.M. (1978). Climate of the Permian in Australia: the invertebrate fauna. Palaeogeogr.Palaeoclimat.Palaeoecol., 23: 33-46.

Dott, R.H., Jr. (1961). Squantum "tillite", Massachussets. Evidence of glaciation or mass-movement. Geol.Soc.Am.Bull., 72: 1289-1306.

Dott, R.H., Jr. (1963). Dynamics of subaqueous gravity depositional processes. Am.Assoc.Petrol.Geol.Bull., 47(1):1289.

Dreimanis, A. (1956). Steep Rock boulder train. Proc.Geol.Assoc.Canada, 8: 28-70.

Dreimanis, A. (1976). Tills: their origin and properties. In: Leggett, R.F., editor, Glacial Till. Spec.Publ.N° 12, Royal Soc.Canada, 11-490, Ottawa.

Dreimanis, A. (1979). The problems of water lain till. In: Schluechter, Ch., editor, Moraines and Varves, A.A.Balkema,Rotterdam.

du Toit, A.L. (1921). The Carboniferous Glaciation of South Africa. Trans.Geol. Soc.South Africa, 24: 188-227.

Fairbridge, R.W. (1970). An Ice-Age in the Sahara. Geotimes, 15(6): 18-20.

Flint, R.F. (1971). Glacial and Quaternary Geology, 2nd edition, Wiley, New York.

Frakes, L.A.; Amos, A.J. & Crowell, J.(1969) Origin and stratigraphy of Late Paleozoic diamictites in Argentina and Bolivia. In: Gondwana stratigraphy, IUGS Symposium, UNESCO, 2: 821-843.

Frakes, L.A. & Crowell, J. (1967). Facies and paleogeography of Late Paleozoic Lafonian diamictite, Falkland Islands. Geol.Soc.Am.Bull., 78: 37-58.

Frakes, L.A. & Crowell, J. (1969). Late Paleozoic Glaciation: I. South America. Geol.Soc.Am.Bull., 80: 1007-1042.

Frakes, L.A. & Crowell, J. (1970). Late Paleozoic Glaciation: II. Africa, exclusive of the Karroo Basin. Geol.Soc.Am. Bull., 81: 2261-2286.

Frakes, L.A. & Crowell, J. (1975). Characteristics of modern glacial marine sediments -application to Gondwana glacials. In: Campbell, K.S.W., editor, Third

Gondwana Symposium, 1973, Canberra. Australian University Press, 373-380.

Frakes, L.A.; Figueiredo Filho, P.M. de & Fulfaro, V. (1968). Possible fossil eskers and associated features of the Paraná Basin, Brazil. Jour.Sed.Petrol., 38(1):5-12

Frakes, L.A.; Mathews, D. & Crowell, J. (1971). Late Paleozoic Glaciation: III. Antarctica. Geol.Soc.Am.Bull., 82: 1581-1604.

Folk, R. (1975). Glacial deposits identified by chattermark trails in garnets. Geology, 3: 473-475.

Folk, R. (1977). Glacial deposits identified by chattermark trails in garnets. Reply. Geology, 5: 249.

Folk, R. & Ward, W. (1957). Brazos River Bar, a study in the significance of grain-size parameters. Jour.Sed.Petrol., 27(1): 3-27.

Gilbert, G.K. (1905). Crescentic gouges on glaciated surfaces. Geol.Soc.Am.Bull., 17: 303-316.

González, C.R. (1972). La Formación Las Salinas, Paleozoico Superior de Chubut, Argentina. Partes I y II. Asoc.Geol.Argentina Rev.,27(1-2).

González, C.R. (1980). Sobre la presencia de "glendonita" en el Paleozoico Superior de Patagonia. Asoc.Geol.Argentina Rev., 35(3): 417-420.

González, C.R. (1981). Pavimento glaciario en el Carbónico de la Precordillera. Asoc.Geol.Argentina Rev., 36(3): 262-266.

Gosh, P.K. & Mitra, D. (1969). Recent studies on the Talchir Glaciation in India. In: Gondwana stratigraphy, IUGS Symposium, UNESCO, 2: 537-550, Paris.

Gravenor, C. (1980). Chattermarked garnets and heavy minerals from the Late Paleozoic glacial deposits of Southeastern Brazil. Can.J.Earth Sci., 17(1): 156-159.

Gravenor, C. & Gostin, V. (1979). Mechanisms to explain the loss of heavy minerals from the Upper Paleozoic tillites of South Africa and Australia and the Late Precambrian tillites of Australia. Sedimentology, 26: 707.

Grey, K.; van de Graak, W. & Hocking, R.M. (1977). Precambrian stromatolites as provenance indicators in the Permian Lyons Fm., Carnavon Basin. Ann.Rep.W. Australia Geol.Surv., 1976, 70-2.

Haaf, E. ten (1956). Significance of convolute lamination. Geol.en Mijnbouw, 18: 188-194.

Hambrey, M. & Harland, W. (1979). Analysis of Pre-Pleistocene glacigenic rocks: aims and problems. In: Schluechter, Ch., ed., Moraines and Varves, 271-275, A.A.Balkema, Rotterdam.

Hambrey, M. & Harland, W. (1981). Criteria for the identification of glacigenic deposits. In: Hambrey, M. & Harland, W., editors, "Earth's Pre-Pleistocene glacial record", Cambridge University Press, 4: 14-17, Cambridge, U.K.

Harland, W.; Herod, K. & Krinsley, D. (1966). The definition and identification of tills and tillites. Earth Sci.Rev., 2: 225-256.

Haughton, S.H. & Frommurze, H. (1936). The geology of Warmbad District, South West Africa. Mem.Geol.Surv.South Africa, 2, South West Africa Series.

Heath, D.C. (1972). Geologie van die Sisteem Karroo in die gebied Mariental Asab, Suid-wess-Afrika. Mem.Geol.Surv.S.Africa, 61.

Holdsworth, G. (1973). Ice calving into the proglacial Generator Lake, Baffin Island, NWT, Canada. Jour.Glaciology, 12: 235-250.

Konecki, M.; Dickins, J. & Quinlan, T.(1958). The geology of the coastal area between the lower Gascoyne and Murchison Rivers, Western Australia. Rep.Bur.Min.Resour.Geol. Geophys.Australia, 37.

Krinsley, D. & Donahue, J. (1968). Environmental interpretation of sand grain surface textures by electron microscopy. Geol.Soc.Am.Bull., 79: 743-748.

Kurtz, D.D. & Anderson, J.B. (1979). Recognition and sedimentologic description of recent debris flow deposits from the Ross and Wedell Seas, Antarctica. Jour.Sed. Petrol., 49 (4): 1159-1170.

Landim, P. & Berrios, M. (1972). Distinção de tilitos dentre os mistitos do Subgrupo Itararé. Rev.Bras.Geoc., 2(4): 270-279.

Landim, P. & Frakes, L. (1968). Distinction between tills and other diamictons based on textural parameters. Jour.Sed.Petrol., 42: 89-101.

López Gamundi, O. (1982). Litofacies y paleoambientes de la Formación Hoyada Verde, San Juan, Argentina. CONICET, unpubl.rpt.

Martin, H. (1964). The directions of flow of the Itararé ice sheets in the Paraná Basin, Brazil. Bol.Paran.Geog., 10-15: 25-76.

Martin, H. (1981a). The Late Paleozoic Dwyka Group of the South Kalahari Basin in Namibia and Botswana and the subglacial valleys of the Kaokoveld in Namibia. In: Hambrey,M. & Harland, W., editors, "Earth's Pre-Pleistocene glacial record", Cambridge University Press, 61-66, Cambridge, U.K.

Martin, H. (1981b). The Late Paleozoic Dwyka Group of the Karasburg Basin, Namibia. In: Hambrey, M. & Harland, W., editors,"Earth's pre-Pleistocene glacial record", 67-70, Cambridge University Press, Cambridge, U.K.

Mills, H.H.(1977). Differentiation of glacier environments by sediment characteristics: Athabasca Glacier, Alberta, Canada. Jour. Sed.Petrol., 47(2): 728-737.

Rocha-Campos, A. (1967). The Tubarao Group in the Brazilian portion of the Paraná Basin. In: Bigarella, J.J.; Becker, R. & Pinto, I. editors, "Problems in Brazilian Gondwana Geology", 27-102, Curitiba.

Rocha-Campos, A. (1976). Direction of movement of Late Paleozoic glaciers in Angola (Western Africa). Boln.Inst.Geoc., Univ. São Paulo, 7: 39-44.

Rocha-Campos, A.(1981). Late Paleozoic tillites of the Sergipe-Alagoas basin, Rondonia and Matto Grosso, Brazil. In: Hambrey, M.& Harland, W., editors, "Earth's Pre-Pleistocene glacial record", 838-841, Cambridge University Press, Cambridge, U.K.

Rocha-Campos, A.; Farjallat, J.E. & Yoshida, R. (1969). Crescentic marks on a Late Paleozoic glacial pavement in Southeastern Brazil. Geol.Soc.Am.Bull., 80(6):1123-1126.

Rocha-Campos, A.; Oliveira, M.E.de; Santos, P.R.dos & Saad, A. (1976). Boulder pavements and the sense of movement of Late Paleozoic glaciers in central eastern Sao Paulo, Paraná Basin, Brazil. Inst.Geoc., Univ.Sao Paulo, 7: 149-160.

Rocha-Campos, A. & Santos, P.R.dos (1981). The Itararé Subgroup, Aquidauana Group and San Gregorio Formation, Paraná Basin, south eastern South America. In: Hambrey, M. & Harland, W., editors, "Earth's Pre-Pleistocene glacial record, Cambridge University Press, Cambridge, U.K.

Schemerhorm, L.J. (1966). Terminology of mixed coarse-fine sediments. Jour.Sed. Petrol., 36: 831-835.

Schemerhorm, L.J. (1974). Late Precambrian mixtites: glacial and/or nonglacial. Am.J.Sci., 274: 673-824.

Schmidt, D.L. & Williams, P. (1969). Continental glaciation of Late Paleozoic Age, Pensacola Mountains, Antarctica. In: Gondwana Stratigraphy, IUGS Symposium, UNESCO, 2: 617-649. Paris.

Schreuder, C.P. & Genis, G. (1974). Die Geologie van die Karasburgse Karrookom. Ann.Geol.Surv.S.Africa, 10(1973/4): 7-22.

Schwarzbach, M. (1964). Criteria for the recognition of Ancient Glaciations. In: Nairn, A.E.M., editor, "Problems in Paleoclimatology". Interscience, London.

Smith, A. (1959). Structures in the stratified Late Glacial clays of Windermere, England. Jour.Sed.Petrol., 29(3): 447-453.

Smith, A. (1963a). Evidence for a Talchir (Lower Gondwana) Glaciation: striated pavement and boulder bed at Irai, central India. Jour.Sed.Petrol., 33(3): 739-750.

Smith, A. (1963b). Striated pavements beneath basal Gondwana sediments in the Ajay River, Bihar, India. Nature, London, 198: 880.

Stratten, T. (1969). Preliminary report on a directional study of tillites in the Karroo Basin of South Africa. In: Gondwana Stratigraphy, IUGS Symposium, UNESCO, 2: 741-762. Paris.

Stratten, T. (1970). Tectonic framework of sedimentation during the Dwyka period in South Africa. In: Haughton, S.H., editor,

284

Second Gondwana Symposium, Cape Town.
C.S.I.R., Pretoria, 483-490.

von Brunn, V. & Stratten, T. (1981). Late
Paleozoic tillites of the Karroo Basin
of South Africa. In: Hambrey, M. &
Harland, W., editors, "Earth's Pre-
Pleistocene glacial record", 71-79,
Cambridge University Press, Cambridge, U.K.

Walker, R.G. (1978). Deep water sandstone
facies and ancient submarine fans: models
for exploration for stratigraphic traps.
Am. Assoc. Petrol. Geol. Bull., 62 (6):
932-966.

Wentworth, C.K. (1936). An analysis of the
shapes of glacial cobbles. Jour.Sed.
Petrol., 6 (2): 85-96.

Winterer, E. (1964). Late Precambrian
pebbly mudstone in Normandy, France:
tillite or tilloid?.In: Nairn, A.E.M.,
editor, "Problems in Paleoclimatology".
Interscience, London.

Glacial events in Carboniferous sequences
of Western Argentina

ALFREDO J.CUERDA
Universidad Nacional de La Plata, Argentina

ABSTRACT

On the basis of stratigraphical and lithological analysis of Carboniferous sequences
exposed in the Pampean Ranges (Tupe and
Tuminico sections) and the Precordillera
(La Chilca section) of Western Argentina,
evidence for a glacial event of that age
is discussed in this paper.
The glacial evidence includes strata of
argillaceous composition with clastic
inclusions of varied size which are interpreted as "dropstones", faceted and striated pebbles and polished and striated
surfaces.
These strata are connected to levels with
elements of the *Rhacopteris* flora, which
suggest a Middle to Upper Carboniferous age
for this glacial event of Western Argentina.

INTRODUCTION

The first notice about glacial deposits
interbedded in the Carboniferous sequences
of Western Argentina was provided by
Keidel (1916). This author pointed the
existence of "morainic" sediments in the
Andean Precordillera region, between Río
Jáchal and Río Mendoza. In a later paper
(Keidel 1922) he recognized two outcrop
belts, in regional terms:

(1) western belt, between Barreal (San Juan)
    and Uspallata (Mendoza), 120 Km-long.
(2) eastern belt, between Trapiche (La
    Rioja) and Cerro Pelado (Mendoza),
    270 Km-long.

He also extended the reach of the glaciation to the Pampean Ranges domain, discussing Carboniferous sequences previously
described by Bodenbender (1902) and Hausen
(1921).
In general terms, the western belt deposits
are related to marine sequences with invertebrate faunas which have been studied by
Cowper Reed (1949), Harrington (1938) and
Amos (1957, 1971). The eastern belt units,
on the contrary, are of terrestrial origin,
containing a fossil flora (Kurtz 1895, 1896;
Frenguelli 1941, 1946; Archangelsky 1971;
Arrondo 1972).
After Keidel's work, the investigations
conducted by Du Toit (1927, 1937), Keidel
& Harrington (1938), Heim (1945), Cuerda
(1945) and more recently by González (1981a,
1981b) have extensively confirmed the
existence of the referred glacial accumulations, particularly those of the western
belt.
However, other authors have doubted about
the glacial origin of these deposits
(Fossa-Mancini 1943). This investigator
sustained that those sedimentary rocks and
their related structures considered as
"tillites" and "glacial pavements" could be
interpreted as subaqueous deposits later
modified by tectonic activity, i.e. "the
supposed ice pavements and many of the
supposed tillites are Paleozoic aqueous
sediments which have locally acquired
that appearence under the action of dias-
trophic forces" (Fossa-Mancini 1943).
Other critical observations have been
formulated by Frakes, Amos & Crowell (1969)
who believed that the so-called "tillites"
were in fact diamictites, genetically
related to submarine-transport processes
rather than glacial processes (i.e., "diamictites which internal structure indicates a
genesis through processes of submarine
mass-movement rather through glaciation").
Finally, a short review of the distribution
and lithological characteristics of the
Upper Paleozoic glacial deposits of Argentina has been provided by Amos & López
Gamundi (1978).

RECENT GEOLOGICAL INVESTIGATIONS

The Carboniferous sequences of Western Argentina may be correlated with any of the following paleogeographic units:
(1) Paganzo Basin
(2) Calingasta-Uspallata Basin
In both cases, the sedimentary sequences are characterized by distinctive lithological features.

(1) PAGANZO BASIN (Azcuy & Morelli 1970)

This area of continental deposition (Figs. 1 and 2) extended over 150,000 Km², approximately and includes the following geological provinces (Azcuy & Morelli 1970):
(a) Famatina Range
(b) Pampean Ranges (partly)
(c) Andean Precordillera (partly)

The units which form the Carboniferous sequences are of continental origin with a thickness of up to 2600 m.
In the Central-Western and Northern portions of the Basin, thin marine beds with brachiopods and bivalves are interbedded.
On the basis of the lithofacies distribution isopach maps and stratigraphic relationships

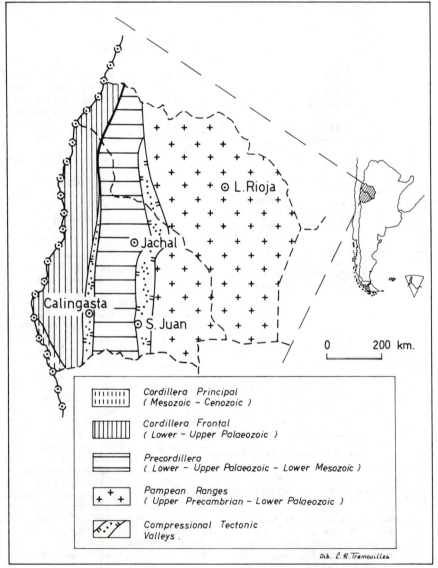

Cordillera Principal
(Mesozoic - Cenozoic)

Cordillera Frontal
(Lower - Upper Palaeozoic)

Precordillera
(Lower - Upper Palaeozoic - Lower Mesozoic)

Pampean Ranges
(Upper Precambrian - Lower Palaeozoic)

Compressional Tectonic
Valleys .

Dib. C. R. Tremouilles

Figure 1. Geological sketch of Central-Western Argentina, based on the distribution of geological provinces.

288

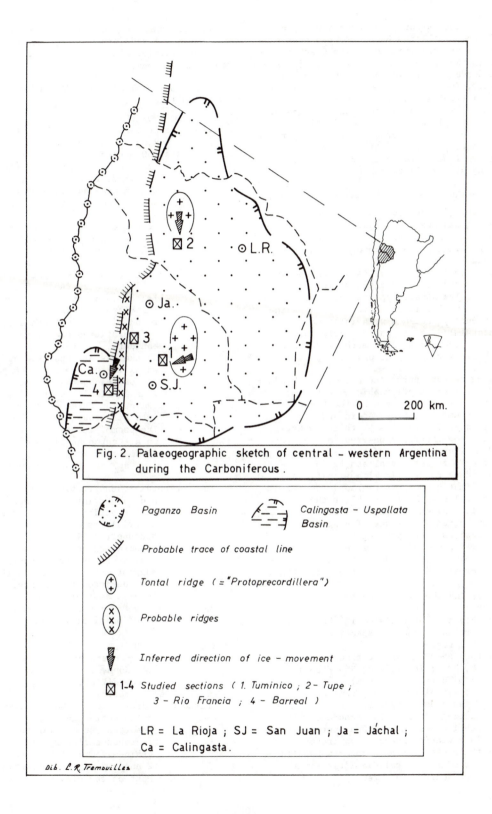

Fig. 2. Palaeogeographic sketch of central – western Argentina during the Carboniferous.

Paganzo Basin

Calingasta – Uspallata Basin

Probable trace of coastal line

Tontal ridge (= "Protoprecordillera")

Probable ridges

Inferred direction of ice – movement

1-4 Studied sections (1. Tuminico ; 2 - Tupe ; 3 - Rio Francia ; 4 - Barreal)

LR = La Rioja ; SJ = San Juan ; Ja = Jáchal ; Ca = Calingasta.

Dib. L. R. Tremouilles

between the Carboniferous sedimentary cover and the Crystalline Basement, positive elements (dorsals) which delimitate sub-basins within the Paganzo Basin have been inferred. The dorsals of Villa Unión-Tupe (Umango Range), Tuminico (Valle Fértil Range) and Tontal (Tontal Range) integrate the "Protoprecordillera", an ancient paleogeographic-paleogeologic unit of Paleozoic times, which is relevant for the present paper. The Tontal Range also provides the southwestern limit between the Paganzo and Calingasta-Uspallata basins.

(2) CALINGASTA-USPALLATA BASIN (Amos 1972)

This unit is located west and southwest of the Paganzo Basin and extends over two geological provinces: Precordillera (partly) and Cordillera Frontal, both in the Andean domain.
The Carboniferous deposits of this basin are related to two Pacific marine transgressions. The strato-type is found at Barreal (San Juan) and it is composed of a marine sedimentary sequence with interbedded continental layers. A gentle angular uncomformity divides these 2000 m-thick deposits into two stratigraphic units.

During recent years, the author has conducted detailed stratigraphic work on three Carboniferous sequences of the Paganzo Basin (Tuminico, Río Francia and Tupe profiles). The results of these investigations, together with observations performed in 1946 on the strato-type deposits of the Calingasta-Uspallata Basin, have enabled us to delimit with significant approximation the stratigraphic position of the glacial beds, establish their main lithological characteristics and give a chronological reference. The stratigraphic successions corresponding to the four profiles previously mentioned have been depicted in Figures 3 and 4, with perhaps an over-simplification of their stratigraphic details.
Concerning the glacial units, it should be pointed out that:
(a) they are intercalated in the lower portion of their respective stratigraphic sequences
(b) they show comparable thickness (Tuminico, 30 m; Tupe, 30 m; Río Francia, 58 m; Barreal, 20-50 m)
(c) the lithology is very similar in Paganzo Basin sequences and they are distinguished by the abundance of pebbly mudstones with dropstones formed in a glaciolacustrine environment
(d) the rocks of the Calingasta-Uspallata Basin are typical diamictites with glacigenic structures.

The Barreal beds are included in a moderately tectonically-deformed belt, whereas the other profiles are part of simpler, homoclinal tectonic structures.

LITHOLOGICAL AND STRATIGRAPHICAL DESCRIPTION OF THE PROFILES

(1) TUMINICO PROFILE
    Location: western border of Valle Fértil Range (San Juan)
    Thickness: 30 metres
    Lithology: a sequence composed of pinkish-grey, feldspar-rich, medium-grained sandstones, shales and slightly limonitic shales.
        At the base of the profile, greenish-grey siltstones which include metamorphic-rock clasts scattered all over the unit (=dropstones). The clast size ranges from 1-2 cm to 50 cm in diameter. The clayey matrix has been found filling in the smallest irregularities of the major boulders.
        Rhythmites are observed upwards in the sequence, composed of feldspar sandstones with parallel and ondulitic lamination which grade upwards into siltstones and shales. The psammitic bodies carry an allochtonous flora of *Rhacopteris sp.*, *Ginkgophyllum sp.* and *?Botrychiopsis sp.*
    Environmental reconstruction: fluvial and lacustrine environments, with contribution of coarser materials, probably ice-transported. The study of lithofacies and paleocurrent indicators as cross-bedding suggests that the sediments were derived from an area located eastwards.
    Age: Middle to Upper Carboniferous, based on plant remains.

(2) RIO FRANCIA PROFILE
    Location: 45 Km north of the city of San Juan, in the Andean Precordillera domain.
    Thickness: 58 metres
    Lithology: dark green, pebbly shales, silty shales and shales. Unsorted, subangular to subrounded clasts, with grain-size range between 2 and 600 mm in diameter. Most of the clasts are of

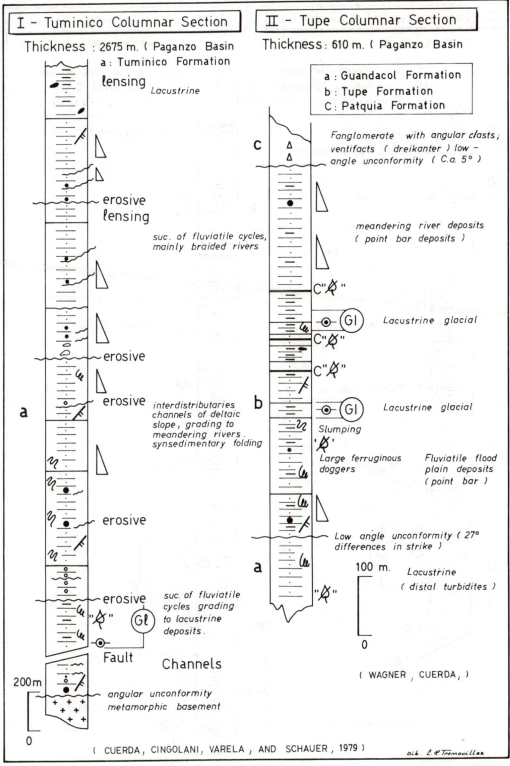

I – Tuminico Columnar Section

Thickness : 2675 m. (Paganzo Basin

a : Tuminico Formation

lensing *Lacustrine*

erosive
lensing

*suc. of fluviatile cycles,
mainly braided rivers*

erosive

erosive

a

erosive *interdistributaries
channels of deltaic
slope, grading to
meandering rivers.
synsedimentary folding*

erosive

erosive *suc. of fluviatile
cycles grading
to lacustrine
deposits.*

Fault   Channels

200m

*angular unconformity
metamorphic basement*

0

(CUERDA, CINGOLANI, VARELA, AND SCHAUER, 1979)

II – Tupe Columnar Section

Thickness : 610 m. (Paganzo Basin

a : Guandacol Formation
b : Tupe Formation
C : Patquia Formation

C

*Fanglomerate with angular clasts;
ventifacts (dreikanter) low –
angle unconformity (C.a. 5°)*

*meandering river deposits
(point bar deposits)*

C"∅"

GI   *Lacustrine glacial*

C"∅"

C"∅"

b

GI   *Lacustrine glacial*

Slumping

'∅'

*Large ferruginous   Fluviatile flood
doggers   plain deposits
                      (point bar)*

*Low angle unconformity (27°
differences in strike)*

a

100 m.   *Lacustrine
(distal turbidites)*

"∅"

0

(WAGNER, CUERDA,)

Dib. L. P. Tremouillas

Fig. 3

291

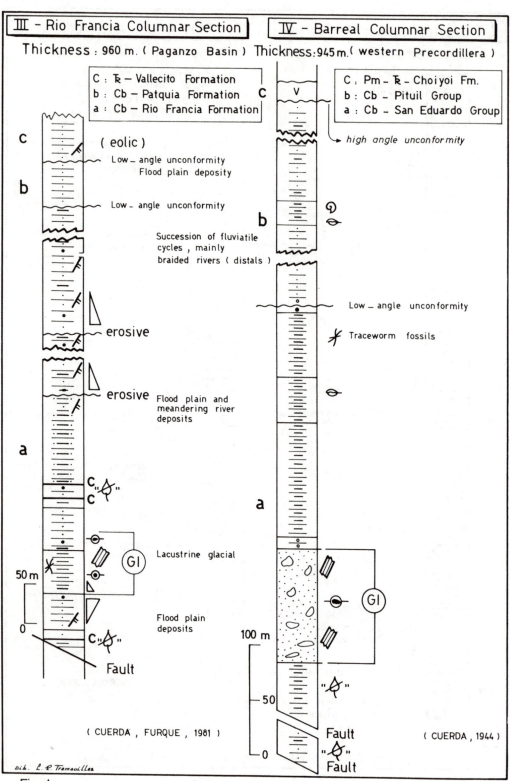

Fig. 4

▯=▯	shales
▯·▯	sandstone, fine - grained
▯•▯	sandstone, coarse to microconglomerate
▯∘▯	conglomerate
▯⚲▯	cross - bedded
▯ᴜ▯	synsedimentary folding
▯ᴜ▯	ondulitic lamination
— C	coal
"✿"	vegetal fossils, determinable.
·✿·	vegetal fossils, non- determinable.
-⊙-	dropstones
Gl	Glacial
▱	striae - glacial pavement
⊕	invertebrate marine fossils
✕	trace worm fossils
◿	fining up strata
◺	fining down strata
v	volcanic rock (Rhyolite)

References to the stratigraphic columns of Figures 3 and 4.

metamorphic rocks but there are also some graywacke pebbles. To the upper levels, this sequence is connected with Carboniferous beds with *Calamites peruvianus* Gothan.

Environmental reconstruction: lacustrine with ice-transported coarser material (dropstones).

Age: Middle to Upper Carboniferous.

(3) TUPE PROFILE

Location: southernmost end of the Villa Unión Range (Pampean Ranges)

Thickness: 30 metres

Lithology: a sequence of light-green, silty sandstones, siltstones and shales. Subangular to sub-rounded, up to 0.60 m in diameter metamorphic clasts are scattered in the unit (=dropstones). One triangular, faceted, parallel-striated clast has been found.

On the basis of the stratigraphic relationships of the Carboniferous beds with the Crystalline Basement, a positive paleogeographic element has been inferred northwards from this locality and is considered as the provenance area.

Environmental reconstruction: lacustrine, with ice-transported, coarser material (dropstones).

Age: Upper Carboniferous (Stephanian-Westfalian), on the basis of a floristic association with *Rhacopteris ovata* McCoy, included in overlying coal beds.

(4) BARREAL PROFILE

Location: 4 Km east of Barreal, San Juan. Western border of the Andean Precordillera.

Thickness: variable, between 20-50 m.

Lithology: unsorted, massive diamictite, containing blocks up to 1 m^3 in a wacky, dark green matrix. In the upper levels, González (1981a) has pointed the existence of glendonite which reveals a proximal marine environment. This observation is confirmed by the finding of marine gastropods (*Neoplatyteichum sp.*, *Perusvira sp.*, *Glabrocingulum sp.*) by the present author in those beds described by González (1981a).

González (1981a) has recognized here a 200 m-long glacial pavement. The clasts are triangular,

faceted and show parallel striations (84%) and crossed-striated pattern (16%). A north to south ice-flow direction has been suggested by González on the basis of the interpretation of structures as crescentic gouges and nailheads.

Environmental reconstruction: continental, above tidal reach, and marine, near the coast.

Age: Upper Carboniferous (Namurian-Westfalian), in beds corresponding to the *Levipustula* zone (Amos & Rolleri 1965).

CONCLUSIONS

The Carboniferous sequences of Western Argentina provide evidence of a glacial event which took place in the second half of the period.

The resulting glacial deposits are of two types:

(1) pebbly shales, siltstones and shales in the Paganzo Basin

(2) diamictites, with glacial pavements and other glacigenic structures, in the Calingasta-Uspallata Basin.

In the first group, the paleoenvironment corresponds to ice-contact water bodies, namely glacial lakes, related to ice-marginal conditions of the ice masses descending from neighbouring dorsals. In the second type, the environment was mixed, marine-continental, within the littoral belt and the nearby shore. The existence of fossil marine shells and plant remains suggests a Namurian-Westfalian age for the glacial episode and it may be considered contemporaneous of the well-known glacial epochs of the Bolivian, Chaco-Paranaense and Central Patagonian basins (Amos 1972; Frakes 1979).

At the continental scale, these events are related to the Upper Paleozoic Glaciation which covered part of the Gondwana Super-Continent.

REFERENCES

Archangelsky, S. (1971). Las tafofloras del Sistema de Paganzo en la República Argentina. An.Acad.Bras.Cienc., 43: 67-88.

Amos, A.J. (1957). New syringothyrid brachiopods from Mendoza, Argentina. Jour.Paleontol., 31(1): 99-104.

Amos, A.J. (1958). Some lower Carboniferous brachiopods from the Volcan Formation, San Juan, Argentina. Jour.Paleontol., 32(5): 838-845.

Amos, A.J. (1971). The Carboniferous fauna

of Argentina, new data. An.Acad.Bras. Cienc., Supl., 43: 19-22.

Amos, A.J. (1972). Las cuencas carbónicas y pérmicas de la Argentina. An.Acad.Bras. Cienc., Supl., 44: 21-36.

Amos, A.J. & Rolleri, E.O. (1965). El Carbónico marino del valle de Calingasta-Uspallata (San Juan-Mendoza). Boletín de Informaciones Petroleras, 368: 1-23, Buenos Aires.

Amos, A.J. & López Gamundi, O. (1978). Las rocas glacígenas del Paleozoico Superior de la Argentina. Acta Geologica Lilloana (Supl.), 14: 111-113, Tucumán.

Arrondo, O.G. (1972). Síntesis del conocimiento de las tafofloras del Paleozoico Superior de Argentina. An.Acad.Bras.Cien., Supl., 44: 37-58.

Azcuy, C.L. & Morelli, J.R. 1970. Tectonic and sedimentary characteristics of the Gondwana sequences in northwestern Argentina. 7. The Paganzo Basin. Sec.Gondwana Symp., South Africa, pp. 241-246.

Bodenbender, G. (1902). Contribución al conocimiento de la Precordillera de San Juan y Mendoza y de las Sierras Centrales de la República Argentina. Bol.Acad.Nac. Cienc.Córdoba, 17: 203-261.

Cowper Reed, F.R. (1949). Upper Carboniferous fossils from Argentina. In: A geological comparison of South America with South Africa. Carnegie Inst., Publ. 381, pp. 129-149, Washington.

Crowell, J.C. & Frakes, L.A. (1975). The Late Paleozoic Glaciation. In: K.S.W. Campbell, editor, Gondwana Geology, Australian National Univ., Canberra, A.C.T., pp. 313-331.

Cuerda, A.J. (1945). Estratigrafía y tectónica al este de Barreal, Provincia de San Juan. Unpublished PhD Thesis, Museo de La Plata, La Plata, Argentina.

Cuerda, A.; Cingolani, C.A.; Varela, R.; & Schauer, O.C. (1979). Depósitos carbónicos en la vertiente occidental de la Sierra de Valle Fértil, Provincia de San Juan. Asoc.Geol.Arg.Rev., 34(2): 100-107.

Cuerda, A.J. & Furque, G. (1981). Depósitos carbónicos de la Precordillera de San Juan. Parte 1 - Comarca del Cerro La Chilca (Río Francia). Asoc.Geol.Arg.Rev., 36(2): 187-196. Buenos Aires.

Du Toit, A.L. (1927). A geological comparison of South America with South Africa. Publ.Carnegie Inst., 381, pp.157, Washington.

Du Toit, A.L. (1937). Our wandering continents. Oliver & Boyd, 332 pp., Edinburgh.

Fossa-Mancini, E. (1943). Supuestos vestigios de glaciaciones del Paleozoico en la Argentina. Rev.Mus.La Plata (N.S.), I, Geol., 10: 347-404, La Plata.

Frakes, L.A. (1979). Climates throughout geologic time. Elesevier, 267 pp., Amsterd.

Frakes, L.A.; Amos, A.J. & Crowell, J.C. (1969). Origin and stratigraphy of Late Paleozoic diamictites in Argentina and Bolivia. Gondwana Stratigraphy, IUGS, Symp., Buenos Aires, UNESCO Earth Sci., pp. 821-843.

Frenguelli, J. (1941). Sobre la flórula carbonífera de Agua de los Jejenes, San Juan, conservada en el Museo de La Plata. Not.Mus.La Plata, 6(36): 459-478, La Plata.

Frenguelli, J. (1943). Acerca de la presencia del género Rhacopteris en el Paganzo I de Villa Unión, La Rioja. Rev.Mus.La Plata, N.S., 2(12): 11-47, La Plata.

Frenguelli, J. (1946). El Carbonífero argentino según sus floras fósiles. Asoc.Geol.Arg.Rev., 1(2): 107-115.

González, C.R. (1981 a). Pavimento glaciario en el Carbónico de la Precordillera. Asoc.Geol.Arg.Rev., 36(3): 262-266.

González, C.R. (1981 b). El Paleozoico Superior marino de la República Argentina. Bioestratigrafía y Paleoclimatología. Asoc.Paleontol.Arg.Rev., Ameghiniana, 16 (1-2): 51-65, Buenos Aires.

Hausen, H. (1921). On the lithology and geological structure of the Sierra de Umango area, Province of La Rioja, Argentina. Act.Acad.Aboensis, Mathem., et Phys., 1 (4): 1-138, Abo.

Heim, A. (1945). Observaciones tectónicas en Barreal, Precordillera de San Juan. Rev.Mus.La Plata, N.S., 2, Geol., 16: 267-286, La Plata.

Keidel, J. (1916). Memoria de la Dirección General de Minas, año 1914. In: Anales Ministerio de Agricultura, Sec.Geol., 11(4): 75-79, Buenos Aires.

Keidel, J. (1922). Sobre la distribución de los depósitos glaciarios del Pérmico. Bol.Acad.Nac.Cienc.Córdoba, 25: 239-267.

Keidel, J. & Harrington, H.J. (1938). On the discovery of Lower Carboniferous tillites in the Precordillera of San Juan, Western Argentina. Geol.Mag., 75: 103-1029, London.

Kurtz, F. (1895). Contribuciones a la Paleophytología argentina. Rev.Mus.La Plata, 6(1): 119-124, La Plata.

Kurtz, F. (1896). Recent discoveries of fossil plants in Argentina. Geol.Mag., 4(3): 338-349, London.

The supposed glacial sediments of the Lower Gondwana
in the Aguas Blancas area (Oran, Province of Salta), Argentina

CÉSAR R.CORTELEZZI
CIC de la Provincia de Buenos Aires, La Plata, Argentina

JOSÉ SOLÍS
Dirección de Minas de la Provincia de Salta, Argentina

ABSTRACT

Sedimentary rocks of the Tarija Formation, Carboniferous, of the Aguas Blancas area, Northern Argentina, are studied at a classic locality.
These sedimentary rocks are composed of so-called "tillites", with interbedded lenses of siliceous sandstones. The authors consider that these diamictites are not glacially originated, according to their petrographical composition and depositional characteristics.
Sedimentological studies on both lithological types have been performed, with special interest in their coarser fractions, so as to make their characterization and to establish if they were deposited by mud flows.

This study has been performed in the Serranía del Divisadero, a part of the Subandean Ranges of the Province of Salta, Northern Argentina.
The studied exposures are located to the north of the village of Aguas Blancas (lat. 22°S; long. 64° 21' W), near the international boundary with Bolivia, about 14 Km along the right embankment of the Río Bermejo, on the road to the village of Los Toldos (Figure 1).
The stratigraphy of the Subandean Ranges in this boundary zone is well known from many studies, especially those connected with oil geology investigations.
According to Mingramm and others (1978), the stratigraphy of this area can be summarized as shown in Table 1.
Within this area, the present authors have studied the Tarija Formation (Carboniferous), which forms the core of the Sierra del Divisadero Anticline. In general terms, the Tarija Formation has

uniform lithological characteristics. Two types of rocks are present: (1) the so-called, dark gray "tillites", and (2) light coloured sandstone lenses.
Frakes and others (1969) and Amos & López Gamundi (1981), together with Mingramm and others (1978) have considered that the former should be better called "diamictites" until they are petrographically studied in detail and its supposed glacial origin, which most of the geologists previously working in this area had favoured, is clearly established or rejected.

TABLE 1

STRATIGRAPHY OF THE AREA UNDER STUDY

Tertiary continental sedimentary rocks		
San Telmo Formation	Mandiyutí	PENNSYLVANIAN
Las Peñas Formation	Group	
Tarija Formation	Machareti	MISSISSIPPIAN
Tupambi Formation	Group	
uncomformity		
BARITU FORMATION		

The "tillites" constitute the majority of the exposures. They are dense, consolidated rocks, with a sandy-conglomeradic fraction distributed irregularly within abundant silty-clayey matrix. The coarser fraction

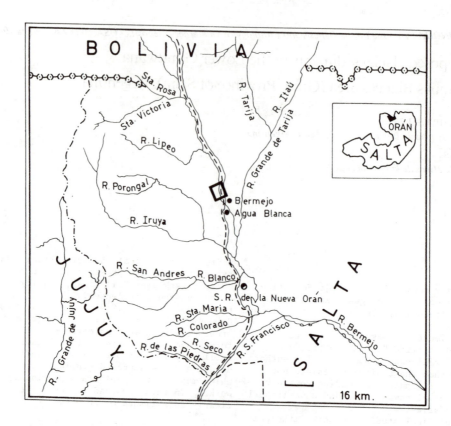

Figure 1. Location map. Exposure area shown by a black, rectangular frame.

is composed of clasts of varied size, chaotically disseminated within the finer matrix. . The homogeneity of these deposits is only broken by 3-4 m-thick sandstone lenses.

The sandstones are dense, fine-to-medium grained, with occasionally some conglomeratic intercalations. Bedding planes observation is hindered by jointing.

The section studied has an approximate length of 1 Km. The north-westernmost 200 metres belong to Upper Tertiary Sandstones. These deposits are non-consolidated, fine-to-medium grained, yellowish brown sandstones which appear in 0.50 - 2.00 m thick beds. These rocks are interbedded with thin reddish claystones. These beds are heavily jointed and a major, regional fault bonds these sandstones with the Tarija Formation (Figure 2).

The following 700 m belong to the Tarija Formation which starts with 50 m of steeply-dipping, not too dense, well-bedded sandstones; then, dark gray diamictites with pink granite, quartzitic and volcanic clasts, follow. Well-bedded, light-coloured sandstone lenses are intercalated. At 900

m, another fault puts together the Tarija Formation against the Tertiaty Sandstones, which are similar to those to the NW. Some samples have been studied in thin sections, whereas others were sieved using series of 0.5 φ-interval sieves, from -1.0 φ (2.0 mm) to 4 φ (0.062 mm).

These samples were collected in the sandstone lenses, between 200-250 m and 345-375 m from the beginning of the section to the NW. These rocks are unimodal sandstones with moderate sorting.

The mineralogy of these sandstones is as follows:

Quartz: 79%; subrounded clasts, with crystalline inclusions; most of them have cataclastic structure with wavy extinction.

Feldspar: 10%; subrounded clasts. K-feldspar is predominant (orthoclase and microcline) and it appears slightly altered.

Zircon: 1%; rounded clasts.

Magnetite: 0.5%, subrounded clasts, partially altered into hematite.

Rock fragments: 0.5%, rounded clasts of volcanics with porphyritic texture, very altered. Chert clasts are predominant,

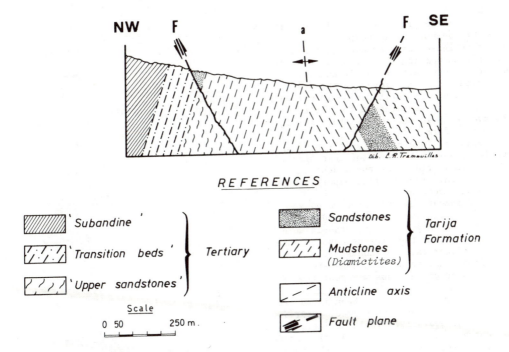

REFERENCES

Subandine	
Transition beds	} Tertiary
'Upper sandstones'	

Scale

0 50   250 m.

Sandstones	} Tarija Formation
Mudstones (Diamictites)	
Anticline axis	
Fault plane	

Figure 2. Schematic section of the Tarija Formation and structural relationships.

subrounded, with microcrystalline structure, partially stained with hematite.
Samples between 375-480 m are very massive
sandstones, with poorly defined SS planes.
The principal component is quartz (about
95%); most of it is cataclastic with wavy
extinction. Feldspar crystalline grains are
fresh, with microcline, orthoclase and
plagioclase, in order of abundance (4%).
The rest of the mineral grains (1%) are
subrounded clasts of zircon + magnetite,
altered into hematite. Matrix is formed by
clasts mineralogically similar to those of
the larger-size fraction, with a slightly
higher proportion of magnetite and hematite.
Cement is clayey and scarce. The rock is
classified as a sub-graywacke.
Other samples of the same section are
lighter, not so massive, with a larger
matrix content. Clasts are subrounded and
the smaller ones are subangular; cement is
scarce. In these rocks, a higher number of
clasts belong to granite, quartzite and
andesite.
These sandstones and mudstones become more
gravelly in some sections. Some clasts of
these diamictites are striated. The striae
are not too deep and few are parallel.
We think that they may have a tectonic origin or, at least, non glacial. However,
according to Amos & López Gamundi (1981),

who have studied other exposures further
South in the area of Tartagal, these striae
are in fact of glacial origin, but the
clasts are considered as retransported from
true tillites which were once lying to the
NE and E of this area.
Petrographic description of the samples
agrees with the terminology used by Spalletti (1980) for diamictites. The morphology
of some clasts from the diamictites was
studied. Clasts are equant and sphericity
is very high (from 0.62 to 0.87). They are
also slightly flattened, with shape factor
between 1.218 and 1.774, one clast exceptionally reaching 2.119, and most are rounded.
In most of the clasts the form is not typically glacial. Although no petrofabric
studies have been made, it was observed
that there is no preferred orientation for
the larger clasts.
Samples from 480-750 m are similar to those
already described but with a smaller proportion of larger clasts, being classified as
gray massive wackes.

CONCLUSIONS

The rocks under study are called "diamictites" in this paper due to their heterogeneous grain-size: between 40 % to 85 % are
sand grain-size, 10 % to 40 % are clays and

5 % to 35 % are gravel stones. The diamictites are polymictic in composition.
Sandstone lenses between diamictites are
common, with varying thickness. This alternation of beds seems to be frequent in
most diamictites mentioned in the geological literature. Stratification is not well
defined and it is rather difficult to
observe, even in sections without strong
tectonic deformation.
The rocks are interpreted as the product of
mudflows, due to the lack of thin stratification, indicating that these deposits
belong to the front or middle portions of
the density current.
Deformed sandstones are interpreted as
unstability indicators. The absence of
conglomerates suggests that gravels were
dispersed within the flowing mass because
of their lower value of cohesion.
The directional structures suggests a
predominance of the mass movement from W
to E.
Glacial pavements have not been found and
the striae are poorly preserved, superficial and non-parallel. These striations
may correspond to a former glacial episode,
and the clasts were included in the muddy
flowing mass, thus preserving some of the
striae from erosion.
From the above exposed considerations, we
suggest that these diamictites are not
tillites and that the isolated finding of
striae should not be taken as a proof of
ancient glaciation.

ACKNOWLEDGMENT

The authors thank Professor Arturo Amos
for comments on the manuscript and the
English translation.

REFERENCES

Amos, A.J. & López Gamundi, O. (1981). Las
    diamictitas del Paleozoico Superior de
    la Argentina, su edad e interpretación.
    VIII Congr.Geol.Arg., Actas, III: 11-58.
Frakes, L.A.; Amos, A.J. & Crowell, J.C.
    (1969). Origin and stratigraphy of Late
    Paleozoic diamictites in Argentina and
    Bolivia. Gondwana Stratigraphy Symp.,
    IUGS, UNESCO, Buenos Aires. UNESCO Earth
    Sci., 2: 821-843.
Mingramm, A.G.; Pozzo, A.; Russo, A. &
    Cazau, L.B. (1979). Sierras Subandinas.
    II Simposio Geología Regional Argentina,
    Acad.Nac.Cienc.Córdoba, 1: 95-137,
    Córdoba.
Spalletti, L.A. (1980). Diamictitas en pa-
    leoambientes sedimentarios en secuencias
    silicoclásticas. Rev.Asoc.Geol.Arg.,
    Serie B, N°8, 124-149, Buenos Aires.

General

Remanent magnetism in glacial tills and related diamictons

DON J.EASTERBROOK
Western Washington University, Bellingham, USA

ACQUISITION OF REMANENT MAGNETISM IN POLYMODAL SEDIMENTS

Since the early systematic measurements of remanent magnetism of sediments (McNish and Johnson, 1938; Ising, 1942; Benedict, 1943), use of detrital remanent magnetism (DRM) has become well established. These and other studies demonstrated the important parameters for acquisition of DRM: (1) ambient magnetic field, (2) magnetic minerals in the sediment, (3) hydrodynamic forces, (4)particle size distribution, and (5) Brownian movement (Johnson et al, 1948; Griffiths et al 1960; Nagata 1961, 1962; Rees, 1961; King and Rees, 1966; and Stacey, 1972).

The intensity of magnetization of a sample depends upon the strength of the earth's magnetic field at the time of magnetization and the magnetic particles which carry the DRM (usually magnetite). Most of the samples measured in this study contain relatively high amounts of magnetite which produces strong sample intensities, generally in the range of 10^{-3} to 10^{-5} oersteds.

Hydrodynamic forces may exert mechanical torques on magnetic particles which can cause grain rotation. In addition, certain types of deposits, such as glacial till, are deposited in the presence of shear stresses which may mechanically align magnetic grains. The effect of such mechanical forces on grains becomes increasingly important with larger sized particles. Gravitation may also play a role in determining particle inclination in water-laid sediments. As an inequant grain settles to the bottom of a water column, if one end of the particle touches bottom before the other, it may rotate the grain into a shallower inclination than that of the ambient field. Declinations are usually not appreciably affected when this happens. Although hydrodynamic and gravitational grain rotation may thus produce errors in DRM, Rees (1965) and King and Rees (1966) have found that small magnetite grains orient essentially parallel to the ambient magnetic field. The critical particle diameter beyond which viscous flow or shear may significantly disturb the DRM is estimated to be 10-180 microns (King and Rees, 1966). For silt-sized particles, the critical limit is probably 40-50 microns (Keen, 1963). Thus, only silt-sized and smaller particles are likely to retain a reliable DRM.

The principal remanence in sediments is carried in single and pseudo-single domain grains. Single domain behavior has been observed for elongate grains up to 17 microns (Evans et al, 1968) but grains larger than that probably behave as multidomain particles. If a sufficiently large number of grains of single or pseudo-single domain magnetite grains occur in a sediment, a highly coercive remanence is produced. The lower size limit of magnetic grains is controlled by Brownian movement which randomizes DRM, so particles smaller than 0.1 microns in diameter are not important in determining the remanence of a sediment (King and Rees, 1966).

If a sufficient amount of silt and clay is present in a sediment, it may retain a stable magnetic remanence even though the sediments are poorly sorted. Magnetic grains lying within the critical range of particle sizes attain a DRM, whereas those larger or smaller than the critical size produce a random component of magnetization which effectively cancels itself out. This appears to be the case in the diamictons studied in this investigation.

Till units sampled include the type Nebraskan, Elk Creek, Nickerson, Cedar Bluffs, and Clarkson tills, Nebraska, and unnamed tills from David City, Nebraska, Hartford, South Dakota, and Messena, Iowa,

REMANENCE IN DIAMICTONS DEPOSITED IN THE ABSENCE OF SHEARING STRESS

Pleistocene glaciomarine drift provides a good example of how a poorly sorted sediment may retain a stable remanence. Many samples of such sediments, which are abundant in the Puget Lowland of Washington, were studied in this investigation.

The glaciomarine drifts studied consist of diamictons deposited from floating ice in a marine environment (Easterbrook, 1963, 1964, 1969). Melting of the floating ice releases englacial debris which rains down on the sea floor below where it accumulates as a poorly-sorted, till-like sediment (Fig. 1).

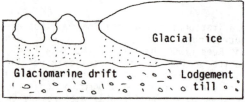

Figure 1. Depositional environments of glaciomarine drift and till.

Measurement of particle size distribution in glaciomarine drifts shows an average of about 70% silt-clay. The glaciomarine drift also has higher voids ratios than glacial till, and lower bulk densities (Easterbrook, 1964). As shown on Fig. 2, both the glaciomarine drift and till samples have appreciable pore space now and undoubtedly had even more at the time of deposition, the significance of which bears on the mechanism of acquisition of magnetic remanence discussed later in this paper. Evidence that the glaciomarine drift was deposited in situ by floating glacial ice is detailed in Easterbrook (1963, 1964, 1982).

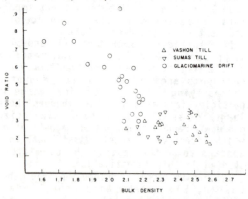

Figure 2. Voids ratios and bulk densities of till and glaciomarine drift

Oriented samples of glaciomarine drift were taken by pushing, or gently tapping, one-inch plastic cylinders vertically into the moist sediment which generally posed little resistance to penetration. The cylinders were oriented in place with a Brunton compass, then extracted, and labelled. The open ends of the cylinders were sealed with paraffin to prevent drying of the samples which might lead to shrinkage and possible rotation of the sediment inside the cylinders during measurement of the remanence. The transparent plastic cylinders allowed visual inspection of sediment samples to check for any distortion of material inside which might have taken place during sampling. All samples showing any signs of visible sediment disruption were discarded.

Stability of remanent magnetism

Remanent magnetization of several hundred specimens was measured with a Schonstedt SM-1A spinner magnetometer. Stepwise a.f. demagnetization, using Schonstedt GSD-5 tumbling demagnetization equipment, was carried out on selected samples of three ages of glaciomarine drift, Everson (11-13,000 years B.P.), Possession (70-90,000 years B.P.), and Double Bluff (150-250,000 years B.P.). Most samples were demagnetized in steps of 50, 100, 150, 200, 300, 400, and 600 oersteds (oe). Some samples were demagnetized to 800 oe. The glaciomarine drifts measured all showed exceptionally stable remanence. Figure 3 shows typical plots of equal area stereonet projections of a.f. demagnetized samples of glaciomarine drift of each age. Declination and inclination values of each sample remain tightly clustered even at demagnetization intensities up to 600 oe. Alpha-95 values (radius of the cone of 95% confidence) for the samples shown in Figure 3 were; 2.2° and 2.8° for two adjacent samples of Double Bluff Drift; 5.3° for Everson Drift; and 2.1°, 2.2°, and 3.6° for three adjacent samples of Possession Drift. The close grouping of points indicates that high levels of demagnetization are required before significant randomizing takes place. Measured declination and inclination values for each demagnetization step shown in Figure 3 are given in the appendix.

Many other samples of glaciomarine drift were also step demagnetized and exhibited stability of remanence similar to the examples shown here.

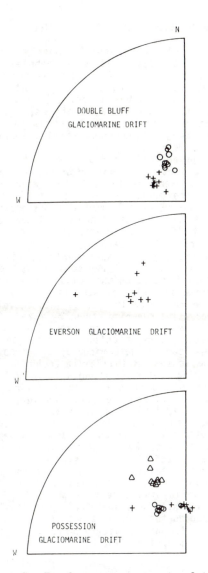

Figure 3. Equal area stereonet plots of glaciomarine drift a.f. demagnetized to to 600 oe.

Computer-generated orthogonal projections of magnetization vectors (Zijderveld plots) were also used to evaluate stability of the remanence of the glaciomarine drifts. The upper curve in Figure 4 shows progressive changes in direction of declination with increasing peak intensity of the demagnetizing field. The lower curve shows progressive changes in inclination with each demagnetization step. Direction of magnetization changes very little through demagnetization intensities of 600 oe, indicating stable magnetization.

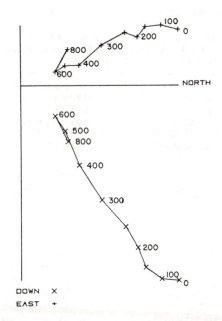

Figure 4. Orthogonal projection of magnetization vectors (Zijderveld plot) for glaciomarine drift a.f. demagnetized to 800 oe.

Samples of glaciomarine drift have high coercivities. 50% of the natural remanent magnetism (NRM) remains in many samples under intensities up to 300 oe (Fig.5).

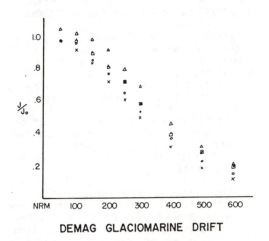

DEMAG GLACIOMARINE DRIFT

Figure 5. Decrease in sample intensity (J/Jo) with increasing demagnetization intensity.

305

Reproducibility

Multiple contemporaneous samples were collected laterally along the same horizon at several sites to test for reproducibility of declination and inclination. The natural remanent magnetism (NRM) for 10 samples from a single horizon of glaciomarine drift clusters tightly (Fig. 6) and remains so after a.f. demagnetization of 200 oe. Alpha-95's are 5.2° for NRM's and 5.4° for samples demagnetized at 200 oe.

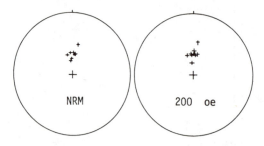

Figure 6. Equal area stereonet plots of multiple samples along a single horizon of Possession glaciomarine drift.

These stability and reproducibility tests demonstrate that poorly sorted glaciomarine drifts retain magnetic remanence which records the declination and inclination of the ambient magnetic field at the time of deposition, in spite of substantial amounts of coarse material.

MECHANISM OF ACQUISITION OF REMANENT MAGNETIZATION IN GLACIOMARINE DIAMICTONS

The stable, reproducible remanent magnetism in the glaciomarine diamictons studied is acquired in a subaqueous environment during deposition, and perhaps shortly thereafter. The silt/clay size, remanence-carrying magnetite grains are free to rotate into alignment with the earth's magnetic field during the time they settle through the water column to the sea floor and also have freedom to respond to magnetic torques while the sediment remains water-saturated. Fixing of the DRM in the deposits is believed to occur as pore fluids are expelled during sediment compaction.

Although both the glaciomarine drifts and tills which were studied are diamictons with similar particle size distribution, the mechanism of DRM acquisition is believed to differ. The glaciomarine drift attains a significant portion of its DRM by settling through a column of water, whereas the tills do not.

Glacial till consists of a poorly sorted mixture of clay, silt, sand, pebbles, and cobbles. At least three origins are commonly recognized: (1) lodgement till deposited from basal ice, (2) flow till deposited from ice as mudflows, and (3) ablation till deposited as glacial ice melts away to leave englacial and superglacial debris.

Lodgement till is deposited at the base of glaciers where motion of the ice over its bed exerts mechanical shear which produces preferred orientation of elongate particles. Flow till may also possess preferred alignment of elongate particles as a result of viscous motion. Particles in ablation till are randomly oriented.

Paleomagnetic measurements of till in this study are believed to be from lodgement till, based on the following evidence: Elongate particles of sand size and larger may be mechanically rotated into the plane of pervasive shearing stress by movement of a glacier over its base. The alignment of magnetic grains oriented by such shearing forces can be determined by anisotropy of magnetic susceptibility, a technique for measuring preferred orientation of a large number of multidomain magnetic minerals, usually elongate magnetite (Fuller, 1964; Rees, 1961, 1965; Stupavsky and Gravenor, 1975; Stupavsky et al, 1974). Such a magnetic fabric, which is largely independent of the DRM in a sample, may be characterized by three axes which define an ellipsoid. Anisotropy measurements of till and glaciomarine drift made by R.Crandall (MS thesis) indicate that the tills have significantly less spherical susceptibility ellipsoids than the glaciomarine drifts, ie the tills have a microfabric induced by glacial shearing (Fig. 7).

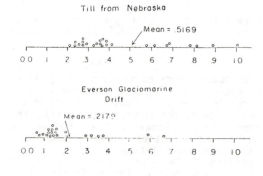

Figure 7. Anisotropy of magnetic susceptibility of till and glaciomarine drift.

Because of such orientation of grains by mechanical forces, tills have long been discounted as possible recorders of ancient magnetic fields. The results of more than 500 measurements made on tills by the author between 1971 and 1982 demonstrates that glacial tills are indeed reliable carriers of DRM which records the earth's magnetic field at the time of deposition.

Most of the till samples studied in this investigation were taken from cores. The cores were stored in plastic bags to retain their field moisture content and sampled by inserting plastic cylinders one-inch in diameter into the centers of one-inch slices of core. Sampling of only the core centers minimized any potential sediment distortion due to drilling. The ends of the plastic cylinders were sealed with paraffin to prevent any disturbance of the sample during magnetic measurements.

Magnetic mineralogy

Samples of till and glaciomarine drift were sieved and heavy minerals were extracted in tetrabromoethane. Magnetite was then removed with a hand magnet.

The glaciomarine drift contained large amounts of magnetite, probably because of the granitic provenance of the glacial debris in the sediment. The tills studied were from Nebraska, Iowa, and South Dakota where glaciers were characterized by a more varied provenance and contained somewhat less magnetite than the glacio-marine drifts of Washington. Although no attempt was made to identify other minerals which might have contributed to the remanence of the sediments, the abundance of magnetite and the strong coercivity of the samples suggest that magnetite is mainly responsible for the DRM.

Stability of remanent magnetism in glacial till

Samples of glacial till from widely separated localities in Nebraska, Iowa, and South Dakota were subjected to stepwise a.f. demagnetization at peak intensities of 50, 100, 150, 200, 300, 400, and 600 oe. A few were demagnetized to 800 oe. Figure 8 shows stereonet plots of three demagnetized till samples. All are from unoriented cores so declinations are relative to the initial arbitrary orientation selected before spinning the sample in the magnetometer. Only inclination values are oriented correctly.

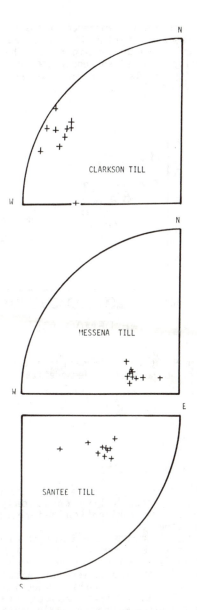

Figure 8. Equal area stereonet plots for Clarkson till (730), Messena till (750), and Santee till (877).

Declination and inclination for the samples deviate very little upon demagnetization. Sample 750 shows the tightest clustering of declination and inclination with an alpha-95 of 3.6°. The alpha-95 of sample 730 is 5.3° and sample 877 is 8.0°.

307

The decrease in remanence intensity with increasing a.f. intensity of three till samples is shown in Figure 9. The a.f. intensity required to reduce the remanence intensity to 50% (coercivity) is generally about 250 to 300 oe.

RELATIVE DECLINATION

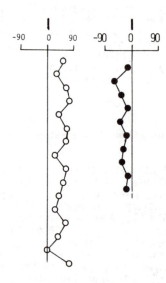

INCLINATION

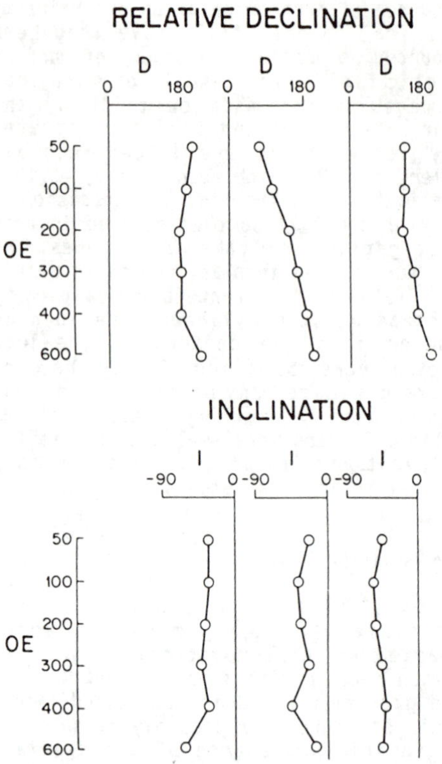

Figure 10. Demagnetization curves of three adjacent samples of Elk Creek till, Neb.

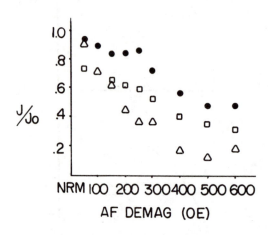

Figure 9. Decrease in sample intensity with increasing a.f. demagnetization.

Reproducibility

Because till samples were collected from cores, reproducibility could only be measured for inclination and relative declination.

Multiple samples were taken at closely spaced intervals from each till. A typical example is shown in Figure 10 in which three adjacent samples were a.f. demagnetized stepwise. All relative declination and inclination curves follow nearly identical paths with very little deviation under increasing demagnetization intensity, indicating good magnetic stability. All three samples have reversed polarity and thus their magnetization cannot be caused by the present normal magnetic field.

Ten adjacent samples of till from David City, Nebraska are all reversely magnetized (Fig. 11), indicating that the till did not acquire it's DRM in the present normal magnetic field.

Seventeen adjacent samples of Nickerson till from Nebraska give consistent inclination values (Fig. 11), averaging 50°.

Figure 11. Inclination values of replicate samples of Nickerson till (left) and till from David City, Nebraska (right).

Data from two samples of Nickerson till collected from an outcrop in Nebraska are shown in Figure 12. Both declination and inclination remain relatively constant through peak demagnetizing intensities up to 600 oe. A small low-coercivity component is removed after 50 oe demagnetization. The plot of relative sample intensity vs. demagnetizing intensity shows that the samples have high coercivity. Fifty percent of the sample intensity remains after demagnetization intensities of more than 400 oe.

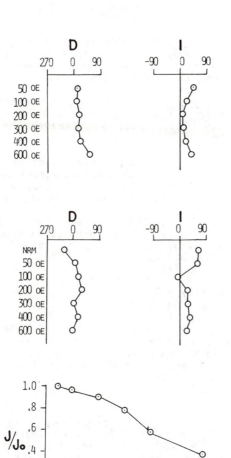

Demagnetization intensity

Figure 12. Step demagnetization curves for replicate samples of Nickerson till.

MECHANISM OF REMANENCE ACQUISITION IN TILL

Data from hundreds of measurements of remanent magnetism in multiple till units in Nebraska, Iowa, and South Dakota demonstrate that a stable remanence representing the earth's magnetic field at the time of deposition is preserved in certain types of till, even in the presence of pervasive shearing stress from glacial overriding. Gravenor et al (1973) came to a similar conclusion for the Port Stanley till along the north shore of Lake Erie in Ontario, Canada. However, interpretation of depositional environments in that area has been complicated by recent evidence that the origin of some previously recognized "tills" are actually subaquatic glacio-lacustrine sediments.

Gravenor et al (1973) considered that glacial till could acquire a remanence because "there is a layer of water at the ice-sediment interface" and "particles would _fall_ through a water layer or a slurry composed of water and suspended rock and mineral fragments."

A different mechanism for acquisition of magnetic remanence in till was proposed by Easterbrook (1977). Close examination of the bases of glaciers in Iceland, Alaska, Washington, Austria, Switzerland, France and elsewhere discloses no such water or slurry layer through which a particle could fall, but instead water-saturated till so "gooey" that a person quickly sinks into it up to knee level even though it supports the weight of the glacier. Small (silt size or less) magnetic grains do not "fall through" the till--they orient themselves parallel to the earth's magnetic field in situ between other grains in the till. They are able to do this because pore spaces are filled with water which carries a portion of the glacial load and allows the magnetic grains sufficient freedom to rotate into alignment with the earth's magnetic field. Because fluids do not transmit shear stresses, the pore water does not transmit glacial shear stress from the overriding ice which might otherwise interfere with the magnetic alignment (Fig. 13). Larger

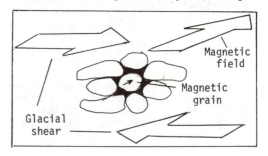

particles are affected by glacial shearing and may become oriented relative to the maximum and minimum stress directions, but the magnetic remanence is carried in the smaller size fraction which is "protected" from the shear stress by pore fluids and remains free to orient parallel to the magnetic field (Easterbrook, 1977, 1978a, 1978b). This effect is aided by the fact that magnetite, which is usually mostly responsible for the DRM in tills, generally consists of equant crystals which minimize effective alignment by shearing stress.

The DRM thus acquired in the till, becomes fixed when enough pore water has been expelled to restrict any further grain rotation if the magnetic field changes. The key factors in this concept are (1) adequate amounts of silt-clay particles in the till matrix, (2) thorough fluid saturation of pore spaces, and (3) subsequent expulsion of pore fluids. When these conditions are met, the small remanence-carrying magnetic grains rotate into alignment with the earth's magnetic field even if the sediment is deposited in the presence of glacial shear. That this is in fact the case is shown by examples cited above of consistent, stable, reliable reversed remanent magnetism preserved in tills whose anisotropy of magnetic susceptibility indicates mechanical shear-alignment of its larger elongate grains.

Not all tills can be expected to carry a reliable magnetic remanence. Only those with a relatively high silt-clay matrix deposited under the conditions suggested above are likely to yield good data.

CONCLUSIONS

Stable, reliable, magnetic remanence in glacial till and glaciomarine diamictons records the earth's magnetic field at the time of deposition as a result of magnetic alignment of silt-clay sized grains free to rotate in water-saturated sediment beneath glaciers, opening up the possibility of direct correlation of glaciations on a world-wide scale.

REFERENCES

Benedict, E.T., 1943, A method of determination of the direction of the magnetic field of the earth in geological epochs: Amer. Jour. Sci., v.241, p.124-129.

Easterbrook, D.J., 1963, Late Pleistocene events and relative sea level changes in the northern Puget Lowland, Washington: Geol. Soc. Amer. Bull., v.80, p.1465-1484.

_____,1964, Void ratios and bulk densities as means of identifying Pleistocene tills: Geol. Soc. Amer. Bull., v. 75, p.745-750.

_____, 1969, Pleistocene chronology of the Puget Lowland and San Juan Islands, Washington: Geol. Soc. Amer. Bull., v.80, p.2273-2286.

_____, 1973, Paleomagnetism and geochronology of the late Pleistocene: International Quat. Assoc. Congress, abs., Christchurch, New Zealand, p.86.

_____, 1977, Paleomagnetic chronology and correlation of Pleistocene deposits: Geol. Soc. Amer., abs., Seattle, Wash., p.961-962.

_____, 1978a, Paleomagnetism of glacial tills: Symposium on genesis of glacial deposits, International Quat. Assoc. Comm., abs., Zurich, Switzerland.

_____, 1978b, Paleomagnetism of glacial till and the Brunhes/Matuyama boundary: Geol. Soc. Amer., abs., Toronto, Canada, p.395.

_____, 1982, Physical criteria for distinguishing glacial deposits: Amer. Assoc. Pet. Geol. Memoir 31, p.1-10.

_____, and Boellstorff, J., 1981, Age and correlation of early Pleistocene glaciations based on paleomagnetic and fission-track dating in North America: Internation Geol. Corr. Prog., Rept. 6, p.72-82.

_____, and Othberg, K.O., 1976, Paleomagnetism of Pleistocene sediments in the Puget Lowland, Washington: International Geol. Corr. Prog. Rept. 3, p.189-207.

Evans, M.E., McElhinny, M.W., and Gifford, A.C., 1968, Single domain magnetite and high coercivities in a gabbroic intrusion: Earth and Planetary Letters, v.4, p.142-146.

Fuller, M.D.,1964, A magnetic fabric in till: Geol. Mag., v.99, p.233-237.

Gravenor, C.P., Stupavsky, M., and Symons, D.T., 1973, Paleomagnetism and its relationship to till deposition: Can. Jour. Earth Sci., v.10, p.1068-1078.

Griffiths, D.H., King, R.F., Rees, A.I., and Wright, A.E., 1960, The remanent magnetism of some recent varved sediments: Proc. Roy. Soc. London, v.59, p.359-383.

Ising,G.,1942, On the magnetic properties of varved clay: Arkiv. Mat. Astron. Fysik, v.29, p.1-37.

Johnson, E.A., Murphy, T., and Torreson, O.W., 1948, Pre-history of the earth's magnetic field: Terr. Mag. Atmos. Elec., v.53, p.340-372.

Keen, M.J., 1963, The magnetization of sediment cores from the eastern basin of the north Atlantic Ocean: Deep-sea Res., v.10, p.607-622.

King, R.F., and Rees, A.I., 1966, Detrital magnetism in sediments: an examination of some theoretical models: Jour. Geophys. Res., v.71, p.561-571.

McNish, A.G., and Johnson, E.A., 1938, Magnetization of unmetamorphosed varves and marine sediments: Terr. Mag., v.43, p.401-407.

Nagata, T., 1961, Rock magnetism: Maruzen Ltd., Toyko, 350 p.

_____, 1962, Notes on detrital remanent magnetization of sediments: Jour. Geomag. and Geoelect., v.14, p.99-106.

Rees, A.I., 1961, The effect of water currents on the magnetic remanence and anisotropy of susceptibility of some sediments: Geophys. Jour., v.5, p.235-251.

_____, 1965, The use of anisotropy of susceptibility in the estimation of sedimentary fabric: Sedimentology, v.4, p.257-271.

Stacey, F.D., 1972, Role of Brownian motion in the control of detrital remanent magnetization of sediments: Pure and Applied Geophys., v.98, p.139-145.

Stupavsky, M., and Gravenor, C.P., 1975, Magnetic fabric around boulders in till: Geol. Soc. Amer. Bull., v.86, p.1534-1536.

_____, Symons, D.T., and Gravenor, C.P., 1974, Paleomagnetism of the Port Stanley till, Ontario: Geol. Soc. Amer. Bull., v.85, p.141-144.

APPENDIX

Symbols: Demag = peak demagnetization intensity
Dec = declination
Incl = inclination
Jx10 = sample intensity
ASD = angular standard deviation for measurement

DEMAGNETIZATION DATA FOR FIGURES

A. Figures 3, 4, and 5.

DOUBLE BLUFF GLACIOMARINE DRIFT

Demag	Dec	Incl	J x 10	ASD
NRM	330.8	64.0	5.6 -5	4
50	331.0	67.8	5.2 -5	4
100	335.2	69.4	4.4 -5	4
150	334.1	68.4	3.8 -5	4
200	327.5	70.2	3.3 -5	4
250	328.9	70.8	2.8 -5	3
300	341.9	73.6	2.1 -5	4
400	341.5	64.7	2.0 -5	4
500	343.8	61.2	1.2 -5	4
600	340.8	62.4	8.7 -6	4
NRM	297.5	73.6	5.5 -5	6
50	304.6	73.5	5.0 -5	6
100	301.8	72.0	4.4 -5	7
150	296.5	72.1	3.5 -5	7
200	299.7	71.5	3.0 -5	6
250	304.1	70.0	2.6 -5	7
300	295.5	71.7	2.0 -5	8
400	316.7	70.4	1.5 -5	8
500	294.1	79.0	1.1 -5	10
600	302.8	67.3	7.3 -6	10

EVERSON GLACIOMARINE DRIFT

NRM	334.6	46.0	0.9 -3	
100	330.8	44.1	0.7 -3	
150	324.6	42.3	5.6 -4	
200	329.2	39.3	4.5 -4	
250	325.3	39.2	3.3 -4	
300	335.0	30.3	2.1 -4	
350	340.0	26.5	1.9 -4	
400	307.0	16.5	1.3 -4	

POSSESSION GLACIOMARINE DRIFT

Demag	Dec	Incl	J x 10	ASD
NRM	6.8	69.1	5.9 -5	6
50	4.3	68.5	6.1 -5	6
100	3.0	67.5	6.0 -5	6
150	0.2	67.2	5.7 -5	6
200	359.0	67.2	5.4 -5	7
250	357.9	65.8	4.7 -5	7
300	356.2	66.7	4.1 -5	7
400	354.6	66.5	2.7 -5	6
500	343.4	64.8	1.9 -5	6
600	0.6	65.6	1.2 -5	5
NRM	341.3	48.4	6.9 -5	5
50	338.6	50.2	6.8 -5	5
100	337.7	50.8	6.6 -5	5
150	335.7	52.7	6.1 -5	6
200	338.3	51.2	5.5 -5	5
250	334.0	49.5	4.9 -5	5
300	334.3	50.9	3.9 -5	5
400	341.6	51.2	2.7 -5	6
500	336.9	42.5	1.8 -5	6
600	339.4	38.6	1.3 -5	6
NRM	327.6	65.0	1.3 -4	8
50	329.8	65.2	1.3 -4	8
100	330.0	63.9	1.2 -4	9
150	327.2	64.2	1.1 -4	8
200	331.5	64.7	9.4 -5	8
250	332.5	65.2	7.9 -5	7
300	323.0	65.4	6.4 -5	7
400	330.5	64.3	4.0 -5	8
500	326.4	64.7	2.3 -5	8
600	326.7	60.8	1.4 -5	8
800	354.2	66.2	6.1 -6	6
NRM	348.7	72.7	1.2 -4	7
50	347.2	72.3	1.1 -4	8
100	339.4	72.0	1.1 -4	7
150	341.4	72.1	9.7 -5	7
200	349.2	72.4	8.8 -5	7
250	339.2	70.5	7.4 -5	6
300	339.1	72.8	6.2 -5	7
400	331.5	76.1	4.2 -5	5
500	4.0	73.8	2.6 -5	6
600	329.2	78.1	1.6 -5	4
800	356.0	68.3	8.4 -6	5

B. Figure 6. Replicate samples of Possession glaciomarine drift.

Demag	Dec	Incl	J x 10	ASD
NRM	9.9	45.2	1.2 -4	5
200	8.0	42.7	9.1 -5	2
NRM	9.0	61.4	8.6 -5	3
200	13.3	59.9	6.2 -5	1
NRM	7.7	59.7	1.3 -4	3
200	6.7	58.0	9.1 -5	3
NRM	4.3	59.7	1.0 -4	3
200	3.2	57.2	7.2 -5	2
NRM	3.2	60.4	1.1 -4	1
200	7.0	59.5	7.5 -5	2
NRM	2.9	57.5	1.8 -4	3
200	2.6	55.4	1.3 -4	4
NRM	342.3	63.2	2.3 -4	4
200	343.5	61.3	1.8 -4	3
NRM	350.1	74.9	1.4 -4	2
200	354.5	73.8	9.7 -5	3
NRM	355.5	67.0	1.4 -4	2
200	356.8	66.9	1.1 -4	3
NRM	354.8	58.3	1.5 -4	2
200	354.9	59.1	1.0 -4	1

CLARKSON TILL, NEBRASKA

Demag	Relative Dec	Incl	J x 10	ASD
NRM	283.8	40.1	3.8 -6	10
50	290.1	40.8	3.4 -6	10
100	292.0	42.2	2.7 -6	8
150	295.8	42.2	2.3 -6	8
200	295.9	46.5	1.7 -6	10
250	291.1	45.2	1.4 -6	11
300	295.2	38.2	1.4 -6	12
400	292.2	53.5	6.5 -7	26
500	311.3	64.4	4.5 -7	45
600	290.9	43.4	6.9 -6	25

MESSENA TILL, NEBRASKA

Demag	Dec	Incl	J x 10	ASD
NRM	128.5	77.3	7.9 -6	2
50	112.6	69.7	6.0 -6	4
100	107.9	66.2	5.4 -6	4
150	108.6	64.1	5.2 -6	7
200	111.1	62.6	5.1 -6	4
250	107.4	61.5	4.5 -6	4
300	113.3	63.1	4.1 -6	3
400	115.1	62.4	3.2 -6	4
500	101.1	63.6	2.7 -6	6
600	119.8	57.9	2.5 -6	7

SANTEE TILL, NEBRASKA

Demag	Relative Dec	Incl	J x 10		ASD
NRM	135.5	73.7	1.1	-5	2
50	136.9	72.3	8.8	-6	3
100	133.6	73.8	5.9	-6	4
150	145.1	71.7	4.3	-6	7
200	113.1	70.8	3.2	-6	4
250	125.2	73.1	2.5	-6	6
300	129.5	74.8	1.6	-6	7
400	182.8	61.0	1.1	-6	18
500	86.9	50.6	5.0	-7	24
600	300.2	73.8	4.4	-7	35

ELK CREEK TILL, NEBRASKA

Demag	Relative Dec	Incl	J x 10	
50	208.4	-33.1	6.8	-6
100	196.1	-31.8	7.2	-6
200	174.8	-39.2	7.1	-6
300	178.7	-43.7	6.2	-6
400	187.6	-32.6	5.4	-6
600	228.6	-63.3	5.0	-6
50	138.1	-46.5	5.7	-6
100	136.1	-55.8	6.5	-6
200	133.6	-51.6	6.9	-6
300	158.2	-45.1	7.6	-6
400	168.2	-41.5	8.5	-6
600	199.5	-45.6	7.3	-6
50	74.2	-23.5	6.0	-6
100	111.5	-38.3	4.5	-6
200	143.6	-32.4	2.8	-6
300	166.0	-22.6	3.1	-6
400	191.0	-45.4	4.5	-6
600	204.3	-17.9	3.6	-6

D. Figure 11 Inclination values

NICKERSON	DAVID CITY
47.9	-18.0
55.8	-62.5
32.7	-37.8
66.0	-15.9
77.4	-43.0
40.9	-20.1
68.6	-31.4
64.3	-35.4
27.9	-16.2
66.3	-21.7
56.2	
41.2	
26.2	
63.0	
37.1	
75.3	

313

Environments and soils of holocene moraines and rock glaciers, central Brooks Range, Alaska

JAMES M.ELLIS * & PARKER E.CALKIN
State University of New York at Buffalo, USA

1 ABSTRACT

Field measurements and multivariate analyses of over 50 cirque glaciers confirm that Neoglacial moraines without ice cores occur in cirques with low input of head-wall debris and extensive direct-radiant energy during the ablation season. Most moraines are cored with glacial ice, having formed in environments with either minimal solar energy and variable input of supraglacial debris or high inputs of both debris and solar energy. Neoglacial transition zones upslope of rock glacier tongues are equivalent to glacier-cored moraines but are ~100 m lower in altitude. A few form where as little as 1% direct radiation is received and vertical head-wall exposures exceed 400 m.

Cirque-glacier deposits in granites were more extensively glacier cored than those studied in sedimentary terrain. Their cirque environments differ respectively in solar energy received (65 + 20% versus 85 + 10%), potential debris supply as measured by height of bedrock in cirque cliffs (245 + 135 m versus 130 + 90 m), and ELA depression during Neoglacial maxima (130 + 60 m in contrast with 70 + 35 m).

Soil development helps differentiate these Holocene deposits in sedimentary terrain. Thin organic horizons and oxidation to depths of 15 cm occurs in <400 years on Neoglacial moraines. On moraines lichenometrically dated ~2000 B.P. the organic horizons reach 3 cm thicknesses, A horizons extend to depths of 10 cm and oxidized C horizons to 20 to >50 cm. Rock glacier tongues, downslope of Neoglacial transition zones, had upper surfaces stabilized by early Holocene time based on relative-dating criteria including soil pH values of 4.8-6.4. These are intermediate between values of 7.5-8.0 for freshly deposited till and 4.7-5.4 for late Pleistocene moraines downvalley.

2 INTRODUCTION

Cirque glaciers across the central Brooks Range have similar dimensions and orientations, yet varying environments in sedimentary and granitic terrains promote formation of different types of glacial deposits. The objectives of this study are to physically differentiate and explain the setting of alpine glacial deposits by 1) completing an environmental and morphological analysis of Neoglacial deposits, and 2) developing a preliminary chronosequence of soil development on moraines and rock glaciers. Both objectives complement our on-going glacial chronologic studies (Calkin & Ellis 1980). This paper also considers the validity of combining chronologies found on various landforms to delineate glacial and climatic variations through time in this most northerly mountain mass of western North America.

The Brooks Range is an east-west trending mountain system lying above the Arctic Circle which has been repeatedly glaciated since at least early Pleistocene time (Fig. 1; Hamilton 1977). Most valleys in the central Brooks Range were free of Pleistocene ice by the beginning of Holocene time (Hamilton & Porter 1975). Three field areas within the central Brooks Range were studied. The Atigun Pass area was chosen for the main portion of the work because of easy access to a large number of glaciers located along the trans-Alaska oil pipeline corridor (Fig. 2; Ellis & Calkin 1979). The centrally-located Anaktuvuk

* Present address: Gulf Oil E&P Company, Central Exploration Group, P.O.Box 36506, Houston, TX 77236, USA

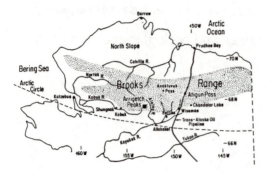

Figure 1. Location map of the central Brooks Range, northern Alaska. Cirque glaciers and their deposits near Atigun Pass, Anaktuvuk Pass, and the Arrigetch Peaks form the basis of this study.

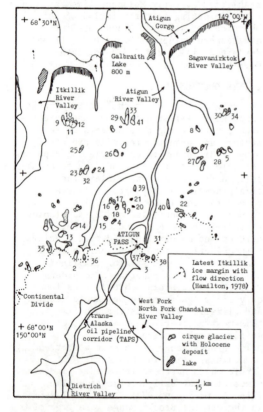

Figure 2. Location map of the Atigun Pass area, east-central Brooks Range. Forty-one cirque glaciers and their deposits were mapped in the field with 1-12 identified as non ice-cored moraines (M), 13-34 as ice-cored (Mg) and glacier-cored moraines (MG), and 35-41 as glacier-cored rock glaciers with Neoglacial transition zones (TRGC) up-slope of early Holocene rock glacier tongues.

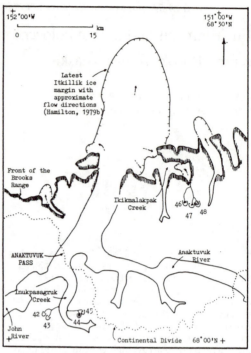

Figure 3. Location map of the Anaktuvuk Pass area, central Brooks Range. Cirques 42-47 were empty and of late Pleistocene age or had rock glaciers without exposed glacier cores. A glacier-cored moraine (MG) was mapped in cirque no. 48 and used in this study.

Pass is significant because of previous reconnaissance work undertaken on deposits of cirque glaciers (Fig. 3; Detterman et al. 1958; Porter 1966). Glaciers in the Arrigetch Peaks (Fig. 4), the third area studied, are of special interest because of their different climatic and lithologic setting (Hamilton 1965; Ellis et al. 1981).

Some of the basic questions considered within the objectives of this study are as follows:

a) Which environmental factors best discriminate between different types of glacier deposits?

b) Are there physical differences between cirque-glacier deposits in the three study areas which may cause significant differences in chronologies?

c) Which soil properties are consistent with lichenometry and other dating techniques as age indicators in sedimentary terrain?

Preliminary studies of cirque- and glacier-distribution patterns have been undertaken by the U.S. Geological Survey (1978) for the Brooks Range (scale 1:250,

316

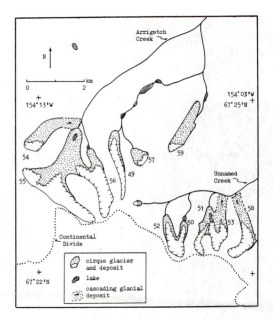

Figure 4. Location map of the granitic
Arrigetch Peaks, west-central Brooks Range.
Moraines without cores of ice (M, 49-51),
moraines cored with glacier ice (MG, 52-
58), and a glacier-cored rock glacier with
an upslope Neoglacial transition zone
(TRGC, 59) were mapped.

000) and by Porter (1966) in the Anaktuvuk
Pass area. In addition, more detailed
studies of glaciers and their downslope
deposits have been done for the Atigun
Pass area (Ellis & Calkin 1979) and the
Arrigetch Peaks (Ellis et al. 1981). However, to our knowledge no environmental or
soil analysis has been done on cirque
glacier moraines and rock glaciers in the
valley heads of the central Brooks Range.

Four types of cirque glacier deposits are
found in the central Brooks Range (Fig.
5A-D). Three out of four cirque glaciers
examined are associated with moraines
which are cored with ice (Mg, Fig. 5B).
In some cases glacial ice is exposed
beneath a thin cover of debris (MG, Fig.
5C). The remaining cirque glacier moraines
either have formed without ice cores (M,
Fig. 5A) or are situated as glacier-cored
transition zones (TRGC, Fig. 5D) upslope
of rock glacier tongues. The crescentic
ridges of transition zones are of Neoglacial age in the central Brooks Range
(Ellis 1982). In contrast, the rock glacier tongues are of early Holocene age and
show evidence of en mass downvalley movement.

3 GEOGRAPHICAL SETTING AND CLIMATE

The three study areas (Figs. 2-4) are above
the boreal spruce tree line and within the
zone of continuous alpine permafrost
(Ferrians 1965). Vegetation above tree
line, at 600 to 700 m altitude, consists
of shrubby tundra with alder and dwarf
birches which gives way to a mixed, her-
baceous-dwarf tundra vegetation at higher
altitudes (Brown 1980: 36). Bouldery,
cirque glacier moraines are unvegetated
when located near receding ice margins;
however, lichens, algae, mosses, and
grasses can cover these deposits and rock
glaciers to varying degrees.

The cirque glaciers are defined as subpolar because firn is saturated with water
during the ablation season. In addition,
internal ice temperatures were measured as
about -1°C in an upper cirque of a glacier
farther east in the Brooks Range (Orvig &
Mason 1963). The glaciers are cold because meltwater streams meander on their
surfaces (and do not descend to the bed)
(Paterson 1969: 178).

3.1 Atigun Pass area

Most of the higher peaks and glacierized
cirques in this region are composed of
siliceous conglomerates and sandstones or
quartzites of the resistant Devonian
Kanayut Conglomerate (Brosgé et al. 1979).
These peaks rise to altitudes of 2300 m.
The northernmost portion of this region is
made of a thick sequence of the crystalline Lisburne Limestone. Less resistant,
phyllitic Hunt Fork Shale dominates immediately south of the Continental Divide
(Fig. 2). At least 130 cirque glaciers
lie within this region; 97% of these occur
north of the Divide where the severe arctic
climate of the North Slope dominates.

Climatic data collected along the trans-
Alaska oil pipeline system (TAPS) since
1975 (Brown 1980) indicate temperatures
above freezing from May through September
in the Pass area, but low insolation during
the rest of the year allows temperatures
to reach -45°C. The mean annual temperature at Atigun Pass (1440 m) is estimated
about -14°C. Glacier snow pits (Bruen
1980) and TAPS data suggest annual precipitation at Atigun Pass ranges between 400
and 700 mm of which about 50% is snow.

3.2 Anaktuvuk Pass area

Anaktuvuk Pass is at an altitude of only
650 m (Fig. 3); there are fewer glacierized
cirques here than around the higher Atigun
Pass area. Field work was undertaken in
cirques supported by the Kanayut Conglomerate. Climatic data are meager, but
Porter (1966) estimated that the mean an-

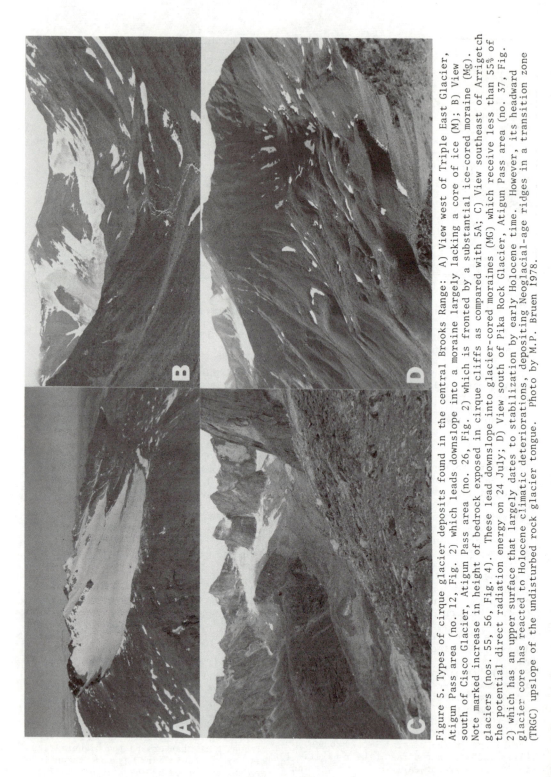

Figure 5. Types of cirque glacier deposits found in the central Brooks Range: A) View west of Triple East Glacier, Atigun Pass area (no. 12, Fig. 2) which leads downslope into a moraine largely lacking a core of ice (M); B) View south of Cisco Glacier, Atigun Pass area (no. 26, Fig. 2) which is fronted by a substantial ice-cored moraine (Mg). Note marked increase in height of bedrock exposed in cirque cliffs as compared with 5A; C) View southeast of Arrigetch glaciers (nos. 55, 56, Fig. 4). These lead downslope into glacier-cored moraines (MG) which receive less than 55% of the potential direct radiation energy on 24 July; D) View south of Pika Rock Glacier, Atigun Pass area (no. 37, Fig. 2) which has an upper surface that largely dates to stabilization by early Holocene time. However, its headward glacier core has reacted to Holocene climatic deteriorations, depositing Neoglacial-age ridges in a transition zone (TRGC) upslope of the undisturbed rock glacier tongue. Photo by M.P. Bruen 1978.

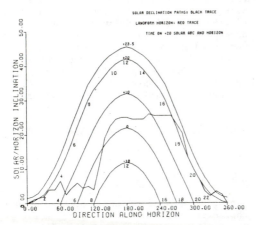

Figure 6. Topographic horizon of Grizzly Glacier (no. 3, Fig. 2) superimposed on paths of the sun at lat. 68°N for different solar declinations (-10° ≅ 24 February or 20 October, 0° ≅ 21 March or 23 September, +10° ≅ 16 April or 28 August, +20° ≅ 21 May or 24 July, +23.5° ≅ 21 June). In this study the +20° declination was used for terrain screening calculations as it represents a typical part of the ablation season. The hours of the day are plotted in two-hour increments along the trace of the +20° solar path. Sunrise(s) and sunset(s) are determined by following the horizon trace from left to right. When the horizon crosses and goes inside the solar path it means the sun has just appeared above the horizon ("sunrise"). When the horizon trace crosses and goes outside the sun path the sun is blocked from view ("sunset"). In this example, the sun initially appears at 1.6 hrs of the 24 hr day and radiation is received at the moraine until 16.9 hrs when the sun is blocked. The sun reappears at 18.9 hrs and shines on the deposit until 22.2 hrs into the day. Sunshine duration totals 18.6 hours on 24 July; this converts to 93% of the potential energy available to a horizontal, unscreened surface at this latitude.

nual temperature and annual precipitation at the pass was about -10°C and 300 mm. Only one cirque was mapped with an exposed headward glacier (no. 48, Fig. 3); the data for the deposit was combined with the 41 ice masses mapped at Atigun Pass (Fig. 2).

3.3 The Arrigetch Peaks

The Arrigetch Peaks occupy an area of about 110 km² on the south flank of the west-central Brooks Range (Fig. 4) and consist of granitic orthogneiss of Cretaceous age (Nelson & Grybeck 1980). The area is characterized by nearly vertical slopes. These protrude as much as 1000 m above the more erodible limestone, shale, and schist of the surrounding terrain, forming peaks to altitudes of 2150 m.

The moisture regime of the Arrigetch area appears to be transitional between the wetter maritime conditions of the western Alaskan coast and the relatively dry continental climate of interior Alaska (Ellis et al. 1981). Coastal storms commonly invade from the west via the broad Noatak and Kobuk valleys (Fig. 1). The sparse climatic data and widespread cover of Trentepholia iolithus algae on the granitic moraines (Ellis et al. 1981) suggest the Arrigetch Peaks is slightly warmer and wetter than the more easterly Atigun Pass area.

4 METHODS

Altitudes were taken from 1:63,360 and 1:250,000 topographic sheets which display contour lines at 100 ft (30 m) and 200 ft (60 m) intervals, respectively. Over 50 cirque glaciers and their deposits were analyzed in the field for area, altitude, aspect, topographic horizon, receipt of direct radiation energy, and potential supraglacial-debris supply as measured by height of bedrock exposed above the glacier in surrounding cirque cliffs. All were planimetrically reconstructed to their previous shape during Neoglacial maxima based on surficial geologic and lichenometric mapping (Ellis 1982). Equilibrium-line altitudes (ELA's) for Neoglacial maxima were determined with an assumed accumulation area to ablation area ratio of 2:1 or a ratio of accumulation area to total glacier area (AAR) of 0.67. ELA depression values were calculated from present mean glacier altitudes to the ELA's reconstructed for Neoglacial maxima.

The shadowing effect of surrounding cirque cliffs and mountainous terrain on the different types of glacial deposits was analyzed in the field by plotting the horizon at 015° increments (24 horizon inclinations per survey site) with a Brunton compass attached vertically to a tripod. Each landform's horizon was superimposed upon the sun's 24 hour path at +20° declination (~24 July) to determine times of sun appearance and disappearance (Fig. 6). This

Table 1. Variables used in multivariate analysis of glacial landforms.

Variable	Explanation
NSUN	Direct radiation energy received (%)
NELA	Altitude of the ELA during Neoglacial maxima (m)
NHEAD	Height of bedrock exposed in cirque cliffs (potential supraglacial debris supply, m)
NLAT	Latitude of glacial landform (°)
NDROP	Amount of ELA depression as measured from present mean glacier altitude to altitude attained during Neoglacial maxima (m)
NAREA	Area involved during Neoglacial maxima (km^2)

solar path (List 1951) was chosen to best characterize the glacial ablation period.

Superimposing the landform's horizon upon the sun's path, as it would be on 24 July, establishes the duration of direct radiation received at each site (Fig. 6). The amount of direct radiation energy received at each landform during these times of sun appearance was then calculated (Ellis 1982), and the sum compared to that measured at lat. 68°N on unscreened horizontal surfaces under clear skies (Kondratyev 1973: 304-305). This measured amount (~590 cal cm^{-2} day^{-1}) is treated as the potential unscreened energy and assigned a value of 100%. The comparison of the actual amount of energy received to this potential energy provides a measure of the reduction in direct solar energy due to screening by surrounding terrain. Albedo of the moraines was considered constant. A similar study on one valley glacier has been carried out in the northeastern Brooks Range with a theodolite (Wendler & Ishikawa 1974).

Increasing slope inclination in a more northerly aspect also reduces solar insolation (Ellis et al. 1981, Table 2). The effects of aspect, slope, and terrain screening were combined to provide a measure of the total receipt of direct radiation energy at each cirque glacier deposit during a typical part of the ablation season.

Morphological and environmental variables (Table 1) were analyzed in order to detect those that were most capable of distinguishing between the different types of cirque glacier moraines (Fig. 5A-D). The results of linear discriminant analysis (Klecka 1975) on 43 sedimentary and 11 granitic moraines are presented in this paper. The technique indicates the relative contribution of each variable (Table 1) to discriminate between the previously defined groups of glacial deposits (M, Mg/MG, TRGG) and between moraines in sedimentary and granitic terrains.

Twenty-three soils on moraines and rock glaciers that date from 0 to ~12,500 B.P. were examined in the Atigun Pass region for pH, color, and horizon thicknesses. These properties are useful for relative-dating of glacial deposits in other alpine environments (Mahaney 1974). The soil pits were located on crests of moraine and rock glacier lobes where drainage is unimpeded during the thaw season. Munsell dry colors were used in the chronosequence. pH was determined in the field by colorimetric methods. The soil profiles were given an approximate age based on lichenometric, geometric, or radiocarbon correlations.

5 GLACIERS, NEOGLACIAL MORAINES, AND ROCK GLACIERS

5.1 Cirque glaciers

Average slopes of 17° and lengths of 740 m characterize the cirque glaciers in the Atigun Pass area (Ellis & Calkin 1979); average slopes of 16° and lengths of 1280 m were measured for 11 glaciers in the granitic Arrigetch Peaks (Ellis et al. 1981). The Atigun and Anaktuvuk glaciers average ~1800 m in mean altitude and are all above 1500 m while those in the west-central Arrigetch have mean altitudes about 300 m lower. Glaciers across the central Brooks Range are very strongly oriented north-northeast to minimize insolation, demonstrating marginal conditions and highly significant climatic control on glacierization during Holocene time. Relatively debris-free areas of glaciers average ~0.4 to 0.5 km^2; however, they range from 0.1 to over 2.0 km^2 (Table 2). Cirque glaciers averaged only ~0.65 km^2 in area during Neoglacial maxima.

5.2 Moraines without cores of ice

Moraines which lack substantial ice cores (M) have subtle relief and are the most stable type of cirque glacier deposit (Fig. 5A). Those in sedimentary terrain receive ~94% of the potential direct radiation energy on 24 July while those in the granitic Arrigetch gain only ~85% (Table 3). In both terrains these deposits have low topographic horizons (minimal shading) and little bedrock exposed in cirque cliffs (Table 2) as compared with ice- or glacier-cored moraines.

Table 2. Area and altitudinal summary of present and Neoglacial maxima cirque glaciers.

Type lithology no. examined	Present glacier area (km²)	Neoglacial maxima area (km²)	Exposed headwall (m)	Neoglacial maxima ELA (m)	ELA depression (m)
M, Mg, MG, TRGC					
All (n = 53)	0.45 ± 0.40	0.64 ± 0.64	153 ± 109	-	82 ± 49
Sedimentary (n = 42)	0.44 ± 0.37	0.58 ± 0.42	129 ± 89	1730 ± 85	70 ± 35
Granitic (n = 11)	0.49 ± 0.50	0.87 ± 0.83	246 ± 134	1340 ± 30	128 ± 62
M					
All (n = 15)	0.58 ± 0.44	0.79 ± 0.54	72 ± 55	-	78 ± 43
Sedimentary (n = 12)	0.61 ± 0.49	0.82 ± 0.60	58 ± 49	1795 ± 90	57 ± 24
Granitic (n = 3)	0.45 ± 0.08	0.71 ± 0.23	100 – 180	1295 – 1375	148 ± 47
Mg					
All (n = 18) —Sedimentary—	0.36 ± 0.20	0.53 ± 0.26	148 ± 70	1740 ± 50	79 ± 35
MG					
All (n = 12)	0.55 ± 0.54	0.91 ± 0.86	190 ± 122	-	119 ± 57
Sedimentary (n = 5)	0.52 ± 0.49	0.77 ± 0.67	100 ± 0	1715 ± 50	97 ± 54
Granitic (n = 7)	0.55 ± 0.61	1.00 ± 1.00	247 ± 125	1345 ± 30	135 ± 58
TRGC					
All (n = 8)	0.28 ± 0.36	0.49 ± 0.43	250 ± 139	1640 ± 80	40 ± 40
M, Mg, MG					
All (n = 45)	0.48 ± 0.40	0.72 ± 0.57	137 ± 95	-	89 ± 48
Sedimentary (n = 35)	0.47 ± 0.37	0.59 ± 0.42	112 ± 70	-	75 ± 35
Granitic (n = 10)	0.53 ± 0.50	0.92 ± 0.84	226 ± 122	1345 ± 30	139 ± 53

5.3 Ice-cored moraines

Ice-cored moraines (Mg) lack a visible ice core, yet have marked relief and sharp-crested fronts in the downvalley direction (Fig. 5B; Østrem 1971). They can grade into zones without ice cores along continuous ridges. Ice-cored moraines were only mapped in the Atigun Pass area (Fig. 2); in the other two field areas glacier cores were clearly exposed beneath the till. Considerable overlap in planimetry and terrain screening values (Tables 2, 3) between these moraines and those visually-cored with glaciers (MG, Fig. 5C) substantiates grouping them together in subsequent statistical analyses. Therefore, all ice cores are considered to be made of glacier ice.

5.4 Glacier-cored moraines

Cirque glacier deposits in the granitic Arrigetch Peaks are more extensively cored with glacier ice than those studied in sedimentary terrain. Glacier-cored moraines (Mg/MG) in sedimentary terrain ac-quire ~83% of the potential solar energy. In contrast, eight measured in granite receive only ~61% of the potential energy (Table 3). The potential debris supply available to glacier-cored moraines is greater in the deeper cirques of the Arrigetch where 245 + 125 m of bedrock is exposed in cirque cliffs as compared with 140 + 65 m for similar deposits in sedimentary terrain.

5.5 Neoglacial transition zones of glacier cored rock glaciers

Eight Neoglacial transition zones (TRGC) were mapped; they average 250 + 140 m of bedrock exposed in cirque cliffs. Terrain screening measurements demonstrate mean horizons of ~20° and average receipts of direct radiation about 73%. However, a few transition zone moraines form where as little as 1% solar energy is received and/or vertical headwall exposures exceed 400 m. In the east-central Brooks Range, glaciers fronted by transition zones (TRGC) that have partially overridden down-slope rock glacier tongues are ~100 m lower in altitude than those glaciers fronted by only morainal loops (M, Mg, and MG; Ellis

Table 3. Terrain screening and exposure of glacial landforms.

Type of deposit lithology no. of sites	Site altitude (m)	Duration (hr)	Potential energy blocked (%)	Exposure gain or loss (%)	Total direct energy received (%)	Mean horizon (°)	Maximum horizon (°)
M							
All (n = 17)	1623 ± 201	17.1 ± 3.1	-6.6 ± 5.9	-1.3 ± 1.6	92.1 ± 6.2	13.5 ± 4.4	26.4 ± 5.6
Sedimentary (n = 14)	1687 ± 126	17.9 ± 2.9	-4.8 ± 4.5	-1.4 ± 1.7	93.8 ± 5.2	12.0 ± 3.2	24.2 ± 3.1
Granitic (n = 3)	1267 ± 29	13.8 ± 1.4	-14.8 ± 4.1	-0.7 ± 0.6	84.5 ± 4.6	20.3 ± 1.7	36.7 ± 0.8
Mg							
All (n = 20) -Sedimentary-	1675 ± 66	12.6 ± 2.2	-15.3 ± 7.2	-2.5 ± 2.2	82.3 ± 7.7	18.2 ± 2.4	30.7 ± 3.4
MG							
All (n = 17)	1429 ± 215	11.4 ± 3.5	-24.1 ± 15.1	-2.4 ± 1.8	73.5 ± 15.6	20.5 ± 5.2	36.1 ± 10.2
Sedimentary (n = 9)	1608 ± 77	13.9 ± 2.2	-12.8 ± 6.1	-2.3 ± 1.7	84.9 ± 6.2	16.4 ± 3.3	28.0 ± 5.1
Granitic (n = 8)	1227 ± 106	9.5 ± 2.2	-36.7 ± 11.4	-2.5 ± 2.1	60.8 ± 12.6	25.2 ± 1.5	45.2 ± 5.5
TRGC							
All (n = 16)	1490 ± 174	10.6 ± 3.8	-25.5 ± 25.4	-1.9 ± 1.8	72.6 ± 24.9	20.4 ± 4.9	33.4 ± 7.1
Sedimentary (n = 14)	1541 ± 100	11.4 ± 3.0	-19.2 ± 16.3	-2.1 ± 1.9	78.8 ± 16.1	19.1 ± 2.9	31.5 ± 4.8
Granitic (n = 2)	1000 - 1270	8.4 - 1.5	-41.6 - -99	-1 to 0	57.4 to 1	23.7 - 34.5	44.0 - 50.5

& Calkin 1979).

5.6 Reconstructed ELA's

ELA's reconstructed for Neoglacial maxima with AAR = 0.67 average 1755 + 55 m for cirque glaciers fronted by morainal deposits (M, Mg, MG) and 1640 + 80 m for those leading into rock glacier transition zones (TRGC) in the Atigun Pass area. Moraines without cores of ice (M) have the highest, steady-state ELA's averaging 1795 + 90 m. These east-central ELA's are significantly higher than those found in the Arrigetch Peaks where ELA's for glacier-cored moraines averaged 1345 + 30 m.

The lowering of ELA from the mean altitude of present glaciers fronted by moraines (excluding transition zone deposits) was 75 + 35 m in the Atigun/Anaktuvuk region and 140 + 55 m in the more westerly Arrigetch. For glaciers fronted by the transition zone moraines, an ELA lowering of only ~40 m occurred during Neoglacial maxima (Table 2).

6 DISCRIMINATION OF NEOGLACIAL DEPOSITS

A trend of decreasing solar energy received (92% to 73%) and increasing head- and sidewall horizons (13° to 20°) generally characterizes the gradation (Fig. 7)

from non ice-cored moraines (M) through glacier-cored moraines (MG). However, measurements on Neoglacial transition zones (TRGC) of rock glaciers are scattered in a plot of radiation versus horizon (Fig. 7). A more continuous transition from M through MG to TRGC is revealed when receipt of direct radiation energy is plotted against estimated height of bedrock in cirque cliffs (Fig. 8). This figure also demonstrates that moraines without ice cores (M) are formed in sedimentary terrain where potential debris is less than 150 m high in cirque cliffs and greater than 83% of the potential solar energy is received. The marked environmental contrast between the cirques of the Arrigetch moraines and those in sedimentary terrain is well shown in Figure 8.

Moraines without ice cores, glacier-cored moraines and transition zones were distinguished with linear discriminant analysis using four of six environmental and morphological variables (Table 1). Analysis 1 (Fig. 9) utilized latitude (LAT), Neoglacial maxima ELA (NELA), direct radiation energy received (NSUN), and potential debris supply (NHEAD). It indicates the importance of the Neoglacial maxima ELA in controlling the type of moraine deposited. The higher this steady-state ELA the more likely moraines without ice cores (M) will be deposited. The

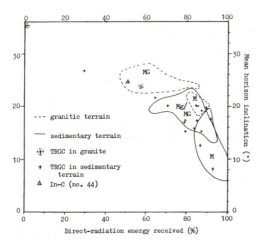

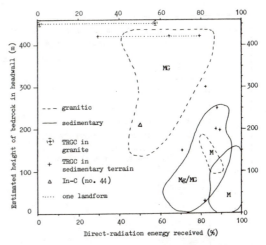

Figure 7. Plot of horizon inclination versus percent of direct radiation energy received for moraines without cores of ice (M), moraines cored with glacier ice (Mg/MG), and glacier-cored rock glaciers with transition zones (TRGC). The deposits are distinguished according to lithology in the central Brooks Range: sedimentary terrain of Atigun and Anaktuvuk and granitic terrain of the Arrigetch Peaks.

Figure 8. Plot of estimated height of bedrock exposed in cirque cliffs (potential debris supply) versus percent of direct radiation energy received for moraines without cores of ice (M), moraines cored with glacier ice (Mg/MG), and glacier-cored rock glaciers with transition zones (TRGC). These deposits are distinguished according to lithology in the central Brooks Range: sedimentary terrain of Atigun and Anaktuvuk and granitic terrain of the Arrigetch Peaks.

higher ELA's imply localization of glaciers in cirques with lower topographic horizons, less supraglacial debris and greater solar inputs. TRGC and MG deposits have considerable overlap; however, a few transition zone moraines have markedly lower solar inputs and Neoglacial ELA's, and higher potential debris supply.

Analysis 2 (Fig. 10) used amount of ELA lowering (NDROP) and area involved during Neoglacial maxima (NAREA) in place of the geographically-dependent NELA and NLAT, which were used in Analysis 1 (Fig. 9). This second analysis clearly discriminates non ice-cored moraines from the glacier-cored varieties. In addition, there is extensive intermixing of glacier-cored moraines and those of transition zones suggesting more similarities than differences.

Potential debris supply (NHEAD) is the most important environmental factor in determining the type of deposit formed; the less bedrock exposed in cirque cliffs the more likely a moraine without an ice core will be formed. A moraine deposited in an environment with a high receipt of solar energy (92%), which favors formation of non ice-cored moraines, will possess a substantial glacier core if a high debris input (~250 m) from cirque walls is available. In those glacierized cirques of the Arrigetch Peaks where input of debris is

relatively low (150 m), a reduced input of solar energy (53%) also results in deposition of glacier-cored moraines. The pooled within-groups correlation matrix showed little correlation (-0.04, -0.02) of these two environmental factors when discriminating the three moraine groups (M, Mg/MG, TRGC) in the same lithologic setting (Ellis 1982). The correlation coefficients in Figure 10 indicate minimal depressions of ELA during Neoglacial maxima favor TRGC and, for reasons unknown, moraines without cores of ice.

A third discriminant analysis (Fig. 11) was applied between the 43 sedimentary and 11 granitic cirque deposits. The variables are non-geographical and the same as used in the second analysis (NSUN, NHEAD, NDROP, NAREA). The amount of ELA depression from present mean glacier altitude is by far the most important discriminant between the two terrains. Of secondary importance are the marked differences in solar energy received and potential debris supply, both of which may be explained by granitic terrain supporting deeper cirques than sedimentary bedrock. This discriminant analysis indicates a slight negative correlation (-0.38) between the two factors, energy received and

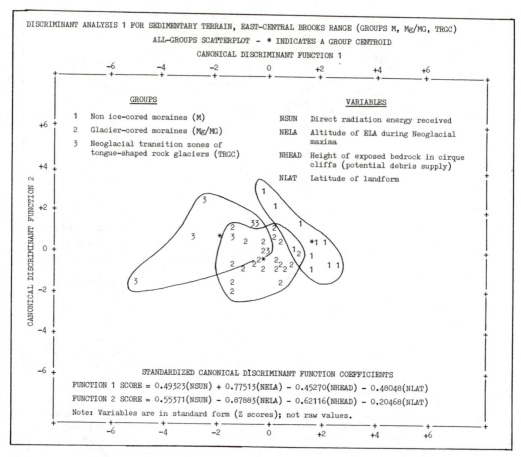

Figure 9. Scatterplot of Linear Discriminant Analysis 1 for glacial deposits in sedimentary terrain with morainal groups (M, Mg/MG, TRGC), variables, and standardized coefficients for discriminant variables shown. These coefficients indicate the relative contribution of each variable to the function. The closer the coefficient is to ± 1.0, the larger the relative contribution of the variable to discriminating between the groups. Here the ELA for Neoglacial maxima (NELA), a geographically-dependent variable, is the most important discriminant.

height of bedrock exposed in cirque cliffs.

7 SOILS OF MORAINES AND ROCK GLACIERS

Soil development in the sedimentary terrain of the Atigun Pass area helps differentiate moraines and rock glaciers. Tills sampled at the contact of receding glaciers or from unweathered zones at depth far beyond the ice margin yielded pH soil values from 7.5-8.0 and color chromas of ~2 (Fig. 12). Thin organic horizons of ~1 cm thickness and oxidation depths to 15 cm develop in less than 400 lichenometric years. On moraines licheno-metrically dated at ~2000 B.P., organic horizons reach 3 cm thicknesses, A horizons extend to depths of 10 cm with pH values of ~6.2, and oxidized C horizons occur from 20 to >50 cm. Thicker A horizons, development of a B horizon, chromas of ~3, and soil pH values of 4.8-6.4 help distinguish early Holocene rock glacier tongues from moraines located upslope in Neoglacial transition zones.

Soils on Pleistocene valley drift dated at ~12,500 B.P. (Hamilton 1979a) are more uniformily well developed than those on rock glaciers. They have pH values ranging from 4.7-5.4. In northern Atigun Valley (Fig. 2), these late Pleistocene

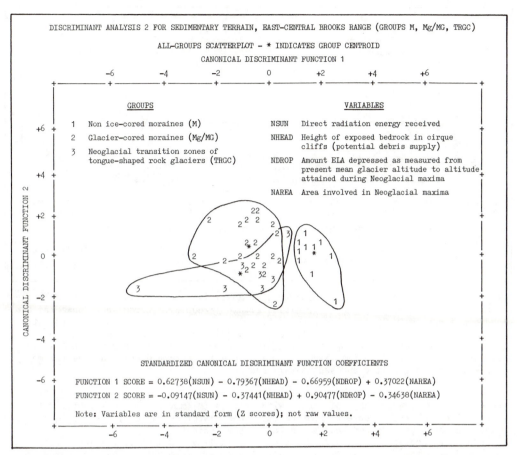

Figure 10. Scatterplot of Linear Discriminant Analysis 2 for glacial deposits in sedimentary terrain with morainal groups, variables that should be relatively independent of geographic location, and standardized coefficients given (see Fig. 9). Potential supraglacial debris supply (NHEAD) is the most important variable discriminating the morainal groups.

moraines typically have organic horizons to 7 cm depth, A horizons from 20 to 25 cm thicknesses, and color B horizons with chromas to 4. On the south side of the Continental Divide two soil pits on valley moraines had pH values from 5.0 to 5.4 in the upper solum; color chromas here were also ~4. More caliche build-up was observed to the south than to the north of the Continental Divide. Calcareous loess derived from extensive exposures of limestone along the south flank of the east-central Brooks Range may cause this calcium carbonate build-up.

8 DISCUSSION

Moraines without cores of ice occur in relatively unique cirque environments which are characterized by minimal heights of bedrock exposed in cirque cliffs, extensive receipt of direct radiation energy, and low topographic horizons. The higher the cirque, the higher the probability that the glacier will form non ice-cored moraines.

Moraines cored with ice form in cirques with variable inputs of solar energy and debris; however, increased supraglacial debris is the most important environmental factor promoting preservation of glacier cores. The widespread occurrence of glacier-cored moraines and rock glaciers emphasizes the major contribution supraglacial debris has in deposition of alpine tills of the central Brooks Range.

Active, tongue-shaped rock glaciers in

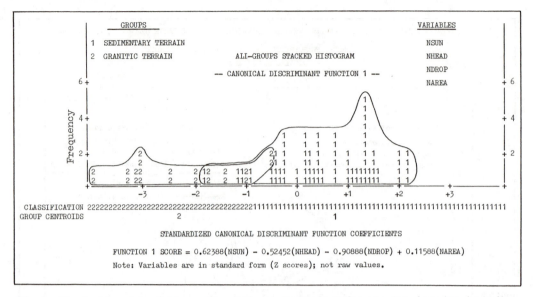

Figure 11. Scatterplot of Linear Discriminant Analysis 3 for glacial deposits in sedimentary terrain (group 1) and in granitic terrain (group 2). Cirque-environmental and landform-morphological variables used in Figure 10 are also used in this analysis to minimize geographical dependence. Although potential debris supply and solar energy received (NHEAD, NSUN) are important factors, the differences in amount of ELA depression during Neoglacial maxima (NDROP) are most significant in discriminating the moraines formed in granitic and sedimentary bedrock.

the study areas are probably all cored with glacial ice. However, those with headward glacier cores presently exposed were able to respond to Neoglacial climatic deteriorations due to their relatively higher altitudes. This has resulted in deposition of glacier-cored morainal ridges within Neoglacial transition zones (TRGC). These zones have not completely overrun their downslope, early Holocene rock glacier tongues (Fig. 5D).

Glaciers presently fronted only by morainal deposits (M, Mg, MG) are ~100 m higher than those glaciers leading downslope into transition zone moraines. In addition they averaged greater ELA depressions during Neoglacial maxima. These higher-altitude glaciers apparently obliterated or incorporated any previous early Holocene rock glacier lobes emanating from their cirques.

The most significant environmental factors favoring formation of transition zones and preservation of downslope rock glacier tongues are lower glacier altitudes and increased bedrock exposed in cirque cliffs. The overall physical similarity between glacier-cored deposits (Mg, MG, TRGC) enhances their combined use in a glacial chronology; however, moraines without ice

cores (M) may provide a slightly different history because of their unique cirque environments.

Whalley (1974) believes ice-cored rock glaciers form rather than ice-cored moraines where there are substantial cliffs above the glacier and a plentiful supply of debris (see also Currey, 1969). Johnson (1980) proposes that debris supply is not a factor in glacier-cored, rock glacier placement; a small confined cirque is regarded as the optimum site. Our study supports the importance of supraglacial debris.

The environmental factor that most effectively distinguishes between moraines of the granitic Arrigetch and those of the more easterly Atigun and Anaktuvuk areas is the amount of ELA depression during Neoglacial maxima. Why the ELA lowering is greater in the west-central Brooks is uncertain; however, it may be related to 1) inconsistencies in determining ELA by the planimetric method or 2) the effects of increased annual precipitation, temperature, and cloudiness on glacier dynamics in the Arrigetch. Terrain screening and potential debris supply also discriminate between moraines in the deeper granitic cirques and those in sedimentary terrain

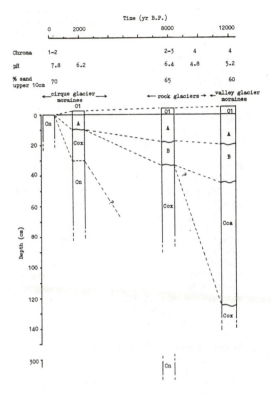

Figure 12. Generalized schematic of soil development from 0 to ~12,500 B.P. on moraines and rock glaciers of the Atigun Pass area, east-central Brooks Range. The parent material is till derived from sedimentary rocks. An oxidized C horizon is designated by "Cox" and a seemingly-unweathered C horizon by "Cn" (Birkeland 1974). B horizons are largely "color B's". Development of a calcium carbonate build-up apparently depends on influx of calcareous loess.

Because of their differing cirque environments, the timing of Neoglacial fluctuations may vary between the west- and east-central Brooks Range. The glacial history developed thus far for these regions (Ellis et al. 1981; Ellis 1982) is based largely on lichenometry; it may be too imprecise to detect such differences.

In the sedimentary terrain of the central Brooks Range, soil evolution on moraines and rock glaciers is typified by pH declining from 7.5-8.0 to 4.7-5.4, color chroma increasing from ~2 to ~4, and marked development of soil horizons over a ~12,500 year interval. Chelating agents, associated with the widespread lichen

population, may have a significant role in this weathering trend. The colorimetric pH values measured in this study compare favorably with values of 4.5 to 5.0 recorded in the upper solum of late Pleistocene soils in valleys of the northeastern Brooks Range (Brown 1966). The rate of soil pH change reflects northern Alaska's harsh climate as compared with coastal southern Alaska (Crocker & Major 1955; Ugolini 1968) and the Yukon Territory (Jacobsen & Birks 1980). Soil pH on glacial drift in these latter two areas declines from ~8.0 to 5.0 and ~6.0 in 75 and 200 years, respectively. Our preliminary use of soils as a relative-dating tool in the central Brooks Range demonstrates strong potential for distinguishing surfaces on alpine moraines and rock glaciers and determining time since deposition.

9 ACKNOWLEDGEMENTS

This work was supported by U.S. National Science Foundation grants through the Research Foundation of State University of New York. We wish to thank B. Whalley and P.W. Birkeland for encouraging us to examine the environments and soils of alpine moraines and rock glaciers. The assistance of A. Roy and L.D. Ellis in multivariate analysis and computerization of data is gratefully acknowledged.

10 REFERENCES

Birkeland, P.W. 1974, Pedology, weathering, and geomorphological research. New York, Oxford University Press.
Brosgé, W.P., H.N. Reiser, J.T. Dutro, & R.L. Detterman 1979, Bedrock geologic map of the Philip Smith Mountains quadrangle, Alaska 1:250,000. U.S. Geol. Surv. Misc. Field Studies Map MF-879 B.
Brown, J. 1966, Soils of the Okpilak River region, Alaska. U.S. Army Cold Regions Research and Engineering Laboratory Research Report 188, Hanover, New Hampshire.
Brown, J. 1980, The Road and its Environment. In J. Brown & R.L. Berg (eds.), Environmental engineering and ecological baseline investigations along the Yukon River-Prudhoe Bay haul road. U.S. Army Cold Regions Research and Engineering Laboratory Report 80-19, p. 3-52, Hanover, New Hampshire.
Bruen, M.P. 1980, Past and present climatic regimes of a cirque glacier and rock glaciers, Atigun Pass, Alaska, Geol. Soc.

Am., Abs. with programs, 12(2):27.

Calkin, P.E. & J.M. Ellis 1980, A lichenometric dating curve and its application to Holocene glacier studies in the central Brooks Range, Alaska, Arctic and Alpine Res. 12:125-140.

Crocker, R.L. & J. Major 1955, Soil development in relation to vegetation and surface age at Glacier Bay, Alaska, J. Ecology, 43:427-448.

Currey, D.R. 1969, Neoglaciation in the southwestern United States. Ph.D. dissertation, University of Kansas, Lawrence, Kansas.

Detterman, R.L., A.L. Bowsher, J.T. Dutro 1958, Glaciation on the Arctic Slope of the Brooks Range, Alaska, Arctic, 11:43-61.

Ellis, J.M. 1982, Holocene glaciation of the central Brooks Range. Ph.D. dissertation, State University of New York at Buffalo.

Ellis, J.M. & P.E. Calkin 1979, Nature and distribution of glaciers, Neoglacial moraines, and rock glaciers, east-central Brooks Range, Alaska, Arctic and Alpine Res, 11:403-420.

Ellis, J.M., T.D. Hamilton, P.E. Calkin 1981, Holocene glaciation of the Arrigetch Peaks, Brooks Range, Alaska, Arctic, 34:158-168.

Ferrians, O.J., Jr. 1965, Permafrost map of Alaska 1:2,500,000, U.S. Geol. Surv. Misc. Field Studies Map I-445.

Hamilton, T.D. 1965, Comparative photographs from northern Alaska, J. Glaciol, 5:479-487.

Hamilton, T.D. 1977, Late Cenozoic stratigraphy of the south-central Brooks Range. In K.M. Johnson (ed.): The U.S. Geological Survey in Alaska - Accomplishments During 1977. U.S. Geol. Surv. Circ. 772-B:36-38.

Hamilton, T.D. 1978, Surficial geology of the Philip Smith Mountains quadrangle, Alaska 1:250,000, U.S. Geol. Surv. Misc. Field Investigations Map MF-879-A.

Hamilton, T.D. 1979a, Radiocarbon dates and Quaternary stratigraphic sections, Philip Smith Mountains quadrangle, Alaska, U.S. Geol. Surv. Open File Report 79-866.

Hamilton, T.D. 1979b, Surficial geologic map of the Chandler Lake quadrangle, Alaska 1:250,000. U.S. Geol. Surv. Misc. Field Studies Map MF-1121.

Hamilton, T.D. & S.C. Porter 1975, Itkillik glaciation in the Brooks Range, northern Alaska, Quat. Res., 5:471-497.

Jacobsen, G.L., Jr. & H.J.B. Birks 1980, Soil development on recent end moraines of the Klutan Glacier, Yukon Territory, Canada, Quat. Res., 14:87-100.

Johnson, P.G. 1980, Glacier-rock transition in the southwest Yukon Territory, Canada, Arctic and Alpine Res, 12:195-204.

Klecka, W.R. 1975, Discriminant analysis. In N.H. Nie et al. (eds.), Statistical package for the social sciences (Second Ed.), p. 434-467. New York, McGraw-Hill.

Kondratyev, K.YA. (ed.) 1973, Radiation characteristics of the atmosphere and the earth's surface, p. 269-311. New Delhi, Amerind Publishing Co. Pvt. Ltd.

List, R.J. (ed.) 1951, Smithsonian meterological tables (Sixth Ed.), Smithsonian Misc. Collections. 114, Publication 4014:504-512.

Mahaney, W.C. 1974, Soil stratigraphy and genesis of Neoglacial deposits in the Arapaho and Henderson cirques, central Colorado Front Range. In Mahaney, W.C. (ed.), Quaternary environments-proceedings of a symposium, p. 197-240. York University - Atkinson College Geographical Monograph 5, Toronto, Canada.

Nelson, S.W. & D. Grybeck 1980, Geologic map of the Survey Pass quadrangle, Alaska 1:250,000, U.S. Geol. Surv. Misc. Field Studies Map MF-1176-A.

Orvig, S. & R.W. Mason 1963, Ice temperatures and heat flux, McCall Glacier, Alaska, Comm. of Snow and Ice, Inter. Assoc. Sci. Hydrology Publ. 61:181-188.

Østrem, G. 1971, Rock glaciers and ice-cored moraines, a reply to D. Barsch. Geog. Ann., 53A:207-213.

Paterson, W.S.B. 1975, The Physics of Glaciers. Oxford, Pergamon Press.

Porter, S.C. 1966, Pleistocene geology of Anaktuvuk Pass, central Brooks Range, Alaska, Arctic Inst. North Am. Tech. Paper 18.

Ugolini, F.C., 1968, Soil development and alder invasion in a recently deglaciated area of Glacier Bay, Alaska. In G.M. Trappe et al. (eds.), Biology of alders, p. 115-140. U.S. Forest Service, Pacific NW Forest and Range Experiment Station, Portland, Oregon.

U.S. Geological Survey 1978, Brooks Range glacier inventory 1:250,000 (unpubl. data). Available from: World Data Center - A for Glaciology (Snow and Ice), Institute of Arctic and Alpine Research, University of Colorado, Boulder.

Wendler, G. & N. Ishikawa 1974, The effect of slope, exposure, and mountain screening on the solar radiation of McCall Glacier, Alaska: a contribution to International Hydrological Decade, J. Glaciol., 13:13-25.

Whalley, B. 1974, Rock glaciers and their formation: as part of a glacier debris-transport system, University of Reading Department of Geography Geographical Paper 24.

Stratigraphy of the glacigenic deposits in Northern James Ross Island, Antarctic Peninsula

JORGE RABASSA
Universidad Nacional del Comahue, Neuquén & CIC de la Provincia de Buenos Aires, La Plata, Argentina

ABSTRACT

Six drift units which are considered to be related to five main glacial episodes have been recognized in Northern James Ross Island, Antarctic Peninsula.
Marine shells have been radometrically dated and they enable us to suggest Wisconsin, Lower Holocene, Upper Holocene and Present ages for the four younger drifts, respectively. The glaciers which deposited these units were of the mountain ice-sheet type and peninsular origin for the two oldest drifts. The four remaining units were generated by outlet glaciers which were occupying valleys and with a locally-derived debris load. These ice-bodies were, at least partly, wet-based; that is, the temperature at their basal portion was at or above the pressure-melting point.
Several raised marine beaches show the existence of isostatic up-lifting of the island, which could have probably reached mean annual values of 2 to 4 mm. Part of this up-lifting could have been related to glacioisostatic recovery due to ice-cover loss.

INTRODUCTION

James Ross Island (lat. 63°45' - 64°20' S; long. 57°05' - 58°30' W) is the largest of the islands in the archipelago of that name which is located east of the northernmost tip of the Antarctic Peninsula. Other islands which are forming this archipelago are: Marambio (Seymour) Island, Vega Island, Cerro Nevado (Snow Hill) Island and some smaller ones.
The highest point of James Ross Island is Mount Haddington (1620 m a.s.l.), the summit of an ancient volcanic cone of Pliocene-Pleistocene age. This volcano extends over marine and continental Cretaceous sedimentary rocks, which form the outcropping

bedrock of this island. Older metamorphic and granitic rocks do not outcrop here but they do it so only westwards of the Prince Gustav Channel, in the Antarctic Peninsula itself (Figure 1).
James Ross Island is located in the precipitation shadow due to the transversal position of the mountaineous ranges of the Antarctic Peninsula with respect to the more humid winds coming from the west and southwest. Thus, although most of its surface is ice-covered (81%, Rabassa and others 1981) there are areas where bare rock is exposed. These areas have been chosen for on-going glacial geology investigations under the GEOGLA Project of the HIELOANTAR Program, Argentine Antarctic Institute (IAA). These bedrock-exposed surfaces are under permafrost conditions with an active layer of 0.80 m, in summer time and at sea level.
The geology of the Antarctic Peninsula may be consulted in Andersson (1906) and Grikurov (1978), among many others. Works on glacial geology and stratigraphy of this area, and related topics, are not so numerous and John & Sugden (1971), Sugden & John (1973), Sugden & Clapperton (1977, 1980a, 1980b), Clapperton and others (1978), Elliot and others (1975), Zinsmeister (1979, 1980), and Elliot (1981), among others, should be mentioned.
For James Ross Island in particular, the classic paper by Bibby (1966) is the most relevant. Recently, Malagnino and others (1981) and Rabassa and others (1981, 1982) have discussed geomorphological and glaciological aspects.
The main purpose of this paper is to provide a preliminary stratigraphic scheme for the glacigenic deposits in this island and several radiocarbon dates performed on marine shells from some of these units. These data suggest a tentative chronology for the latest glacial events which took

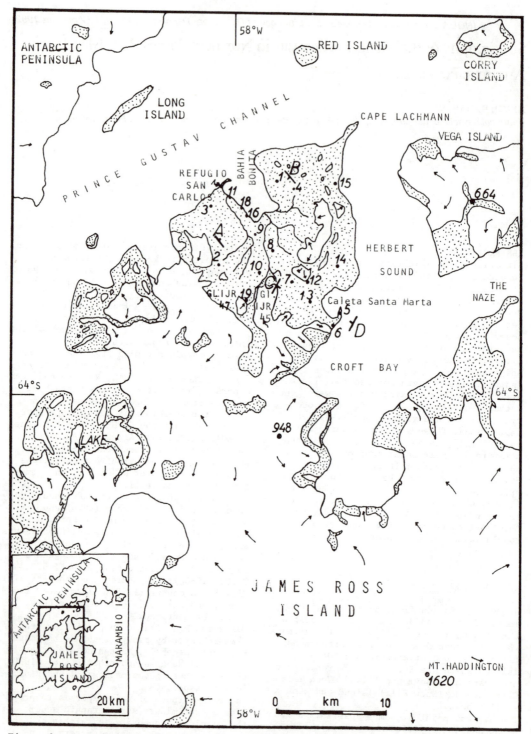

Figure 1.   Location map. Dotted areas: ice-free areas of James Ross and other islands.
Localities 1 to 19, see text. A-B and C-D, approximate geographical location
of schematic cross sections of Figure 2; ⊙ 1620, approximate elevation, m a.s.l.

place in this island.

STRATIGRAPHY

During summer field seasons 1980, 1981 and 1982 many observations, profiles and geological-geomorphological mapping were performed in the area of Bahía Bonita (Brandy Bay) and Caleta Santa Marta (Sta.Martha Cove), Bahía Croft, together with reconnaissance flights over the rest of the island. These observations enabled us to prepare the following stratigraphic scheme (Table 1)

1. James Ross Volcanics

This unit is composed of andesites, basalts, basandesites, mesosilicic to acid ignimbrites, volcanic breccias and agglomerates of varied lithology, tuffs and cinerites. We have also found erratics of a diamictite (within the Bahía Bonita Drift) which is assumed to come from this unit, although no outcrops have been found yet. It is a very dense, compact rock with basaltic, andesitic (locally-derived) and granitic, metamorphic (exotic) pebbles in a sandy, cineritic matrix. Some of the pebbles are striated and penecontemporaneously-deformed laminated

TABLE 1
STRATIGRAPHY OF THE GLACIGENIC DEPOSITS

Lithostratigraphic units	Lithology	Age
Presently-forming Drift	(Locally-derived lithology) Ablation till Shear till Supraglacial debris	Present (XIX and XXth. centuries)
----------------- Glacier recession --
IJR-45 Glacier Drift	(Locally-derived lithology) Ice-cored morainic debris Lodgement till Ablation till Glaciofluvial and glaciomarine associated deposits	Upper Holocene 1590 ± 60 yr.^{14}C B.P.
------------------ Uplift (glacioisostatic recovery?) and erosion -----------------------
Bahía Bonita Drift	(Locally-derived lithology) Subaerial and submarine ablation till Lodgement till Kame deposits Submarine subglacial delta deposits Glaciomarine deposits with dropstones	Lower Holocene 5175 ± 205 5270 ± 120 yr.^{14}C 5580 ± 110 B.P. 5915 ± 233
------------------- Uplift (glacioisostatic recovery?) and erosion -----------------------
Caleta Santa Marta Drift	(Locally-derived and exotic lithology) Ablation till Flow till Glaciomarine deposits with dropstones Glaciomarine melt-out till Deltaic deposits Kame deposits	Upper Pleistocene (Wisconsin) $34115 \pm 1110/975$ yr.^{14}C B.P.
----------------- Uplift (glacioisostatic recovery?) and erosion -----------------------
Refugio San Carlos Drift	(Exotic and locally-derived lithology) Lodgement till Ablation till Scattered erratics	Upper Pleistocene(?) Middle Pleistocene(?) (Illinois?)
----------------- Uplift (glacioisostatic recovery?) and erosion -----------------------
Meseta Lachmann Drift	(Exotic lithology, rarely locally-derived) Submarine melt-out till (*) Lodgement till (*) Scattered erratics without matrix	Middle Pleistocene(?) Lower Pleistocene(?)
	(*) found only in Marambio Island to the moment	
----------------- Erosional uncomformity --

TABLE 1 (continued)

Lithostratigraphic units	Lithology	Age
James Ross Volcanics	Volcanics, tuffs, tephras, volcanic breccias, ignimbrites, etc. Diamictite (tillite?)	Pliocene to Pleistocene
------------------- Angular uncomformity ---		
"Snow Hill Series" "Prince Gustav Series"	Continental and marine epiclastic sedimentary rocks	Upper Cretaceous

silt and fine sand layers are also found within the matrix. From a genetic point of view, this diamictite could have been deposited in fact either by glaciers coming from the Antarctic Peninsula or by non-glacial agents, such as volcanic mudflows related to the evolution of Mount Hadding-ton. Its proper interpretation will undoubtedly need more detailed studies.
The age of these unit has been established as Pliocene to Quaternary, based on radiometric dating (Grikurov 1978).

2. Meseta Lachmann Drift
The surface of the basaltic flows related to the James Ross Volcanics has been partially preserved as structural levels. Northeast of Bahía Bonita, at the northernmost extent of Meseta Lachmann (Lachmann Crags) and over one of such levels (Bibby Point, 280 m a.s.l.; Figure 1, Locality 1 –those sites mentioned in the rest of the paper are also referred to the same figure–; Figure 2), volcanic and granitic-metamorphic (exotic) erratic boulders have been found, some of them being up to 3.0 m in diameter (Figure 3 ; Figure 4, Section A-B).

Figure 3. Granitic boulder (exotic) which indicates a provenance from the Antarctic Peninsula, with no matrix. The boulder is embedded into congelifracted basaltic clasts (Meseta Lachmann Drift). The site corresponds to Locality 1, see Figure 1).

These boulders exhibit polished and striated surfaces and they appear scattered over a large area but no matrix has been found. The boulders are uncomformably lying over basaltic rocks of the James Ross Volcanics. The basalts are severely affected by mechanical weathering and periglacial processes.
Other deposits which may be correlated with these are found at 235 m a.s.l., as isolated erratics with little or no matrix overlying basalts and other volcanics (Figure 1, Locality 2; Figure 2, outcrops at the extreme left of the photograph; Figure 4, Section A-B).
The lithology of these boulders highly resembles that of the erratics found within the till plain of Marambio Island (Rinaldi and others 1978; Elliot and others 1975; Malagnino and others 1981); a tentative, preliminary correlation is suggested for both units.
At Marambio, a severely-eroded lodgement till cover overlies melt-out submarine till. The largest erratics are heavily striated and they appear included in a greenish, sandy-clayey matrix, probably derived from the erosion of Cretaceous

Figure 2. Bahía Bonita, Refugio San Carlos area. The uppermost level is composed of lava flows of the James Ross Volcanics, more than 250 m a.s.l. The intermediate level is formed by Refugio San Carlos Drift (around 100 m a.s.l.). Near sea level, deposits of the Bahía Bonita Drift. In the distance, the Antarctic Peninsula.

332

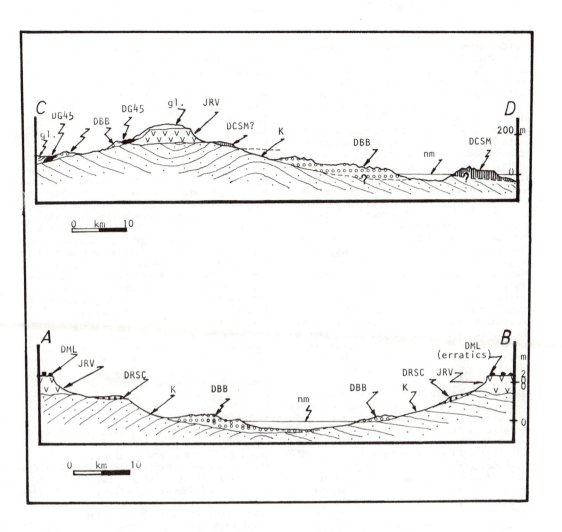

Figure 4. Schematic sections illustrative of the spatial distribution of the glacigenic
deposits in Northern James Ross Island. See approximate location of the sections
in Figure 1.
References: K, Cretaceous; JRV, James Ross Volcanics; DML, Meseta Lachmann
Drift; DRSC, Refugio San Carlos Drift; DCSM, Caleta Santa Marta Drift; DBB,
Bahía Bonita Drift; DG45, IJR-45 Glacier Drift; gl, glacier; nm, sea level.
The topographic position, horizontal and vertical scales, structural position
of the Cretaceous beds and the stratigraphic relationships are approximate and
tentative; they are presented just for descriptive purposes.

sedimentary rocks, with scattered marine
shells fragments. These shells are interpreted as contemporaneous with the melt-out
till.
On the Meseta Lachmann boulders, crustose
lichens (*Rhizocarpon geographicum* and other
taxa) up to 10 mm in diameter have been
found. These sizes are unexpectedly small
considering the age of these deposits.
However, the ecological conditions above
250 m a.s.l. in this island are probably

unsuitable for the growth of these lichens
or, alternatively, the referred surface has
only recently been liberated from its snow
cover due to climatic ascent of permanent
snow line.

3. Refugio San Carlos Drift
Glacigenic deposits have been found on
topographic heights located inmediately
WSW of Refugio San Carlos (Bahía Bonita,
Locality 3; Figure 2), between 100-150 m a.

s.l., thus being more than 100 m below the Meseta Lachmann Drift (Figure 4, Section A-B). These deposits are composed of ablation till, with very loose, sandy matrix and large granitic-metamorphic (exotic) boulders with polished faces. Some of these boulders are striated and reach a maximum diameter of 0.80 m . This till cover is rather thin, up to 3-5 thick, but it is discontinuous and the Cretaceous bedrock outcrops frequently. No original glacial landforms have been preserved. In those areas where the till cover is interrupted, isolated erratic boulders without matrix are found. *Rhizocarpon geographicum* colonies up to 138 mm in diameter have been measured on these units.

The Cretaceous sedimentary rocks include here coarse fluviatile conglomerates with very-rounded, equidimensional clasts. Partly, these pebbles have been incorporated to the overlying till but they are easily distinguished from the true till stones, more angular and faceted.

In front of the Prince Gustav Channel, at the uppermost portion of the cliff (100 m a.s.l., approximately), the glacial sediments are eroded by marine action, forming raised beaches with open-work reddish gravels, slightly oxydized.

Other deposits probably correlating with these glacigenic units are found nearby the col between Bahía Bonita and Cape Lachmann (Locality 4, 160 m a.s.l.; Figure 4, Section A-B). They are composed of lodgement till, 1-2 m thick, with heavily striated, volcanic, granitic and metamorphic boulders and cobbles in sandy-clayey matrix. These units are interpreted as remnants of a severely eroded ground moraine.

4. Caleta Santa Marta Drift

A peculiar landform superficially resembling a spit is found partially closing Caleta Santa Marta (Sta.Martha Cove, Locality 5). However, the study of its sedimentary profile has shown that it is composed of glacigenic deposits, partially submarine. These deposits extend from their visible base at sea level up to 15 m a.s.l. They are truncated by a planar surface with homogeneous granometric-size gravels which is interpreted as a raised marine beach (Figure 4; Section C-D). Melt-out till, submarine ablation till, kame and submarine kame deposits, glacimarine deposits with dropstones, delta deposits and marginal lake sediments have been recognized within these glacial beds.

Between 10.40-10.90 m, many marine shells of *Laternula elliptica* have been found, which form a single bed with most shells complete and some still articulated.

Thus, according to our knowledge of the present ecology of this species (Zulma A. de Castellanos, Museo de La Plata, written communication),we assume that they are in life position or with very restricted transportation. One sample of these shells has been dated in 34,115 $\pm$ 1110/975 years ^{14}C B.P. (Table 2).

The deposits in this locality are tentatively correlated (Figure 4, Section C-D) with: (a) partially eroded terminal moraines which appear southwards along the Croft Bay shore (Locality 6; 30 m a.s.l.) and (b) lodgement till and ablation till deposits which are found at the col between Bahía Bonita and Caleta Santa Marta (Locality 7; 130 m a.s.l.), where they have been observed as a discontinuous, partially eroded cover with large boulders (1.80 m in diameter), N 90° striations and *Rhizocarpon geographicum* crustose lichens up to 168 mm in diameter. These deposits at the col would have been formed by a transfluent glacier moving from W to E.

The existence of submarine drift and terminal moraines at or near sea level are considered as the fundamental criteria to distinguish these deposits from the Refugio San Carlos Drift whose remnants and the marine beaches excavated on them are located at higher elevations.

5. Bahía Bonita Drift

This denomination is proposed for the most conspicuous and more extended glacigenic deposits of the areas under study.

The Bahía Bonita Drift is composed of lodgement till, ablation till, flow till, sands and silts of submarine deltas, kame deltas and glaciofluvial plain deposits, some of them submarine, and glaciomarine silty clays with dropstones. These dropstones would have been contributed either by icebergs coming from local glaciers and the ice-shelf probably occupying then the Prince Gustav Channel, or directly by basal melt-out of floating tongues of one or both origins mentioned.

The lithological content of the Bahía Bonita Drift varies towards the inner portion of the island, showing a progressive diminution of exotic (granites and metamorphics)pebbles till they disappear completely.

The exotic boulders reach 1.0 m in diameter (Locality 9) whereas there are locally-derived volcanic breccias and andesites that are larger than 10 m in diameter. These breccia boulders should have been supraglacially transported, mainly because striae are lacking and also, they would have disintegrated under englacial- or subglacial transport, due to their poor consolidation. Additionally, they are forming the ablation till

cover that overlies the ground moraine
(Figure 5).
In the area of Bahía Bonita, the deposits
are interpreted, based on their geomorphic
distribution, as remnants of the sedimentary
body of the debris-covered Glacier IJR-45
(Rabassa and others 1981, 1982) when it was
extending over almost the entire glacial
valley between Meseta Lachmann and the Cretaceous rocky divides topped by the Refugio
San Carlos Drift (Figure 4, Section A-B).
These deposits are forming a basal till
plain (Locality 8), lateral morainic ridges
(Locality 9), probably a frontal moraine
(?) near Refugio San Carlos, drumlins,
drumlinoid forms and flutings (Figure 5,
Locality 10) and submarine kame-deltas
(Locality 11).
Other terminal morainic arcs are found near
the front of smaller glaciers, i.e., the
regenerated mountain-slope glacier of the
col between Bahía Bonita and Bahía Croft
(Locality 12; Figure 4, Section C-D), the
morainic ridges along the inner shore of
Caleta Santa Marta inland from the Caleta
Santa Marta Drift outcrops (Locality 13;
Figure 4, Section C-D), in Andersson Point
(Locality 14) and near the shore at Cape
Lachmann (Locality 15) where whale bones
have been found at 12 m a.s.l. lying on
marine terraces cut on these moraines.
During this glacial epoch, the terminal
position of the IJR-45 Glacier was below
sea level as it is shown by numerous find-
ings of marine shells, most of them belonging to *Yoldia eightsii*, although some
shells of *Laternula elliptica* and one
specimen of *?Adamussium sp.* have also been
obtained. Most of the material is fragmentary, either due to transportation or periglacial processes, but in several sites
around Locality 16 (Figure 1), complete and
even articulated shells have been recovered.
These shells have been radiometrically dated
between 5175 and 5915 years ^{14}C B.P.
(Table 2).
The colonies of *Rhizocarpon geographicum*
reach up to 19 mm in diameter, but these
values are much smaller inlandwards.
The upper surface and sides of the moraines
at Bahía Bonita also exhibit the effects of
marine erosion at several topographic levels
between 5 and 35 m a.s.l. The uppermost
remnants of these marine beaches are located
in the inner part of the bay whereas those
near sea level are found at Refugio San
Carlos.

6. IJR-45 Glacier Drift
This unit is composed of lodgement till,
ablation till and glaciofluvial deposits,
partially terraced. Glaciomarine beds located over erosional levels on the sides

Figure 5. Drumlins and drumlinoid forms,
Bahía Bonita Drift. See the large volcanic
breccia boulders, superglacially transported, which form part of an ablation till
cover on the ground moraine. At the back,
the IJR-45 Glacier and the deposits of the
IJR-45 Glacier Drift. Drumlin formation
was generated by right-to-left flowing ice.

of the Bahía Bonita Drift moraines are tentatively correlated with this unit, because
of their geomorphological continuity with
the glaciofluvial plains.
The sediments of the IJR-45 Glacier Drift
are found near the margin of the present
glacier of that name (Rabassa and others
1981) and other smaller ice bodies (Figure
4, Section C-D; Figure 5), forming terminal
moraines, till plains and ice-cored moraines.
It is also included in this unit the thick
debris cover which is still over stagnant,
dead and thermokarst-facies ice bodies.
Small lateral moraines and basal till are
seen in front of IJR-45 Glacier (Locality
17), inside and at a lower altitude respect
to the Bahía Bonita Drift moraines. The
supraglacial debris cover in the ice-cored
sections in front of this glacier is up to
2.0-2.5 m thick over more than 90 m of
dead ice. This supraglacial debris is composed of pebbles and boulders of breccias
and volcanics up to 8 m x 6 m x 4 m in
size, in a sandy-clayey matrix, and it has
been concentrated by melting and, probably,
by shearing up from the base of the glacier.
On the debris surface, some melting, collapse depressions, some of them up to 200-
300 m in diameter, and many other smaller
ones are found. Into the larger hollows,
true thermokarst lakes have formed but
the smaller depressions hold shallow meltwater ponds where algae, fungi and lichen
remains accumulate. A sediment sample from
one of these depressions has been investigated, fungi spores, re-transported Cretaceous pollen and colloidal organic matter

have been found in it (Calvin J Heusser &
Linda E Heusser, New York University,
written communication). The colonies of
Rhizocarpon geographicum reach here a maximum diameter of 13 mm in the outer moraines, though in the ice-cored moraines are
just dots or are even totally absent. In
other minor ice bodies, the IJR-45 Glacier
Drift is forming ice-contact morainic
belts, smaller than the Bahía Bonita Drift
moraines and at an inner position with
respect to them.
A marine shell sample coming from the glacimarine deposits which are associated
to this unit (Locality 18, 2.50 m a.s.l.)
has been dated in 1590 ± 60 years ^{14}C B.P.
(Table 2).

7. Presently-forming drift
This informal name is suggested to those
glacigenic deposits which are forming nowadays. They are ablation till, shear till
and supraglacial debris, assuming that
other glacigenic deposits may be also forming underneath present glaciers.
A part of these sediments is also accumulating due to sliding, collapsing, supraglacial stream action or along the margin
of the glaciers. Shear-till formation was
observed at the edge of IJR-45 and IJR-47
glaciers (Localities 17 and 19, Figure 1;
Figure 6). In these ice bodies, it partly
forms evolving morainic belts which are
basically generated by up-shearing of subglacial debris and also, by supraglacially-
derived clasts which slide along the glacier surface. This drift is distinguished
from the IJR-45 Glacier Drift because the
former is found only in active glacial
landforms whereas the latter is a part of
inactive landforms or reactivated glacial
features due to ice-push.

Figure 6. Detail of shear-till at IJR-47
Glacier, Locality 19 of Figure 1. Ice is
still present in the sedimentary mass.
The palimpsest structure in the matrix is
attributed to the up-shearing process

RADIOCARBON DATINGS

The analyzed samples, characteristics,
stratigraphic provenance, acting laboratory
and results obtained are presented in Table
2. The radiocarbon ages shown in this table
have been conventionally referred in years
before present (B.P.). They have not been
corrected neither for δ^{13}C nor for the
reservoir effect, as it has been suggested
instead by Broecker (1963), Sugden & John
(1973) and Sugden & Clapperton (1980b),
among others.

TABLE 2
RADIOCARBON DATINGS

Sample N° and Labor. number	Material and Lithostrati-graphic unit	Age (years ^{14}C B.P.)	δ^{13}C (‰)
38 Hv-11004	Marine shells (*Yoldia eight-sii*) Glaciomarine deposits (rai-sed marine beach) associa-ted to the IJR-45 Glacier Drift	1590 ± 60	+2.6
41 A Hv-11003	Marine shells* Bahía Bonita Drift	5175 ± 205	+0.6
41 B I-12201	Marine shells* Bahía Bonita Drift	5270 ± 120	−0.33
27 I-12200	Marine shells* Bahía Bonita Drift	5580 ± 110	−0.43
107 LP-76	Marine shells* Bahía Bonita Drift	5915 ± 233	−
119 Hv-11002	Marine shells (*Laternula elliptica*) Caleta Santa Marta Drift	34115 ± 1110 975	−0.50

Note: Hv, ^{14}C und ^{3}H Laboratorium, Niedersächsisches Landesamt für Bodenfor-
schung, Hannover, West Germany; I,
Teledyne Isotopes, Inc, New Jersey,
USA; LP: LATYR, Museo de La Plata,
La Plata, Argentina.
* all shells belonging to *Yoldia
eightsii*

GEOLOGICAL HISTORY

It is not possible to determine yet the actual characteristics and areal extent of the glaciation in this island during the James Ross Volcanics episode. Although the Antarctic Continent was already glaciated in the Pliocene (Kvasov & Verbitsky 1981) and thus it is possible that the ice cap was extending over marginal sectors of the Antarctic Peninsula, no conclusive evidence has been found yet to delimit the ice position in this area and for this geological epoch.

The first glaciation recognized for this archipelago formed the Meseta Lachmann Drift, of which only isolated, granitic and metamorphic erratic boulders without matrix and located above 275 m a.s.l., have been preserved. Probably, these erratics may be tentatively correlated with lodgement till and submarine melt-out till found at Marambio Island (Rinaldi and others 1977; Malagnino and others 1981; Elliot 1981), composed of a thin till cover with large, heavily striated boulders, most of them of Peninsular provenance, included in clayey matrix with scattered marine shells fragments. Malagnino and others (1981, p. 887) described these sediments and provided a list of 8 foraminifera species which would confirm its submarine genesis, though some of these forms could be reworked out of the Cretaceous or Marie Tertiary of the region. These authors suggested that the Marambio Island Plateau is due to marine abrasion. This opinion does not agree with our own observations: wave action seems to have been effective only along its margins. The plateau would be instead a ground moraine, affected by periglacial processes. These glacial deposits were generated by a large, several hundreds of metres thick ice-field, which would have extended from the Peninsula towards the edge of the submarine platform (Sugden & Clapperton 1977). The sedimentation conditions at James Ross Island are not well understood yet: it can only be inferred that lodgement till was formed at the base of a, in parts at least, temperate glacier. At Marambio Island, instead, the deposition of basal till took place, partly, over glaciomarine, sub-glacial sediments, probably generated by the floating edge of an ice-shelf. It is also possible that the sequence at Marambio Is. would represent a single glacial event characterized by an advancing grounding-line over its own glaciomarine proximal deposits, either by the increase of ice thickness or by the sea-level lowering. Elliot (1981) depicted the Marambio Drift as "Upper Tertiary Drift" (see the map legend in his paper) but no explanation for this chronological assumption is provided.

The age of the glaciation which formed the Meseta Lachmann Drift is undoubtedly pre-Wisconsin, because of the morphological relationships with landforms and deposits of younger glaciations, for example, the Caleta Santa Marta Drift whose Wisconsin age may be inferred from the radiocarbon dating: 34,000 years ^{14}C B.P. (Table 2). The scarcity of volcanic boulders in the Marambio Drift suggests that ice streams coming from the Peninsula were moving perhaps peripherically to the James Ross Island which would have kept smaller local glaciers of less competence against the large Peninsula glaciers-

This maximum glaciation phase, whose age is still uncertain but probably Lower to Middle Pleistocene (John & Sugden 1971), would have been followed by an episode of more restricted glaciation (interglacial?) in which the glacioisostatic recovery (partially tectonic?) of Marambio Island, in particular, and of the other islands would have taken place. Raised beachs at 275 m a.s.l. in the Shetland Islands (John & Sugden 1971, Sugden & Clapperton 1977) would correspond to this partial deglaciation. Marambio Island would have then remained definitively isolated from the Peninsular ice-sheet; no younger glaciations, even of local character, have been detected and this is basically interpreted as a result of its peculiar climatological situation.

The following ice readvance would have taken place during the Middle to Upper Pleistocene (Illinois? Lower Wisconsin?), when a glaciation phase triggered the extension of large outlet glaciers which modelated deep valleys as the Prince Gustav Channel or the Admiralty Sound. These glaciers covered part of the present submarine platform but affected only the margins of the larger islands, where they would have been confluent with local glaciers. Thus, the sediments of the Refugio San Carlos Drift would have been formed, which at their type locality have a large content of Peninsular erratics but at Lachmann Col (Locality 4, 160 m a.s.l.), they exhibit also a significant proportion of locally-derived volcanic boulders. The modelling phase of the largest troughs was followed by another episode of partial deglaciation with beach development which are now raised up to 100 m a.s.l. or more. The age of this phase and the subsequent beach elaboration is still unknown but undoubtedly it pre-dates the Upper Wisconsin.

In this later period, the glaciation had different characteristics with development of local ice-caps over the islands. These ice-caps generated the sediments of Caleta Santa Marta Drift and expanded towards the marine environments, suggesting that Bahía Croft, as an example, was already existing 34,000 years ^{14}C ago. This interpretation would be coincident with observations of Sugden & Clapperton (1980 a, b) and Clapperton & Sugden (1981), concerning with the lack of major expansions of the Western Antarctica ice-sheet in this epoch. However, this interpretation is contrary to the evidence presented by Elverhøi (1981), who has suggested an expansion of the Filchner-Ronne Ice Shelf within the Weddell Sea for this period. This apparent contradiction is explained by Elverhøi (1981) with the following argument: "different sections of the edge of the ice-sheet could have moved out of phase".

The shells found in the deposits of Caleta Santa Marta correspond to *Laternula elliptica*. This is a typically Antarctic species which presently lives in sandy-clayey environments of the submarine shelf, sometimes at -50 m or more (Zulma A. de Castellanos, Museo de La Plata, written communication). The sedimentological characteristics of these deposits and the finding of this bivalve suggest that its accumulation took place under subglacial conditions or near the edge of a floating tongue.

Its lithological content indicates that part of the material is exotic but it could have been provided either by Peninsular glaciers or by re-working of pre-existing drifts.

Finally, the Holocene glaciations show development of a main phase of outlet glaciers which occupied valleys (Bahía Bonita Drift). These glaciers appear almost entirely disconnected of the Peninsular glaciers. The available evidence would also suggest that, during this epoch, no glacier transfluence would have taken place through the Bahía Bonita-Caleta Santa Marta Col (Locality 7) as it had happened before; both ice bodies in their respective valleys would have remained independent due to thickness loss of the ice-cap. The lithology of Bahía Bonita Drift indicates that its provenance is essentially local, but there are also exotic materials which could have been re-worked from the Refugio San Carlos Drift outcrops, as it is suggested by clast counts performed along the moraines of Bahía Bonita. The marine shells found in these sediments would indicate that they were generated partly in coastal or shallow-water environments, between 0 and -25 metres (Zulma A. de Castellanos,

Museo de La Plata, written communication), having been raised to 35-40 m a.s.l. The four datings obtained on *Yoldia eightsii* shells are suggestively coincident (Table 2), indicating that this local outlet glaciers phase took place between 5000-6000 years ^{14}C B.P.

The sea level in this epoch should have been relatively high, because it is coincident with the Hypsithermal Interval (Deevey & Flint 1957), when the melting of the large Northern Hemisphere continents ice sheets provoked a sharp glacioeustatic rise up to several metres above present sea level. This could have contributed perhaps to accelerate the partial deglaciation of the island due to collapsing of the floating ice tongues or those nearby the sea shore. This was not in fact the situation during the genesis of Caleta Santa Marta Drift, because 35,000 years B.P. the sea level was perhaps at -60 m (Bloom and others 1974). The glacial episodes during the Upper Holocene cannot be separated from those of the Lower Holocene and they indicate that the glaciers were progressively reducing their size till they reached their present dimensions.

The deposits of the IJR-45 Glacier Drift suggest that they would have formed either during phases of such recession or by partial readvances. One of these events would have taken place around 1600 years ^{14}C B.P., accepting that the raised marine beach deposits (+2.5 m) worked out on the slopes of the moraines of Bahía Bonita would correlate with the IJR-45 Glacier Drift, basically considering the geomorphic continuity between these beaches and the glaciofluvial terraces related to this drift.

The period of extensive negative mass-balances seems to have been almost continued since then till nowadays, because the drift accumulation takes place presently under such glaciological conditions, characterized by the abundance of debris-covered or semi-covered glaciers.

CONCLUSIONS

The stratigraphy of the glacigenic deposits of the Northernmost James Ross Island shows that several glacial episodes took place between the Early Pleistocene (?) and the Holocene.

These glaciations have in common a progressive reduction of the Peninsular ice-sheet extent and thickness. This reduction was followed, also progressively, by the growth of local ice-caps, probably not so because of a real increase of local ice thickness but by an easier flow and expansion due to the absence or restriction of glaciers coming from the Peninsula. The general trend of these glacial episodes seems to have been towards a gradual diminution of ice-mass volume and it would indicate that Wisconsin glaciation would have been more reduced than, at least, one more extensive, previous one.

These glacial episodes alternated with epochs of partial glacier recession which permitted the development of marine beaches, isostatically raised later. This elevation would have been partly a glacioisostatic recovery. It is not possible for the moment to separate clearly this recovery from true tectonic ascent, which could have existed being this a tectonically active region (Grikurov 1978). Nevertheless, the relationship between the elevation of raised marine beaches and their estimated ages (Table 2) suggests that the recovery rate -which is assumed as essentially glacioisostatic- would have been close to 2 mm/year since the Wisconsin or 4 mm/year approximately, if the Holocene beaches are taken into account.

Finally, our studies have shown the existence of striated lodged boulders, glacial pavements, drumlins and drumlinoid forms, and basal, melt-out glacigenic deposits. These characteristics are indicators that, at least partly, wet-based, temperate glaciers were responsible for their genesis in the past, both those glaciers coming from the Peninsula as those of local origin. These pressure-melting point seem to continue in present glaciers.

The relative chronology herein proposed for the described lithostratigraphic units is based on the morphostratigraphic field evidence. It has also been confirmed both by the radiometric dating and by the spatial distribution on the drift bodies of crustose-lichen colonies' maximum diameter, such as *Rhizocarpon geographicum* and other species.

Thus, lichenometry appears as a promising technique for relative and even semi-quantitative dating of glacially-generated deposits and erosional surfaces in this area. However, there is apparently a major applicability restriction for this technique when dealing with areas above 200 m a.s.l., because the ecological conditions could be unfavourable for the growth of these colonies.

ACKNOWLEDGEMENTS

The author wish to express his sincere gratitude to the following persons and institutions:

To the Instituto Antártico Argentino and the Director of the Glaciology Section, Ing Pedro Skvarca, for the organization of the Summer Field Seasons of 1980, 1981 and 1982.
To the Fuerza Aérea Argentina for the logistic support of these campaigns.
To Dr Trevor J H Chinn (New Zealand, Ministry of Public Works and Development), Lic. Alejandro Dillon, Lic. Daniel Cobos, Agr. Fernando Piccolella, Mr. Sigfrido Rubulis and Mr. Luis Bertani, who took part of the field trips.
To IANIGLA (CONICET, Mendoza) for the authorization for the participation of Lic. Cobos and Agr.Piccolella.
To Dr Aníbal Figini (LATYR, Museo de La Plata) and Dr M A Geyh (^{14}C und ^{3}H Laboratorium, Niedersächsisches Landesamt für Bodenforschung, Hannover, West Germany) for the radiometric dating of shell samples and the discussion of the results. Other samples were analyzed at Teledyne Isotopes Inc, USA, thanks to a grant from IAA.
To Prof.Dr. Zulma A. de Castellanos (Museo de La Plata, Zoology Department) for the study of marine bivalves and the comments on their ecological significance.
To Professors Calvin J Heusser and Linda E Heusser (New York University, USA), for the palynological and micropaleontological analysis of samples.
To Dr P Redón (Valparaíso, Chile) for the specific determination of lichens.
To Prof.Dr. Francisco Fidalgo (Museo de La Plata) for the critical review of the manuscript.
The ideas and opinions exposed in this paper do not necessarily coincide with those of the investigators mentioned above.

REFERENCES

Andersson, J.G. (1906). On the geology of the Graham Land. Bull.Geol.Inst.Univ. Upsala, 7: 19-71, Upsala.
Bibby, J.S. (1966). The stratigraphy of part of north-east Graham Land and the James Ross Island Group. British Antarctic Surv., Sci.Report 53, Cambridge.
Bloom, A.; Broecker, W.S.; Chapell, J.M.A.; Matthews, R.K. and Mesolella, K.J. (1974) Quaternary sea level fluctuations on a tectonic coast: new ^{230}Th/^{234}U dates from the Huon Peninsula, New Guinea. Quaternary Research, 4: 185-205, Seattle.
Broecker, W.S. (1963). Radiocarbon ages of Antarctic materials. Polar Record, 11 (73): 472.
Clapperton, C.M. & Sugden, D.E. (1981). Late Quaternary glacial history of George

VI Sound Area, West Antarctica. TISAG
(Third International Symposium on Antarctic Glaciology), Abstracts, Columbus.

Clapperton, C.M.; Sugden, D.E.; Birnie, R.
V.; Hanson, J.D. & Thom, G. (1978). Glacier fluctuations in South Georgia and
comparison with other island groups in
the Scotia Sea. in: Antarctic Glacial
History and World Paleoenvironments, E.M.
van Zinderen Bakker, editor, p. 95-104,
A.A.Balkema, Rotterdam.

Deevey, E.S. & Flint, R.F. (1957). Post-
glacial Hypsithermal Interval. Science,
125: 182-184.

Elliot, D.H. (1981). Glacial geology of
Seymour Island. Antarctic Journal of the
U.S., 1981 Review, p. 66-67.

Elliot, D.H.; Rinaldi, C.; Zinsmeister, W.
J.; Trautman, T.A.; Bryant, W.A. & Del
Valle, R. (1975). Geological investigations on Seymour Island, Antarctic Peninsula. Antarctic Journal of the U.S., 10
(4): 183-186.

Elverhøi, A. (1981). Evidence for a late
Wisconsin glaciation of the Weddell Sea.
Nature, 293: 641-642.

Grikurov, G.E. (1978). Geology of the Antarctic Peninsula. Amerind Publishing Co.,
New Dehli, 140 pp.

John, B.S. & Sugden, D.E. (1971). Raised
marine features and phases of glaciation
in the South Shetland Islands. British
Antarctic Survey Bull., 24: 45-111.

Kvasov, D.D. & Verbitsky, M.Ya. (1981). Causes of Antarctic glaciation in the Cenozoic. Quaternary Research, 15: 1-17.

Malagnino, E.C.; Olivero, E.B.; Rinaldi,
C.A. & Spikermann, J.P. (1981). Aspectos
geomorfológicos de la isla Vicecomodoro
Marambio, Antártida. VIII Congr.Geol.Arg.,
San Luis, Actas, t.2, p. 883-896.

Rabassa, J.; Skvarca, P.; Bertani, L. &
Mazzoni, E. (1981). Glacier inventory of
James Ross and Vega Islands, Antarctic
Peninsula. TISAG (Third International
Symposium on Antarctic Glaciology), Abs-
tracts, Columbus, and Annals of Glaciology,
vol.3, in press, Cambridge.

Rabassa, J.; Skvarca, P.; Bertani, L. &
Mazzoni, E. (1982). El inventario de glaciares de las islas James Ross y Vega y
algunos aspectos geomorfológicos asocia-
dos. Contribuciones Científicas del IAA,
vol.6, in press, Buenos Aires.

Rinaldi, C.A.; Massabie, J.; Morelli, J.;
Rosenman, H.L. & Del Valle, R. (1978).
Geología de la Isla Vicecomodoro Marambio.
Contribuciones Científicas del IAA, 217,
37 pp., Buenos Aires.

Sugden, D.E. & John, B.S. (1973). The ages
of glacier fluctuations in the South
Shetland Islands, Antarctica. Paleoecology
of Africa, the surrounding islands and

Antarctica, 8: 139-159, A.Balkema,
Cape Town.

Sugden, D.E. & Clapperton, C.M. (1977).
The maximum ice extent on island groups
in the Scotia Sea, Antarctica. Quaternary
Research, 7: 268-282, Seattle.

Sugden, D.E. & Clapperton, C.M. (1980 a).
Glacial history of the Southern Antarctic Peninsula. AMQUA Sixth Biennial Meeting, Abstr. with Program, 1980, p. 188,
Orono, Maine.

Sugden, D.E. & Clapperton, C.M. (1980 b).
Nature, 286: 378-381.

Zinsmeister, W.J. (1979). Coastal erosion
on Seymour Island, Antarctic Peninsula.
Antarctic Journal of the U.S., 14(5):
16-17.

Zinsmeister, W.J. (1980). Marine terraces
of Seymour Island, Antarctic Peninsula.
Antarctic Journal of the U.S., 15(5):
25-26.

A model of valley bottom sedimentation during climatic changes in a humid alpine environment

D. VAN HUSEN
Technical University Vienna, Austria

1. ABSTRACT

Thick sequences of fluvial and mass wastage sediments were deposited in the main valleys of the eastern Alps during the early Würm (approximately 70,000 to 25,000 yrs. B.P.). The deposition of the various types of sediments appears to have been controlled by climatic fluctuations which are reflected by the characteristics of the sediments and the incorporated pollen. A model early Würm valley sedimentation for the humid eastern Alps is proposed.

2. INTRODUCTION

Many sediment sequences from the first extensive phase of the Würm have been mapped and described by the author. The sediment is glacially deformed and is commonly covered by lodgment till, and therefore represents the remains of valley fill deposited before the main event of the last glaciation.

Many of the sequences investigated appear to show similar variations in their horizontal and vertical stratigraphy as well as in their pollen record. Using this data, the author has attempted to reconstruct the depositional and climatic history of the area.

3. GENERAL DESCRIPTION

3.1 Sediments

The sediment sequences consist predominantly of coarse sediments laterally interfingered with thick units of sandy and clayey silts. Provenance indicates that much of the coarse material was deposited as alluvial cones built into the main valleys by tributary drainages. The interfingering units for the most part show a significant interbedding of clayey and sandy layers whose formation depended on the current velocity between the cones. The thick clayey layers are slack water deposits associated with flood ponding in the main valley while the coarser sandy units are assigned to periods of more swiftly flowing channelized water (van Husen, 1981).

The thickest complete sequence of fine-grained sediments in the eastern Alps is the banded clay of Baumkirchen near Innsbruck, Tyrol (Fliri, 1973). Here more than 100 meters of sediment shows a varve-like stratification which indicates deposition in standing water. Water depth was rather shallow, as indicated by trace fossils on the bedding planes of the banded clay. The shallow water depth indicates that the dam was aggrading at approximately the same rate as the valley bottom (Fliri, 1973). Sedimentation rates probably averaged 5 cm per year (Bortenschlager, 1978).

There are two possibilities for further sedimentation on top of the fine-grained units. In the first case, the sediment becomes progressively coarser through the addition of sand and gravel (Fig. 1). The sedimentary features also show a change from the varve-like bedding associated with the stagnant or slowly running water to cross bedding associated with swifter river currents. The new phase of coarser accumulation is the result of sedimentation adjusting itself to a debris overloaded river, causing accumulation across the entire valley floor. Lodgment till is generally found on top of the coarse sandy layers often with only a thin transition zone.

Figure 1.

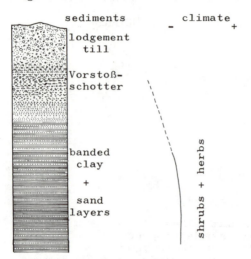

Figure 1. Sediment cycle showing "coarsening upward" development during cooling in early Würm.

In the second case, progressively greater amounts of organic sediments are found interbedded with the clay layers (Fig. 2). The organics probably formed in poorly drained areas on the ancient flood plain of the river. Overlying the interbedded clays and organics, a thick layer (up to 3 m) of slate coal, mixed with wood pieces, can often be found. The lateral extent of these coal layers generally does not exceed 100-200 m. All of these sediments are covered by lodgment till.

3.2 Climatic Record

Climatic fluctuations in the sequences, as indicated by pollen analysis, is paralleled by changes in sedimentation. The pollen record shows a cool period with vegetation in existence on the valley bottom during the deposition of the thick, fine-grained sediment sequence (van Husen, 1981; Draxler and van Husen, 1978). A high percentage of wood vegetation is indicated on the bottom or on lower slopes while grasses covered the higher slopes. The average temperature recorded by shrub-tundra pollen at the Baumkirchen banded clay site is approximately 5°C below present day temperatures (Fliri, 1973). High sedimentation rates probably account for

the surprisingly low pollen concentration when compared with vegetation composition in these sediments.

The beginning of high organic content in the sediment is preceeded by a pronounced climatic amelioration (Fig. 2). The pollen concentration also shows a increase, indicating a distinct decrease in the rate of sedimentation.

The climatic ameliorating gave rise to a forest which is recorded in the slate coal horizon. The forest patches developed in the flooded land surrounding old river branches in the inundation area. The slate coal was formed from wood peat high in carex. Renewed cooling ended organic sedimentation and caused a new phase of detrital sedimentation to begin.

4. PROCESS OF DEPOSITION

The results briefly described above permit a reconstruction of sedimentary processes and conditions in the main valleys of the eastern Alps during the climatic change of the early Würm.

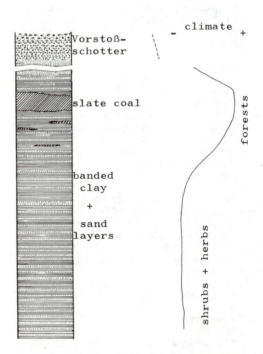

Figure 2. Sediment cycle showing high organic sedimentation associated with warmer phases in the early Würm.

Phase A

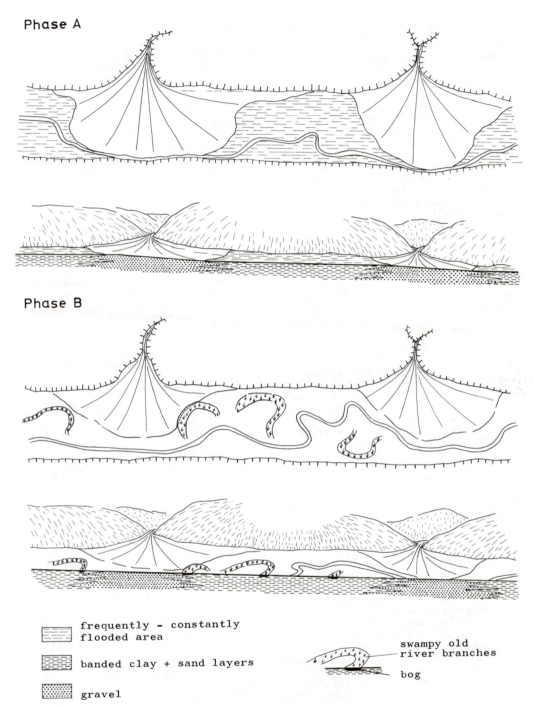

Phase B

frequently - constantly flooded area

banded clay + sand layers

gravel

swampy old river branches

bog

Figure 3. Model showing different depositional environments in the early Würm. Phase A - rapidly prograding alluvial cones block main valley and cause deposition of fine-grained flood material in intercone areas. Phase B - reduced sediment input from tributary valleys allows through fluvial drainage on extensive alluvial floodplains.

During cooler periods, high mountain slopes lost their vegetation or retained only a thin cover of vegetation. Tributary streams resultingly became overloaded with congelifraction debris. The growth of alluvial cones was considerable as the consequence of torrential flow and mud flows. The cones were able eventually to dam the river of the main valley thus causing the gradient to become unbalanced (Fig. 3; Phase A). Flooding deposited the fine-grained sequences between cone basins. The rate of sedimentation depended on the growth of the cones and was accelerated by the high input of debris. A pronounced climatic amelioration allowed vegetation to develop on the valley sides. This reduced the input of debris to the cones so that they aggraded less rapidly and consequently were not so important to the drainage conditions of the main valleys. As a result, the rivers had a constant gradient along their entire course. Erosion and sediment transport kept pace with cone aggradation.

The lower sediment input produced a very slow aggradation of the valley bottom (Fig. 3; Phase B). During this time, meander cutoffs were formed and filled with plants to form slate coal. The model explains the limited extent of these coal deposits. In this phase, sedimentation conditions must have approximated present day conditions.

Because of the lack of complete sections, the number of climatic, and hence sedimentologic cycles, which occurred in the early Würm of this area is not known.

After the last cool period (documented by the banded clay of Baumkirchen) the climate continued to cool. The result was a very high rate of sedimentation due to overloading of even the main rivers with debris. The sand-rich coarse gravels, named Vorstoßsschotter, are the product of increasing congelifraction and outwash production downstream from the advancing glaciers. The glaciers then covered these gravels with a layer of lodgment till. These glaciers formed a connecting net of ice streams at the end of the Würm in the eastern Alps.

5. REFERENCES

Bortenschlager, I. and S., 1978, Pollenanalytische Untersuchungen am Bänderton von Baumkirchen (Inntal, Tirol), Z. Gletscherk. Glazialgeol. 14: 95-103.

Draxler, I. and van Husen, D., 1978, Zur Einstufung innerwürmzeitlicher Sedimente von Ramsau/Schladming und Hohentauern (Steiermark), Z. Gletscherk. Glazialgeol. 14: 105-114.

Fliri, F., 1973, Beiträge zur Geschichte der alpinen Würmvereisung: Forschungen am Bänderton von Baumkirchen (Inntal, Nordtirol), Z. Geomorph. N. F., Suppl. 16: 1-14.

van Husen, D., 1981, Geologisch-sedimentologische Aspekte im Quartär von Österreich, Mitt. Österr. geol. Ges., 74/75: 197-230.

General sediment development in relation to the climatic changes during Würm in the eastern Alps

D.VAN HUSEN
Technical University Vienna, Austria

1. INTRODUCTION

Würmian sediment complexes in the eastern Alps consist of thick fluviatile sequences, which fill valley bottoms, lacustrine sediment sequences found in large ancient lakes, moraines and associated outwash terraces, and extensive mass movement and congelifraction units. Recent investigations and detailed mapping have led to a reconstruction of the pattern of sedimentation in time and space, and have suggested interrelationships between climate, hydrologic conditions and sedimentation within the area.

2.1 EARLY PHASE

Radiocarbon dating (van Husen, 1981) has demonstrated that many of the sediment sequences in the Eastern Alps are older than the Late Würm glacial advance. Van Husen (in press) has also demonstrated a clear relationship between sediment type and climate. According to the model proposed, the fine-grained sediment sequences on valley floors accumulated between alluvial cones during periods of cool climatic conditions when shrubs and herbs were the major vegetation of the valleys and uplands. Due to sparse vegetative cover, the alluvial cones grew rapidly due to torrential fluvial activity and mudflows emanating from the unprotected uplands, which caused aggradation of the entire valley bottom area.

With a warming of the climate, slopes became better protected by vegetation, and debris input to tributaries decreased. As a result, the main river could form a more constant gradient as the damming effect of the cones ceased. During this time many meander cut offs developed which flooded and filled in with organic debris. Valley floor conditions at this time were approximately the same as those of today.

Recent investigation within the Eastern Alps show a cool period at the beginning of the Würm (Fliri, 1978, van Husen, 1982). This cool period, which cannot currently be subdivided, was followed by a warmer period during the middle Würm, represented by several slate coal deposits and other organic sediments. The warm period lasted until approximately 30,000 BP after which accelerated cooling again took place. This led to the development of a connecting net of valley glaciers in the Eastern Alps.

2.2 PHASE OF FINAL CLIMATIC DECAY

The facies development of this phase is best documented in the profile of Baumkirchen (Furi, 1973). The lower part of this sequence consists of banded clays deposited in a shallow lake of constant depth as evidenced by trace fossils on bedding planes. A constant rise of the dam of the lake is indicated by the uniformity of the sediment sequence. The rate of sedimentation in a well investigated horizon was approximately 5 cm per year (Bortenschlaser, 1978). The aggradation of the valley bottom correlates well with the reconstructed climate, which indicates shrub-tundra in the vicinity of the lake. This resulted in unprotected slopes leading to greater sediment input to the tributaries and the resulting rapid aggradation of the alluvial cones (van Husen, in press).

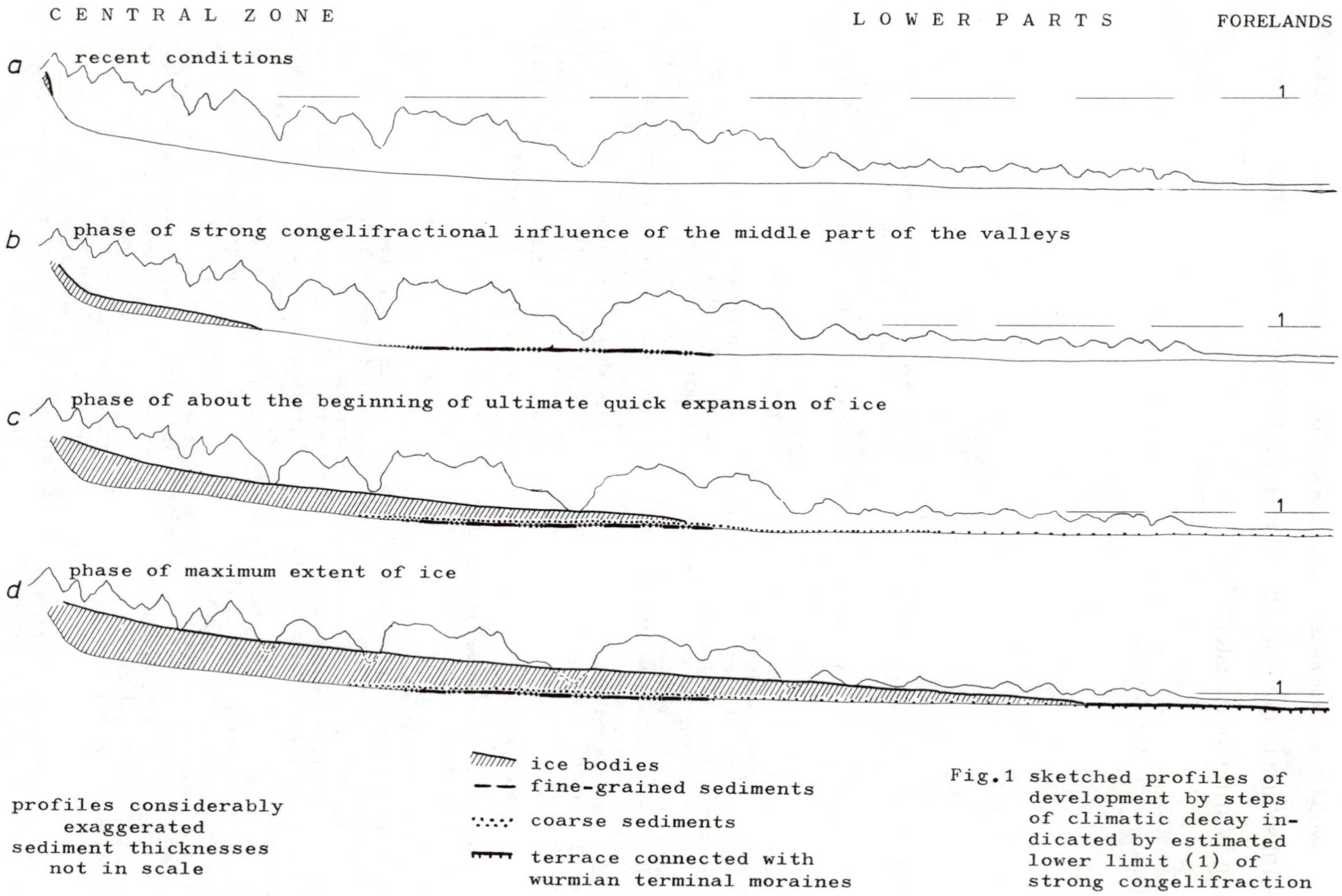

C E N T R A L   Z O N E   L O W E R   P A R T S   FORELANDS

a recent conditions

b phase of strong congelifractional influence of the middle part of the valleys

c phase of about the beginning of ultimate quick expansion of ice

d phase of maximum extent of ice

profiles considerably
exaggerated
sediment thicknesses
not in scale

Fig.1 sketched profiles of
development by steps
of climatic decay in-
dicated by estimated
lower limit (1) of
strong congelifraction

ice bodies
fine-grained sediments
coarse sediments
terrace connected with
wurmian terminal moraines

346

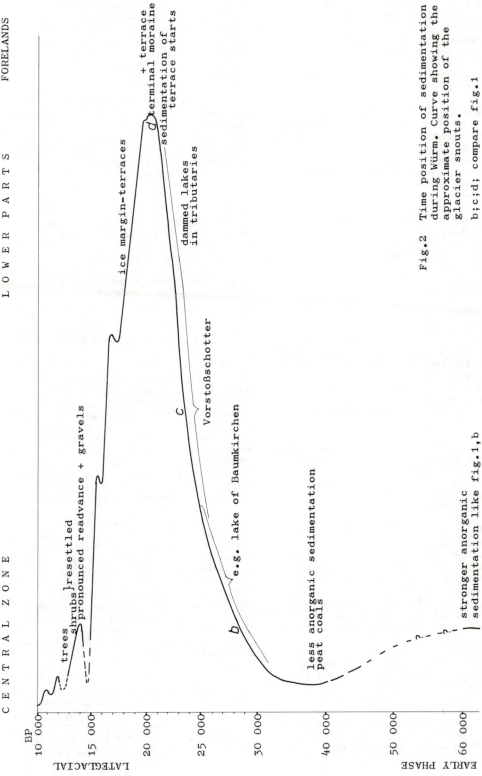

Fig.2 Time position of sedimentation during Würm. Curve showing the approximate position of the glacier snouts.
b;c;d; compare fig.1

347

Due to a strong acceleration of climatic cooling as time progressed, the sedimentation processes also accelerated. At the same time the rivers of main valleys became overloaded with large amounts of coarse sediments, so that sand, and later gravel, began to accumulate across the entire valley bottom.

The top units of these coarse gravels are composed of increasing amounts of congelifraction debris, as shown by increasing amounts of locally derived subrounded and subangular gravel, indicative of short transport distances. These units are in turn overlain by a thin transition zone and lodgment till. The gravels, named "Vorstoßschotter" are genetically related to the final climatic decay culminating with the Würm maximum, and form terraces up to 100 m thick along the main valleys of the Eastern Alps. They are found only in the main valleys, because ice had already filled the major tributary valleys.

In the central portion of the Eastern Alps the ice streams started to interconnect to form a net of ice streams. The ice surface rose rapidly because the ice streams dammed one another, especially in the valley narrows. The accumulation area thus enlarged disproportionately quickly with accelerated glacier growth; this was especially marked towards the end of the advance (Fig. 1:d).

The advance and growth of ice masses in the valleys of unglaciated lower parts of the Eastern Alps was so fast that, in spite of the immense sediment influx, small lakes developed and existed for a fairly long time in the dammed tributaries. These lakes then filled in with clayey silts in non-stratified layers several centimeters thick. The fine-grained layers are often interbedded with sandy layers which extend over the bottoms of entire lakes and indicate a strong current in these lakes. Complete filling of the lakes by prograding delta forsets did not occur due to the constantly rising water level of the lake.

During this phase of ice growth, the non-glaciated lower portions of valleys were aggraded, due to the increased load of glacial outwash and congelifraction debris (Figure 1:c). This aggradation temporarily dammed the non-glaciated tributary streams downstream from the ice margin. However, as the zone of active congelifraction expanded into these areas

of the Eastern Alps, increased debris input from unvegetated slopes caused the tributary streams to aggrade as well. Consequently, debris from tributaries became increasingly important as a sediment source for terraces in the main valleys.

The expansion of the glacial and periglacial zone due to climatic decay not only led to peripheral migration of considerable fine and coarse sedimentation, and formation of terraces but also to the displacement of periglacial phenomena such as rock glaciers and debris flow outward to the lower marginal zones of the Eastern Alps. During the maximum extent of glaciation a connecting net of ice streams existed which also covered the higher passes between the drainage areas. The Würm glacial maximum is marked by pronounced terminal moraines, with terraces graded to them. The valleys behind the terminal moraines are often covered with a continuous till layer, especially in the wider parts.

2.3 LATEGLACIAL PERIOD

The most sudden and conspicuous change in sedimentation took place at the beginning of the Lateglacial Period. Following a minor retreat from the terminal moraine, a rapid deglaciation started. During retreat, large masses of englacial and unconsolidated morainic material as well as larger areas of thawed permafrost soil became available for transport. Slope instability after ice retreat also added to rock and loose debris available through mass movements.

Since plenty of meltwater was available, the available sediment was transported and distributed across the valley floor. In this environment, the many lakes existing on the margins of inactive ice bodies acted as sediment traps. Of the presumed large sediment-infillings of these lakes, only small remnants remain. The lacustrine in-fillings are often seen as staircase-terraces, indicating a step by step filling of the lakes on the ice margin.

Sedimentation into these short-lived basins occurred in the form of deltas, with associated bottomset beds and extensive dropstones. A high rate of sedimentation is documented by turbidity current deposits and subaquatic slumps of

348

the foreset beds which disturb the bottom
set beds by compression or folding of less
consolidated fine-grained sediments. Only
a relative time scale of the development
of the staircase-terraces is possible,
with even relative dating impossible on
the widely distributed ones because their
growth depended on the short term
hydrological fluctuations in the stagnant,
melting ice bodies rather than on a
retreating glacier tongue. During the
first Lateglacial phase the glaciers
melted back to approximately one third of
their original length (Fig. 2). This was
followed by two short readvances of the
ice streams, which oscillated at their
margins without a distinct variation of
the type of sediment deposited in front of
the ice. During oscillation of the ice
front, previously deposited sediments were
occasionally overridden as evidenced by
morainic material overlying disturbed
sediments.

Another retreatal phase occurred,
followed by a pronounced readvance which
deposited distinct terminal moraines and
terraces graded to them in almost all of
the valleys. This indicates a significant
climatic cooling. During this time,
meltwater debris transport in the valley
bottoms increased; and congelifraction
increased due to destruction of hillslope
vegetation cover.

The effect of this climatic
deterioration was confined to the higher
valleys in the central zone of the Eastern
Alps. Following this short cooling event,
a pronounced amelioration occurred which
initiated the final migration of grasses
as well of trees, causing a stabilization
of debris valley bottom and slope, so that
the region of major sediment source and
congelifraction shifted to the highest
peaks. During this time, mass movements
were the only active contributors of
sediment in the lower portions of valleys.
These conditions approximate those of
today in the area. This final climatic
amelioration was interrupted twice by
short glacial advances during which
congelifraction was again important in the
upper reaches of the main valleys.

3. REFERENCES

Bortenschlager, I. & S. 1978,
    Pollen-analytische Untersuchungen am
    Banderton von Baumkirchen (Inntal,
    Tirol), Z. Gletscherk. Glacialgeol.
    14:95-103, Innsbruck.

Fliri, F. 1973, Beitrage zur Geschichte
    der alpinen Wurmverisung: Forschungen am
    Banderton von Baumkirchen (Inntal,
    Nordtirol), Z. Geomorph. N.F., Suppl.
    16: 1-14, Berlin.

Fliri, F. 1978, Die Stellung des
    Bandertonvorkommens von Schabs
    (Sudtirol) in der alpinen
    Wurm-Chronologie, Z. Gletscherk.
    Glazialgeol., 14: 115-118, Innsbruck.

Husen, D. van, 1981, Geologisch-
    sedimentologische Aspekte im
    Quartar von Osterreich, Mitt. Osterr.
    Geol. Ges. 74/75: 197-230, Wien.

Husen, D. van, 1982, Zur Ausbildung and
    Stellung der wurmzeitlichen
    Sedimente im unteren Gailtal, Z.
    Gletsherk. Glazialgeol. 16: 85-97,
    Innsbruck.

Husen, D. van, A Model of Valley Bottom
    Sedimentation during Climatic Changes in
    a Humid Alpine Environment, this volume.

The origin and movement of rock glaciers

G.SERET
Université de Louvain-la-Neuve, Belgium

ABSTRACT

Rock glaciers are mobilized in the same way as the real glaciers, that is by the resulting forces of the load of ice and debris on slopes. The movements within the rock glacier are demonstrated by a series of shear planes with an up-slope dip. Rock fragments are transported to the surface along these shear planes and on the surface of the rock glacier the shear planes become visible due to ridges and furrows more or less parallel to each other, comparable to the shear moraines of glaciers. The frontal arch of the rock glacier is of the same origin. - The ice in a rock glacier results from compaction and refreezing of avalanche-snow, from refreezing of the infiltrated interstitial water and from the formation of lenses of segregation ice. The igher content in fine matrix in the basal zones of the rock glacier allows a proportional increase in ice formation there. A certain minimum ice content is needed to produce the necessary plasticity for a mobilisation of the rock glacier. The matrix is concentrated in the bottom layers by percolation of meltwater and rain. In addition, frost-heaving brings the larger components to the surface. This process accentuates the sorting mechanism. - The origin of the matrix is essentially associated with the microgelifraction as is demonstrated by the high abundance of fractured grains of quartz, feldspar and mica < 2 μ as identified by X-ray analysis. - (Abstract arranged by the Editors.)

RESUME

Les glaciers rocheux sont mobilisés comme les vrais glaciers, grâce aux pressions exercées par la charge des glaces et débris accumulés à l'amont, sur le versant, au pied de falaises abruptes. Le mouvement se traduit par une série de plans de cisaillement à pendage amont, le long desquels sont remontés les constituants internes. En surface, ces plans déterminent la formation de rides et dépressions grossièrement parallèles, comparables aux moraines de cisaillement des glaciers. L'arc frontal a la même origine. Le contenu en glace résulte de la compaction et du regel des neiges d'avalanche, du regel d'eaux intersticielles infiltrées, et de la formation de lentilles de glace de ségrégation. Dans les couches basales des glaciers rocheux, la teneur plus élevée en matrice fine permet un accroissement progressif de la proportion de ces glaces internes. Lorsque celles-ci sont suffisamment abondantes, elles confèrent à l'ensemble la plasticité nécessaire à la mobilisation. La matrice est concentrée dans les couches inférieures par le lavage et l'infiltration des eaux de pluie et de fonte de neige. En outre, le "frost-heaving" remonte vers la surface les éléments les plus grossiers, ce qui accentue encore le tri granulométrique. L'origine de la matrice est essentiellement liée à la microgélivation, comme le montre l'abondance des débris de quartz, feldspath et micas de moins de 2 μ, identifiés aux rayons X.

INTRODUCTION

The study of some rock glaciers in the French Alps and Canadian Arctic shows that they are the result of a particular type of mass movement. Three main processes are involved:

i. Production within the rock glacier of a fine matrix and a granulometric sorting which concentrates this matrix in the inner part of the rock glaciers.

ii. Frost heaving which moves the coarser debris near the surface with a maximum projection plane for the majority of fragments dipping up glacier.

iii. Downslope creep due to a compressionnal overthrusting along shear planes within the rock glaciers, at the level of sufficient ice - enrichment. This movement is quite comparable the one of temperate glaciers. It results from load pressures of material accumulated upslope transmitted on the ice within the rock glacier: compacted snow of avalanches, interstitial regelation ice and segregation ice.

DESCRIPTION OF ROCK GLACIERS

Rock glaciers show a gentle downslope concavity and a tongue shaped form. Their heads lie at the bottom of cliffs from where, downstream, the value of the slope decreases progressively, until a maximum of 7° is attained. The total length can reach 1 km.

The surface is marked by a net of ridges and furrows particularly in the lower part of the rock glaciers. These ridges and furrows are generally parallel and downstream, convex. But this general disposition, is frequently interrupted in detail by meandering, irregular and coalescing pits and holes.

Statistical analysis of the distance between the top of the ridges and the bottom of the furrows gave a mean of 2.40 m, with a maximum of 6.8 m. The upstream side of the ridges is a steep reverse slope. The surface is etched by small cracks and moulins.

The cover of rock glaciers is a sorted debris mantle, with only coarse fragments at the surface, without matrix. These features - form, shape, detailed relief and pavement of coarse blocks - are the main elements of identification of rock glaciers (Capps 1910). More complete descriptions were made by Wahrhaftig and Cox (1959), Lliboutry (1955 and 1965), Barsch (1971), Østrem (1971), White (1971), Potter (1972), Serrat (1979).

The tongue shaped form, their position at the bottom of cliffs, the accumulation of coarse fragments at the surface are believed by several authors to result from rockfalls, solifluction coulees and avalanche deposits. Therefore, their particular morphological lobate form, with oriented, sinuous and coarsely parallel ridges and furrows, appears to be the surface expression of an overthrusting along shear planes within the rock glaciers. It has to be distinguished from other kinds of slope processes. The parallel ridges and furrows are believed to support comparison with the net of surficial crevasses of a temperate glacier, due to the compressional shearing.

ORIGIN OF THE SORTED COARSE DEBRIS OF ROCK GLACIERS

Statistical measures of the debris mantle of rock glaciers show a dominant vertical disposition of the coarser blocks. It appears mainly as though the broken down gives elongated parallelipedic fragments. In this case there is also perceptible a slight parallelism in the long axis of the blocks at the top of the ridges. This disposition appears quite clearly on the surface of the active rock glacier on the eastern side of the Lanserlia (Vanoise, Savoy Alps, France which is surprisingly interpreted as a solifluction coulee by Marnezy 1977). Bedrock is here a dolomitic limestone whose splitting supplies quite elongated parallelipedic blocks, heterometric enough for statistical analysis of their long axis. For the two rock glaciers of the Ribon river (Arcelle, Haute Maurienne, Savoy Alps, France) in the "Schistes lustrés" formation (metamorphic marlstone), this statistical orientation is less perceptible. The debris are too homometric for significant results. In the inactive rock glaciers of the Gladstone valley (138^o03 W - 61^o23 N) and the active one of the Talbot Creek (138^o06 W - 61^o28 N) (Yukon Territory, Canada) the granitic bedrock fragments were also too homometric for meaningful statistical analysis of the distribution of the blocks.

The fragments of the rock glacier mantle undergo an intensive frost splitting

until they reach granular size. This frag-
mentation supplies sand and silt, which
are quickly washed to the inner part of
the rock glacier. The macropermeability of
the coarse block cover permits the immediate infiltration of the meltwater and rain.
A few remnants of the fine material are
deposited with organic matter in some of
the small noles on the surface of rock
glaciers. Granulometric analysis of these
fines shows a composition of clay size
(less than 0.002 mm) between 3% and 19%
(French Alps). On the active Talbot Creek
rock glacier (Yukon), the microfrost
splitting of the granitic bedrock supplies
also clay-size particles. X-ray analysis
of this clay provides mainly micrograins
of quartz and feldspar, and some particles
of illite, vermiculite and chlorite. Clay
mineral content of the total $< 2\ \mu$ particles constitutes less than 20%. Nevertheless, the granite is here quite fresh,
as analysed by petrographic microscope.
Therefore the macropermeability of a rock
glacier mantle does not allow persistance
of fine material near the surface. - Except
for some lichens and a few sparse herbs,
the colonisation by vegetation is not
possible on active rock glaciers. The
mechanism itself prevents the development
of plants, as we shall see.

INNER COMPOSITION OF ROCK GLACIERS

To cut an exposure across rock glaciers is
hard to achieve. The presence of blocks
several tons in weight on the surface
would require large machines only taken
with considerable difficulty on to active
rock glaciers. Nevertheless, Wahrhaftig
and Cox (1959), and Potter (1972) refer
to a few such exceptional cuts.

Natural outcrops take place when the
rock glaciers have collapsed at points
where they are hanging with teir downward
tongue reaching the top of an abrupt slope.
The active rock glacier of Talbot Creek
(Yukon and the two in the Ribon valley,
Savoy) collapsed and clearly showed the
enrichment of fines below the surface,
particularly of clay size particles in the
lower levels (Fig. 1). - Material may be
partly frozen and cemented by interstitial
ice. There are also two other types of
internal ice: large blocks of white ice
with black debris, and lenses of pure
segregation ice. White ice of rock
glaciers is similar to white ice of
valley-glaciers, and results from incorporation and compaction of snow of avalanches. Black debris is also mainly integrated from rock falls and avalanches.
A mass of white ice seen in an outcrop
was 1 m thick and 6 m long.

Lenses of pure segregation ice are
horizontal and thin with a size range of
between a few mm and several cm. Crystals
of ice are parallel and vertical. These
lenses are concentrated some metres below
the surface of rock glaciers, where the
fine interstitial material becomes abundant. The lenses of ice coat large stones
which are progressively upheaded.

In short, a vertical section in a rock
glacier shows:
- a lack of matrix at the surface, with
  coarse fragments dominant in the mantle,
  and an enrichment of fines below, first
  in sand and silt and in clay lower down.
- three kinds of ice: interstitial ice
  resulting from the freezing of infiltration water, masses of white ice due
  mainly to compacted snow of avalanches,
  lenses of pure segregation ice responsible for frost-heaving of large stones.

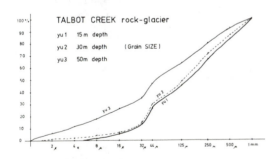

Fig. 1 Grain size analysis on sediments
of the Talbot Creek rock, glacier

ORIGIN AND MOBILITY OF ROCK GLACIERS

Snow is unstable on steep cliffs and frequently mobilized by avalanches. It spreads and accumulates at the bottom of the cliffs in white compacted masses containing fragments of debris mainly due to frost splitting and rockfalls. Frequently repeated, the processses accumulate several tens of metres of coarse material mixed with compacted snow. At high altitudes and/or latitudes, these accumulations are sheltered from a complete summer melting. From year to year, the mass - balance is positive. Summer hot days provide melt water shich at night freezes and causes microfrost splitting of the blocks in the mantle leading to a supply of fine sandy-silty material. The surficial meltwater and rain of summer storms percolate through the rubble of the cover, sorting the material and concentrating the finer grains as interstitial matrix below the surface. The freezing of this infiltrated water provides ice which cements the matrix. The coarse debris mantle shelters the inner ice of rock glaciers from the direct insolation and constitues a thermic insulation body. The volume and the temperature of infiltrated waters determine the depth and number of freeze/thaw cycles. In summer rainstorm precipitation carries the heat deeper, increases the microfrost splitting and thus the amount of fine matrix, and by refreezing increases the amount of ice content under the cover of coarse blocks. Then, in the inner part of rock glaciers, the depth of the freezing/thaw level fluctuates. An increase of fine matrix in this inner zone results in more capillary water and the formation of segregation ice lenses. Frost-heaving may occur. The latter increases the sorting of material in the rock glaciers. Large blocks heave up and, on the other hand, fine matrix is lowered. This enrichment in matrix and ice near the bottom results in a physical state of the lower part of rock glaciers more and more similar to those of true glaciers. The masses of accumulated material - rocks and compacted ice - on the higher slopes determines the slant component of the gravity along the bottom-gradient of the rock glacier. These pressures acting on the inner ice supply compressive stresses of plasticity and shear deformation. By such a mechanism a rock glacier may become mobile.

A rubble sheet characterizes the upper part of a rock-glacier, but downstream lies a progressive network of ridges and furrows, mainly parallel and convex. This reticulation is equivalent of the surficial ogives of a true glacier.

The steep convex ridge at the frontal extremity of a rock glacier is an equivalent of a frontal moraine.

FINAL REMARKS

As assumed by Lliboutry (1955-1965), White (1971) and Potter (1972) it is probable that some rock-glaciers developed initially on dead ice of a cirque glacier, or other buried ice bodies. Avalanches and rock falls would then accumulate and increase internal pressures sufficient to create mobility. These remnants of former glaciers promote the capacity to flow, but do not necessarily constitue the beginning of a rock glacier. Serrat (1979) reports "the greater abundance of fine matrix in the rock glaciers developed from schistes" (p.96) and that "the importance that the presence of fine matrix may have had in their origin (of rock glaciers) is very problematic" (p.96).

This author made an important observation. In fact, the schist promotes the micro-frostsplitting, which then provides an abundant fine matrix. Segregation ice lenses may develop and then sorting increases by frost-heaving. Schist with its cleavage promotes an ideal environment for the development of rock glaciers.

CONCLUSION

Rock glaciers result from a combination of several processes:

- burying and accumulation of avalanches and rock falls at the foot of a cliff, where the slope becomes more gentle. The accumulation must reach several tens of metres in thickness. It may occasionally occur on remnants of former glacier ice bodies.
- Surficial snow melting provides fine silt and sand by nocturnal frost-splitting, especially in summer. Meltwater and rain infiltrate the coarse block-mantle, washing silt and sand which sift downward and concentrate below the surface.
- The freezing of these infiltrated waters increases the volume of inner ice: occasional remnants of former glaciers, blocks of compacted avalanche snow, interstitial ice cement and segregation ice lenses. Interstitial and segregation ice mainly form in the matrix layers.
- Active frost-heaving of rubble occurs in the segregation ice levels. It emphasizes the importance of sorting, and the fine matrix enrichment in the inner part of rock glaciers.
- Movement is involved by compressive internal stresses mainly due to the overlying weight of the material accumulated on the slope above. The component forces are applied parallel with the slope of the rock glacier and induce plasticity and shearing motion. The main mobility occurs into the inner ice-rich levels. Here the type of movement is similar to that of true glaciers.

Ridges and furrows of rock glaciers are believed to be the surface expressions of internal overthrusting. As in true glaciers they appear to express the net of shearing planes.

ACKNOWLEDGMENT

Financial assistance for field work in Yukon Territory (Canada) was partly provided by the Belgian F.N.R.S. Logistical support in the field was kindly shared by our Colleague Prof. M.A. Geurts of Université d'Ottawa.

REFERENCES

Barsch, D. 1971, Rock glaciers and ice-cored moraines, Geogr. Ann. 53A:203-206.

Capps, S.R. 1910, Rock glaciers in Alaska, Journ.Geol. 18:359-375.

Lliboutry, L. 1955, Origine et évolution des glaciers rocheux, C.R. Acad.Sc. Paris, 240:1913-1915.

Lliboutry, L. 1965, Traité de Glaciologie, Masson (ed). I & II, 708.

Marnezy, A. 1977, Glaciers rocheux et phénomènes périglaciaires dans le vallon de la Rocheure (Massif de la Vanoise), Rev.Géogr.Alpine 65:147-167.

Østrem, G. 1971, Rock glaciers and ice-cored moraines, a reply to D. Barsch, Geogr. Ann. 53:207-213.

Potter, N.Jr. 1972, Ice-cored rock glacier - Galena Creek, northern Absaroka Mountains. Wyoming, Geol.Soc.Amer.Bull. 83: 3025-3058.

Serrat, D. 1979, Rock glacier morainic deposits in the eastern Pyrenees. In Moraines and varves, Ch. Schlüchter (ed). A.A. Balkema, 93-100.

Wahrhaftig, C. & Cox, A. 1959, Rock glaciers in the Alaska Range, Geol.Soc. Amer.Bull. 70:383-436.

White, S.E., 1971, Rock glacier studies in the Colorado Front Range. 1961 to 1968, Arctic and Alpine Research. 3:43-64.

The age of tills: Problems and methods

J.-M.PUNNING & A.RAUKAS
Academy of Sciences, Tallin, ESSR, USSR

ABSTRACT

From stratigraphical point of view
in Quaternary section with complicated structure tills are of special
interest as they enable to correlate
lithologically similar horizons in
vast territories. The study of tills
gives also the basis for the compilation of stratigraphic schemes, the
motivation of which is greatly dependent on the reliability of chronological evidence. The present paper
discusses the suitability of different direct and indirect methods,
such as palynologic, thermoluminescent and sulphurmetric ones, for the
dating of tills, and presents some
concrete examples on their studies
in the north-west of the European
part of the Soviet Union, in the
Caucasus, in Finland and in Spitsbergen archipelago.

1 INTRODUCTION

The age of tills is generally determined by bedding conditions, by
their position with respect to interglacial or interstadial deposits.
Unfortunately the latter are rather
uncommon. Besides, the majority of
unconsolidated intermorainic organic
deposits was strongly crushed by
advancing glaciers of succeeding
glaciations, whereas well preserved
lake and bog deposits are far from
being always of primary bedding.
Very often they are imbedded as erratics in younger sediments as a
result of which the apparent age of
tills considerably increases.

2 THE STUDY OF REDEPOSITED POLLEN

In recent years the study of redeposited spores, pollen and fauna
has found ever growing use in dating
of tills since they give the basis
for the elucidation of bedding conditions of interglacial deposits,
enable to determine the ways of the
advance of glaciers and the age of
tills. This method takes advantage
of the supposition that quantitative
ratio of spores and pollen in deposits varies with interglacials. Thus,
in the North Europe relatively warm
and moist Holsteinian (Likhvian)
interglacial was characterized by
considerable spread of conifers
whereas that of broad-leaved trees
was rather limited. The climatic
optimum was accompanied by wide distribution of alder and restricted
spread of hazel. There occurred also some "indicative" species, such
as Osmunda claytoniana, Azolla fi-
liculoides, Ilex aquifolium, Picea
sect. Omorica, Tilia tomentosa and
others, which are lacking in younger, Eemian (Mikulian) deposits.

Eemian interglacial was considerably warmer than Holsteinian and it
can be subdivided into thermoxerotic
and thermohygrotic stages. Its first
half was warm and dry, the second
half - warm and moist. Eemian interglacial was characterized by a distinct appearance and culmination of
broad-leaved trees: at first oak and
afterwards elm, linden and hornbeam;
the most important "indicative"
species were Osmunda cinnamomea,
Carpinus betulus and Brasenia pur-
purea. Unlike Holsteinian interglacial both alder and hazel were

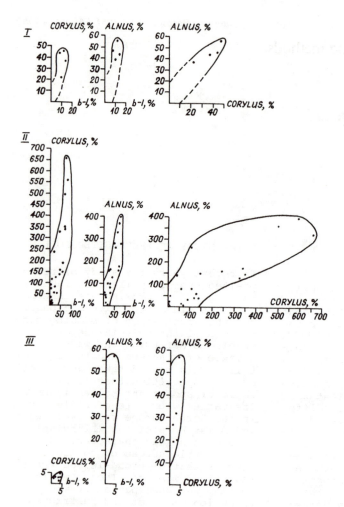

Figure 1. Variograms of the thermofilic plant pollen in Suur-Pranli section, North-Estonia (compiled by E.Liivrand): I - Upper (Weichselian, Valdaian) till with abundant pollen from Eemian (Mikulian) interglacial deposits; II - Eemian (Mikulian) intermorainic deposits; III - Lower (Middle Pleistocenian) till with abundant pollen from Holsteinian (Likhvian) interglacial deposits.

widely distributed in Eemian interglacial.

Proceeding from the assumption that there were only two interglacials with the above differences between them the study of spores and pollen stored in tills provides valuable evidence about their age. However, it should be mentioned that more reliable results are obtained through the compilation of variograms, proposed by V.Grichuk and introduced into practice by E.Liivrand (1969). Variograms (fig.1) represent a graphical description of quantitative ratios of the most typical interglacial components, e.g. alder, hazel and broadleaved species. For the above purposes coordinates are drawn for two components and the pollen content expressed in percents for these two components is carried on the diagram. The result is a body of dots which forms a specific configuration. The shape of the latter varies not only with both interglacials but also with tills assimilated the deposits of one or another inter-glacial.

However, in spite of seemingly high theoretical motivation the above method evokes no special trust. Up to the present the exact number of interglacials and their palyno-logic characteristics are not un-ambiguously elucidated, and in ad-dition, the content of pollen com-prised in tills is also influenced by local peculiarities and abundant interstadial deposits, palynologic characteristics of which are not yet sufficiently studied. At the same time the method of fluorescent microscopy which enables to separate pollen grains of different age on the basis of their physical quali-ties is still in the stage of its development.

3 LEACHING OF CARBONATES

Already for some decades use has been made of technique enabling to date surficial deposits through the determination of the depth of the leaching of carbonates. The method works well with tills of different glaciations the age of which can be easily established also by other methods (e.g. geomorphologically). However, it cannot be applied to dating of stadial and facial tills as the results obtained depend on primary carbonate content and matrix of tills, and also on local climatic and geomorphological conditions. This method is also unsuitable for dating subsurface till horizons.

4 THERMOLUMINESCENT DATING

At present the thermoluminescent (TL) method has claimed special attention due to wide distribution of its study objects - natural quartz and feldspars in glacial deposits. How-ever, it must be noted that physical grounds of the method are poorly elaborated yet and the microproces-ses occurring in crystals can be modelled at an empirical level only. Besides, there usually arises a number of specific complications. Here one has to face such problems as the quenching of light sum in the course of formation and redeposition of tills, and the migration of natu-ral exciters, i.e. natural radio-active isotopes during the time that has passed since the accumula-tion of tills.

Although there are abundant data in literature on modelling natural processes in order to study their influence on quenching of the pri-mary light sum, there is still no guarantee that the models compiled work in natural conditions. We elab-orated criteria for the estimation of the reliability of TL dates. Our selection of dates is based on the following grounds: (1) accurate preparation of the study object, TL measurement and mathematical proces-sing of the results; (2) check of TL dates by comparing them with chronological, paleontological and geological evidence; (3) dating of a geological section by series. The lack of age inversion evidences about the reliability of dates.

We attempted to estimate the suit-ability of sediments belonging to different genetical types for TL studies. From all types of Quater-nary deposits tills proved the most complicated objects for TL dating because of their highly variable genesis (Dreimanis, 1976, a.o.). As known, they may accumulate in sub-aerial and subaqual conditions, under moving and dead ice, before the front of the glacier, under the glacier, and also as a result of surface ablation. Besides, the ad-vancing glacier enriches continuous-ly with the material of underlying rocks, which have considerably dif-ferent thermoluminescent properties.

N.Sudakova and V.Iljitchev (1974) suppose that the quenching of the primary light sum is due to high pressure in glacial covers, however, this suggestion evokes some suspi-cion. Experimental evidence shows that even one-axial pressure, many times exceeding the pressure in gla-ciers, has only unremarkable influ-ence on the stored light sum. In mineral grains the "zero-point" is experimentally realized only after pounding and crushing in a mortar. The evidence obtained show that the ultraviolet radiation should be re-garded as the main factor responsi-ble for the quenching of the stored light sum (Vlasov et al., 1978). Its influence on basal till is rather insignificant. Therefore we are of opinion that on dating tills one should proceed not from tills but from intermorainic (fluvioglacial, marine, etc.) deposits, where the mechanism of quenching of the primary light sum has acted more completely. Although the results obtained through the dating of tills by improved technique (Hütt et al.,1977, a.o.) in the Institute of Geology of the Academy of Sciences of the Estonian SSR are reasonable from geological point of view as shows regular increase in ages with the depth (Kajak, Raukas, Hütt, 1981), the obtained data, as a whole, are in bad correlation with supposable geological ages of tills.

At the same time the dates on intermorainic deposits seem entirely reliable. Here a series of dates for Spitsbergen archipelago may serve as an illustration (Troitsky et al.,1979). In Spitsbergen not only marine and glacio-marine deposits but also tills served as study objects. TL dates were compared with the results obtained by other geological methods. TL and ^{14}C dates showed a good agreement. TL dates were also in good correlation with amino acid determinations (Boulton, 1979) performed by G.H. Miller (fig.2).

In the Arctic where the transport of sediments by water, and accumulation take place mainly in summer during the polar day, the intense ultraviolet radiation is the main factor responsible for the quenching of progenetic light sum stored by minerals.

The sections of North Finland are of particular interest since they serve as key sections for the investigation of the glacial dynamics in Middle Weichselian (Würm, Valdai) period in Europe. We studied Kauvonkangas section in North Finland (samples were collected and presented by H.Hirvas). According to K.Mäkinen (1979) peat formed in this section in the course of long-lasting Middle-Weichselian (Peräpohjola interstadial). K.Korpela (1969) was the first who distinguished the above interstadial on the basis of paleobotanic, lithologic and geochronologic evidence. Our results (fig.3) are in good agreement with ^{14}C dates and contribute to the reliability of the Upper Pleistocene stratigraphic scheme compiled by Finnish geologists (Hirvas, Kujansuu, 1979).

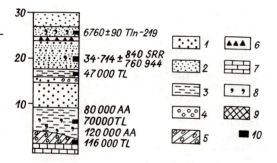

Figure 2. Section in Quaternary deposits in Billefjorden (Spitsbergen) and dates by [14]C (Tln,SRR), TL and amino acid (AA) methods. 1 - shingle, 2 - sand, 3 - silt and mud, 4 - boulders, 5 - till, 6 - slope deposits, 7 - bedrock, 8 - marine molluscs shells, 9 - peat, 10 - sites for sampling.

Lower Weichselian (Lower Valdaian) till was dated by us from Osinovskoje section in Arkhangel District. Evidence obtained through the complex study of this area show that in Late Pleistocene there were two stages in the activization of North European ice cover. TL dates for this region are in good agreement with geological imaginations about cyclic changes in summer radiation, according to which minimal summer temperatures occurred 116 000,72 000 and 22 000 years ago. Also $\delta^{18}O$ variations in skeletons of foraminifers from deep-water bottom deposits (Shackleton, Opdyke, 1976) prove the reliability of TL dates. The above shows that the usage of TL method is promising in the complex with other methods for the dating of Quaternary deposits. Geological prerequisitions and correct modelling of the conditions in which the study object deposited are of consequence for the interpretation of TL dates.

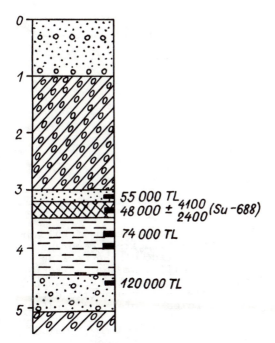

Figure 3. Section in Quaternary deposits Kauvonkangas (North Finland). For the legend see fig.2.

5 THE PECULIARITIES OF DATING TILLS IN MOUNTAINOUS AND ARCTIC REGIONS

In recent years our institute has performed extensive studies on the evolution of glaciation in arctic and mountainous areas in different regions (Spitsbergen, Polar Urals, Caucasus, etc.) with the application of specific methods for dating tills. The selection of optimal techniques takes usually advantage of the methods applied to the study of tills of lowland continental glaciations elaborated in the course of tens of years. However, they still have their specifics and cannot be mechanically transferred to the mountain glaciers. At the same time some methods which are widely used in mountains and in the regions of recent glaciation, e.g. lichenometry, are unsuitable for the study of tills of ancient continental glaciations.

6 SULPHURMETRIC METHOD

On dating tills in mountain glaciers wide use is made of lichenometric method proposed by R.Beschel (1950). This method is based on the consideration of maximal diameter of the thallium of lichens, and assumes that the growing rate of lichens is constant. However, further studies showed that it diminishes with increasing age of lichens (Beschel, 1961; Webber, Andrews, 1973; Andrews, Barnett, 1979, etc.). In addition, the rate of growing is dependent upon climatic conditions (Keith-Pitman, 1973). All the above complicates the interpretation of lichenological data.

Sulphurmetric method (Punning, Punning, 1978) is a good supplement to lichenometric one. It is based on the determination of sulphur content in lichens and assumes that: (1) lichens begin to develop on moraines immediately after their formation; (2) the concentration of sulphur compounds is constant in the atmosphere of polar and high-mountainous regions located far from the areas of human influence; (3) the rate of the accumulation of sulphur in lichens is constant. The investigations performed in different mountainous regions showed that many species of lichens, including our study object Rhizocarpon geographicum, become available for sampling 15-20 years after their appearance on substratum. Atmospheric background of sulphur-containing compounds was studied on the basis of ice cores from Spitsbergen and Severnaya Zemlya archipelagoes. It is established that during the past thousand years the concentration of sulphur in the atmospheric precipitations only slightly deviates from its average value. The assumption about the constancy of the sulphur accumulation by lichens was proved by the study of different series of lichens collected from end-moraine ridges of known age. In the Polar Urals and the High Caucasus the rate of the accumulation of sulphur by Rhizocarpon geographicum and Parmelia centrifuga proved to be constant throughout the whole study period (up to 740 years ago).

The concentration of sulphur in lichens may serve as a basis for the correlation of different forms from

the areas with the similar conditions of the accumulation of precipitations and similar microclimate.
For the establishment of the age of
different moraine ridges covered by
lichens one should have at least one
reference complex dated by other
methods.

For example, we used sulphurmetric method while dating moraine
ridges in front of the Halde glacier
in the High Caucasus. This glacier
is distributed in the upper reaches
of the Inguri River and flows down
to the altitude of 2400 m. In front
of the glacier there lies a series
(6-7) of moraine ridges, formed as
a result of short oscillations on
the background of continuous degradation of the glacier. We collected
samples of Rhizocarpon geographicum
for the determination of sulphur.
Samples were taken from the crest of
the ridge along the axis of the valley. A boulder marked in 1961 and
moraine ridges formed in 1911-1913
and 1980 served as bench marks. Determination of sulphur in these samples showed that the concentration
of sulphur grows linear with the age
of lichens. This reguliarity enabled
us to date two ancient moraines,
situated 2-2.5 km (fig.4) from the
present margin of the Halde glacier.
Our experiment shows that complex
application of lichenometric and
sulphurmetric methods can be applied
to the dating and correlation of
different stages of the development
of mountain glaciers.

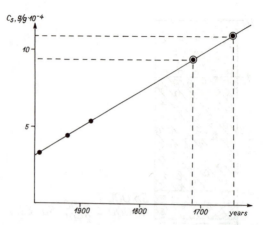

Figure 4. Concentration of sulphur
in lichens Rhizocarpon geographicum
from the surface of the moraine
ridges of different age in front of
the Halde glacier (the Caucasus)
Big circles - sulphurmetric dates,
small circles - reference erratic
(1961) and the ridges dated by
lichenometric methods by N.Golod-
kovskaya (oral announcement).

7 DATING OF ALLOCHTONOUS MATERIAL

As a rule, the material needed for
^{14}C and TL methods is of rather limited distribution in mountainous and
arctic regions. Still, radiocarbon
method is of consequence while
dating tills in mountain glaciers.
It has been applied to dating of
tills from some localities in the
Alps (Patzelt, 1974) and other regions. By this method we have dated
some stages in the evolution of
glaciation in Spitsbergen archipelago (Punning et al., 1976).

In connection with the application
of different dating methods the
problem of interpretation of the results obtained serves some attention. The ages by lichenometric and
sulphurmetric studies are naturally
a bit younger than the real ones,
however, dating of over- and underlying deposits contributes also to
the uncertainity of real ages. The
advance and retreat of the glacier
is always accompanied by abrasion
and washing out of a part of underlying deposits. Undoubtedly, driftwood and shells of molluscs imbedded in tills lied primarily on the
coast or on the bottom of an ice-
free water basin, and only after
some time they were seized by a
glacier and redeposited. In this
case radiocarbon dates indicate the
time of the previous retreat of the
glacier.

In the Arctic where in Holocene
the snow boundary lay considerably
low, and in some valleys the glaciers almost reached the sea, the
data on the age of marine terraces
may be used for indirect dating of
tills. Such investigations are
especially promising in Spitsbergen
archipelago. The dynamics of verti-

cal movements of the Earth's crust is well known (Schytt et al., 1967; Salvigsen, 1978; Punning, Troitsky, 1980). Knowing the rate of the uplift of the Earth's crust it is possible to elucidate the minimal age of tills subjected to marine abrasion or covered by deposits. This approach is justified only in single regions, and the elucidation of the rate of the uplift of Earth's crust takes labour-consuming studies.

8 CONCLUSIONS

The methods used for dating tills are rather limited in number and at present none of them can be regarded as an universal one. Unfortunately, in geological practice one quite often meets with cases when one or another method or dates are generalized without critical approach, and used as a basis for the compilation of complicated paleogeographic and stratigraphic schemes. On the other hand, sometimes some single incorrect dates have arisen suspicion against the method as a whole. New physical and isotope-geochemical methods and elaborated old classic methods (paleobotanic, etc.) considerably enlarged and deepened our knowledge not only of the physical age of deposits but also of the paleogeographic situation in the past. However, only complex application of methods and accurate study of the genesis of studied deposits eliminate potential incorrect conclusions which are unavoidable while using only one method.

9 REFERENCES

Andrews, J.T. & D.M.Barnett, 1979, Holocene (Neoglacial) moraine and proglacial lake chronology, Barnes Ice Cap, Canada, Boreas, 8:341-358.

Beschel, R.E. 1950, Flechten als Altermasstab rezenter Moränen. 2. Gletscherkunde und Glazialgeologie, 1:152-161.

Beschel, R.E. 1961, Dating rock surfaces by lichen growth and its application to glaciology and physiology (lichenometry), Geology of the Arctic, 2:1044-1062.

Boulton, G.S. 1979, Glacial history of the Spitsbergen archipelago and the problem of a Barents Shelf ice sheet, Boreas, 1979, 1:31-57.

Dreimanis, A. 1976, Tills: their origin and properties. In: Glacial Till. An inter-disciplinary study. The Royal Society of Canada, Special Publ., 12:11-49.

Hirvas, H. & R.Kujansuu, 1979, On glacial interstadial and interglacial deposits in Northern Finland, In: Proj. 73-1-24, Quatern. Glaciat. in the Northern Hemisphere, Rept., Prague, 5:146-164.

Hütt, G., J.-M.Punning & A.Smirnov, 1977, TL dating technique in geology, Ann. of Acad. Sci. ESSR, v. 26, Chemistry. Geology, 4:284-288 (in Russian, summary in English).

Kajak, K., A.Raukas & G.Hütt, 1981, Experience obtained through the study of tills of different age in Estonia by thermoluminescent method, In: Pleistocene geology of the NW of the USSR. Geological Institute of Kola Branch of Acad. Sci. USSR, Apatites, 3-13 (in Russian).

Keith-Pitman, G.T. 1973, Lichenometrical study of snow patch variation in the Frederikshab district, Southwest Greenland, Grønlands Geol. Undersøgelse. Bull. 104:1-31.

Korpela, K. 1969, Die Weichsel-Eiszeit und ihr Interstadial in Peräpohjola (nördliches Nordfinnland) im Licht von Submoränen Sedimenten, Ann. Acad. Sci. Fennicae, Ser.A.III, Geol.-Geogr. 99:108.

Mäkinen, K. 1979, Interstadiaalen turvekerrostuma Tervolan Kauvonkankaalla, Geologi, 31:82-87.

Patzelt, G. 1974, Holocene variations of glaciers in the Alps, Colloq. Intern. du C.N.R.S., 219:51-57.

Punning, J.-M., L.Troitsky & R.Rajamäe, 1976, The genesis and age of Quaternary deposits in the eastern part of Van Mijenfjorden, West Spitsbergen, GFF, 98:343-347.

Punning, J.-M. & K.Punning, 1978, Concentration of sulphur in lichens as the criterion of their age, In: Lichen indication of the environmental conditions, Tallinn, 75-79 (in Russian, summary in English).

Punning, J.-M.K. & L.S.Troitsky, 1980, The uplift of earth's crust in Spitsbergen archipelago in Holocene, In: Quaternary Geology and Geomorphology, Remote Sensing, Moscow, 76-79 (in Russian, summary in English).

Salvigsen, O. 1978, Holocene emergence and finds of pumice, whalebone and driftwood at Svartknausflya, Nordaustlandet. Norsk Polarinstitutt, Arbok 1977, Oslo, 217-228.

Schytt, V., G.Hoppe, W.Blake, Jr. & M.G.Grosswald, 1967, The extent of the Würm glaciation in the European Arctic, Intern. Assoc. of Sci. Hydrogeol. I.U.G.G., Bern, 79:207-216.

Shackleton, N.J. & N.D.Opdyke, 1976, Oxygen isotope and paleomagnetic stratigraphy of Pacific core V28-239 late pleistocene to the latest pleistocene, Geol. Soc. Am. Memoir. 145:449-464.

Sudakova, N.T. & V.A.Iljitchev, 1974, Age of tills in Jaroslavl District, In: Engineer-geological study of tills, Jaroslavl, 64-69 (in Russian).

Troitsky, L., J.-M.Punning, G.Hütt & R.Rajamäe, 1979, Pleistocene glaciation chronology of Spitsbergen, Boreas, vol.8, 4:401-407.

Vlasov, V.K., O.A.Kulikov, N.A.Karpov & Ju.A.Petrov, 1978, On the problem of "zero-point" in thermoluminescent dating, In: Abstracts of the conference "The application of the results of the study on the luminescence of minerals in geology", Tallinn, 26-28 (in Russian).

Webber, P.J. & J.T.Andrews, 1973, Lichenometry, a commentary, Arc. and Alp. Res., 5:295-302.

A quaternary pollen profile from the upper valley of Cochabamba (Bolivia)

K.GRAF
University of Zürich, Switzerland

1 INTRODUCTION

The Andes are well known as an imposing topographical and climatological barrier on
the Western border of South America. The
Cordilleras include many longitudinal valleys. These areas were settled early and
used for agriculture because of their fa-
vourite climate in relation to the lowlands
and the high mountains. As examples of such
areas, we mention Bogotá in Columbia, Quito
in Ecuador, Junín, Ayacucho and Cuzco in
Peru, La Paz and Cochabamba in Bolivia. Man-
y of these interandean depressions contain
profound deposits from volcanic, glacial,
lacustrine or fluvial origin, which may
give us wide information about their geomorphological genesis. The present paper
deals with the topographical, vegetational
and climatological conditions in the Valley of Cochabamba during the Pleistocene.

The incentive to this study was given
by bore hole drillings (Driller type Cy-
clone R-43, proyecto integrado U.N./Geobol
1974) down to a depth of 260 m which were
realized in March 1974 by a common project
of U.N. and Geobol (Geological Survey of
Bolivia). The groundwater in detected gravels was the main supply for irrigation purposes; the clays were most favourable for
a scientific study of ostracodes, pollen
grains and spores (=sporomorphs). Some results about ostracodes identification and
frequency are published by Purper & Pinto
(1980), treating the 250 m deap bore profile from Wasa Mayu (17°32'S/65°49'W,
2750 m), above all. Breman has described
them in a manuscript, too. We discuss the
results from pollen analysis of a profile
extracted near the farm Cala Conto (17°34'
S/ 65°56'30"W, 2700 m).

Fig.1 The locality Cala Conto is situated
on 2700 m approx. in the centre of the plain
of Valle Alto de Cochabamba. An irrigation
canal forms the boundary of the cornfield,
and where that farmer is standing, a bore
hole of 260 m depth was drilled for seeking
ground water. With the mountains in the
background we can make out terraces half
way up the hills, and (bright) fluvial cones
at the basis of the slope.

2 GEOMORPHOLOGICAL AND CLIMATICAL CONDITIONS

The Valley of Cochabamba is composed of mainly two neighbouring basins which have an
extension of 20 x 40 km, each. They are
surrounded by mountain chains, reaching
maximum elevations at Mount Tunari (17°17'
S/66°24'W, 5100 m). The basins were never
glaciated in the Pleistocene, but Mount
Tunari was. Hydrographically both basins
are connected by a narrow steep step
("angostura"), and they make part of the
Amazonas system. Many brooks are flowing
from the narrow mountain valleys down to
the Upper Valley and have deposited there
alluvial cones at elevation intervals of

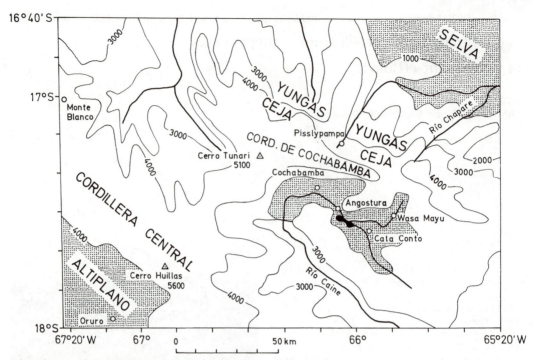

Fig. 2 Map of the Cochabamba Valley and surrounding mountains. Plains are indicated by a dotted pattern, lakes and rivers black, altitudes in m.

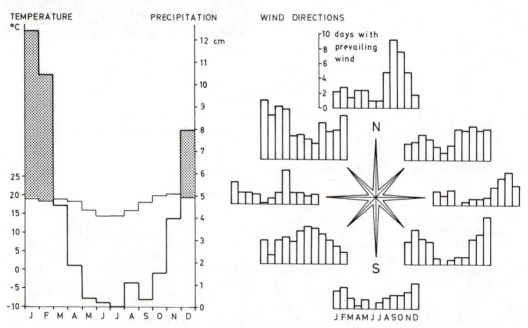

Fig. 3 Climate diagram of Cochabamba (17°23'9"S/66°9'35"W, 2555 m). Humid months are dotted, and the mean monthly temperature is indicated by a finer line than the mean monthly precipitation. Sources: Yearbooks/Anuarios Meteorológicos (1970-1974) and Ministerio de Agricultura (1967).

100 - 200 m. The slopes bordering the basin are often subdivided by rocky terraces which probably indicate former lake levels at 2750 - 2800 m.

The climate of this interandean region is arid to semiarid with 9 arid months (March till November, see fig. 3; aridity after Lauer 1960: 234). Due to the isolation of the area there are very special wind conditions. Frequently prevailing winds during the summer months (October till April, e.g. the wet season) are from the northwest (from Mount Tunari), north and northeast. As an exception there are frequent winds from SW during the dry season (extremely April till September). So we can conclude that the wind system is caused by the Cordillera de Cochabamba, specially in the wet season which is related to vegetation growth and pollen production. So, the main part of pollen, transported a long distance, originates from that eastern border of the Andes, e. g. from the "Ceja" or the "Yungas" (see the following chapter).

3 THE PRESENT VEGETATION

As we have seen, winds from the Northeastern slope of the Andes are frequent and influence the spread of pollen in the Cochabamba Valley. So it is useful to know the vegetation in the whole transsect from the interandean basin to the external slopes of the Andes. Our description is based on the observations of Herzog (1923: 168-197) and on our own herbarium collected in December 1974 and January 1975 in the Chapare Valley (17°10'S/65°30-40'W, 2600-3600 m) and in the area between Cochabamba and the Mount Tunari. All the plants mentioned with their species name are from our own herbarium. We gratefully acknowledge the indentifications by the team of Patricia Holmgren, the New York Botanical Garden, Bronx. In particular we mention plants which are well represented in our pollen diagram.

In the Cochabamba Valley, we find a dry vegetation between 2500 and 3200 m, with Amaranthaceae (Guilleminea densa), Anacardiaceae (Schinus), Bromeliaceae (Deuterocohnia, Dyckia, Puya), Cactaceae (Cereus, Echinocactus, Opuntia), Compositae (Baccharis dracunculifolia, Gynoxys, Mutisia viciaefolia, Proustia), Convolvulaceae (Evolvulus frankenioides), Euphorbiaceae (Acalypha plicata, Croton boliviensis), Gramineae, Labiatae (Salvia orbignaei), Leguminosae

(Acacia, Psoralea; Cassia after Cárdenas, herbarium Nr. 4431), Loganiaceae (Buddleia), Rosaceae (Dodonaea, Kageneckia), Solanaceae (Nicotiana, Solanum pallidum) and Umbelliferae (Eryngium).

On the interandean slopes of the mountains, there is a shrub belt between 3200 and 3800 m. One frequently finds Campanulaceae (Siphocampylus), Compositae (Achyrocline, Baccharis, Erigeron, Hieracium, Perezia, Senecio, Viguiera pazensis, Werneria), Gentianaceae (Gentiana, Halenia), Leguminosae Astragalus, Lupinus), Rosaceae (Acaena cylindrostachya, Alchemilla, Polylepis), Saxifragaceae (Escallonia), Solanaceae (Solanum) and Violaceae (Viola).

The high andean herbaceous belt (3800-4600 m) is characterized by Alstroemeriaceae (Bomarea zosteraefolia), Caryophyllaceae (Cerastium), Compositae (Gynoxys psilophylla, Hypochoeris meyenianus, Lucilia conoidea, Perezia pinnatifida, Senecio, Werneria), Gnetaceae (Ephedra), Gentianaceae (Gentiana), Geraniaceae (Geranium), Gramineae, Malvaceae (Nototriche cf. flabellata) and Plantaginaceae (Plantago australis). In the area of small mountain lakes there is the water fern Isoëtes, Halorrhagaceae (Myriophyllum), Hydrophyllaceae (Phacelia secunda) and Juncaceae (Distichia).

On the Eastern slope of the Andes the climate is much wetter what we note above all from 3400 m downwards with a distinctly different vegetation from the interandean one. Especially Polylepis is lacking here (after Herzog 1923: 183), an observation which is only right for this study area in the Northeast of the Cochabamba Valley. We treat two vegetation belts of the Eastern slope, the cloud forest ("Ceja") and the upper mountain forest (the upper "Yungas").

The "Ceja" lies between approx. 2800 and 3400 m and contains a lot of trees and shrubs such as Betulaceae (Alnus), Compositae (Barnadesia, Gynoxys), Cunoniaceae (Weinmannia), Myrtaceae (Eugenia after Cárdenas, herbarium Nr. 5987; Myrteola microphylla) and Podocarpaceae (Podocarpus). We find also Campuanulaceae (Centropogon, Siphocampylus), Geraniaceae (Geranium bangii), Melastomataceae (Brachyotum, Miconia, Tibouchina bicolor), Myricaceae (Myrica), Oxalidaceae (Oxalis), Solanaceae (Poecilochroma) and Verbenaceae (Duranta).

The upper "Yungas" include the warm, wet belt between 1500 and 2800 m. Frequently there are growing specimens of trees of Alnus, Juglandaceae (Juglans after Herzog 1923: 195/196), Salicaceae (Salix),

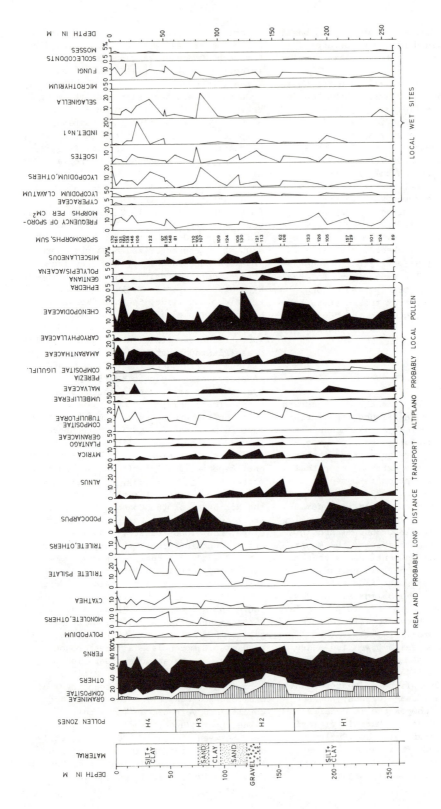

Fig. 4   Profile H
Cala Conto   2700 m
17° 34' S / 65° 56' 30" W   K. Graf

368

Tiliaceae (Luehea) and tree ferns (Cyathea). Here further members apart of the tipical plants of the "Ceja" are Araliaceae (Oreopanax), Euphorbiaceae (Acalypha), Malvaceae (Abutilon, Valeriana), Phytolaccaceae (Phytolacca octandra), Solanaceae and Urticaceae (Cecropia).

With regard to the frequent occurence of spores in our pollen diagram, we especially want to mention some ferns. As stated above, the tree fern Cyathea finds its upper limit at approx. 2600 m in the upper "Yungas". Other ferns with trilete spores generally avoid the high andean herb belt (with some exceptions as Cheilanthes, Ctenopteris, Jamesonia). They include Gleichenia and Pteris of humid regions, but also ferns with very different demands of humidity. So within the Sinopteridaceae, Cheilanthes, Notholaena and Pellaea (Herzog 1923: 151/169) grow on dry fluvial cones in the Cochabamba Valley, while Eriosorus (called Gymnogramme by Herzog 1923: 189/196) and Pityrogramma prefer the wet regions of the upper "Yungas" and the "Ceja". Polypodium and other ferns with monolete spores also grow in many various ecological conditions.

With all this general knowlegde about the distribution of the present vegetation, we will try to interpret the pollen spectra in the following chapter.

4 DISCUSSION OF THE PALYNOLOGICAL RESULTS

4.1 Description of the pollen diagram Cala Conto (Fig. 4)

There were 260 samples with intervals of 1 m in the profile Cala Conto, and we chose 31 samples for a detailed pollen analysis. The study was quite difficult because of the generally bad conservation and rareness of the sporomorphs. So there was a relatively big amount of no identifiable pollen grains. This was also due to the fact that above all they are of Pleistocene age and could not always be compared with the postglacial ones of our earlier studies. Radiometric date fixations are not at our disposal.

In Fig. 4 the percentages of Gramineae are shown by hatching. The amounts are white with the ferns on the left side, the compositae in the middle and the specimens of wet localities on the right side. Within this group there are the fern allies, the water fern Isoëtes, "Indet. No. 1" (probably a megaspore), spores of fungi or

their fruit bodies (Microthyrium after Van Geel 1972: 270-272) and moss spores (Marchantiaceae after Boros & Járai 1975: 78 ff.). As a special case we habe included the mandibles of crabs, too (Scolecodonts after Tasch & Shaffer 1961: 371). We distinguish four zones in the diagram.

Zone H 1: 260-165 m

In the lowest pollen zone we note high percentages of tree pollen (Podocarpus, Alnus) and trilete spores. Also Malvaceae and Polylepis/Acaena are quite frequent. On the other hand, Myrica and Plantago are rare, both plants of the cloud forest and of bare stony sites. Equally there are only few Lycopodium, Selaginella and Isoëtes. The sporomorph spectrum is very abundant and provided with a relatively high sporomorph frequency which shows the favourable conditions of sedimentation in the former lake.

Zone H 2: 165-105 m

This is the most conspicuous pollen zone of the whole diagram. Various taxa reach their maximum amounts such as Gramineae, Myrica, Plantago, Chenopodiaceae, Gentiana and Polylepis/Acaena. Also well represented are Alnus, Compositae tubuliflorae and Amaranthaceae now. On the contrary Podocarpus and ferns (with the exception of Isoëtes) are characterized by an intense decrease. Apart from Podocarpus and Malvaceae extremely many pollen grains are found in this section, and generally few spores only. It seems that the bad conditions of growth favoured a pioneer vegetation.

Zone H 3: 105-55 m

Again we note high percentages of pollen grains of Podocarpus and trilete spores, like in zone H 1. Zone H 3 is characterized by the slow retreat of Alnus, Myrica and Gramineae. From now on pollen of Polylepis/Acaena is practically absent, and on the other hand Umbelliferae and Malvaceae are relatively frequent. There are a lot of Lycopodium and Selaginella indicating locally humid conditions. In the course of the sedimentation of gravels and sand following, during H 2 and H 3, the lake must slowly have silted up. Ostracodes (Heterocypris, Limnocythere after Purper & Pinto 1980), which are always found from the base of the profile, are now lacking completely from 65 m upwards. Probably the upper clay does not correspond to a lake sediment, but rather a fine material transported by wind ("Loess"). Remnant lakes and swamps can still be considered to be in the centre of the basins.

Zone H 4: 55 - 0 m

In the uppermost pollen zone Amaranthaceae and Chenopodiaceae get a dominant role. However, Podocarpus is reduced again to minimum percentages like in zone H 2. An unexpected thing is now the low amounts of Gramineae pollen and the frequent ones of monolete spores. This means a fundamental change in the sporomorph spectrum, in connection with an increased importance of the local vegetation. Therefore the hygrophile plants of the local vegetation are well documented, too.

4.2 Interpretation of the palynological data

Our interpretation is based on the assumption that periods of intense local pollen production change with accentuated periods of pollen transported a long distance. Especially in the Valley of Cochabamba we have to consider variable conditions of sedimentation and transport by wind or water.

Let us suppose the case of an increasingly cold climate, first. As the basins are ± entirely surrounded by mountains, a migration of plants downwards to warmer regions was very difficult. They could not cross the Cordillera, and so depressions of temperature had a very strong influence on the vegetation. Also an increase of dryness had complex influences, because it could cause the disappearance of lakes or the partial dessication of the basins, at least. This enables a pioneer vegetation to develop locally and to produce a lot of local pollen, which reduces the importance of pollen transported a long distance.

In the case of increasing temperatures our interpretation is inverse. Then, the cloud forest belt with its abundant vegetation reaches up to higher sea levels, and therefore comes nearer to the basins leading to an increase of the long distance transport. If the increasing humidity takes more importance on the evaporation than the warmth of climate, large areas of the basins can be overflooded. Also this fact causes a dominance of the pollen transported a long distance.

With this model of interpretation we are going to explain the changes of vegetation and climate within the pollen diagram Cala Conto. We suppose that a 260 m deep profile can not only originate from the last 10.000 years, and so it does not only represent a postglacial sedimentation. There is a Pleistocene diagram the geomorphological facts of which resemble to those from Bogotá - Ciudad Universitaria (van der Hammen et al 1973: 24/25).

Zone H 1 is characterized by a great many sporomorphs transported a long distance. Only a sparse local vegetation with Gramineae , Compositae, Malvaceae and Chenopodiaceae has developed. Presumably the "Ceja" has come very near to the Cochabamba Valley and has caused high percentages of tree pollen (Podocarpus, Alnus, Polylepis). This warm phase corresponds to an interglacial period. It scarcely dates from the Pliocene because of the rareness of possibly tertiary sporomorphs from the areas of Cochabamba/ Pisslypampa and Cerro Rico de Potosí (19° 40'S/65°40'W) such as Gleichenia, Lauraceae (Mespilodaphne), Leguminosae (Acacia, Caesalpinia, Cassia, Inga, Mimosa), Moraceae (Coussapoa), Myrtaceae (Myrcia), Nyctaginaceae (Pisonia) or Rutaceae (Pilocarpus), quoted in Berry (1922: 157 and 1939: 45-56).

In zone H 2 we note an intensely changed pollen spectrum. Although the long distance transport seems to be reduced only a little, the influence of the local vegetation becomes dominant with Gramineae, Compositae and many pioneer plants (Amaranthaceae, Chenopodiaceae, possibly Myrica and Plantago). Podocarpus has lost its importance which may be due to dry, cold conditions of climate. The beginning upfilling of the lake basin with gravel is probably a consequence of a wet climate during the upper half of H 2, which causes a higher water level of the rivers and consequently a higher transport energy. There must have been a pleniglacial period, connected with the largest glaciar advance in the mountains. The present climatic snowline, lying presumably at 5400 m, then suffered a depression to 4600 m, in a minimum (Furrer & Graf 1978: 453/454). These are estimated data which arise in the larger context of the whole Bolivian Andes. There, approximately, the present lower limit of structural soil normally corresponds with the snowline during the last ice age. This one does not differ much in the ice age next to the last. Moreover we have observed former glacial cirques at 4600 m on Mount Tunari. In similar dimensions, the andean vegetation belts may have been shifted.

H 3 reflects similar conditions of dominant long distance transport like H 1. But there are only small amounts of Alnus, a typical tree of wet river banks. This allows the conclusion that this warm period was no more as humid as H 1. However, Pteridophyta with trilete spores and Myrica are quite frequent, and so we assume a corresponding climate with semihumid conditions. We cannot conclude precisely, if H 3 results from an interglacial or only an interstadial period.

Zone H 4 is difficult to interpret. Notably the Gramineae are reduced. There must have taken place an intense effect of wind, what can be deducted from the deposits of "Loess" and from the big quantities of sporomorphs transported a long distance. But also pioneer plants are frequent (Chenopodiaceae, Caryophyllaceae, Amaranthaceae). All this may indicate a dry and very cold steppe climate belonging to the last ice age. Eventually the postglacial period is not represented in this diagram due to erosional processes, or at least it can not be made out palynologically.

As a whole the pollen diagram Cala Conto resulted to be Pleistocene and to reflect an interesting sequence of warm and cold periods. Relatively few, in relation to postglacial diagrams, there occur Caryophyllaceae, Gentiana, Lycopodium clavatum and Cyperaceae. With the other components, the sporomorph spectra are divers and very similar to those from postglacial, high andean peat bogs (Graf 1979, 1981). This is certainly effected by their similar origin as pollen transported a long distance. H 1 and H 3 correspond to warm periods with a similar climate like today. If H 3 only meant an interstadial period, H 2 and H 4 would be allocated to the last ice age. The profundity of sediments in H 3, however, rather points to a longer and more intense warm period. In this case we have to consider H 1 and H 3 as interglacial periods, while H 2 and H 4 correspond to the two last ice ages. Moreover, H 2 is divided into a first part with dry climate and a second, wet period. Following H 4, the postglacial period is not to delimit clearly. We expect that erosion by wind or human actions have reduced these youngest fine-grained sediments.

Acknowledgments

Our field studies were realized with the support from the Swiss National Fundation and the Geological Survey of Bolivia (Geobol). Mr R.P. Müller (Wädenswil) assisted in the text editing, Mr W.Schwarz (Geographical Institute of the University of Zurich) in the chart drawing. The samples were prepared in the Laboratory of the same institute by Mr B.Kägi. Fern studies were realized in colaboration with Dr C.U.Kramer (Botanical Institute of the University of Zurich). We thank them all for their support.

Bibliography

Ahlfeld, F. 1972. Geología de Bolivia, p. 1 - 190. La Paz: Ed. "Los Amigos del Libro".

Anuarios Meteorológicos 1970-1974, p. 1 - approx. 150. La Paz: Ministerio de Transportes y Comunicaciones, Dep. de Meteorología.

Berry, E.W. 1922. Pliocene fossil plants from eastern Bolivia. Bol. The John Hopkins Univ., Baltimore, Studies in Geology 4: 145 - 203.

—   1939. The fossil flora of Potosí, Bolivia. Bol. The John Hopkins Univ., Baltimore, Studies in Geology 13: 9 - 67.

Breman, E. Manuscr.: Paleoecología de los ostácodos de pozos perforados en el Cuaternario de Cochabamba, Bolivia, p. 1 - 10.

Cárdenas, M., Herbarium material from the University "San Simón", Cochabamba.

Furrer, G. & K. Graf 1978. Die subnivale Höhenstufe am Kilimandjaro und in den Anden Boliviens und Ecuadors. Erdwiss. Forschung, Akad. der Wiss. und der Lit. 11: 441 - 457.

van Geel, B. 1972. Palynology of a section from the raised peat bog "Wietmarscher Moor", with special reference to fungal remains. Acta bot. neerl. 21: 261 - 284.

Graf, K. 1979. Untersuchungen zur rezenten Pollen- und Sporenflora in der nördlichen Zentralkordillere Boliviens und Versuch einer Auswertung von Profilen aus postglazialen Torfmooren. Habil. Univ. Zürich: 1 - 104.

—   1981. Palynological investigations of two Post-glacial peat bogs near the boundary of Bolivia and Peru. J.of Biogeogr. 8: 353 - 368.

van der Hammen, T., J.H. Werner & H. van
    Dommelen 1973. Palynological record of the
    upheaval of the northern Andes: A study
    of the Pliocene and lower Quaternary of
    the Columbian Eastern Cordillera and the
    early evolution of its high-andean biota.
    Rev. of palaeobotany and palynology 16:
    1 - 122.
Lauer, W. 1960. Klimadiagramme. Erdkunde
    14/3: 232 - 242.
Ministerio de Agricultura 1967. El clima
    del Valle de Cochabamba. La Paz: Bol.
    climatológico 2: 1 - 25.
Proyecto integrado U.N./Geobol 1974,
    Manuscr.: Informe final del pozo BC-23
    Cala Conto.
Purper, I. & I.D. Pinto 1980. Interglacial
    ostracodes from Wasa Mayu, Bolivia. Porto
    Alegre: Pesquisas 13: 161 - 184.
Tasch, P. & B.L. Shaffer 1961. Study of
    scolecodonts by transmitted light. Micro-
    paleontology 7: 369 - 371.
Urquidi, G. 1954. Monografía del departe-
    mento de Cochabamba, p. 1 - 366. Cocha-
    bamba: Impr. "Tunari".

Differentiation of morainic deposits based on geomorphic, stratigraphic, palynologic, and pedologic evidence, Lemhi Mountains, Idaho, USA

DAVID R.BUTLER
Oklahoma State University, Stillwater, USA

CURTIS J.SORENSON & WAKEFIELD DORT Jr.
University of Kansas, Lawrence, USA

1 INTRODUCTION

1.1 Background

Proposed chronologies of alpine glacial sequences in the Canadian Rocky Mountains differ markedly from those that have been compiled for the Central and Southern Rockies of the United States. The Canadian record includes two Holocene advances - the Cavell of the last few centuries and the less extensive Crowfoot which occurred shortly(?) before 6,600 BP - and three Pleistocene episodes - the Late Portage Mountain, culminating near 12,000 BP; the Early Portage Mountain of around 22,000 BP; and an undated, though much older, Early Advance (Luckman and Osborn, 1979; Rutter, 1976).

The chronology that has been developed for the Rocky Mountains of Wyoming and Colorado documents a greater number of Pleistocene and Holocene glaciations. The late Pleistocene record contains two Bull Lake advances and at least three during Pinedale time (Richmond, 1965). Holocene glaciations include the Satanta Peak Advance, culminating prior to 9,200 BP; the Ptarmigan Advance occurring between 7,250 and 6,600 BP; and three Neoglacial advances, the Triple Lakes (5,000-3,000 BP), the Audubon (2,400-900 BP), and the Arapaho Peak (350-100 BP) (Benedict, 1981).

1.2 Objectives

The preceding paragraphs illustrate the major differences between recognized sequences of glacial advances in the Canadian versus the Southern/Central American Rockies. The Northern Rockies of the United States, particularly in Idaho and Montana, bridge the geographic gap between these two regions, yet have largely been ignored in studies of the Quaternary records of alpine areas. Recent exceptions include detailed work in the Missoula Hills of western Montana (Elison, 1981), the Pioneer Range of central Idaho (Evenson et al., 1979 and 1981), and the Lemhi Range in east-central Idaho (Knoll, 1977). The Lemhi Range in particular offers the potential for significant data bearing on the glacial chronology of western North America.

It has been suggested that "regions characterized by low intensity of glacierization will, in concert, provide the greatest detail of climatic and geologic fluctuations during the Pleistocene history of the North American Cordillera (because) land forms or materials attributable to earlier episodes of glaciation, or to interglacial intervals, have not been removed in entirety during times of later advances" (Dort, 1977, iii).

The location of the Lemhi Range intermediate between those alpine areas studied extensively in Canada and in the United States provides an optimum environment for investigation of glacial land forms and materials and for reconstruction of the late Quaternary glacial chronology. The additional advantage of low intensity glacierization in this range assures preservation of much of this record.

This paper examines the glacial moraines in a small mountain valley in the east-central Lemhi Range. The study was initiated to examine and interpret unusual stratigraphic features exposed in a large terminal moraine, and to reconstruct the late Quaternary glacial history, a reconstruction aimed at resolution of differences between the Canadian and American alpine glacial chronologies.

1.3 Location and Geography

The specific study area is a valley situated on the eastern slope of the Lemhi Range about 25km south of the town of Leadore, Idaho (Figure 1). Drainage runs northerly into the Lemhi River. We have unofficially named this location the Mountain Boy Valley after a small mining operation located there.

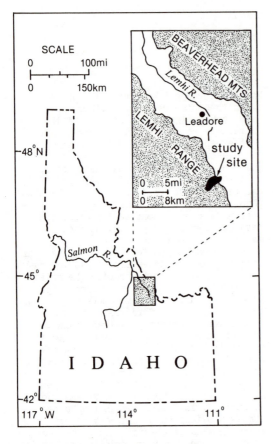

Figure 1. Location of study site.

The valley heads in a small, though fairly well developed cirque, the floor of which has an elevation of approximately 2,800m. Mountain Boy Valley extends 3km northeasterly from the cirque headwall to its debouchment onto the broad alluvial flank of the Lemhi Valley. A massive terminal moraine is present at this point. Elevation of the valley floor here is about 2,300m, providing an average valley gradient of almost 175m/km.

2 LATE QUATERNARY GLACIAL HISTORY

2.1 Number and Distribution of Moraines

Segments of 37 separate and distinct moraines were identified and mapped in Mountain Boy Valley. Morphology, slope stability, soil development, and vegetation cover of these land forms vary greatly. In general, the larger moraines are also the older ones and are distributed mainly in the lower 1.5km of the valley. Largest of all is the terminal moraine at the valley mouth (Figure 2).

Younger, generally smaller moraines are located in the upper half of the valley. The younger the moraine, based on many relative-age dating criteria (such as boulder weathering, boulder implantation, soil development, steepness of slope, stability of deposits, and vegetative cover), the rockier and less stable its surface and the less the cover of vegetation.

2.2 Pleistocene Glacial Events

Geomorphic, stratigraphic, pedologic, morphologic (cross-cutting or over-riding), and palynologic evidence suggests that at least six late Pleistocene glacial advances occurred in Mountain Boy Valley. This number can be compared with the accepted sequence of three advances in southern Canada and at least five in the American Rockies.

Two Bull Lake and one early Pinedale advance are recorded by superimposed tills in the Mountain Boy terminal moraine described below. Later Pinedale advances are distinguished on the basis of geomorphic evidence in the form of moraines which were over-ridden and truncated by subsequent readvances.

The third major Pinedale glacial advance deposited an arcuate moraine about 300m upvalley from the main terminus. Behind this blockade, fine sediments accumulated to a minimum explored thickness of 5m. A radiocarbon date of 10,130±500 BP (Beta 3659) for an organic horizon at a depth of 160cm provides a minimum date for this third advance. (The large standard deviation for the date is a consequence of the extremely small amount of organic matter in the dated horizon.) Later Pinedale advances, represented by moraines situated farther up the valley, probably also occurred prior to 10,000 BP, but none of these was accurately dated. Figure 3 illustrates the stratigraphic position of the unit dated at 10,130 BP.

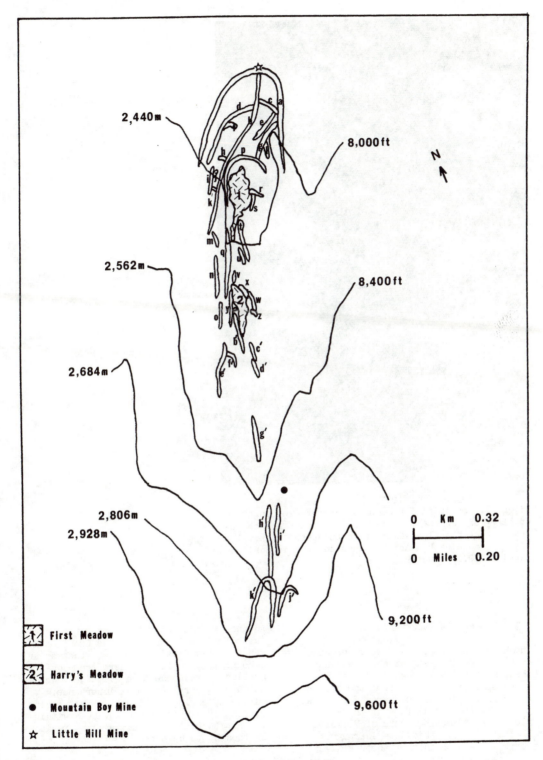

Figure 2. Distribution of moraines, Mountain Boy Valley.

375

Figure 3. Stratified fine sediments exposed in trench
upvalley of moraine P (Figure 2). Another trench revealed that the fine sediments overlap the proximal
slope of moraine P. Arrow points to organic horizon
dated at 10,130±500 BP.

2.3 Holocene Glacial Events

The Mountain Boy glacial record of Holocene time begins with an early advance
that built a small, tightly curved moraine approximately 1km from the cirque
headwall. Charcoal recovered from an exposure in the distal face of this moraine
yielded a radiocarbon date of 7,100±120
BP (Beta 2163). This date is roughly
correlative with that of the Ptarmigan

Advance in the Colorado Front Range
(Benedict, 1981) and with the Crowfoot Advance in the Canadian Rockies (Luckman and
Osborn, 1979). It is, therefore, suggestive of response to a climatic fluctu-
ation of clearly regional importance.
Furthermore, it is noteworthy that radiocarbon dates place these advances within
the Altithermal interval, a time previously thought to have been warm and dry.

The glacial and mass-wasting record of Neoglacial time in Mountain Boy Valley more closely resembles that of the Colorado Rockies than the Canadian pattern. An advance believed, on the basis of several relative-age dating methods, to be of early Neoglacial age may correlate with one of the Triple Lakes pulses of Benedict (1981). Later events, represented in Mountain Boy Valley solely by mass-wasting debris and land forms, appear to be at least roughly correlative with the Audubon and Arapaho Peak stades.

3 SPECIFIC MORAINAL FEATURES

3.1 Stacked Tills of the Terminal Moraine

The terminal moraine in Mountain Boy Valley is a massive, looped ridge from which lateral extensions prolong the land form upvalley for almost half a kilometer. This moraine is about 75m high at its point of greatest relief near the axis of the valley. On the basis of surficial characteristics discernible on aerial photographs, this had been previously classified as "Pinedale moraine" (Dort, 1962).

Bulldozer cuts made in conjunction with mining activities are located at several positions on the terminal and western parts of the moraine complex. These excavations expose a sequence of three apparent stacked or superimposed tills, the lower two units being separated from the uppermost till by a carbonate-rich paleosol (Figure 4). This paleosol (B in Figure 5) clearly delineates a major break between the time of deposition of the two

older, underlying tills and the younger till above. However, no such clear pedologic marker separates the two lower tills from each other (C and D in Figure 5). Rather, these two tills meet at a wavy, though distinct planar contact.

Are these two lower tills representatives of two separate glacial advances, or are they instead ablation till overlying ground-moraine till (Drake, 1971; Rabassa and Aliotta, 1979) or supraglacial till overlying subglacial till (Boulton and Eyles, 1979)? There is ample evidence that they do indeed represent two different, distinct glacial advances.

The lower two tills have relative-age characteristics that support a Bull Lake age for each unit. However, the similarities end there. Table 1 records several of the characteristics of these two lower tills, as well as the carbonate paleosol and the uppermost till. The analysis for each unit represents an average for samples collected from four exposures along the moraine crest. Of particular note is the combined silt/clay percentage for Till C, which argues against a supraglacial origin. Supraglacial tills normally have combined silt/clay percentages of less than 15% (Boulton and Eyles, 1979).

The few fragments of diorite and andesite scattered through the lowest till, unit D, are extremely rotten, crumbling at touch. Fragments of the same rock types present in unit C are somewhat more coherent, though not at all approaching the soundness expected of these lithologies in a deposit of Pinedale age. Quartzite clasts in unit D are coated with bulbous accumulations of carbonate caliche 3-4mm

Table 1. Characteristics of the Stacked Tills and Paleosol

Sample	% Sand	% Silt	% Clay	% Organic Matter	Color	Assigned Age
Till A	46.5	28.3	25.2	4.70	10YR6/6	Early Pinedale
Paleosol B	43.4	39.7	25.9	7.03	10YR8/3	Pinedale/Bull Lake Interglacial
Till C	48.0	34.5	17.5	1.54	2.5Y7/4	Late Bull Lake
Till D	57.5	27.6	14.9	2.15	10YR6/4	Early Bull Lake

A
B
C
D

Figure 4. Superimposed tills in terminal moraine. (See Figure 5 for identity of units.)

thick. Similar clasts in unit C have coatings only about 1mm thick. (A sample of 30 clasts was examined in each unit at four separate exposures.) Particle size data, degree of weathering of igneous rocks, and thickness of carbonate coatings on quartzite clasts all suggest that unit C is considerably younger than unit D and therefore the two units represent tills deposited by separate glacial advances.

Table 2. Type and Percentage of Main Pollen Forms Present (Methodology in Butler, 1982)

Sample	Main Pollen Types Present	Percentage of Total
Till A	Larix	61
	Pinus	13
	Pseudotsuga	9
	Gramineae	8
Paleosol B	Gramineae	89
	Chenopodiaceae	3
	Artemisia	2
Till C	Pinus	43
	Gramineae	16
	Artemisia	13
	Picea	5
	Abies	2
Till D	Pinus	34
	Gramineae	26
	Picea	8
	Chenopodiaceae	8
	Artemisia	4

Mining debris
___

Unit A
Pinedale till

___

Unit B
Paleosol

___

Unit C
Younger Bull Lake till

___

Unit D
Older Bull Lake till

___

Figure 5. Superimposed tills in lateral extension of terminal moraine.
Mining debris cover is thick here, but almost non-existent in previous figure. Thickness of till and paleosol units varies, but can be
clearly traced from the exposure of Figure 4 to the exposure seen here.

Table 2 records additional differences
among the three sedimentary units and the
paleosol. Pollen was extracted from each
of these horizons. At least 300 grains
from each sample were identified and counted in the Palynology Laboratory, Department of Geography, University of
Kansas. There are significant variations
in the plant communities represented.
For example, note that Larix is common in
the uppermost till but is absent from the
lower tills. Gramineae percentages vary
widely, although, excepting the paleosol,
they increase through the older sediments.

3.2 Areal Extent of Terminal Moraine Till
Units

Dissection of the terminal moraine ridge
by several bulldozer cuts permits lateral

tracing of sedimentary units through a
distance of almost half a kilometer
(Figure 5). Although thicknesses of individual units vary somewhat, all three
till units and the paleosol can be
identified in every exposure. Their
physical characteristics are remarkably
constant.

4 CONCLUSIONS

The presence of three tills of demon-
strably differing characteristics and inferred ages in a vertical sequence within
a single sharp-crested moraine ridge
apparently is unusual in published descriptions of glaciated areas in the
Rocky Mountains. The ability to extract
and analyze pollen from these tills pro-

vided an independent line of evidence, separate from strictly physical character-istics, which confirmed the distinctive-ness of the three till units. The tech-niques used in this study to recover pollen from non-lacustrine sediments can in the future be applied to discriminate among glacial units in other areas.

The late Quaternary glacial chronology of Mountain Boy Valley contains elements of sequences previously compiled for both the Canadian and American Rockies. This Idaho chronology may serve to reconcile records observed in widely separated re-gions. The location of the Lemhi Moun-tains apparently allowed them to be af-fected by glacial-climatic episodes occur-ring both to the north and to the south.

Finally, the detailed morainic, strati-graphic, and palynologic records which have been analyzed in Mountain Boy Valley reinforce the opinion of Dort (1977) cited earlier. The Lemhi Mountains cer-tainly were characterized by low in-tensity glacierization when compared with either the Canadian or central American Rockies. However, this low level of glacial activity allowed preservation of the many moraines and a detailed glacial stratigraphic succession from which a com-plex history has emerged.

5 ACKNOWLEDGEMENTS

Financial assistance from the National Science Foundation, Doctoral Dissertation Research Grant EAR-8112516; the Geological Society of America, Research Grant 2587-80; and the University of Kansas Graduate School', Dissertation and Summer Fellow-ships, is gratefully appreciated. Dr. William C. Johnson contributed valuable advice and assistance in the field and on various aspects of this project. Assist-ance in the field was also provided by J. Butler, M. Butler, B. Hall, K. Millington, R. Sewell, and M. Winter. Figure 1 was provided by the Kansas Uni-versity Cartography Service.

6 REFERENCES

Benedict, J.B. 1981, The Fourth of July Valley: Glacial geology and archeology of the timberline ecotone. Ward, Colorado, Center for Mountain Arche-ology.

Boulton, G.S. & N. Eyles 1979, Sedimenta-tion by valley glaciers; a model and genetic classification. In C. Schlüchter (ed.), Moraines and varves, p. 11-23. Rotterdam, A.A. Balkema.

Butler, D.R. 1982, Late Quaternary glacia-tion and paleoenvironmental changes in adjacent valleys, east-central Lemhi Mountains, Idaho. Unpublished doctoral dissertation, Department of Geography, University of Kansas.

Dort, W., Jr. 1962, Multiple glaciation of southern Lemhi Mountains, Idaho: preliminary reconnaissance report, Tebiwa 5: 2-17.

Dort, W., Jr. 1977, Foreward. In K.M. Knoll, Chronology of alpine glacier stillstands, east-central Lemhi Range, Idaho, p. iii. Pocatello, Idaho State Museum of Natural History.

Drake, L.D. 1971, Evidence for ablation and basal till in east-central New Hampshire. In R.P. Goldthwait (ed.), Till: a symposium, p. 73-91. Columbus, Ohio State University Press.

Elison, M.W. 1981, Relative dating of major moraines of the southeast Jocko Valley, Northwest Geol. 10: 20-31.

Evenson, E.B., T.A. Pasquini, R.A. Stewart & G. Stephens 1979, Systematic provenance investigations in areas of alpine glaciation: applications to glacial geology and mineral explora-tion. In C. Schlüchter (ed.), Moraines and varves, p. 25-42. Rotterdam, A.A. Balkema.

Evenson, E.B., J.F.P. Cotter & J.M. Clinch 1981, The central Idaho glacial model. In E.B. Evenson (ed.), Symposium and field trip on the genesis and lith-ology of morainic deposits in an alpine environment, abstracts and selected re-prints and handouts, 30p. Bethlehem, Pennsylvania, Lehigh University.

Knoll, K.M. 1977, Chronology of alpine glacier stillstands, east-central Lehmi Range, Idaho. Pocatello, Idaho State Museum of Natural History.

Luckman, B.H. & G.D. Osborn 1979, Holo-cene glacier fluctuations in the middle Canadian Rocky Mountains, Quat. Res. 11, 52-77.

Rabassa, J. & G. Aliotta 1979, Sediment-ology of two superimposed tills in the Bariloche Moraine (Nahuel Huapi Drift, Late Glacial), Rio Negro, Argentina. In C. Schlüchter (ed.), Moraines and varves, p. 81-92. Rotterdam, A.A. Balkema.

Richmond, G.M. 1965, Glaciation of the Rocky Mountains. In H.E. Wright, Jr. & D.G. Frey (eds.), The Quaternary of the United States, p. 217-230. Princeton, Princeton University Press.

Rutter, N.W. 1976, Multiple glaciation in the Canadian Rocky Mountains with special emphasis on northeastern British Columbia. In W.C. Mahaney (ed.), Quaternary stratigraphy of North America, p. 409-440. Stroudsburg, Pennsylvania, Dowden, Hutchinson & Ross, Inc.

Glacier and moraine inventory on the eastern slopes
of Cordón del Plata and Cordón del Portillo, Central Andes, Argentina

LYDIA E.ESPIZÚA
IANIGLA, CONICET, Mendoza, Argentina

ABSTRACT
Three-hundred and eighty seven glaciers
and snow patches above 1500 m a.s.l. have
been inventoried within the Río Tunuyán
Basin, a 2423 Km^2 watershed with dry, continental climate. The inventory has been
carried out by means of aerial photointerpretation at 1:50,000 scale and field observations. The World Glacier Inventory outlines, with some modifications and additional information have been followed. The
glacierized area is 144.03 Km^2, 40% of
which corresponds to uncovered ice and 60%
to covered ice. Both uncovered and covered
ice show a predominant SE trend. The minimum altitude is 4500 m a.s.l. for uncovered
ice and 3400 m a.s.l. for covered ice.
In comparing this inventory with those of
nearby basins, it is observed that uncovered
ice increases towards the S, whereas covered
ice increases northwards.

INTRODUCTION
This inventory has been conducted at the
Argentine Institute of Nivology and Glaciology (IANIGLA - CONICET) and it is a contribution to the World Glacier Inventory (WGI),
international program coordinated by the
Temporary Technical Secretariat (TTS), ETH-
Zentrum, Geography Department, Zürich, Switzerland. The aim of this project is to provide a global knowledge of snow and ice as
water source and the characteristics of the
ice bodies of the world. Our own program
intends to do it so in the Argentine Central
Andes.

GEOGRAPHY
The inventoried zone lies on the eastern
flank of the Cordón (=range) del Plata and
Cordón del Portillo, a part of the Río
Tunuyán Basin, between lat.33°00'S and lat.
33°45'S, and 69°15'-69°40' of longitude
West of Greenwich. It covers an area of
2423 Km^2, approximately, above 1500 m a.s.l.

The western limit is given by the higher-
most ridges of the Cordillera Frontal, with
NNE trend. This ridge is called Cordón del
Plata to the north, with elevations above
5000 m a.s.l., and a maximum at Cerro del
Plata with 6100 m a.s.l. The Cordón del
Portillo is somewhat lower, with most sum-
mits above 4400 m a.s.l. and a maximum at
Cerro Pirca (5597 m a.s.l.) (Figure 1).
The valleys draining the Cordón del Plata
have a general SE trend; they are, from N
to S: Arroyo Negro, Arroyo Guevara, Arroyo
Cuevas, Río de la Carrera, Río Chupasangral,
and Río Santa Clara. Southwards, the streams
are NE-oriented: Río de las Tunas, Arroyo
Novillo Muerto and Arroyo Barraquero. From
Cordón del Portillo, the valleys descend
towards the E, being the most important
Arroyo Pircas and Arroyo Manzano. All these
streams reach the piedmont zone where they
are used for irrigation and water-supply.
The piedmont zone shows the Quaternary
sedimentary filling of a tectonic graben,
which has descended more than 1000 m
(Polanski 1963).
The climate of the Inventory region is
essentially of the continental type, with
large daily variations, raising from december to march above 0°C. However, sunrise
frosts are frequent. During the rest of the
year, the mean temperature keeps below
freezing-point.
Precipitation above 4000 m a.s.l. oscillates
between 400 mm and 600 mm a year (Polanski
1964) and it falls mainly in winter as snow
and hail. In other areas below that altitude, precipitation is more reduced in winter and rain may occasionally fall in
summer.
In the studied area, there are three gaging
stations of the Departamento General de
Irrigación of the Province of Mendoza
(Figure 2):
(a) Río Santa Clara (2150 m a.s.l.; lat.
    33°15'S, long. 69°30'W): 25 years of

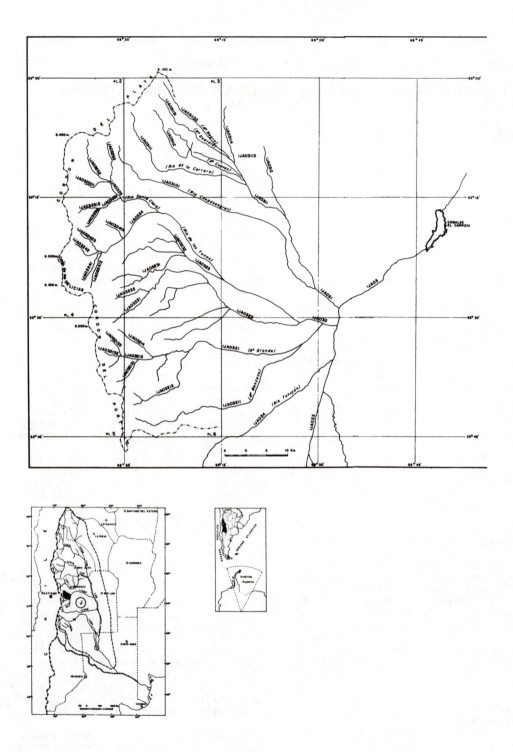

Figure 1. Location map. Scales in kilometers.

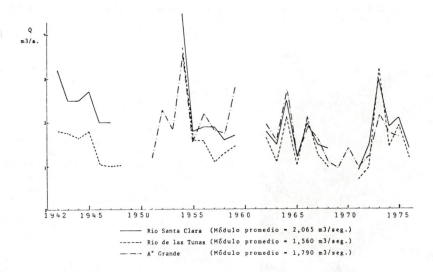

Q
m3/s.

| 1942 | 1945 | 1950 | 1955 | 1960 | 1965 | 1970 | 1975 |

——— Rio Santa Clara (Módulo promedio = 2,065 m3/seg.)
------ Rio de las Tunas (Módulo promedio = 1,560 m3/seg.)
—·—·· A° Grande (Módulo promedio = 1,790 m3/seg.)

Figure 2. Mean annual discharge of three streams in the study area (1942-1976).
Data provided by the Dirección General de Irrigación.
Note: "módulo promedio" = mean discharge.

discontinuous record (1942-1976).
25 yr.-mean annual discharge: 2.065
m³/sec.
Maximum mean annual discharge:
4.496 m³/sec (1953-1954)
Minimum mean annual discharge:
0.877 m³/sec (1970-1971)
(b) Río de las Tunas (2150 m a.s.l.; lat.
33°18'S; long.69°28'W): 25 years of
discontinuous record (1942-1976).
25 yr.-mean annual discharge:
1.560 m³/sec
Maximum mean annual discharge:
3.701 m³/sec (1953-1954)
Mimimum mean annual discharge:
0.680 m³/sec (1970-1971)
(c) Arroyo Grande (1700 m a.s.l.; lat.
33°36'S; long.69°23'W): 23 years of
discontinuous record (1950-1975).
23 yr.-mean annual discharge:
1.790 m³/sec
Maximum mean annual discharge:
3.554 m³/sec (1953-1954)
Mimimum mean annual discharge:
0.950 m³/sec (1968-1969)

METHODOLOGY
Photointerpretation was done on 1:50,000
scale stereopairs of March-April 1963. Due
to the lack of suitable cartography enlargem-
ents of Landsat imagery were used as base
maps. The drainage network and water-divide
position were mapped on them. Later, these
drainage maps were enlarged to 1:50,000
scale and the observed ice bodies and
perennial snow fields were superimposed on

them. Elevation data were obtained (1)from
Instituto Geográfico Militar topographic
sheets at 1:50,000 and 1:100,000 scales,
and (2) directly in the field with a Pauli
althimeter.
Due to the reconnaissance characteristics
of the cartography errors up to $\pm$ 100 m
may be expected in the altitudinal data.
Several field determinations have been
performed to establish the scale of the
base maps. Error has been found to be
always in excess and smaller than + 8 %.
Photointerpretation was extended over an
area of 2423 Km², approximately, from the
summits down to the 1500 m-contour line.
The glacierized area is 144.03 Km² (6 % of
the total area of the basin). Surface data
were obtained with polar planimeter. The
smallest snow and ice bodies were measured
with a millimetric grid instead, to avoid
the large planimeter error.
The drainage network has been codified
following the recommendations of the TTS
for WGI (Müller and others 1978). All the
observed ice bodies were numbered from the
mouth or outlet of the basin, proceeding
clockwise upstream. The obtained values
have been presented in Fortran data sheets.

CLASSIFICATION OF THE ICE AND SNOW BODIES
The photointerpretation techniques have
enabled us to recognize the following ice-
body types:
(1) Uncovered ice
It includes perennial ice masses, free of
debris cover. Morphologically, they may be

383

divided into valley glaciers, mountain glaciers, glaciarettes and snow-fields (Figures 3 and 4).

Figure 3. De la Carrera Glacier, Cordón del Plata. Valley-glacier with uncovered ice (H) in its sources, and in the terminal portion of the tongue, debris-covered ice in facies of: ice-cored moraine (M), thermokarst (T), active rock glacier (A) and inactive rock glacier (I). Aerial photograph of 1963.

The ice exposed in glacier tongues is found only above 4500 m a.s.l. It has not been possible to establish firn line position due to the lack of suitable contrast in the whitish tones of the stereopairs. The lowest perennial snow patches were observed (January 1982) in the Cordón del Portillo above 3775 m a.s.l., and in the Arroyo de la Carrera, above 4500 m a.s.l. (February 1982).

(2) Covered ice

The existence of a debris cover on the glacier surface is a very common feature in the Central Andes of Argentina and Chile. The angular clasts provided by mechanical (congelifraction) weathering are transported along the valley slopes down to the glacier surface.

It is possible to distinguish certain facies in relation to altitude and debris-cover thickness (Corte 1976a, 1976b, 1978; Corte and Espizúa 1978). It may happen that all or just some of the facies are present in a single glacial system. The interfacies limits are transitional and normally it is difficult to set up a definite line between them. We must accept that any delimitations on these ice bodies are partly subjective.

2.1. Ice-cored moraines

Under the term "ice-cored moraines", we include morainic deposits which appear marginally to small glaciers near glacierization limit (Østrem 1964, 1970, 1971) and those which are associated with valley glaciers at lower altitudes (Corte & Espizúa 1981). In both cases, a gentle slope and lack of a definite flow pattern are remarkable characteristics (Figures 3 and 4). There are other bodies in which these features are rather uncertain and may be

Figure 4. Arroyo Negro Glacier, Cordón del Plata. Valley-glacier with uncovered ice in its sources (H), and in the terminal portion of the tongue debris-covered ice in facies of: ice-cored moraines (M), thermokarst (T) and active rock glacier (A). Aerial photograph of 1963.

considered as transitional between ice-cored moraines and rock glaciers. According to Barsch (1977), the transition between glacier and rock-glacier takes place at the snout of the glacier when the morainic-debris thickness is large enough (20 – 30 m) and permafrost conditions are established. The ice-bodies of this region have been classified as one or another type in terms of the dominant characteristics. In the study area, the ice-cored moraines are located above 4300 m a.s.l.

2.2. Thermokarst

Thermokarst is considered an ablation facies which develops on ice with a thin debris-cover (Popov 1956; Corte 1960; Kachurin 1962; Clayton 1964; Aleshinskaya and others 1972).
It is characterized by circular hollows up to 200 m in diameter and asymmetrical slopes, with a gentler sun-facing side (Figure 5). Thermokarst facies are very common in the Central Andes of Argentina and in the study area they are found above 4000 m a.s.l. (La Carrera Glacier, Figures 3 and 5).

2.3. Rock glacier

Rock glaciers are lobe- or tongue like, generally slowly moving debris and ice bodies.
Genetically, they may be divided in:

(a) ice-cored rock glaciers: debris-covered ice bodies, related to the terminal portion of a glacial system. They show well-defined flow structures, end in an abrupt front and lack vegetation (Figures 6 and 7). Following Barsch (1977), their movement is due to plastic deformation of basal layers related to tensional stress caused by debris + ice weight and the surficial boulder layer.
We include in this type the "cirque floor rock glaciers" (Outcalt and Benedict 1965), "ice-cored rock glaciers" (Potter 1972), "debris-covered glaciers" (Corte 1976a, 1976b) and "moraine rock glaciers" (Johnson 1978). In the studied area, they are found above 3700 m a.s.l. (La Carrera Glacier, Cordón del Plata, Figure 6; Manantiales Glacier, Cordón del Portillo, Figure 7).

(b) Talus rock glaciers (with interstitial ice): these glaciers are not related to glacial systems but are the result of debris and snow sliding and avalanches in valley slopes (Wahrhaftig & Cox 1959; Outcalt & Benedict 1965; Potter 1972; Corte 1976a; Johnson 1978). The talus rock-glaciers form in periglacial environment with abundant debris supply and permafrost conditions. In general, they are small bodies (300-1000 m long), with lobed or tongue-like morphology and flow alignment (longitudinal and transversal archs and troughs) due to the strong slopes. Several coalescent bodies may appear on the same slope if the avalanche troughs are close enough to each other (Figure 8).
These glaciers show two strikingly differentiated layers: the upper one, thin, composed of large boulders and a steep front; the lower one, thicker, with finer debris and a gentler (33°-38°) terminus (Figure 9). In the study

Figure 5. De La Carrera Glacier, Cordón del Plata. Debris-covered ice in thermokarst facies. The debris which covers the ice core shows a 30-40cm thickness. The surface is highly irregular, with circular depressions. Elevation: 4050 m a.s.l. Photograph taken looking NE, by C.Aguado (February 1982).

area, these glaciers have been observed above 3400 m a.s.l., for example Rock Gla-cier N° 10, Cordón del Portillo Sub-basin, (Figure 9).

Figure 6. De la Carrera Glacier, Cordón del Plata. Uncovered ice can be seen in the sour-ces, thermokarst facies in the central part of the glacier and active- and inactive rock-glacier facies at the snout. Note the smooth, gentle morphology and the low-gradient front that exhibits the inactive rock glacier (I) (left part of the photograph) in relation with the active rock glacier (A). The darker patches over the inactive rock-glacier surface are due to vegetation. Photograph taken looking N by C.Aguado (February 1982).

Figure 7. Manantiales Glacier, Cordón del Portillo. Active rock-glacier with ice core developed in the terminal part of the glacier tongue. See the flow structures on its surface, abrupt front, lack of vegetation. Elevation: 3650 m a.s.l. The glacier ends over morainic deposits. Photograph taken looking N by R. Bottero, December 1981.

Figure 8. Cordón del Portillo. Talus rock-glaciers, active and coalescent, with flow alignments on their surfaces. The front is very steep. Elevation: 4000 m a.s.l. Photograph taken looking E. December 1981.

Figure 9. Cordón del Portillo.
Active talus rock-glacier.
In the profile, two layers
may be seen: an upper one,
thin, with large boulders and
very steep front; a lower one,
with finer debris and gentler
slope. Gradient: 37°. Elevation: 3430 m a.s.l. Photograph
taken looking NW. December
1981.

According to the activity of the tongue,
we have classified them in:
(a) Active rock glacier: with steep front
    and well-defined flow structures
    (Figures 6, 7, 8, 9 and 10)
(b) Inactive rock glacier: this covered-
    ice facies is the outermost one to the
    glacial system, where the ice, if
    still present, has been covered by a
    thick debris layer.
    They appear as collapsed ice bodies
    with a gentle front and vegetation on
    their snouts. They lack hydrological
    significance (Figure N° 6). In this
    area, they are located above 3600 m
    a.s.l.
In the present inventory, it has not been
possible yet to distinguish between
"fossil" and "inactive" rock glaciers.
There are just a few examples of inactive
talus rock-glaciers; they have been recognized by their collapsed morphology and
their smooth, gentle terminus slope.

(3) Pleistocene moraines
This term is used to embrace all glacial
deposits which are presently disconnected
from the ice, due to glacier recession.
They have no hydrological significance but
they have been included in the Inventory
because of their palaeoclimatic importance.
During the Pleistocene, this region of the
Andean Cordillera was heavily glaciated.
The glacial features are clearly visible
in the higher mountains but they disappear
towards the lower areas.
Morainic deposits have been identified in
the indicated sites (Table 1). In this
table, we have included the approximate
lowest elevation of the moraine and the
distance away from the present glaciers,
measured on the maps or aerial photographs.
Based on morphological criteria, these
moraines are generally considered of
Pleistocene age (Figure 11), but those
deposits closer to the glaciers may be,

in fact, of Holocene age. No radiocarbon
dating has been performed yet on these
deposits.

Figure 10. Cordón del Portillo. Active
talus rock-glaciers, located
below avalanche troughs on the
valley sides. Elevation: 3625
m a.s.l. Photograph taken looking NW. December 1981.

Figure 11. Arroyo Grande, Cordón del Portillo. Pleistocenic lateral moraine
along the Arroyo Grande Valley.
Photo taken looking WNW. Dec.1981

387

TABLE 1

LOCATION, ALTITUDE AND DISTANCE TO PRESENT GLACIERS OF PLEISTOCENE MORAINES

Site	Approximate elevation (m a.s.l.)	Distance to present glaciers (m)
Arroyo Negro	2450	14,200
Arroyo Guevara	2400	14,600
Arroyo Cuevas	2500	17,600
Quebrada Norte	2850	12,000
Quebrada del Azufre	2850	9,600
Quebrada Cortaderas	2850	8,500
Río de las Tunas	3250	10,300
Arroyo Barraquero	3500	7,700
Arroyo Arenales	3500	10,000
Arroyo Grande	2200	12,500
Arroyo Guindo	2200 (?)	16,000

DATA ORGANIZATION

For the compilation of this inventory, suggestions and recommendations by Müller and others (1977, 1978) and Espizúa (1980) have been considered.

In the data sheets, the following information has been included:

1. Identification and glacier number
2. Glacier name
3. Glacier geographical co-ordinates
4. Number of drainage basins
5. Topographic sheets
6. Aerial photographs
7. Accuracy
8. Exposed area
9. Ice-cored moraines area
10. Thermokarst area
11. Active rock glacier area
12. Inactive rock glacier area
13. Glacier orientation
14. Facies orientation
15. Mean total glacier length
16. Mean uncovered glacier width
17. Maximum total glacier length
18. Maximum uncovered glacier length
19. Maximum glacier length
20. Minimum glacier length
21. Minimum exposed-area elevation
22. Morphological classification

RESULTS

The percentage distribution of uncovered and covered ice according to its orientation is represented in the azimuthal histograms of Figure 12. The individual ice bodies included in this Inventory are represented in the maps of Figures 13, 14, 15 and 16.

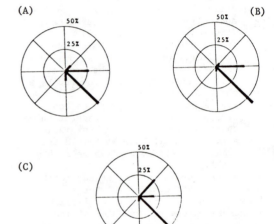
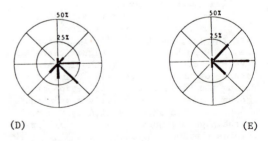

Figure 12. Percentage distribution of uncovered and covered-ice according to its orientation. (A): uncovered ice (100% = 56.61 Km^2); (B): ice-cored moraine (100% = 16.09 Km^2); (C): thermokarst (100% = 11.34 Km^2); (D): active rock glacier (100% = 49.95 Km^2); (E): inactive rock glacier (100% = 7.49 Km^2)

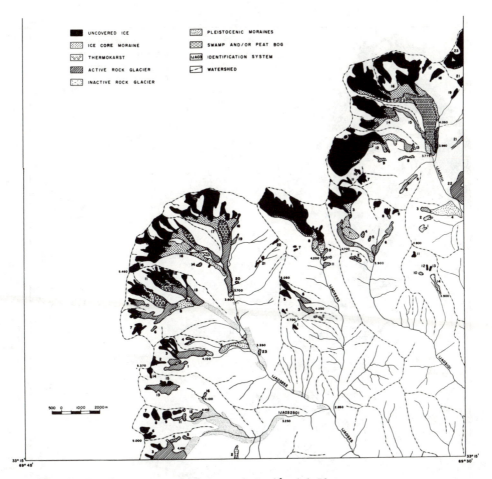

Figure 13. Glacier inventory map of Central Cordón del Plata

(1) Measured areas have been organized in
    Table 2 (in Km^2).

TABLE 2
GLACIER AREAS (Km^2)

Basin	UI	Mo	Th	aRG	iRG	Total area
Arroyo Negro	2.63	0.41	0.68	1.24	–	4.96
Arroyo Guevara	5.82	0.78	–	1.18	–	7.78
Arroyo Cuevas	6.27	2.28	0.10	3.25	0.43	12.33
A° de la Carrera	10.20	3.11	2.35	3.01	1.13	19.80
A° Chupasangral	1.53	0.41	0.02	1.17	0.26	3.39
A° Santa Clara	12.21	4.32	1.33	13.30	2.46	33.62
Río de las Tunas	5.12	0.42	1.46	3.71	0.38	11.09
A° Barraquero	2.02	0.21	0.22	2.14	–	4.59
A° Grande	5.21	2.00	2.12	13.45	1.69	24.27
A° Pircas	3.95	1.92	2.14	4.46	1.15	13.62
A° Manzano	2.38	0.69	1.54	3.74	0.03	8.38
Total area	57.34	16.55	11.96	50.65	7.53	144.03
Percentage of glacierized area	40	12	8	35	5	

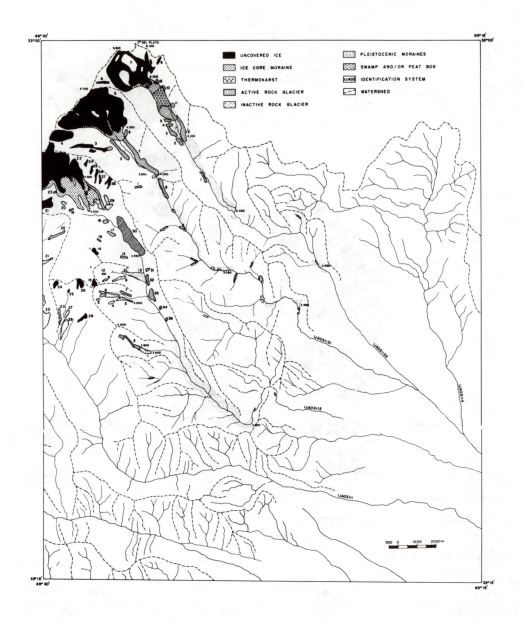

Figure 14. Glacier inventory map of Northern Cordón del Plata area.

References to Table 2
UI: uncovered ice
Mo: ice-cored moraines
Th: thermokarst facies
aRG: active rock glaciers
iRG: inactive rock glaciers

The total glacierized area (144.03 Km^2) is divided into 40% of uncovered ice and 60% of debris-covered glaciers. The basins with larger uncovered-ice glacierization are Arroyo Santa Clara and Arroyo de la Carrera, whereas those for debris-covered ice are Arroyo Santa Clara and Arroyo Grande. Uncovered ice is dominant in the Cordón del Plata within southeast-oriented watersheds.

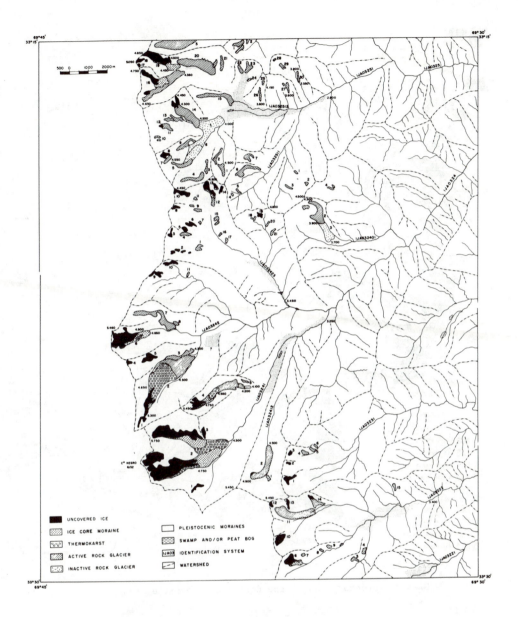

Figure 15. Glacier inventory map of the Cordón de las Delicias area.

(2) Valley glaciers with an area larger
    than 4 Km²

In this Inventory, a total number of 387
ice and snow bodies have been identified
so far, but most of them are of very small
size.
However, the ten larger glaciers exceed
4 Km² in area. Table 3 presents the list
of such glaciers and their location within
the study area.

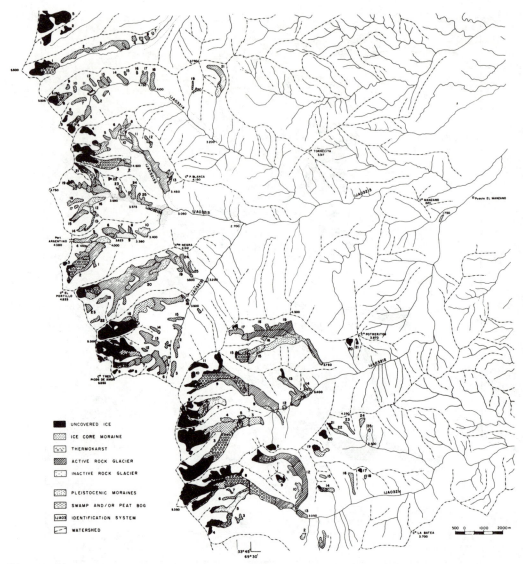

Figure 16   Glacier inventory map of the Cordón del Portillo area.

TABLE 3
VALLEY GLACIERS LARGER THAN 4 KM^2

Basin identification	Glacier number	Total Area	Uncovered-ice area
Arroyo Negro – IJA031132	7	4.20	2.61
Arroyo Guevara – IJA031131	5	5.82	5.10
Arroyo Cuevas – IJA03112	23	7.88	5.30
A° de la Carrera – IJA03111	12	4.45	3.51
A° de la Carrera – IJA03111	16	8.64	4.31
A° Santa Clara – IJA03252	16	4.56	2.09
A° Santa Clara – IJA03252	17	4.65	2.11
Río de las Tunas – IJA03241	2	5.01	2.82
Arroyo Arenales – IJA032131	20	4.19	0.25
Arroyo Pircas – IJA03212	3	4.16	2.03

(3) Orientation
The SE- and E- oriented uncovered ice represents 78% of the total uncovered-ice area (see Figure 12).
91% of the ice-cored moraine area is also oriented to the SE and E. For the thermokarst facies, SE orientation is prevailing with 50.5%, followed by NE and E orientations.
For the active rock glaciers, orientation is scattered in several quadrants, being the SE direction predominant with 31.5%. Inactive rock glaciers have a preferred orientation towards the E quadrant, followed by the NE direction.

(4) Elevation
The exposed ice is developed only above 4500 m a.s.l. The ice-cored moraines appear at somewhat lower altitudes, around 4300 m a.s.l. Thermokarst facies extend well below, till 4000 m a.s.l.
The ice-cored, active rock glaciers are found above 3700 m a.s.l. and the talus rock glaciers, above 3400 m a.s.l. The inactive rock glaciers have lowest elevation at 3350 m a.s.l.

(5) Relationship between basins, glacierized areas and mean annual discharge
In three of the studied basins, discharge data has been obtained for 25, non-consecutive years (see "Geography" section). The relationship between areas and discharge are presented in Table 4.

Arroyo Santa Clara has the largest mean annual discharge, corresponding, as expected, to the larger basin area and larger glacierized area. However, the proposed ratio (Discharge x % Area -glacierized- / Total Area; see Table 4), suggests that Arroyo Grande, with the largest percentage of glacierized area (11.7 % of the total area) has also the largest mean annual discharge per unit area per percentage of glacierized area (i.e., 10.02×10^{-2} $m^3.Km^2.sec^{-1}$).
However, this ratio and the presented results should be taken only as an indicative parameter. It is not possible yet, with the available information, to establish a direct, absolute relationship between discharge, basin area and percentage of glacierized area, because other significant parameters such as precipitation régime, basin orientation, rock and soils permeability, mean elevation, mean slope, etc., should be taken into consideration. Nevertheless, the proposed ratio should be investigated in further detail for these and other basins. It could perhaps provide a good indicator of glacier meltwater contribution to superficial runoff in these mountaineous regions.

(6) Comparison of the glacierization of neighbouring basins
For the purpose of this comparison, only those basins which have highly-similar characteristics have been considered. Thus, in this first approach, eastern portions of the N-S trending Cordillera, in the Río Mendoza and Río Tunuyán basins, have been

TABLE 4
RELATIONSHIP BETWEEN BASINS, GLACIERIZED AREAS AND MEAN ANNUAL DISCHARGE

Basin	Total Basin Area Km^2	Glacierized area Km^2	%	Mean annual discharge (m^3/sec)	Discharge $\dfrac{1}{\text{Area} \times \%A}$
Arroyo Santa Clara	296	33.62	11.4	2.065	7.95×10^{-2}
Río de las Tunas	223	11.09	5.0	1.560	3.50×10^{-2}
Arroyo Grande	209	24.47	11.7	1.790	10.02×10^{-2}

taken for this analysis. The high elevations of these ranges, above 5000 m a.s.l., favours a more extended glacierization.

Data in Table 5 suggest that glacierized area increases southwards and covered-ice is predominant on uncovered-ice northwards.

TABLE 5
COMPARISON OF THE GLACIERIZATION BETWEEN NEIGHBOURING BASINS

Basin	Latitude S	Area (Km^2)	Glacierized area	%UI	%CI
Río Mendoza Basins (Eastern Cordillera del Tigre and Cordón del Plata)	32°15'-33°00'	2526	85.15 Km^2	30	70
Río Tunuyán Basins (Eastern Cordón del Plata and Cordón del Portillo)	33°00'-33°45'	2423	144.03 Km^2	40	60

In Table 6, data concerning the Río Mendoza Basin (Corte & Espizúa 1981) and the Río Atuel Basin (Cobos 1980) are presented.

In both Table 5 and Table 6, % UI and % CI mean percentage of uncovered-ice area and percentage of covered-ice area, respectively.

TABLE 6

COMPARISON OF THE GLACIERIZATION BETWEEN RIO MENDOZA AND RIO ATUEL BASINS

Basin	Latitude S	Total area	Glacierized area	% UI	% CI
Río Mendoza Basin	32°15'-33°15'	6311 Km^2	646.98 Km^2	47	53
Río Atuel Basin	34°26'-35°17'	2965 Km^2	185.93 Km^2	80	20

In the southernmost basin of those presented here, the Río Atuel Basin (Cobos 1980), uncovered ice is by far the most important glaciological component of the glacial systems, whereas the Río Mendoza Basin (Corte & Espizúa 1981) shows a relative equilibrium between covered- and uncovered ice.

In summary, uncovered ice is clearly predominant southwards where, although altitude is lower, there are larger precipitations and firn line is presently at lower altitudes.

Finally, if the distribution of uncovered- and covered ice is a result, mainly, of a climatic pattern, distribution of the glacigenic sediments related to the aforementioned facies should also be, at least partially, climatically controlled, both in the present glaciological systems as in the Pleistocene glacial epochs.

ACKNOWLEDGEMENTS
The author is greatly indebted to the following:
To Lic.Carlos Aguado and Técn. Jorge Suarez and Víctor H. Videla, for the significant information obtained during field work and the permanent collaboration.
To Técn. Rafael Bottero for his field work participation and maps and figures design.
To Agr. Luis Lenzano for the calculation of base map error.
To Dr Arturo Corte for the reviewing of the manuscript and his valuable suggestions.
To all colleagues and members of IANIGLA.
To Dirección General de Fabricaciones Militares for the aerial photographs.
To the Centro de Fotogrametría y Catastro of San Juan National University and the Dirección de Catastro of Mendoza Province for the collaboration in the base map preparation.
To Dr Calvin J Heusser, with special thanks, for his important suggestions and indications during field work in Cordón del Portillo.

REFERENCES
Aleshinskaya, Z.V.; Bondarev, L.G. & Gorbunov, A.P.(1972). Periglacial phenomena and some paleogeographical problems in Central Tian-Shan. Biul.Perygl., 21: 5-13
Barsch, D. (1977). Nature and importance of mass-wasting by rock glaciers in Alpine permafrost environments. Earth Surface Processes, 2: 231-245.
Clayton, L. (1964). Karst topography on stagnant glaciers. Jour.Glac., 5(37):107-112.
Cobos, D. (1980). Inventario de glaciares de la cuenca del Río Atuel (Mendoza). IANIGLA-CONICET, unpubl.rept., 62 pp.
Corte, A. (1976a). The hydrological significance of rock glaciers. Jour.Glaciology, 17: 157-158.
Corte, A. (1976b). Rock glaciers. Biul. Peryglac., 26: 157-197.
Corte, A. (1960). Experimental formation of sorted patterns in gravel overlying a melting ice surface. Biul.Perygl., 8: 65-72.
Corte, A. (1978). Rock glaciers as permafrost bodies with a debris cover as an active layer. A hydrological approach. Andes of Mendoza, Argentina. Third Intern. Permafrost Conference, Edmonton (Canada), p. 350-355.
Corte, A. & Espizúa, L. (1978). Glacier inventory of the Río Mendoza Basin. Intern. Symp.on World Glacier Inventory, Aletsch Center, Switzerland, Abstr., p.8-9.
Corte, A. & Espizúa, L. (1981). Inventario de glaciares de la cuenca del Río Mendoza. IANIGLA-CONICET.
Espizúa, L. (1980). Guía para la compilación y ordenamiento de datos para el inventario de glaciares de la República Argentina: glaciares descubiertos, morenas con núcleo de hielo y glaciares de escombros. IANIGLA-CONICET.
Johnson, P.G. (1978). Rock glacier types and their drainage systems, Grizzly Creek, Yukon Territory. Canadian J of Earth Sci., 15 (19): 1496-1507.
Johnson, P.G. (1980). Glacier-rock glacier transition in the southwest Yukon Territory, Canada.

Kachurin, S.F. (1962). Thermokarst within the territory of the U.S.S.R. Biul. Peryglac., 11: 49-55.

Müller, F.; Caflisch, T. & Müller, G. (1977). Instructions for the compilation and assemblage of data for a world glacier inventory. ETH, Zürich, TTS-WGI, p. 2-19.

Müller, F.; Caflisch, T. & Müller, G. (1978). Instructions for the compilation and assemblage of data for a world glacier inventory: supplement with maps. ETH, Zürich, TTS for WGI, p. 1-7.

Østrem, G. (1964). Ice-cored moraines in Scandinavia. Geografiska Annaler, 46 (3): 282-337.

Østrem, G. (1971). Rock glaciers and ice-cored moraines: a reply to Barsch. Geografiska Annaler, 53 (3-4): 207-213.

Østrem, G. & Arnold, K. (1970). Ice-cored moraines in southern British Columbia and Alberta, Canada. Geografiska Annaler, 52 A (2): 120-128.

Outcalt, S.I. & Benedict, J.B. (1965). Photo-interpretation of two types of rock glaciers in the Río Colorado front range, U.S.A. Jour.Glaciology, 5(42): 849-856.

Polanski, J. (1963). Estratigrafía, neo-tectónica y geomorfología del Pleistoceno pedemontano entre los ríos Diamante y Mendoza. Rev.Asoc.Geol.Arg., 17 (3-4): 127-349.

Polanski, J. (1964). Descripción geológica de la Hoja 25a, Volcán San José, Provincia de Mendoza. Dirección Nacional de Geología y Minería, Buenos Aires, Boletín n° 98, 94 pp.

Popov, A.I. (1956). The thermokarst. Biul. Peryglac., 4: 319-330.

Potter, N. (1972). Ice-cored rock glacier, Galena Creek, Northern Absaroka Mountains, Wyoming. Geol.Soc.Amer.Bull., 83:3025-3058.

Wahrhaftig, C. & Cox, J. (1959). Rock glaciers in the Alaska Range. Geol.Soc.Amer. Bull., 70: 383-436.

Moraines and tills in southwest Scandinavian icesheet

F.GRUBE
Geologisches Landesamt Schleswig-Holstein, Kiel, Germany

The glacigenic sediments in the southwest region of the scandinavian quaternary continental glaciation were investigated with changing intensity and different methods. In the early phases of investigation petrographic studies of boulders as well as of the parent material in their homeland prevailed. The discovery of ice-scratched rocks in Rudersdorf/Berlin led to a marked progress in the quaternary research (Gripp 1975, Kaiser 1975). Bloc boulder enrichments and morphological walls are most important for the mapping of endmoraines. The periglacial ageing was understood and could be stratigraphically used to distinguish between weichselian young moraines and the old moraines of the Saalian (Gripp, 1924). In the present time the point of main effort in the investigation of tills and moraines is lying in the regional documentation and evaluation of the knowledge about glacigenic sediments. At the same time intensified attention is given to the genesis of the sediments of glaciers.

The regional group "Glacigenic deposits in Southwest Scandinavian ice sheet" of the INQUA Commission on Genesis and Lithology of Quaternary deposits is trying to correlate the results of research regarding the tills in the region of the southwest-scandinavian icescape, that is Norway, Sweden, Denmark, North-Germany, DDR, Poland and the Netherlands. It is an important concern of this conference to standardise and to propagate the working methods and technology.

A lot of valuable knowledge has been earned in single excavations and has to be compared. It is not easy to transmit these single dates on greater areas by geological mapping.

To characterize and to identify special stratigrafic and genetic tills all kinds of working methods should be used. Only a few technics have succeeded in daily usage and are applied by many active research workers of the Quaternary. The researches of single persons are in spite of their scientific value not helpful for correlations of various areas.

The fine gravel method, coming from the Netherlands (Zandstra 1978), is now used in northern Germany (Ehlers 1978), in the DDR, Poland and in Denmark and makes it possible to correlate the results in spite of the great areas between them, if the same gravel size is used (3-5 mm). Its easy practicability allows the work of trained assistants under the supervision of experienced scientists. One can differentiate quartz, crystalline, flint, chalk, Paleozoic limestone and other sediment gravels.

The stone countings and indicator boulder analyses presumes a solid knowledge of the bedrock in the homeland of the glacier and of the local geology. Therefore this method can only be employed by a few experts (Meyer 1970, Schlüter 1978, Schuddebeurs 1981). Only characteristic crystalline and sedimentary rocks, which occur very rarely in the homeland of the inlandglacier, are interpreted.

Concerning valley glaciers the usage of this method is hardly possible or even leads to mistakes. G. Richmond and E. Evensen (INQUA-CGLQD-Symposium 1981) pointed out that there are considerable differences in the boulder content of a valley glacier at Livingstone/Idaho, whose lateral-moraines contain on one hand boulders of granite and Paleozoic and on the other hand stones of Tertiary Volcanics. A systematic documentation of the boulder content in a valley glacier results in information about the areas where a fossil valley glacier had been, which often can be much larger than the areas of the current river valleys.

During the exaration and during the transport in the glacier the boulder is strained, broken, rounded and polished. By pressing the so-called "Kehl"boulders are formed, which are known in the archaeological literature as false artefacts as well (Regenhardt 1972 u. 1975).

Sediment-petrographic and geochemical basis datas of tills are very important for an exact definition. But the use of the common methods for the tills are very laborious because of the great amounts of test samples which are necessary and because glacier sediments are commonly far from beeing homogeneous. Practicable homogeneous test methods are missing to get comparable results for the main part of the tills, which consist of clay, silt and sand.

The pebble and boulder content of a till must be measured by a huge experiment. A method for the measurement of the coarser material in a plane of a quadratmeter will be developed. Considering the conventional grain size analysis of a till usually the pebble fractionation is not regarded. Quantity and quality of clay minerals, heavy minerals and geochemical dates depend upon weathering and diagenesis (Grube 1981). A summary of the publications advertised within the CGLD-regional group has been published by Grube (1981).

Resulting from researches in excavations and exposures the percentage of pebbles and boulders seems to be different in elsterian, saalian and weichselian tills. The investigation of heavy minerals is showing positive and negative results for the differentiation of tills. Fundamental sediment petrographical and geochemical researches were realised by Vortisch (1981), Valeton u. Khoo (1979), Höfle & Schlenker (1979), Haldorsen (1981) and Biermann (1980). This lets us hope for a systematical continuation of these researches. They will make it possible to differentiate the tills of the different great glaciers. The calcareous content was usually not investigated during many mechanical routine researches. But preliminary evaluations of the present Ca-analysis (Baermann 1981) show clear differences between the stratigraphic different tills. The differences of the Ca content is so eminent that the content of chalk and Paleozoic limestone can be used as distinctive mark in geological mapping of saalian tills.

Palaeontological methods are rarely used in the till investigation. The practicability of the palaeontological research in the till investigation is limited. But the reworked fossils of the Cambrian up to the Tertiary beds in the scandinavian and the baltic sea have been collected and studied since deceniums. The results have been published in many publications of boulder research (Hucke & Voigt 1967). Marine clays of Eemian interglacial age were reworked by younger glaciers. In Denmark they separate the saalian tills from the weichselians. These clays were investigated by micropalaentological methods. Sporomorphs in tills were evaluated in a similar way. Investigations of chalk foraminiferas are helpful to determine the drift direction of a glacier and the identification of the origin of the chalk (Stenestad, unpublished).

The future petrological researches of the till investigation will be concentrated on the grain size distribution, the content of pebbles and boulders and of clay minerals. The fine gravel analysis should belong to the fundamental investigation of glacial sediments. From the geochemical parameters the determination of the primary Ca content must not be neglected. If these basic datas are known, particular tillbeds like the Niendorfer Moraine of the middle saalian can be used as indicator horizons from Niedersachsen to Hamburg to Schleswig-Holstein up to Denmark.

All these researches should not neglect the extensive secondary alterations which take place after the sedimentation of a till. Chemical, pedological and biological processes change the character of a till. Especially the decalcination by leaching is very important. The chalk is dissolved by the groundwater or precipitated in form of single concretions, stripes or continous banks. Before taking a sample the grade of weathering of a till has to be examined for correlation purposes with unweathered material because the content of silica, chalk and iron changes depending on the degree of the weathering process. The cristallisation of new minerals is typical for diagenesis. The primary fabric and preconsolidation is disturbed above all by periglacial processes like ice wedges, kryoturbation, flows etc. A profound pedological research is necessary for all lithostratigraphical till researches in order to determine the depth of weathering. Glaciers transported praeglacial weathered material, for example fine gravelled boulder. The postglacial weathering can have been caused in an interglacial. In older tills the weathering of several interglacials and interstadials can have effected the glacial sediment. The influence of the groundwater is an important factor for the weathering and diagenesis of tills.

Tills are widely distributed bedrocks and are well known to the construction engineers as building grounds. In many routine researches the grain size, the natural water content and other parameters of soil mechanics of tills were studied, but without considering stratigraphy, stratification, weathering and genesis. Therefore these results are only of limited use for a systematical scientific analysis.

The diversity of tills comprises bloctills as well as clay-silt tills, which are used for the production of brickstones. If conventional laboratorial methods of soil mechanics, which were developed for well sorted sediments, can be used to characterize the geotecnical properties of glacial sediments, is currently tested by Baermann, Ehlers and Grube (1981). Even sampling of till material from drill and excavations is difficult because of the boulder and gravel-content. The squeezing out of the drill-kernels out of the capsules is a reason for many errors as well. The natural water content is almost ever measured in routine researches as well as in scientific ones. The temperature and the air humidity can have a great influence on the sample. The depth of the groundwater table should be known regarding the watersaturation point. The groundwater flow is influenced by fissures. Plastic reactions of the tills are caused by the content of the clay-mineral-fraction. Clay mineral aggregates greater than 2μ are not important for the plastic qualities of glacial material (Keller u. Smolczyk 1982).

The geotechnical qualities of the tills vary according to the petrochemical consistence and in the compaction by inland glaciers. The erratic blocks hamper all construction activities and the setting up of excavations. They delay in a lot of cases the progress of the constructions-especially tunneling, buildings on piles and benthonit walls. Fine sandtills tend to a fabric collaps, to a hydraulic soil subsidence and they cause landslips. On the contrary the till of the Elsterian is very hard and behaves like soft concrete.

The great amount of experiences which are at hand should be documented and interpreted in order to avoid expensive difficulties in excavations and during the construction. The refinement of the till stratigraphy, the improvement of soil-mechanical methods in laboratories and the testings of new geophysical methods will perfect the geotechnical valuation of the tillbeds in order to elevate the security of construction sites in moraines. The geological layering of the Saalian tillbeds is in large areas of Niedersachsen very thin with a thickness from 0,5 m up to more

than 3 m. The Weichselian covertill in the east of Schleswig-Holstein has very often a thickness of about 4 m. On the contrary there are indications that in single areas no till was sedimented by the glacier in the open landscape of northern Germany, for example in Schenefeld-Eggerstedt, as well as in some mountain regions like Norway (Holtedahl, 1975, Garnes & Bergersen 1980). Typical for mountain countries is the spotted lodging of tills in the driftshadow of a valley glacier at the junction of a tributary valley into the main valley.

In Northern Germany two genetic types of terrestrial tills are common due to the rarity of ablation tills. The ground-till (german: Grundmoräne) consists mainly of different material and is composed relativly uniform, depending upon the stratigraphy characterized by specific parameters like fine gravel content, content of chalk and grainsize distribution and is jointed by flat falling shear-planes and "Mohr"fissures. The formation of the soletill is extremly heterogeneous because of the assimilation of local material (Lokalmoräne). The soletill is always jointed and often marked by big units of strange material in various orders of magnitudes. Overridden tills are reshaped structurally. One facies of the soletill is similar to the groundtill but a bit coarser, the fabric of the longaxis of the boulder is mostly excellent. Another type of till is the sheartill. It consists of different materials like till, sand or silt, which is intensely sheared with slideways. The pushtill consists of intensely deformed various material with a thickness from 0,2 up to over 50 m. This sort of till is widespread in Northern Germany. The transported stonematerial is separated sharply from the undeformed underground. The question if the tectonical structures can be compared with the slideways of great landslides, has still to be researched. Flow tills are known from the marginal zone of the inland glaciers. In this area the famous end-moraines like the Hüttener Berge, the Dammer Berge, and Moens Klint were pushed up. Waterlain tills are very rarely in Northern Germany, they are only known from a local region in Eastern Holstein and in Münster (Westfalen). In excavations glacilacustrine layers are intercalated with tillbeds with a cover of groundtill, a soletill is missing.

Glacier sediments are forming a landscape. Great areas in the range of the Southwest Scandinavian glaciers, but also in the Northern Alpes are covered by a

covertill. The characteristic morphology
of moraines is marked by the glacier ice
and by the melt water. By compressing the
wall or endmoraines are modelled, which
often overlook the surrounding landscape
with many dekametres. Fields of drumlin-
hills are especially characteristic for
quaternary areas of ice formation. The
long axis of these drumlins are running
mainly parallel to the drift direction
of the glacier. But the transverse lying
drumlins (ger. Rogenmoraines) are more
numerous than we thought in earlier times.
The relation between praeglacial and syn-
genetic material of rocks is very much
changing in the structure of the drumlins.
Kames are very rare in the landscape of
moraines. Kettleholes result from melting
of buried dead ice or from erosion by
melting-water. They give together with
the drumlin hills the specific character
to the young morain. By deep exaration
long channels are generated, so called
glacielle or glacier-basins with changing
sizes from a little pond up to lakes as
for example the Plöner- or Chiemlake.

By ageing by periglacial processes the
hills are smoothing and the kettleholes
of Saalian age are filling up with glacial
sediments forming the old morain landscape.

This manuscript was translated by Elfi
and Anke Grube, and written by Mrs. Wein-
berg, to all my thanks are due.

References

BAERMANN, A., EHLERS, J. & GRUBE,F.
   (1981): Moränen-Geotechnik -
   Ingenieurgeologische Probleme im Grenz-
   bereich zwischen Locker- und Festgestei-
   nen.-
   DFG-Schwerpunktprogramm
   Bonn-Bad Godesberg
BIERMANN, M. (1980): Sedimentologie und
   Geochemie der Weichsel-Moräne von
   Timmerhorn. -
   unveröffentl. Diplomarbeit
   Geol.-Paläontol. Institut Univ. Hamburg
EHLERS, J. (1978): Feinkieszählungen nach
   der niederländischen Methode im Hamburger
   Raum. -
   Geschiebesammler 12: 47-64, Hamburg
FORCHHAMMER, G. (1835): Danmarks geogno-
   tiske Forhold. - Universitäts-Festschrift
   zum Reformationsfest,
   112 S., Kopenhagen
GARNES, K. & BERGERSEN, O. (1980): Wastage
   features of the inland ice sheet in
   central South Norway. -
   Boreas 9: 251-269, ISSN 0300-9483 Oslo
GRIPP, K. (1924): Über die äußerste Grenze
   der letzten Vereisung in Nordwestdeutsch-
   land. -
   Mitt. geograph. Ges. 36, Hamburg
GRIPP, K. (1975): 100 Jahre Untersuchungen
   über das Geschehen am Rande des nordeuropäischen Inlandeises:
   Eiszeitalter und Gegenwart, 26:
   31-73, Öhringen
GRUBE, F. (1979): Zur Morphogenese und
   Sedimentation im quartären Vereisungsgebiet Nordwestdeutschlands. - Verh.
   naturwiss. Ver. Hamburg, NF 23: 69-80,
   Hamburg
GRUBE, F. (1981): Postsedimentäre Veränd-
   erungen von Gletscherablagerungen. -
   Verh. naturwiss. Ver. Hamburg, 24:
   103-112, Hamburg
GRUBE, F. (1981): Glacigenic Deposits in
   the Southwest Parts of the Scandinavian
   Icesheet. -
   Verh. naturwiss. Ver. Hamburg, 24:
   37-42, Hamburg
HALDORSEN, S. (1981): Grain-size distrib-
   ution of subglacial till and its rela-
   tion to glacial crushing and abrasion. -
   Boreas 10: 91-105, ISSN 0300-9483 Oslo
HÖFLE, H.-C. & SCHLENKER, B. (1979): Das
   Pleistozän in der Kreidegrube Hemmoor
   bei Stade - ein Beitrag zur Quartär-
   stratigraphie im Elbe-Weser-Dreieck. -
   Geol. Jb. A, Hannover (in print)
HOLTEDAHL, H. (1975): The Geology of the
   Hardangerfjord, West Norway. - Norg.
   geol. undersoekelse 323: 87 p. Oslo
HUCKE, K. & VOIGT, W. (1967): Einführung
   in die Geschiebeforschung. - Nederlandse
   geol. Vereniging, 132 S. Oldenzaal
KAISER, K. (1975): Die Inlandeis-Theorie,
   seit 100 Jahren fester Bestandteil der
   Deutschen Quartärforschung: Eiszeitalter
   und Gegenwart, 26: 1-30, Öhringen
KELLER, P. & SMOLCZYK, U. (1982): Ver-
   fahren für die Bestimmung eines Parameters für den Verwitterungsgrad von
   Tongesteinen - Ingenieurgeologische Probleme im Grenzbereich zwischen Locker-
   und Festgesteinen. -
   DFG-Schwerpunktprogramm, Bonn-Bad Godesberg
MEIßNER, R. (1981): Neue Erkenntnisse zur
   Scherwellenseismik - Ingenieurgeologische
   Probleme im Grenzbereich zwischen Locker-
   und Festgesteinen. -
   DFG-Schwerpunktprogramm, Bonn-Bad Godesberg
MEYER, K. D. (1970): Zur Geschiebeführung
   des Ostfriesisch-Oldenburgischen Geest-
   rückens. - Abh. naturw. Ver. Bremen,
   37: 227-246, Bremen
REGENHARDT, H. (1972): Das Kehlgeschiebe,
   eine neue Form der Gletschererosion. -
   Z. Geomorph. N.F. 13: 26-31, Berlin
REGENHARDT, H. (1975): Konvergente Formen
   unter glazigenen Kehlgeschieben und

artifiziellen Nasensteinen Heidelberger
Industrien im Pleistozän Norddeutsch-
lands. -
Mitt. Geol. Inst. Univ. Hamburg 44:
531-536, Hamburg

SJÖRRING, S. & FREDERIKSEN, J. (1980):
Glacialstratigrafiske observationer i
de vestjyske bakkeoer. - Dansk geol.
Foren. Arsskrift for 1979: 63-77,
København

SCHLÜTER, G. (1978): Geschiebezählungen im
Altmoränengebiet von Schleswig-Holstein.-
Geschiebesammler 12: 3-12, Hamburg

SCHUDDEBEURS, A.P. (1981): Die Geschiebe
im Pleistozän der Niederlande. -
Geschiebesammler 13-15: 163-168 ff,
Hamburg

VALETON, I. & KHOO, F. (1979): Die Drenthe-
I-Moräne von Hamburg-Bahrenfeld. -
Verh. naturwiss. Ver. Hamburg 23:
19-38, Hamburg

VORTISCH, W. (1982): Clay mineralogical
studies of some tills in northern
Germany. - Geologica et Palaeontol.
15: 167-192, Marburg

ZANDSTRA, J.G. (1978): Einführung in die
Feinkiesanalyse. - Geschiebesammler 12:
21-38, Hamburg

Glacial processes and disturbance in vegetation richness

ALDO A.BRANDANI
Centro de Geologia de Costas, Mar del Plata & CONICET, Argentina

ABSTRACT

Researchers from several disciplines, such as geomorphologists and glaciologists, make use of present vegetational attributes to assess some related feature within the framework of their own investigations. The vegetation types growing on terminal and lateral moraines are, for example, often considered indicators of the time elapsed from the sedimentation period until the moment of observation. However, most of this biological information is qualitative in nature, frequently concentrated on few conspicuous aspects, therefore limiting its potential usefulness.

The classical theory of ecological succession underlies most of the uses made of the distribution and growth of plants. The theory states that the vegetation develops through different stages, each formed by a particular association of plant species, culminating in a regional, self-perpetuating, climax. Implicit in the theory are the following notions: (1) the intermediate and climax stages of a successional trend are unique for a particular region; (2) these stages are structurally different either in terms of plant species composition, habit of growth, or both; (3) the successional stages are not interchangeable in time, i.e. only one sequence is possible, and (4) any disturbance of some vegetational stage by environmental factors is seen as an external influence, often of catastrophic proportions.

This paper explores the advantages of an alternative approach based on the analysis and quantification of the causal relationships among environmental disturbances and vegetation structural dynamics.

Environmental disturbances are of continuous nature. They vary from chronic and normal to acute and catastrophic. Plant communities interact with those factors and this interaction generates the structural and functional dynamics classically referred to as successional patterns and trends. This paper analyses the nature of the causal environment-vegetation relationships. The disturbances studied here are those related to glacial and periglacial phenomena: cycles of freezing and thawing, formation of moraines, sedimentation on ice, ice-push along river and lake shores, ice storms and cryogenic movement of soils. The disturbances are identified by three quantitative, or potentially quantifiable, parameters: (1) magnitude or intensity; (2) frequency; and (3) predictability. Different biological responses would be expected if, for example, a glaciation is of local or regional scale, if a variation is of seasonal or historical frequency, and whether a disturbance is recurrent and, if so, how regularly repeated is in time. The plant attributes are considered at three different levels: (1) individual morphology; (2) single species population; and (3) ecosystem structure. Among the individual traits explored are the growth habit, phenological strategies, branching patterns, leaf morphology, leaf layering, shade tolerance, and seed size. The population attributes studied are age-classes distributions and survivorship patterns. At the ecosystem level the features studied are the distributions of abundance-dominance, within-habitat and between-habitat diversity of plant species, and gradients of species abundance.

The feasibility of establishing casual relationships among environmental disturbances and vegetational features is analyzed for different regions, with emphasis on the ecosystems of Northern Patagonian Andes. The identification of these relationships can better serve the double purpose of predicting the outcome of natural disturbances in terms of plant community dynamics, and inferring the occurrence of a particular disturbance regime from existing plant attributes in a given region.

INTRODUCTION

Vegetation is essentially dynamic and dynamics are usually initiated by natural disturbances. However, perturbations in plant communities are traditionally defined in terms of external events, often of catastrophic proportions such as fires, landslides, floods, and windstorms (White, 1979). This traditional view is rooted on the classical theory of ecological succession (Clements, 1916 in Golley, 1977; Odum, 1971). Central to this theory is the paradigm of the climax: a self-reproducing, terminal community whose species, seen as the best competitors in a regional flora, are under the control of regional climate. The climax, if not existing already, is reached through the replacement of some plant species by others, starting with the colonization of bare substrate in the most general case. Succession is understood as a sequence of intermediate-seral stages each featuring a characteristic species or group of them, dominating the landscape during a certain period before giving place to later colonizers.

Several concepts are implicit in the succession paradigm. First, it is assumed that for every region there is a given sequence of seral stages leading to a typical and unique climax. At most, succession can be slowed down or continually interrupted by environmental perturbations but the potential climax always exists and will eventually be reached in the long run. It follows that, for a regional climax to be expressed, long periods of "stable" climatic conditions are needed. Second, for every sequential stage- seral and climax- there is a typical species or, more generally, an optimal set of morphological and physiological attributes dominating the landscape. Finally, the replacement in time of a set of species- or attributes- by another does not happen in an haphazard manner. Rather, the successional trend is an orderly and unidirectional, therefore predictable, sequence. In terrestrial successions for example, more shade tolerant species of plants will always replace shade intolerant

ones, and never the other way round. It is widely assumed that climax species are the most shade tolerant, hence the most successful competitors for light, which ensures their persistence in a regional flora (Odum, 1971; Bazzaz, 1979). Even more generally, and always under the traditional belief, it is supposed that to reach a climax community, "inferior" taxa have to be replaced by more "complex" ones. In regions where broadleaf, deciduous, temperate forests are expected as climax (such as Eastern North America) the colonization of unvegetated sites supposedly begins with an ephemeral stage of cryptogams and weedy annual herbs, later joined by perennial herbs and shrubs. Afterwards, saplings of the most invasive trees become installed and from then on a much more protracted phase of succession begins in which canopy closure and the demise of heliophile species are the dominant processes (Wells, 1976).

It is not my intention to present here an extensive account of the theory of succession since detailed reviews can be found elsewhere (Odum, 1971; Drury and Nisbet, 1973; Pickett, 1976). Instead, I would like to discuss some of the applications made by Glaciology, Geomorphology and Palynology of the central postulates of succession which, although abandoned already by the majority of ecologist, are still at the base of much environmental interpretation of landscape dynamics and evolution. Afterwards, I will propose an alternative approach to the study and interpretation of natural community dynamics. Finally, I might support predictive (sometimes quantitative) landscape analysis, multidisciplinary and integrated environmental studies.

At the core of the classical theory of succession lies the contrast between allogenic and autogenic vegetation change (Odum, 1971). Allogenesis is community change driven by environmental change, while autogenesis is community change caused by internal biological activity. This central postulate is critical for the entire theory since, it should be recalled, it was modeled after neo-lamarkian evolution: biological communities were perceived as supra-individual organisms affected in their development by environmental stresses that complicate their form and function (Clements, 1916).

While autogenic development needs the definition of stable environments, it is within such an environment where development toward a terminal climax can happen; environmental variability drives allogenic change and prevents autogenesis.

Palynology has made extensive use of allogenic interpretations of vegetation development, a classical example being the gradual and directional climate modifications inferred from changes in the fossil pollen records reflecting variations in the distribution of plants.

On the other hand, Glaciology and Geomorphology interpret supposedly autogenic changes (such as post-glacial "pond-filling", plant colonization of moraines, deglaciated areas and alluvial deposits) in order to estimate the total elapsed time from the moment the substrate became available to plants. Usually, these estimations involve the aging of plant individuals, especially trees, and the maximum age observed is then considered the minimum elapsed time. Moreover, these calculations are often corrected, by adding up a certain number of years to the minimum time, to account for the time "succession needs" to reach a given stage of plant community development.

Confronted with a growing body of evidence that plant communities usually do not accommodate satisfactorily to the classical theory of succession, the ecologists developed, still within the realm of the theory, a complex set of concepts and terminology to describe vegetation dynamics. It was found that several different communities persisted in a single region; this was called poly-climax. A persistent, non-climax community was named a disclimax; progressively, ersatz climax began to be seen as more common than true regional ones, the last transformed in an almost ideal, seldom attained, stage of vegetation development. In tropical forests, for example, no single climax with one or a few dominant trees exists. Instead, the forest is a mosaic where several hundreds of tree species thrive in different patches, each patch consisting of a particular guild of species. In addition, primary successions (those that, beginning with bare substrate end up producing climax communities) were complemented with secondary, cyclical, recurrent, and reverse successions, to name just a few. The last one, in particular, describes the replacement of a forest community that ends with a herbaceous "climax". Observing the entire theoretical development from a historical perspective, one is reminded of the complicated planetary orbits Ptolemaic astronomers devised in order to explain the erratic movements of celestial bodies in a geocentric universe.

The major drawback of the succession paradigm is its search for first causes of natural systems in either the biotic or abiotic components. This artificial community-environment duality does not, in fact, exist in natural ecosystems (White, 1979). Landscapes may be more appropriately described as systems where evolution resulted in mutual interrelationships among plants and disturbance patterns. This has been particularly well shown for tropical lowland forests. There, most tree species are adapted to environmental conditions existing in gaps formed by the fall of gap-recolonizing canopy trees, that is, gaps tend to breed gaps (Hartshorn, 1978; Orians, in press).

Although the classical theory of succession may induce faulty, misleading or even erroneous interpretations of vegetation dynamics, no new paradigm has entirely replaced it. In what follows I will analyze the mutual interdependence of natural disturbance and vegetation as a causal factor for local and regional flora.

THE DISTURBANCE CONTINUUM

Change of environmental factors is essentially gradual, ranging from relatively minor to major events. Only in the last cases environmental change is perceived as a "perturbation". Wind, for example, causes changes from simple pruning of tree branches to extensive damage by hurricanes (Bormann and Likens, 1979).

No matter its type or ecosystem it affects, every disturbance may be described in terms of three independent parameters: magnitude, frequency and predictability.

For most terrestrial disturbances, intensity is usually correlated to the total area affected, to its duration, or to both. The disturbance frequency is measured as its number of occurrences per unit time. It follows that similar disturbance types may be of different magnitudes and frequencies. Variations of temperature, for example, may be daily, seasonal, secular or commensurate with geological time. In the sense used here, a natural disturbance is any change of environmental conditions that may affect a biological community. Consequently, predictability is defined in a dynamic sense: it is the inverse of the variation around the most probable value of recurrence for any given disturbance. In the above example, daily and seasonal temperature changes would have the highest predictability while annual or secular may be more variable.

Some disturbances are initiated or promoted by biotic components of a community. We already mentioned that successful development of lowland tropical trees is highly dependent upon gap formation, in turn colonized by wind throw prone individual species. This biological facilitation effect has recently been demonstrated as a major factor also for temperate forests, particularly beech-maple (Brever and Merrit, 1978) and balsam-fir (Sprugel and Bormann, 1981) communities.

Although for any given region several different disturbance types are possible, I will concentrate my discussion on those related to glacial and periglacial phenomena acting in montainous landscapes.

ICE PUSH: the destructive effect of ice-push has been noted for the scouring of riversides and lake shores (Lindsey and others, 1961; Raup, 1975). This effect can be similar to recurrent flooding in the sense that creates (and maintains) "longitudinal" gaps along streams to which riparianvegetation is adapted. This disturbance produces a recurrent destruction of the vegetation cover and the consequent enrichment of the ice with organic matter. It also causes in situ deposition of part of the plant material overriden by the moving ice. The type of plant regrowth on the till will depend on the material laid down, its richness in organic matter and whether there was already some vegetation growing on a mantle of ablation debris upon the moving ice as described for supraglacial moraines (Rabassa and others, 1981).

CRYOGENIC MOVEMENT OF SOILS: freeze-thaw cycles and the related soil movement especially in mountain slopes, cause the opening up of new surfaces, destroy plant root systems and whole individuals, and provoke changes in the water table in arctic and alpine tundra (White, 1979). Cryogenic movement of soils shows both diurnal and seasonal variations and can also be a consequence of more lasting and intense regional glaciation dynamics.

MOVING AND MELTING ICE: ice bodies carrying ablation debris are of considerable interest for plant colonization studies because, while being remarkably persistent in time, the substrate is under the chronic disturbance of ice melting and debris reaccommodation to which particular plant communities are adapted (Rabassa and others, 1981). Whether the ice body is stagnant, as in the Malaspina Glacier (Sharp, 1958) or still moving downslope like the Casa Pangue Glacier (Rabassa and others, 1981), the supraglacial sediments maintain true forests, with well developed tree and shrub strata. Whether these forests are reached through seral

replacement (as classical succession predicts) or are the adaptive response to the particular environmental conditions of the ablation mantle, is critical for the estimation of total time of the latest ice advance. In the Malaspina Glacier for example, is the Sitka spruce forest reached through succession from firewood to willow and alder, or is one of the first colonization stages? The estimated time of ice advance would differ significantly according to the answer given, as is particularly clear from existing descriptions (Sharp, 1958).

Malaspina Glacier is a good example of the disturbances described so far because it also encompasses deglaciated areas once covered by a forest overriden by the latest ice advance. The ice push destroyed the forest, depositing great amounts of organic matter in situ, but also transporting some to variable distances. The forest debris are still recognizable tree trunks and stumps (some still rooted), limbs, bark, cones and needles (Sharp, 1958). These deglaciated areas are vegetated in summer, supporting willow and firewood as the most conspicuous woody plants, and the growth is apparently restricted to soil patches with large amounts of organic material from the original forest (Sharp, 1958).

ICE STORMS: they develop in winter when warm fronts reach areas with ground temperatures below freezing. Storms cause limb damage that opens the way for wood attacking insects and fungi, or limit the elevational range of some plant species (White, 1979).

SNOW AND ICE AVALANCHES: they strongly limit the altitudinal range of tree species, and influence the distribution and abundance of morpho-physiological traits adaptive to extreme cold and drought conditions. Entire communities are adapted to these perturbations, among which dwarf forests and *krummholtz* are the most thoroughly known.

ALLUVIAL EROSION, DEPOSITION AND FLOODING: in glacial valleys where river and stream dynamics are regulated by the building up and melting of ice, vegetation shows multiple morphological and physiological adaptations related to flood frequency and intensity, as well as to the importance of alluvial processes.

INDIRECT AND FACILITATED DISTURBANCES: ice masses indirectly produce multiple environmental effects including temperature and precipitation fluctuations, local movement of air masses, and changes of the total surface available for new colonizing biota. Movement of substrate, avalanches and landslides can be facilitated by the disgregation of soils by ice, either massive or interstitial.

THE ADAPTIVE RESPONSE

Regional floras all around the world are rich in disturbance dependent species. I will discuss some adaptations to disturbances while describing few selected examples in each of three categories: morpho-physiological individual traits, population attributes and community features.

INDIVIDUAL TRAITS: seed size and allocation of cotiledonary reserves, root/shoot ratios, branching patterns, apical dominance, leaf layering, reproduction and phenological strategies have all been shown as significant for the adaptation to disturbance regime (Wells 1976; Bazzaz 1979; White 1979; Orians -in press-). The size of seeds affects the colonization probability of a new opened area. Larger seeds would have higher probability of surviving during longer periods of time than smaller seeds, and the seedlings produced by large seeds can grow larger before becoming dependent on their own photosynthesis. These advantages are countered, however, by the probability of being transported to lesser distances from the parental individual. Therefore, it is expected that smaller seeds will be successful colonizers on large areas relatively far away from the seed source. The proportion of energy reserves of the seed allocated to root *versus* shoots varies interspecifically and should affect survival under different environmental conditions. Allocation of more energy to roots than to shoots is expected where competition for nutrients is high, while apical dominance and vertical growth would be higher in those areas where competition for light is critical, e.g. recently liberated substrates, moraines and deglaciated areas, where little changes in height are correlated with high variations in the total radiation received.

Perhaps the most clear set of disturbance-adapted traits is found in flood plain trees, such as *Acer*, *Betula* and *Populus* species. They all have large yearly seed crops, wind and water-dispersed small seeds, fast growth rates, low wood density, short life spans, low shade tolerance, ability to sprout when damaged, and high tolerance to floods. They also have early spring blooming period, early summer seed dispersal, lack of seed dormancy, and short viability periods for seeds.

These traits taken together are clearly advantageous in environments such as river valleys of glacial origin. There, flow patterns and overall flood dynamics are relatively predictable and recurrent, dispersion by water is the major way of spatial distribution, high water table levels are reached in late spring and early summer months to which blooming periods are correlated and successful establishment is dependent on rapid germination and development prior to adverse winter conditions. It is noteworthy that bottomland species, e.g. *Taxodium spp.* and *Nyssa spp.*, attain their best growth on sites with permanent inundation, but their seeds need aereated ground sites above the water level to germinate. This feature indicates the species dependence on a disturbance pattern of shifting inundation, erosion and deposition (White 1979). At least some of these adaptive traits also exist in bottomland *Nothofagus* species, e.g. *N. antarctica*, of the Southern Andes in Argentina (Brandani 1977).

POPULATION ATTRIBUTES: age structure of single species populations, especially trees, is one of the most interesting characters related to disturbance patterns. If a regionally spreading population of trees shows patchily distributed ages, i.e. each patch of forest is of a particular predominant age, a process of cyclical replacement may be acting over the entire region. The similarity of ages within patches would suggest that each patch originated through a relatively synchronous regrowth of trees after the elimination of whatever vegetation was present in the site. If the population shows a mosaic-like structure, with several patches each of a given age, the differences in age between patches reflect therefore the frequency of elimination and regrowth processes. A forest of all-young patches would indicate high frequency of disturbance, while widespread senile stands result from very low recurrence. Furthermore, if at any site the disturbance frequency is lower than the life span of the tree species, the population becomes locally extinct, i.e., a patch with no trees may also indicate disturbance infrequency. A typical example of local extinction of trees is pink cherry; it is dependent on large scale disturbances and disappears from a site if no disturbances are acting there.

The variability of the age differences among patches would reflect the predictability of the pattern of disturbance recurrence. High variability in the differences among predominant ages would highlight low predictability while more similar differences suggest a more regular disturbance.

The magnitude of a recurrent disturbance may be estimated from the regional extent of the mosaic-like community, or from the entire area covered by a population showing age-specific patches.

It should be noted that the pattern described is not species-dependent: it can be studied regardless of the spatial distribution of any particular species, since climate may be limiting the geographical range, as long as the species bears some morphological feature correlated to age such as tree-rings. The differential survivorship of particular "age classes" of a tropical palm, *Astrocarium mexicanum*, is related to the occurrence of a gap -by windthrow or other natural disturbance- within the mature rain forest (Sarukhan 1978). Balsam fir, *Abies balsamea*, forests of the northeastern United States show a pattern initiated by the windthrow of adult trees and the subsequent regeneration of the forest in waves (Sprugel and Bormann 1981). Within one "wave cycle" all age classes, from seedlings to senescent trees can be found ordered sequentially in the downwind direction. *Nothofagus* forests of Chile are apparently maintained by the occurrence of massive disturbance, landslides, windthrow and vulcanism among them (Veblen and Ashton 1978; Veblen 1979). There, disturbances seem to be promoting the occurrence of even aged stands of *Nothofagus pumilio* and *Nothofagus betuloides* in the beech forests. Dendrochronology has proven a successful technique to study disturbance patterns such as climatic changes and water table variations (Isaak and others 1959), and fires (Johnson and Rowe 1975), and will not, therefore, be discussed extensively here. However, it is more limited in scope than the analysis I proposed since it mostly takes advantage of rings and ring-like structures of old trees that withstood past disturbances, and seldom uses the entire age structure of a population for vegetation dynamics analysis.

BIOLOGICAL COMMUNITY CHARACTERISTICS: disturbance may create landscapes that are in dynamic equilibrium with regard to plant species composition and structure. Landscapes contain a continuum disturbance from prone situations to protected ones. Both extremes of this gradient act as a refugia and seed sources, providing the recolonizing elements when conditions change. This is particularly true for postglacial forested landscapes where plant communities growing within a drainage basin range from closed forests to open meadows to fringing and riparian vegetation (Brandani 1977; White 1979). The zonation of vegetation in a basin, and the relative extension of the different communities, are a reflection of the environment and the related disturbance dynamics.

The magnitude of a disturbance is reflected in the spatial extension of disturbance-adapted communities, while recurrence is reflected in the spatial shifting in time of disturbance-adapted, as well as protected site, communities.

Two widely used community parameters, community diversity and concentration of the dominance, are useful for describing the relationship among community structure and disturbance regime.

Community diversity, whether measured as the total number of species present or by some more elaborate index (Pielou 1976), is lower at either end of a disturbance gradient. It is a good indicator of the importance of site disturbances because a community can become adapted to high disturbance frequency or, conversely, at very low frequency there will be a lack of compositional elements adapted to disturbance. It is in the alternation of disturbance and disturbance-free areas that a variety of environmental situations are created facilitating the development of a mixture of species with different strategies, therefore increasing total diversity. Concentration dominance in a community describes the proportion of the most abundant plant species to the next-most-common one and so forth until the entire set of species is exhausted. It gives information similar to the species diversity since the strongest concentration of the dominance will be found at either end of a disturbance gradient. However, it differs from the diversity index in, at least, two aspects: (1) Concentration of the dominance is a function related to the dynamic apportioning of resources by the whole set of a community species, while a diversity index is a synthetic, static, parameter; (2) Changes in species abundance may not be reflected by a diversity index while they will show up in the relation of abundance-dominance. As an example of how community structure and composition are related to disturbance regime, I will discuss the analysis made by Tisdale and others (1966) of vegetation development on deglaciated areas near Mt. Robson, British Columbia. Tisdale and collaborators found that, while studying the vegetation cover of sequential moraines, plant species considered early colonizers according to the classical succession theory were actually more common in moraines of inter- mediate age. The most conspicuous species was *Dryas octopetala*. At the same time, Tisdale and others (1966) found little evidence of succession in the area studied, either in the depressions between moraines or, with even less change, in the alluvial flats above two small lakes. In addition they mentioned that intermoraine areas were covered in part by a mixture of pioneer herbs and shrubs. They also found that shingle flats above the lakes had even less vegetation cover than 60 years before but that the dominant species was the same: *Dryas octopetala*. What looked as a case of "anomalous" succession was in fact a situ- ation where different plant adaptations, as well as total diversity and plant coverage, were associated to a particular disturbance regime. We know now that *Dryas octopetala* shows a pattern of cyclic replacement as a response to freeze-thaw disturbances (Anderson 1967). *Hedysarum* and *Salix* species, characteristically adapted to high soil water changes, were common in inter- moraine areas near Mt. Robson were periodic disturbance by glacial melt waters and runoff exists. Indeed, the majority of species described in the Tisdale study as pioneers are adapted to particular distur- bance regimes. Moreover, differences in diversity and vegetation cover were also related to similar patterns of environmental modifications and not to decadence because, as mentioned in the Tisdale paper, the species growing in areas of sparse vegetation cover did not show any particular sign of stress or decadence; quite the contrary they had well developed root and shoot systems with conspicuous root nodules in legumes (Tisdale and others 1966).

CONCLUSIONS

So far I have sketchily discussed the generalities concerning some relationships among environmental disturbances and vegetation features significant for glacial and periglacial landscapes. Specific situations would necessarily be more complex and with their own particularities. This complexity might arise from: the regional combination of different disturbance patterns; the scale at which disturbance and communities are defined; the mutual interdependence of disturbance parameters, and the co-occurrence of vegetational traits.

Usually, disturbance frequency is inversely correlated with magnitude, e.g. few major glaciations have occurred during the Quaternary, while many small glacial fluctuations have been recorded for a comparable period of time.

Several disturbance types may be associated or facilitated by one another such as insect infestations that may follow ice storms (Ashe 1918, cf. White 1979), or heavy epiphyte load that increases the damage suffered by a woodland during windstorms (Siccama and others 1976). The pattern of disturbance association may be further complicated by the characteristics of each disturbance involved. Low frequency and high intensity vulcanism is associated with more frequent, less intense and relatively more predictable biological disturbance. The relatively very regular pattern of flowering and death of bamboo, *Chusquea spp.*, in beech forests of Southern Andes, glacial cycles and irregular vulcanism, all appear to have a significant importance in the distribution and presence of the forests (Veblen and Ashton 1978; Veblen 1979). Analogously, vegetation attributes, while theoretically independent from one another, are under similar selective forces and therefore only few successful sets of morpho-physiological attributes are found together in a community.

However, the approach followed here seems to be a promising one for the analysis of vegetation dynamics and its relationships with disturbance patterns of environmental variables. Quantitative models relating community attributes to disturbances has already been explored taking into account one disturbance type, fires, acting in an unique region (Johnson and Rowe 1975).

Although more complex models and mathematical functions would have to be derived, especially to simultaneously relate vegetation to the three environmental parameters of intensity, frequency and predictability, they would lead to significant interdisciplinary environmental prediction of landscape dynamics.

REFERENCES

Anderson, D.J. (1967). Studies on the structure in plant communities. IV. Cyclical succession in *Dryas* communities from Northwest Iceland. J.Ecol. 55: 629-635.

Bazzaz, F.A. (1979). The physiological ecology of plant succession. Ann.Rev.Ecol.Syst., 11: 351-371.

Bormann, F.H. and Likens, G.E. (1979). Pattern and process in a forested ecosystem. Springer-Verlag, New York, 253 pp.

Brandani, A.A. (1977). Identification de nichos ecologicos de plantas en un ecosistema de bosque Andino-Patagonico. PhD Thesis, University of Buenos Aires, Argentina 234 pp.

Brewer, R. and Merrit, P.G. (1978). Windthrow and tree replacement in a climax beech-maple forest. Oikos, 30: 149-152.

Drury, W.H. and Nisbet, I.C.T. (1973). Succession. J.Arnold Arbor.Harv.Univ., 54: 331-368.

Golley, F.B., editor (1977). Ecological succession. Benchmark papers in Ecology, vol.5, Dowden, Hutchinson and Ross, Stroudsburg, Pennsylvania, 373 pp.

Hartshorn, G. (1978). Tree falls and tropical forest dynamics. In: P.B.Tomlinson and Zimmermann, M.H., editors, Tropical Trees as Living Systems, pp. 617-638, Cambridge Univ. Press, Cambridge.

Isaak, D.; Marshall, W.H. and Buell, M.F. (1959). A record of reverse plant succession in a tamarack bog. Ecology, 40: 317-320.

Johnson, E.A. and Rowe, J.S. (1975). Fire in the wintering grounds of the Beverley caribou herd. Amer.Midl.Nat., 94: 1-14.

Lindsey, A.A.; Petty, R.O.; Sterling, D.K. and van Asdall, W. (1961). Vegetation and environment along the Wabash and Tippecanoe Rivers. Ecol.Monogr., 31: 105-156.

Odum, E.P. (1971). Fundamentals of Ecology, 3rd. edition. Saunders, Philadelphia, 574 pp.

Orians, G.H. (in press). The influence of tree-falls in tropical forests on tree species richness.J.of Tropical Ecology.

Pickett, S.T.A. (1976). Succession: an evolutionary interpretation. Amer.Nat. 110: 107-119.

Pielou, E.C. (1977). Mathematical Ecology. Wiley, New York.

Rabassa, J.; Rubulis, S. and Suarez, Jr. (1981).Moraine-in-transit as parent material for soil development and the growth of Valdivian Rain Forest on moving ice: Casa Pangue Glacier, Mount Tronador (lat. 41 10'S), Chile. Annals of Glaciology, 2: 97-102.

Raup, H.M. (1975). Species diversity in shore habitats. J. Arnold Arbor. Harv. Univ., 55: 126-165.

Sharp, R.P. (1958). The latest major advance of Malaspina Glacier, Alaska Geograph. Rev., 48: 16-26.

Siccama, T.G.; Weir, G. and Wallace, K. (1976). Ice damage in a mixed hardwood forest in Connecticut in relation to *Vitis* infestation. Bull.Torrey Bot.Club., 103: 180-183.

Sprugel, D.G. and Bormann, F.H. (1981). Natural disturbance and the steady-state in high-altitude Balsam fir forests. Science, 211: 390-393.

Tisdale, E.W.; Fosberg, M.A. and Poulton, C.E. (1966). Vegetation and soil development on a recently deglaciated area near Mount Robson, British Columbia. Ecology, 47: 517-523.

Veblen, T.T. and Ashton, D.H. (1978). Catastrophic influences on the vegetation of the Valdivian Andes, Chile. Vegetatio, 36: 149- 167.

Veblen, T.T. (1979). Structure and dynamics of *Nothofagus* forests near timberline in South-Central Chile. Ecology, 60: 937-945.

Wells, P.H. (1966). A climax index for broadleaf forest: an n-dimensional, ecomorphological model of succession. Proc. First Meeting of the Central Hardwood Forest Conference, Southern Illinois University at Carbondale, pp.31-176.

White, P.S. (1979). Pattern, Process and Natural Disturbance in Vegetation. Bot. Rev., 45: 229-299.

Contributions related to the field excursions:
A. United States: Wyoming and Idaho

INQUA Commission on Genesis and Lithology of Quaternary Deposits: North America Annual Meeting, Wyoming and Idaho 1981 Field trip and symposium

DONALD F.ESCHMAN
University of Michigan, Ann Arbor, MI, USA

GERALD M.RICHMOND
US Geological Survey, Denver, CO, USA

EDWARD B.EVENSON
Lehigh University, Bethlehem, PA, USA

The 1981 INQUA Commission on the Genesis and Lithology of Quaternary deposits annual meeting was held in the Rocky Mountains of Wyoming and Idaho, August 20-30, 1981. Thirty-eight participants from 14 countries including Germany, Austria, Ireland, Sweden, Great Britian, Canada, Norway, Argentina, Spain, Switzerland, Belguim, Estonia, Netherlands and the U.S.A. were in attendance.

Most of the participants at the meeting were members of the INQUA Commission Working Group I: Genetic classification of tills and criteria for their differentiation (A. Dreimanis, President). The meeting was entitled: Symposium and Field Trip on the Genesis and Lithology of Morainic Deposits in an Alpine Environment. This title accurately describes the theme of the field trip and symposium, although a comparison of these deposits with other genetically different diamictons was made at several places.

The first five days of the meeting (Aug. 20-24) were spent in the field examining alpine glacial deposits in the Wind River, Gros Ventre and Teton Ranges and in the Jackson Hole and Yellowstone Plateau areas. A two day symposium (Aug. 25-26) at the University of Michigan's Camp Davis was devoted to the presentation of papers and work group discussions. The final three days (Aug. 27-29) were spent in the field investigating the glacial deposits of the Pioneer, Sawtooth and White Cloud Mountains of southern Idaho.

The symposium was divided into three sessions: (1) active mountain glaciation and mineral exploration (6 papers), (2) Quaternary mountain glaciation (6 papers), and (3) till: genesis, properties and classification (7 papers). Active

discussion followed the presentation of most papers and two evening meetings were held to discuss important points raised during the day and to conduct commission activities (A. Driemanis presiding as chairman).

The field aspect of the meeting, divided into two segments by the symposium, was largely devoted to a demonstration and discussion of the characteristics of alpine tills and associated sediments. Considerable attention was paid to lithology, fabric, provenance, granulometry, clast characteristics (shape, striae, weathering) and the presence or absence of stratification. In addition, characteristics such as surface morphology, soil development, boulder weathering and geomorphologic relationships of the various deposits were used to assign relative age and demonstrate a glacial or non-glacial origin of the sediments under discussion.

The participants gathered in Rock Springs, Wyoming (Fig. 1) on the morning of August 20, and later that afternoon crossed the Green River Basin by bus arriving at Boulder Lake, located at the southwestern foot of the Wind River Range, in time to make several stops. Led by Dr. Gerry Richmond, the group investigated moraines, tills and deformed lacustrine sediments associated with the last glaciation (Pinedale) along the shore of moraine-dammed Boulder Lake, and drumlins at the mouth of Boulder Creek Canyon.

Following a welcoming dinner at Boulder Lake Lodge, Richmond lead a discussion of the history of the late Pleistocene ice-cap on the Wind River Range and showed maps of the piedmont moraines at the mouths of each canyon.

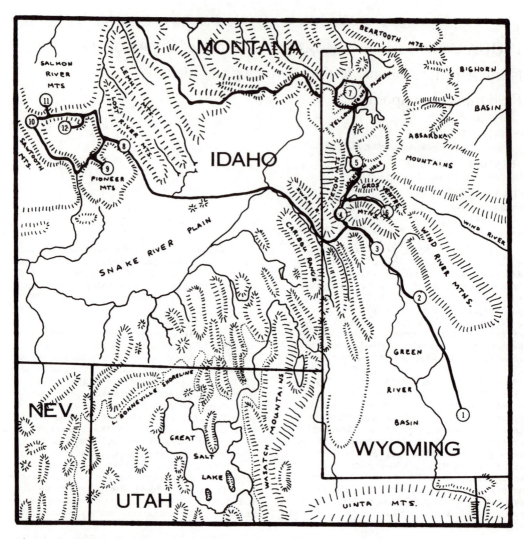

Figure 1   Route of field trip and location of areas discussed in text.
1 - Rock Springs, Wyoming, 2 - Boulder Lake and Pinedale,
3 - Hoback Basin, 4 - Jackson and Camp Davis, 5 - Jackson Hole,
6 - Gros Ventre River, 7 - Yellowstone National Park, 8 - Mackay,
Idaho, 9 - Copper Basin, 10 - Stanley Basin, 11 - Yankee Fork
dredge, 12 - Railroad Ridge.

The morning of August 21 was devoted to the investigation of Bull Lake and Pinedale tills and soils at Fremont Lake (Richmond, 1973) near Pinedale, Wyoming, the type area of the Pinedale Glaciation. The tills of this area are silty, sandy tills derived wholly from crystalline granitic rocks, and form well developed nested moraine sequences (Fig. 2). One of the topics of discussion was the homogeniety of the tills, the absence of a well developed clast fabric, and the complete lack of stratification and other recognizable deformation structures. The absence of these characteristics in a till forming well developed moraines led several participants to question the validity of the extension of genetic criteria developed for lowland tills to alpine deposits--a

414

Figure 2   Ice disintegration features in Pinedale moraines near Halfmoon and Fremont Lakes, Wyoming.

theme which permeated many of the subsequent discussions during the field trip. The afternoon was spent investigating the tills near Cora, Wyoming. These tills, which are composed largely of resistant fine-grained Paleozoic and Mesozoic carbonate and sandstone lithologies, were contrasted with the granitic provenance tills visited in the morning. Here, clasts were glacially shaped with striated "bullet shaped" clasts occurring in abundance. The matrix to clast ratio was much higher and well developed fissility and crude stratification were occasionally observed. This led to discussions of the idea that till composition was very important in the development of many of the criteria used in the investigation of till genesis.

As the group proceeded north over the Green River Basin rim and into the Hoback River drainage, Dr. Don Eschman reviewed geomorphic evidence for extensive piracy of the Green River drainage by the Hoback River, as well as several smaller scale pirated valleys and numerous stream terraces related to the glacial history of the Hoback drainage. The group arrived at Camp Davis, the University of Michigan's field camp and research station, in the late afternoon. This facility served as the base for subsequent days field activity and for the symposium.

Saturday, August 22, was spent investigating the glacial deposits of Jackson Hole, the Teton Range and the Gros Ventre drainage. The characteristics of moraine-forming tills of this area were contrasted and discussed in terms of interpretation of their genesis, and

several stops were particularly informative. One stop, in a well developed morainic ridge in Jackson Hole (Fig. 3), exhibited a till composed entirely of well rounded quartzite cobbles, about 1% of which were striated, in an open, loose matrix of sand sized material. The cobbles were derived from Cretaceous and Paleocene conglomerates that underlie both glaciated and unglaciated areas in the nearby mountains. Much discussion ensued concerning the interpretation of this diamicton as a till as it had very few of the characteristics ordinarily attributed to tills and no characteristics that aided in defining its genetic origin. The argument offered by Evenson was that it was a till composed almost entirely of glacially reworked inwash gravels which were pushed into a morainic ridge and striated by the advancing ice lobe in Jackson Hole. Again, the concept that a till's characteristics are closely related to the material available to the ice was expressed and argued. The next stop was in a moraine deposited around Jenny Lake by ice flowing out of the dominantly crystalline Teton Range. This till was very similar to the crystalline tills visited at Pinedale and had few characteristics helpful in assigning a genetic origin (basal, supraglacial, meltout, etc.) to the deposit. The afternoon was spent in the Gros Ventre Valley comparing tills consisting of both sedimentary and crystalline clasts in a

Figure 3   View of Pinedale moraines on floor of Jackson Hole, Wyoming. Moraines in this area are composed almost entirely of stream rounded quartzite cobbles pushed into morainic loops by a glacier flowing south from the Yellowstone Plateau.

fine-grained clayey matrix derived from Mesozoic shale with local large landslide deposits, especially the famous Gros Ventre landslide. The characteristics by which the landslide deposits could be differentiated from the very similar alpine tills were discussed. A well developed basal (or valley side) lodgment till was visited at the lower end of Upper Slide Lake. This till had a well developed boulder pavement (the boulders of which had faceted, striated, upper surfaces with the striae oriented parallel to the valley side), strong fissility, a well developed clast fabric and a silty matrix and was unanimously accepted as a "typical lodgment till". Evenson and Richmond explained that this till was probably formed by a lobe of the glacier in Jackson Hole moving up the Gros Ventre Valley. Apparently the glacier dammed the valley, impounding the drainage and forming a lake into which the glacier advanced. Glacial reworking and incorporation of the lacustrine sediments provided the fine-grained material necessary to produce a till with most of the characteristics commonly attributed to the lodgment process, again demonstrating the relationship between the material available and the genetic features which develop.

On August 23, the group left Camp Davis for a two day excursion to Yellowstone National Park, much of which lies in a caldera 600,000 years old. A first stop was made along the canyon of Lewis River, which descends the outer rim of the caldera, to observe the characteristics of till composed almost wholly of rhyolite and rhyolite tuff. The till is an unsorted silty sand, consisting of angular grains of quartz and sanidine, and abundant glass shards. The clasts are mostly rough-surfaced, irregularly shaped and angular to subangular. However, a few are smooth rounded pebbles or cobbles of andesite, an erratic derived from the east. Striations preserved on a very few of these andesites are the only criteria demonstrating that the deposit is a till.

After ascending to the caldera rim, the bus descended into the caldera basin, in which Yellowstone Lake is enclosed, held in on its western side by obsidian-rhyolite flows. After a walk of about 1km along the lake shore, the group arrived at Rock Point, where three superposed late Wisconsin tills overlie a lacustrine sand. The tills, though texturally distinct, are all sandy but lack any obvious bedding. Because the source of the glacier was to the southeast across Yellowstone Lake, as indicated by striated erratics of andesite,

discussion centered on whether or not the tills were deposited in an ancestral lake. Further along the lake bluff, at our lunch stop, a silt till containing both angular clasts of rhyolite and rounded clasts of andesite overlies an obsidian rhyolite flow K-Ar dated as 145,000 years old. This till, which is believed to be of late Illinoian age is overlain by varied silts that pass upward into massive silt containing two tephra layers, one dated as about the same age as the underlying rhyolite flow. Above the tephra are more lake silts characterized by undulating and folded bedding and by abundant slickenslides. Infolded in the upper part of these sediments is a massive silt till, obviously consisting of reworked lake silt and containing abundant erratics of andesite. This till was clearly deposited on soft lake sediments disrupted by the pressure of over-riding ice. The glacial sediment was probably not dropped in water, but was deposited as the ice was flowing over nearshore lake bottom sediments.

Returning to the buses, we quenched our thirst at the Lake store before proceeding to the north end of the lake to examine a till-like diamicton overlying, highly folded, varied sediments and overlain by wavy bedded fine lake sand. A few years ago, the diamicton was in contact with a thin organic layer C-14-dated at about 13,000 years BP. A tephra, known to be about 12,000 years old, is present in the overlying sand. The diamicton consists of angular and rounded clasts in a brownish to bluish matrix. Many of the clasts are of hydrothermally altered and cemented lacustrine silt, sand, or gravel. Studies have shown that it is a hydrothermal explosion breccia. Such deposits are formed where subaqueous thermal springs are confined by hydrostatic pressure that is explosively released by a sudden fall in lake level. The elliptical crater produced by this explosion-or explosions- is about a kilometer long and nearly 50m deep.

Our arrival at Canyon Village was delayed in Hayden Valley by a large herd of buffalo crossing the road, and by the reluctance of our colleagues to stop photographing them.

On the morning of August 24 we made an adventurous descent into the upper canyon of the Yellowstones to observe another till-like diamicton. This deposit is overlain by a thick sequence of sediments ranging from late Wisconsin till at the canyon rim, down through older fluvioglacial, interstadial and interglacial sediments, some possibly of pre-Illinoin age. The diamicton is composed of a greenish or brownish gray tuffaceous sandy clay enclosing pebble-to-boulder size angular clasts of fine-grained biotite diorite. The diamicton is tough, massive, nonbedded and unsorted and has been called both a till and a talus; but it is more likely a subaqueous tuffaceous mudflow or subaqueous lithic tuff. It lies within the 600,000 years old caldera and is older than a rhyolite flow dated at about 590,000 years old.

Returning to the canyon rim, we descended again to examine thick tuffaceous lake sediments beneath the diamicton and to observe the characteristics of an underlying slightly welded pumiceous ignimbrite containing numerous rounded clasts of older rhyolite flows, silica-cemented lake sediments, quartzite and granite gneiss incorporated from the ground surface by the incandescent ash flow. Before climbing again to the rim we paused at the brink of the Lower Falls, 97m high, to photograph and gaze at the brilliantly hued Grand Canyon of the Yellowstone before us.

From the Canyon we drove across the Solfatara Plateau to Norris Geyser Basin, which we visited all too briefly before proceeding to our lunch stop along Gibbon River. Here wine flowed like the nearby water, and some dozed as we proceeded on to geyser basins along the Firehole River. Old Faithful very kindly erupted just as we arrived, following which some toured the hot springs and smaller geysers while others retreated to the Inn for liquid refreshment. The trip back to Camp Davis was uneventful.

Tuesday and Wednesday, August 25 and 26, were devoted to the symposium and evening sessions of the working group described above. During a temporary power outage, the group was assembled by Dave Mickelson and Curt Sorenson to partake of the favorite American past time--a baseball game. It rapidly became apparent that activities like catching and batting a baseball are not inate! Don Eschman arranged for a local ranch hand, "Cowboy Bob", to perform (with guitar) for the group at a huge campfire. Here, it rapidly became apparent that our European counterparts were better singers than the native scientists!

Early Thursday morning the bus left for Copper Basin, located midway between Mackay and Sun Valley, Idaho at the head of the Big Lost River. Here the base of operations was the geology field camp of Lehigh University, consisting of a classic log cabin cow camp augmented by "rustic accommodations"--a number of tents, two outhouses and running water (cold and very cold) nestled in the midst of a large morainic complex of Pinedale and Bull Lake age (Fig. 4). This isolated basin was a truly magnificent setting for a group studying alpine glacial deposits. On the drive into the basin, Evenson introduced the complex history of glacial advances and retreats in the several tributary basins, each with its own similar, but slightly different, chronology. As a result of this temporal variance, a glacial lobe occupying one valley system would occasionally dam another valley, only to have the resulting lake suddenly "break out" producing, in at least one case, a flood capable of transporting boulders 3 to 4 meters in diameter far down valley to leave them widely distributed across the gravel terraces (Fig. 5). The details of the history of glacier oscillation in the area have been painstakingly worked out by Lehigh graduate students; their job helped immeasurably by detailed provenance studies that allow all the glacial deposits in the area to be accurately related to their source areas. Evenson briefly reviewed the new (Evenson, et al., in press) stratigraphic nomenclature proposed for the area.

August 28 was spent investigating the tills of the Copper Basin, Wildhorse Canyon

Figure 4   Copper Basin Cow Camp and "dormitories". The large feature at the base of the rainbows is a lateral moraine.

417

Figure 5   Large flood transported boulder on a fluvial terrace. The lithology of this boulder proves that it was transported over 15 Km from its source. It is located 10 Km beyond the maximum limit of glaciation. Note S. Gawarecki for scale.

and the Summit Creek area. At Wildhorse Canyon, Keith Brugger, a Lehigh graduate student, explained how provenance investigations aided in reconstructing ice deployment and transport paths for various lithologies (see Brugger, et al., this volume) and Jim Cotter gave a detailed explanation of the glacial history of the Summit Creek area (see Cotter and Evenson, this volume). The group was later transferred to four-wheel drive vehicles (Fig. 6) to transport them to an active rock glacier near Green Lake (Fig. 7). Saturday, August 29, the last working day of the field trip and Commission meeting,

was spent on a long loop drive over Trail Creek Road, continuing north into the Stanley Basin where granitic tills of the Sawtooth Range were discussed in detail. The group then voyaged east along the Salmon River to the Yankee Fork River (of the Salmon) where they visited an area of the Yankee Fork that had been dredged for gold along an extensive stretch of Pinedale outwash. This extensive area of dredged bouldery spoil (Fig. 8) provided an excellent opportunity to see the internal composition and clast size and shape characteristics of a late Pleistocene

Figure 7   Active rock glaciers above Green Lake.

outwash sequence. The final stop of the field trip was in the White Cloud Mountains near the Livingstone mill and mine. Here the group was again transferred to four-wheel drive vehicles for an ascent to the Railroad Ridge Drift (Fig. 9) described by Ross (1929). Although Ross interpreted this flat, bouldery deposit, situated at 10,000 ft. on the flank of the mountains, as a till of "Nebraskan" age the group was told, and then convinced, that the deposit was actually a huge mass wastage deposit (see Gawarecki and Evenson, this volume). After the long and arduous, but scientifically stimulating, trip to Railroad Ridge the group was returned by bus to the Copper Basin via Challis, Idaho.

Figure 6   The INQUA Commission partici-pants enroute to Green Lake via 4-wheel drive vehicles.

The stay in the Copper Basin camp (elevation 7600 ft.) although all too brief, was probably the high point of the

Figure 8 Yankee Fork dredge and dredge spoil material along the Yankee Fork River, Idaho.

entire meeting for our foreign visitors for it provided them with an opportunity to experience the beauty and charm of the mountain cow country. A final dinner party at the Copper Basin, attended by 22 of the local cowboys and ranchers--all friends of longstanding--highlighted the western alpine experience. Gifts and addresses were exchanged and friendships were started and renewed around a campfire which was later filled with empty beer and wine bottles. Perhaps the reticence we all felt as we prepared to leave the Basin the next morning was also experienced by the bus, for it steadfastly refused to start. As we were 30 miles from the nearest town or telephone, numerous tactics were tried to rectify the problem, but to no avail. Because some of the group had to reach the airport at Idaho Falls, some 125 miles away, by early afternoon they were quickly loaded into the back of an open pickup truck--the only alternative means of transport available. Those who stayed behind, including some of the trip leaders, will long remember the view of some of the most distinguished Quaternarists of all of Europe riding off in the back of a dusty pickup, nestled in and on their luggage, and chauffeured by Ed Evenson along the winding road up the Pinedale lateral moraine!

The bus driver then left in the last available vehicle in hopes of finding help in the form of advice over the phone from his company, 350 miles away in Rock Springs, Wyoming. Before noon, two different pairs of cowboy-mechanics showed up, demonstrating to all how well the grape vine works in an area where everyone is isolated and hence depends on far-flung

neighbors for help when in trouble. A little after noon the cantankerous deisel was finally repaired by Walt Johnson whose name will never be forgotten by any of those still present. Thus, the group gained still further insight into the "real" West and its ways. Even those who have spent many field seasons in the cow country of the Rocky Mountains were once more impressed with the genuineness and warmth of the local "cowboy".

From a scientific point of view, perhaps the salient point learned by all was the real differences between the genetic characteristics of mid-continent tills deposited by large ice sheets and those produced by alpine glaciers. The influence of lithology and particularly that of fine-grained matrix material produced either in proglacial lakes or from communition of sedimentary bedrock cannot be overlooked or overstated. Many of the participants commented on the frequent absence, especially when dealing with tills of crystalline provenance, of many of the features (fissility, striations, shaped clasts, fabric, compactness, etc.) that we take for granted in assigning a genetic term to lowland mid-continent tills. In fact, had it not been for the fact that many of the tills investigated were in distinct moraines, and therefore undeniably

Figure 9 Surface morphology of the Railroad Ridge diamicton. This flat surface is at an elevation of approximately 10,000 ft. The deposit is a diamicton with boulders up to two meters in diameter in a silty, sandy matrix. In the center of the photo the diamicton is exposed in a cirque headwall cut by "post-Railroad Ridge" glaciers.

tills, many of the participants familiar
only with lowland tills would, and some
still may, deny that the deposits were, in
fact, tills.

The organizers would like to thank all
those who worked so diligently to make the
excursion and symposium a success. Laurie
Cambiotti acted as secretary, typist,
travel agent and banker. Without her
superhuman effort this meeting could not
have been conducted. Jim Cotter, Keith
Brugger, Michael Clinch, Jim Bloomfield and
Annette Wagner acted as field assistants,
cooks and drivers as needed. Victor and
Scott Johnson provided four-wheel drive
vehicles and logistic support as did the
Rangers from the Mackay office of the U.S.
Forest Service. The National Science
Foundation provided an underwriting grant
for travel. The citizens of Mackay, Idaho
and the members of the Copper Basin
Cattlemen's Association provided lodging,
friendship and help in our hour of need.
To all these individuals and organizations,
the organizers extend their sincere
appreciation.

REFERENCES

Brugger, K.B., Stewart, R. A. and Evenson,
    E. B. (this volume) Chronology and
    Style of Glaciation in the Wildhorse
    Canyon Area: Idaho.

Cotter, J. F. P. and Evenson, E. B. (this
    volume) Glacial History and
    Stratigraphy of the North Fork of the
    Big Lost River, Pioneer Mountains,
    Idaho.

Evenson, E. B., Cotter, J. F. P. and
    Clinch, J. M., Glaciation of the
    Pioneer Mountains: A Proposed Model
    for Idaho: in Cenozoic of Idaho, R.
    Breckenridge, ed., Idaho Bureau of
    Mines (in press).

Gawarecki, S. L. and Evenson, E. B. (this
    volume), The Railroad Ridge Diamicton:
    A Relict Fan Complex Formerly
    Considered Glacial Till.

Richmond, G. M., 1973, Geologic Map of the
    Fremont Lake South Quadrangle,
    Sublette County, Wyoming U.S.
    Geological Survey Map GQ-1138

Ross, C. P., 1929, Early Pleistocene
    Glaciation in Idaho: in Shorter
    contributions to General Geology: U.S.
    Geol. Survey Prof. Paper 158, p.
    123-128.

Deer Ridge: A pre-Bull Lake multiple till locality in Wyoming

DONALD F.ESCHMAN
University of Michigan, Ann Arbor, MI, USA

GORDON R.BREWSTER
Saskoil, Regina, Saskatchewan, Canada

WILLIAM F.CHAPMAN
Slippery Rock State College, PA, USA

Deer Ridge is a broad flat-topped divide separating the drainages of Shoal Creek and Granite Creek, both being westward flowing tributaries of the Hoback River heading in glaciated canyons of the Gros Ventre Range of northwestern Wyoming. The elevation of the ridge top near its northwestern end (near the mountain front) is approximately 2685 meters. A large head cut of a multiple slump on the southeast (Shoal Creek) side of the ridge exposes a 38 meter thick section consisting of two tills capping the rather poorly consolidated sandstones and siltstones of the Hoback Formation of Paleocene age (Figure 1).

In the field, one can see signs of extensive weathering in both tills, from top to bottom (Figure 2). The base of the Upper Till, largely cemented by secondary calcite, rests directly on top of the slightly red, leached upper part of the Lower Till. The contact between the two tills rises toward the western end of the exposure (Figure 1). The base of the Lower Till is covered with colluvium in most places, although the underlying Hoback Formation was exposed in one place during the summers of 1980 and 1981.

The position of the two tills approximately 400 meters above the level of the modern streams and 120 meters above the crest of a lateral moraine of Bull Lake age bordering the northwest side of Shoal Creek Valley together suggest that both tills are pre-Bull Lake in age. The evidence of extensive weathering of both tills would tend to support this interpretation. If the interpretation of this section is correct, the Deer Ridge exposure is one of the best outcrops of pre-Bull Lake tills in stratigraphic succession in the Wyoming Rockies. This paper reports the work presently completed to better define the nature and extent of weathering of the tills.

To check our interpretation that the data obtained from the analyses of the two tills capping Deer Ridge reflect extensive weathering (i.e. pre-date Bull Lake tills), samples of till of Bull Lake age were also examined. Samples were collected from a small slump scar near the top of a lateral moraine of Bull Lake age 120 meters below the crest of Deer Ridge. The sample analyzed came from a depth of 340 cm below the top of the exposure and moraine crest. The exposed till is calcareous throughout and is light brownish-gray to light olive brown in color (2.5Y 6/2 dry; 2.5Y 5/4 moist). The matrix has a loamy texture, containing approximately 40% sand, 35% silt, and 24% clay and is mildly alkaline in reaction (pH=7.86).

The Deer Ridge Lower Till unit is 19 meters thick. It is also loamy in texture, with the till matrix averaging 46% sand, 35% silt, and 18% clay. The color of this Lower Till unit ranges from brown (7.5YR 5/4 dry; 7.5YR 4/4 moist) near its upper boundary to light gray (2.5Y 7/2 dry; 2.5Y 6/4 moist) at its base. However, the predominant color of the till is yellowish brown to dark yellowish brown (10YR 5/4 dry; 10YR 4/4 moist). The till is leached of carbonate to a depth of one meter. There is some small amount of secondary carbonate present below the one meter depth, evident in the form of white mottling. The matrix of the unleached portion of the till has a pH of 8.00.

The Upper Till is approximately 17 meters thick, with the upper 4 meters leached of carbonate. The lower 13 meters of this till are strongly cemented with calcium carbonate. This strongly cemented K horizon (Birkeland, 1974) and the resultant vertical face has thus far stymied attempts at detailed sampling of the till. To date, only a few "grab" samples, obtained by reaching down from the top of the exposure, have been analyzed. The two

Figure I: The Deer Ridge exposure. Solid line is contact between Upper Till and Lower Till; dashed line is contact of Lower Till and Hoback Formation.

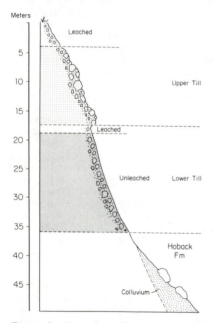

Figure 2: Diagrammatic section of Deer Ridge exposure.

samples analyzed are from depths of 30 cm and 80 cm below the top of the exposure, and thus are from the upper leached zone. The color of the till matrix within the leached zone is light olive brown (2.5Y 5/6 dry; 2.5Y 4/4 moist). The texture is a sandy loam tending toward a loamy sand, with the matrix averaging 68% sand, 22% silt, and 10% clay. There is a very strong contrast between the size distri-

bution, and especially the sand-clay ratio, of the upper and lower till units. The pH values for the two samples analyzed were 7.03 for the uppermost and 7.16 for the lowermost samples respectively.

Till clasts are Precambrian granites and granite gneisses and of Paleozoic and Mesozoic sediments, all of which outcrop in the Shoal Creek drainage basin upstream from Deer Ridge, in and along the front of the Gros Ventre Range. A much higher percent of the clasts in the Upper Till unit are of Precambrian granitoid rocks from the crystalline core of the Gros Ventre Range. One variety of this rock type that is quite common in the till is so weathered throughout the entire thickness of the till that cobble- and boulder-sized clasts have become grus and can be readily picked apart with the fingernail. Three till fabric measurements, made in an attempt to determine the configuration of the glacier which deposited the till (i.e. valley or piedmont) were inconclusive; the three fabric maxima determined were: N 20° W; N 15° E; and N 70°W.

X-ray diffraction (XRD) powder pattern analysis was performed on the coarse silt-sized fraction (50 μm to 20 μm) of samples obtained from the two Deer Ridge till units and from one sample of a nearby Bull Lake till. The mineral grains were randomly mounted on a glass slide using clear nail polish as a mounting medium. From the Lower Till unit, samples 15 cm and 120 cm below the top and 75 cm above the base were analysed. The uppermost sample from this unit revealed only quartz and K-feldspar XRD peaks, while the lower two samples contained quartz and K-feldspar as well as calcite and dolomite. Both calcite and dolomite increase in amount and crystallinity with increasing depth. In the lowermost sample (75 cm above the base) both calcite and dolomite peak heights exceed that of K-feldspar, and the calcite: dolomite ratio is less than 1.

Examination of the XRD traces produced from the coarse silt-sized fractions of the two samples analyzed from the upper Till unit revealed the dominance of quartz and the presence of K-feldspar. Neither calcite nor dolomite were present (as evidenced by XRD peak position) in either of these two samples. By comparison, analysis of the Bull Lake till sample revealed the dominance of quartz and substantial amounts of both calcite and dolomite, with only trace amounts of K-feldspar being present. Unlike the Lower Till unit of Deer Ridge, the calcite: dolomite ratio in the Bull Lake till sample was greater than one.

For the clay mineral analysis, XRD patterns were obtained for the following conditions: Mg-saturated and then air-dried (Mg-ADPO); Mg-saturated and then saturated with ethylene glycol (Mg-GPO); K-saturated and then air-dried (K-ADPO); K-saturated and then heated to 350° for 0.5 hours (K-350°C); K-saturated and then heated to 550°C for 0.5 hours (K-550°C). All samples were mounted according to the suction method of Kinter and Diamond (1956) using the modification proposed by Rich (1969). The amount of clay-sized material on each porous plate was uniform (186 mg), while the thickness and density of the mounted clay were controlled as much as possible in order to obtain a uniform distribution across the porous plate. The calculated density of the clay layer examined was approximately 14 mg/cm^2, producing the optimum thickness corresponding to a starting angle of 35° 2θ (Rich, 1975). It is important to note that for each mounted sample, the amount of clay and the clay density were controlled as much as possible. By controlling the amount of material mounted on each porous plate and its distribution across the plate, both intra- and inter-sample examination of clay mineral XRD characteristics as revealed by peak size, shape, and location was possible. The samples were examined using Ni-filtered CuKα radiation and a Philips diffractometer with a scanning speed of 2°/min. Figure 3 depicts the XRD traces obtained for samples analyzed from the upper and lower till units on Deer Ridge, and of a sample of Bull Lake till.

XRD analysis of the clay-sized fraction of samples from the Lower Till unit (Figure 3 a to c) reveals the dominance of illite (10.04 Å; K-ADPO) in all three samples. It would appear that illite increases in either amount or crystallinity with increasing depth (Figure 3c). Montmorillonite-like expandable min-erals are also present in these samples (Figure 3a to 3c; 16.97Å; Mg-GPO) and they also increase in amount or crystallinity with increasing depth. The asymmetry of the 10.04-Å peak and the broad plateau between 10.04 Å to 14.24 Å following K-saturation would seem to indicate the presence of a poorly ordered vermicu-lite/montmoril-lonite/Al-interlayered vermiculite complex (Figure 3b). The lack of "sharp" clearly-defined XRD peaks in the uppermost sample of the Lower Till unit (Figure 3a) would seem to suggest an almost complete alter-ation/destruction of pre-existing clay-sized minerals.

Clay mineral analyses of the two samples obtained from the Upper Till unit reveal the dominance of illite, but a lack of montmorillonite-like expandable minerals (Figure 3d and 3e). The lack of "sharp" and "clean" XRD peaks in the uppermost sample would seem to suggest an almost complete lack of crystallinity, perhaps induced by the high rates of weathering/alteration which occur at surface environments (Figure 3d). The asymmetry of the 10.04 Å illite peak in the lower-most sample suggests some form of illite alteration, perhaps to vermiculite.

XRD analysis of clay-sized material obtained from the nearby Bull Lake till exposure reveals a mineral suite which contrasts sharply with the clay mineral assemblages identified in the Deer Ridge tills. The Bull Lake till clay mineral suite is completely dominated by mont-morillonite-like expandable clay minerals (16.97 Å; Mg-GPO) with little or no illite being present (Figure 3f).

Based on the preliminary investigation of the two Deer Ridge till units and the comparison with the Bull Lake till, it would appear that Deer Ridge is capped by two pre-Bull Lake tills. Both field and laboratory examinations reveal that both the upper and lower tills are extensively weathered throughout. The clay- and coarse silt-sized fraction mineralogy of the Deer Ridge tills is quite different in both composition and character than the Bull Lake till, the former showing evidence of much more extensive weathering than the latter. Based on the cursory examination completed to date, there is strong evidence to suggest that the Deer Ridge exposure contains two quite distinct till units. Not only are they mor-

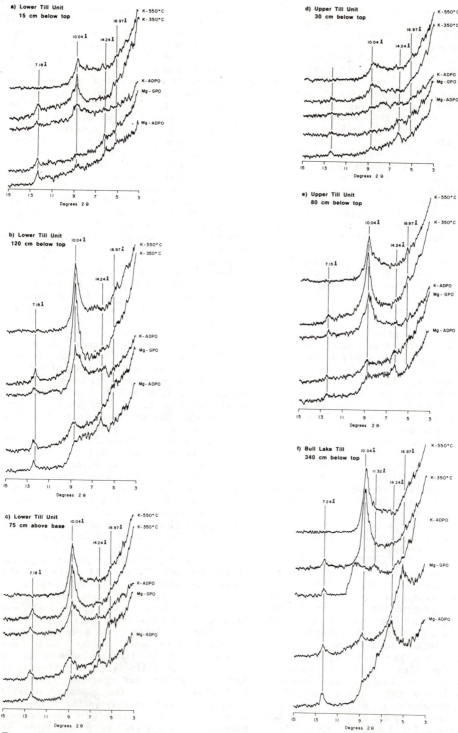

Figure 3: X-ray diffraction traces of samples obtained from the upper and lower Deer Ridge till units and from a nearby Bull Lake till (2 m).

phologically dissimilar, but there are
strong contrasts between both their textures and their mineralogies.

REFERENCES

Birkeland, P. W. 1974, Pedology, weathering and geomorphological research,
   New York, Oxford University Press.
Kinter, E. B. & S. Diamond, 1956, A new
   method for preparation and treatment of
   oriented specimens of soil clays for X-
   ray diffraction analysis, Soil Science
   81: 111-120.
Rich, C. I., 1969. Suction apparatus for
   mounting clay specimens on ceramic tile
   for X-ray diffraction, Soil Science
   Society of America Proceedings 33: 815-
   816.
Rich, C. I., 1975. Amount of clay needed
   for optimum X-ray diffraction analysis,
   Soil Science Society of American Proceedings 39: 161-162.

Glacial history and stratigraphy of the North Fork
of the Big Lost River, Pioneer Mountains, Idaho

JAMES F.P.COTTER & EDWARD B.EVENSON
Lehigh University, Bethlehem, PA, USA

1 ABSTRACT

The drainage of the North Fork of the Big
Lost River includes three principal trunk
streams and their tributaries in the
Pioneer and Boulder Mountains of
south-central Idaho. Deposits of two
extensive glaciations, and one period of
restricted glacier advance (or periglacial
activity), have been mapped and tentatively
correlated within the area. A relative age
(RD) stratigraphy and informal
stratigraphic nomenclature (Birkeland, et
al., 1979) unique to this area was
developed for this study. The two periods
of glacial activity were designated "Kane
advance" (older) and "North Fork advance"
(younger). A still younger phase, largely
a period of periglacial activity, was
described and designated "Neoglacial". The
Kane and North Fork advance deposits, are
correlated with the Copper Basin and
Potholes Glaciations, respectively, of the
Idaho Glacial Model (Evenson, et al., in
press). It is impossible at this time to
correlate the deposits of the study area
definitively with the classic deposits of
the Wind River Range, Wyoming; but by
applying the methods and stratigraphic
constraints used elsewhere in the Rocky
Mountains, the deposits are tentatively
correlated with the Rocky Mountain Glacial
Model (Blackwelder, 1915; Richmond, 1948,
1960a, 1965, 1976; Mears, 1974).

The terminal moraines of the Kane and
North Fork advances can be differentiated
accurately by using morphologic and
pedologic (qualitative) relative age
parameters. Recessional moraines of a
single glaciation cannot be differentiated
by relative age criteria and are subdivided
into "events" on the basis of their
relationship to glaciofluvial terraces and
by down-valley position.

Pebble lithology provenance studies aided
in the reconstruction of the geometry of
the three major mutually independent trunk
glaciers that existed in the area. During
the Kane advance, these three ice streams
coalesced to form one composite valley
glacier. During the North Fork advance,
the maximum extent of glaciation was
considerably less, and only two of the ice
streams coalesced.

2 INTRODUCTION

This project was designed to continue the
mapping of glacial and surficial deposits
in the drainage of the North Fork of the
Big Lost River, south central Idaho.
Surficial deposits in the area,
differentiated by relative age (RD)
techniques, afford an excellent opportunity
to extend the understanding of the glacial
history of the Pioneer Mountains and to
begin to unify this knowledge into a
dynamic stratigraphic model which is
independent of other models used in the
Rocky Mountains. Evenson, et al. (in
press) have called this model the " Idaho
Glacial Model". This model has been
designed so that deposits of a restricted
area such as the North Fork of the Big Lost
River can be; A) assigned to an (local)
informal stratigraphic system (i.e. North
Fork and Kane advances); and B) correlated
to a regional stratigraphic system. If
correlations later prove to be incorrect
they can be changed without having to
change the original stratigraphic
nomenclature. With increased use of this
system the regional significance of the
Idaho Glacial Model increases. The model
remains dynamic, yet independent of glacial
models in regions too distant (e.g. the
Wind River Range, Wyoming) to allow
accurate correlation based solely on RD
criteria.

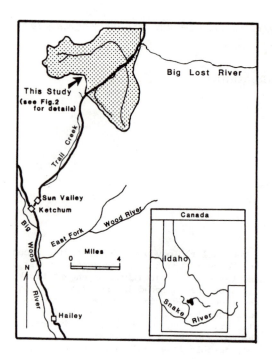

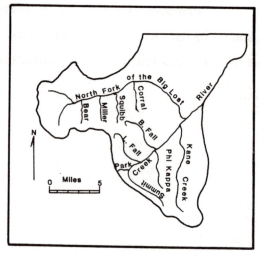

Figure 2 Major glaciated drainages of the study area

Figure 1 Location map of the study area

This paper will discuss the mapping of the surficial geology of the North Fork Region and those RD techniques which proved most effective. In addition the history and philosophy of the Idaho Glacial Model will be reviewed.

3. LOCATION AND BEDROCK GEOLOGY

The study area is located in the northeasternmost portion of the Pioneer Mountains in south central Idaho, north of the Snake River Plain (Figure 1). This area includes approximately 275 square km of the headwaters of the North Fork of the Big Lost River.

The two major trunk streams in the study area are the North Fork of the Big Lost River and Summit Creek (Figure 2). Summit Creek drains northeastward from the divide that separates the Trail Creek and Big Lost River watersheds (Figures 1 and 2). Glaciated valleys tributary to Summit Creek include: Park Creek, Little and Big Falls Creeks, Phi Kappa Creek, and Kane Creek Canyons. (Figure 2). Although there are many streams tributary to the North Fork of the Big Lost River, only the northward flowing canyons (Corral Creek, Miller Creek, Squibb Creek, and Bear Creek; Figure 2) have been glaciated. This is due to the

northwestward decrease of summit altitudes and the unfavorable southward orientation of the catchment areas.

The geomorphologic setting of the study area has been greatly influenced by Quaternary fluvial and glacial activity. Streams in glaciated valleys are often underfit and uplands display classic Alpine-type glaciation features such as cols, aretes, horns, and cirques. Elevations in the study area range from 2360 m along the North Fork, in the northeast corner of the area, to 3955 m above Kane Canyon.

Structurally, the Pioneer Mountains, consist of two northwest-trending domes. These domes are cored by autocthonous Pre-cambrian metamorphic units surrounded by a number of allocthonous, upper Paleozoic clastic lithologies which were emplaced by thrust faulting. During the formation of the Idaho Batholith (Cretaceous) small "satellite" intrusive bodies were emplaced in the western Pioneer Mountains. Stratigraphically overlying all of these units are the Tertiary Challis Volcanics. These volcanics, consisting of interbedded lava and tuffaceous units, probably blanketed the entire region at one time, but subsequent uplift and erosion has resulted in the exposure of older units. Tertiary to Quaternary block faulting is believed to be the cause of this uplift, and the present relief (Umpleby, et al., 1930; Dover, 1966, 1969).

4. PREVIOUS WORK

Umpleby, et al. (1930) and Nelson and Ross (1969) were the first workers to map and subdivide glacial deposits in the Pioneer Mountains. More recently, Dover (1966, 1969) and Dover, et al. (1976) mapped and described, in a generalized manner, the surficial deposits of the area as part of a detailed investigation of the stratigraphy and structure of the area.

The mapping by Lehigh students has been reported in seven unpublished Master's theses. These include mapping and provenance analysis in; the Copper Basin (Wigley, 1976; Pasquini, 1976), Wildhorse Canyon (Stewart, 1977; Brugger, 1983), and the North Fork Drainage (Cotter, 1980; Repsher, 1980), and provenance analysis of glaciofluvial terraces (Pankos, in preparation).

In each of the three glaciated basins (Copper Basin, Wildhorse Canyon and North Fork Basin) the glacial and glaciofluvial deposits were mapped and differentiated using relative-dating techniques (Birkeland, et al., 1979). In addition, models of ice deployment were developed and reported by Evenson, et al. (1979), Cotter, et al. (1981), Evenson, et al. (in press), Repsher, et al. (1980), and Brugger, et al. (this volume).

5. STRATIGRAPHIC APPROACH

Due to the paucity of absolute age dates throughout the Rocky Mountains, relative dating (RD) techniques have been widely used to establish local stratigraphies. RD techniques utilize time dependent characteristics of; landform, rock, and soil modification to differentiate deposits. These techniques assume that at the time of deglaciation a feature is "fresh" and that all modifications indicative of relative age have occurred since that time. As Burke and Birkeland (1979) have discussed, time is not the sole factor in influencing weathering and erosion, therefore, a multiple parameter approach to relative dating increases the reliability of differentiation on a local basis and increases the possibility of obtaining criteria useful for regional correlation. The use of any single method often does not document age difference sufficiently. Also, local factors may result in varying effectiveness of a technique at different localities.

In the North Fork of the Big Lost River we have utilized a multiple RD technique approach. The qualitative and semi-quantitative parameters used include:

1) moraine morphology (number and freshness of ice disintegration hollows, preservation of original morainic form, degree of secondary dissection); 2) downvalley extent of glaciation (older moraines are downvalley or "outside" younger deposits); 3) spatial relationships of moraines and glaciofluvial terraces (terraces are graded to a moraine of the same age and dissect older moraines and terraces); and 4) soil properties (horizonization, color, structure, texture, carbonate concentrations, clay mineralogy and soil chemistry).

With the establishment of increasing numbers of local stratigraphies based on RD techniques (e.g. Richmond, 1948, 1957, 1960a, b, 1962a, 1964b, 1965, 1972, 1976; Moss, 1949, 1951a, b, 1974; Holmes and Moss, 1955; Birkeland, 1964, 1973; Crandell, 1967; Benedict, 1967, 1968, 1973; Madole, 1972, 1976, 1980; Mahaney, 1972, 1974; Carroll, 1974; Currey, 1974; Pierce, et al., 1976; Evenson, et al., 1979; and Colman and Pierce, 1981) problems have arisen with both the over-extension of regional correlations, and the formal (or informal) naming of deposits differentiated by more than one RD parameter. The former problem is caused by regional variation of post-depositional modification processes. As discussed by Birkeland, et al. (1979) the latter problem arises from the fact that the "Code of Stratigraphic Nomenclature" currently makes no provision for defining glacial units on the basis of multiple physical (RD) features.

The American Commission on Stratigraphic Nomenclature (1970) has approved a variety of classifications (i.e.; time-, rock-, and soil-stratigraphic units) for the distinction and correlation of Quaternary deposits, but none of them embrace the multiple weathering parameters used in relative-age dating. In each of the code-approved classifications an individual parameter must be used to define a stratigraphic unit, whereas relative-dating relies on the application of multiple weathering parameters. Our use of multiple relative dating techniques (e.g.; both soil and morphologic features) must therefore result in a mixed or "hybrid" stratigraphic classification (Birkeland, et al., 1979) not formally recognized by the Stratigraphic Code. However, the fact that RD data does not generate a code approved stratigraphic classification does not mean RD based units cannot be named on an informal, and even formal (see Burke and Birkeland, 1979), basis. It is important, however, that care be taken to avoid the proliferation of even informal stratigraphic nomenclature.

We have discussed this problem previously (Cotter, 1980; Cotter, et al., 1981; and Evenson, et al.,in press) and have developed a system in which local, informal names are applied to hybrid RD units ("advances") not recognized by the stratigraphic code. This system, called the "Idaho Glacial Model", was designed to: A) provide local and regional RD based stratigraphic names; B) allow easy correlation of deposits occurring within the individual basins of Central Idaho, but not simultaneously imply absolute age equivalence; C) avoid the incorrect usage of formally defined terms (e.g. till, drift or stade); and D) avoid the problems associated with the extension of the terms "Bull Lake" and "Pinedale" which have been used at various times for morpho-, climato-, soil-, rock- and chrono-stratigraphic designations throughout the Rocky Mountains.

The Idaho Glacial Model then, with continued use, will expand in areal extent yet remain flexible in terms of local correlations. This approach will both avoid the over-extension of regional correlations based solely on RD parameters and facilitate communication on a local level while avoiding conflict with the stratigraphic code.

Admittedly, this system results in the proliferation of stratigraphic terms, but we feel the correlation of local units ("advances") to regional units ("glaciations"; Table I) might alleviate some confusion for those not thoroughly familiar with the literature concerning the area. Using this system it is possible to name and then correlate "hybrid" RD units on a local basis (e.g. the North Fork and Kane advances within the drainage of the Big Lost River) and simultaneously have a set of regional terms (Potholes and Copper Basin glaciations) to which all the local stratigraphic systems can be correlated. This regional chronology may then be correlated to other independent chronologies such as those used in the Cascade Ranges and the central Rocky Mountains.

For this study the deposits of individual glacial events (called "advances") have been assigned informal names: "North Fork (younger) advance" and "Kane (older) advance", and have been correlated to deposits in adjacent basins and to a regional stratigraphic system, the Idaho Glacial Model (Table I; Evenson, et al. in press). In addition, each advance has been subdivided into second order episodes (which we call "events") that are designated by Roman numerals appended to

Idaho Glacial Model Evenson et al in press	North Fork Drainage Cotter, 1980	
Glaciations	Advances	Events
Potholes Glaciation	North Fork Advance	IV III II I
Copper Basin Glaciation	Kane Advance	
Pioneer Glaciation	(Not Recognized)	

Table 1 Stratigraphic Nomenclature of this study and the Idaho glacial model.

the advance name (eg. North Fork I, North Fork II, etc.) If present correlations prove incorrect (e.g. if the Kane advance is not correlative to the Copper Basin glaciation) the correlation can be changed without requiring revision of the local RD based stratigraphic name.

The following is a description of terms (Kane advance and North Fork advance) used in this study.

Moraines of the Kane advance are those which retain only a portion of their original constructional topography, and have weathering characteristics similar to those described at the type Bull Lake section in the Wind River Range, Wyoming (Blackwelder, 1915; Richmond, 1962a). Glaciofluvial deposits of the Kane advance were not recognized. The Kane advance is therefore believed to be the older of the two advances recognized in the study area. Deposits of the Kane advance are tentatively correlated with the Copper Basin glaciation of the Idaho Glacial Model (Table I).

430

Deposits of the North Fork advance are those moraines and glaciofluvial gravels deposited by the second, or younger, glacial advance recognized in the study area. Deposits of this advance are located up-valley, and lower along the valley walls than those of the older Kane advance. Soil characteristics and moraine morphology indicate North Fork advance moraines have been exposed to pedologic development for a shorter period of time, and have been less dissected by fluvial activity subsequent to their deposition, than moraines of the Kane advance. Deposits of the North Fork advance are tentatively correlated with deposits of the Potholes glaciation (Table I).

Informal names rather than letters (e.g. Crandell, 1967) are used because the correlation of letter designated units rapidly becomes cumbersome. In addition neither communication nor convenience is served since letter designations are stratigraphically invalid and can be taken to imply correlation.

6. RESULTS AND DISCUSSION

Figure 3 is a generalized map of the glacial deposits of the North Fork area. The ultimate goal of field work in this area was to develop a model which characterized the history and geometry of glaciation. In addition to detailed mapping of the surficial deposits, field investigations included differentiation of glacial deposits according to age using multiple RD techniques and the determination of till provenance (to reconstruct ice flow patterns and source areas). The combined use of RD techniques (for the development of an RD stratigraphy) and provenance investigations (for ice flow paths) has proven extremely useful in the Rocky Mountains (Richmond, 1960a, b, 1976; Birkeland, et al., 1971; Weber, 1972; Birkeland, 1973, 1978; Mears, 1974; Madole, 1976) including the Pioneer Mountains of central Idaho (Wigley, et al., 1978; Evenson, et al., 1979; Repsher, et al., 1980; Evenson, et al., in press).

6.1 Glacial Geology

The glacial history of the North Fork region included two extensive and distinct periods of glacial advance and deposition, that have been named the "Kane advance" (older), and the "North Fork advance" (younger) (Cotter, 1980; Evenson, et al., in press). During each of these advances, valley restricted glaciers developed in each of the three principal canyons (North

Fork, Summit and Kane). For ease of discussion these three ice streams will be referred to as; the "North Fork", the "Summit Creek", and the "Kane Canyon Lobes"; corresponding to the valley in which they developed.

Wigley (1976), Stewart (1977) and Cotter (1980) have demonstrated that RD techniques can be used to differentiate the glacial deposits in the Pioneer Mountains of Central Idaho. In this investigation the following RD criteria proved most useful.

A) Extent of moraine complexes: Older moraines are found further down valley and higher on the valley sides than younger advances in the same canyon.

B) Grading and cross-cutting relationships of glaciofluvial deposits: glaciofluvial terraces are graded to moraines of the same age, and dissect older deposits. Terraces were particularly useful for the correlation of the deposits of the three separate and mutually independent valley glaciers that occupied the principal valleys.

C) Moraine morphology: morphologic variation in constructional topography, degree of development of drainage, percent undrained hollows, and surface boulder percentage.

D) Till weathering: qualitative pedologic characteristics (soil depths, color, horizonation and ped development) and weathering of till clasts.

As previously discussed, the morphologic and pedologic parameters utilized for RD age differentiation are developed as a function of time and length of exposure to weathering processes. It is inferred, therefore, that the two named advances (Kane and North Fork) represent an older (Kane) and younger (North Fork) episode of glacial activity and till deposition. These advances have been correlated (Evenson, et al., in press) to the Copper Basin (older) and Potholes (younger) glaciations of the Idaho Glacial Model (Table I).

6.2 Pre-Kane Advance Deposits

Both Stewart (1977) and Wigley (1976) have recognized deposits older than the Copper Basin glaciation in adjacent areas of the Pioneer Mountains. (Pioneer glaciation, Evenson, et al., in press). These deposits are; of limited extent, well dissected, highly weathered and located 60-75m higher in elevation than other glacial deposits. However, in the drainage of the North Fork of the Big Lost River, no deposits of this age have been encountered to date. The absence of glaciation deposits in the North

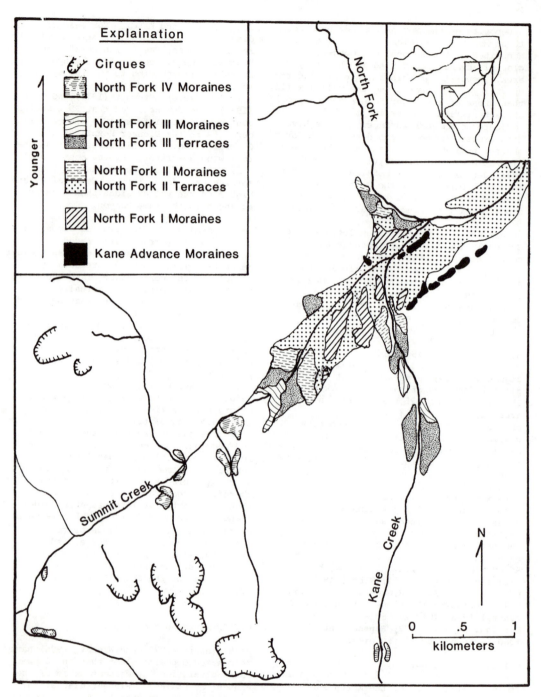

Figure 3. Glacial geology of the North Fork of the Big Lost River.
(North Fork IV moraines and terrace remnants were also
mapped in the headwaters of North Fork Canyon.)

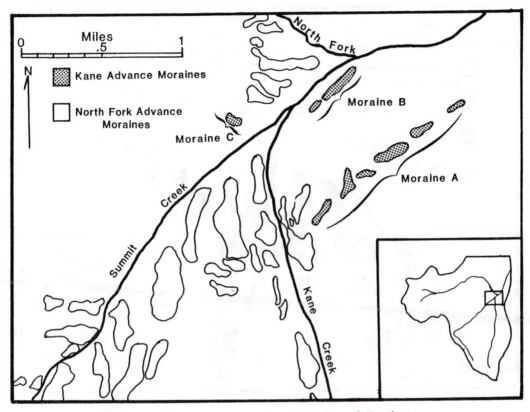

Figure 4 Deposits of the Kane advance. Lettered Moraines
(A, B and C) are discussed in text.

Fork drainage may be due to the removal by subsequent glacial and glaciofluvial erosional processes intensified by the narrow steep-sided valleys of this drainage.

6.3 Kane Advance

The Kane advance is the oldest advance recognized in the study area. Moraines of the Kane advance are; 1) well dissected and washed, 2) within the confines of Summit Creek Canyon, and 3) located down valley from terminal moraines of the North Fork advance. The surficial morphology of Kane advance moraines is extremely subdued. The lack of well defined constructional topography, and the absence of kettles allows differentiation of deposits of this older advance from those of the younger North Fork advance. The moraines assigned to the Kane advance have a much smoother morphology than the moraines of the younger North Fork advance. Nowhere are the moraines of the Kane advance continuous or complete. No terminal moraines have been preserved, and there are no Kane advance glaciofluvial gravels. The geometry of moraines of the Kane advance (Figure 4) indicates that glaciation was more extensive during this period than any subsequent glacial advance.

Figures 3 and 4 show the deposits of the study area assigned to the Kane advance. One deposit (Moraine A, Figure 4), consisting of six mappable lateral moraine remnants, is located northeast of the mouth of Kane Creek Canyon. A second moraine (Moraine B, Figure 4), consisting of two remnants, is located in the center of Summit Creek Canyon, east of the confluence of Summit and Kane Creeks. Also assigned to the Kane advance is a small isolated moraine remnant (Moraine C, Figure 4) on the west side of Summit Creek Canyon, south of the mouth of the canyon of the North Fork of the Big Lost River.

A provenance study was undertaken to determine both the source area for moraines

433

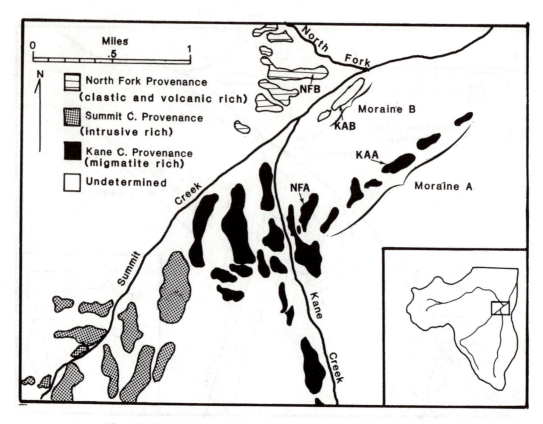

Figure 5 Provenance of Moranic Deposits. KAA, KAB, NFA
and NFB are soil profiles discussed in text.

and the character of parent material (till) for soil analysis. Provenance of moraine deposits was determined by identifying at least fifty pebbles in twenty-five shallow pits within a 3 square km. area surrounding the junction of Kane and Summit Creeks (Figure 5). As shown in Figure 5, the provenance (till pebble lithology) of the "Moraine A" complex is migmatite rich, indicating that the ice which deposited this moraine eminated from Kane Creek Canyon (see Cotter, 1980 for sample site locations, raw data and discussion).

Field studies of soil profiles in tills rich in migmatite lithologies (KAA and NFA, Figure 5) indicate the amphibolite member of the Migmatite Complex in Kane Creek is one of the least resistant lithologies to chemical weathering in the area. The soil profile developed in the "Moraine A" complex (KAA, Figure 5) of the Kane advance has no fresh (see Cotter, 1980, Appendix III) amphibolite boulders remaining in the A- and B-horizons. Those amphibolite clasts that were found in the soil profile

were completely grussified "ghosts". In the soil profile (NFA, Figure 5) of the younger North Fork advance Moraines with the same provenance, more amphibolite boulders were present (3% in provenance samples, see Cotter, 1980, Appendix III) and, for the most part, they are ungrussified. The contrast in the degree of weathering and removal of migmatite clasts in these two profiles indicates that "Moraine A" (Kane advance) has been exposed to weathering for a greater period of time (and is therefore older than the moraine containing NFA).

Moraine A is a lateral moraine of the ice stream eminating from Kane Creek Canyon. This Kane Canyon lobe was deflected during the Kane advance northeastward by the two other ice streams in the study area (see discussion below). The down valley extent (or terminal position) of the Kane Advance could not be determined.

In most cases moraine geometry, flow direction indicators and provenance clearly identified the source area of moraines.

434

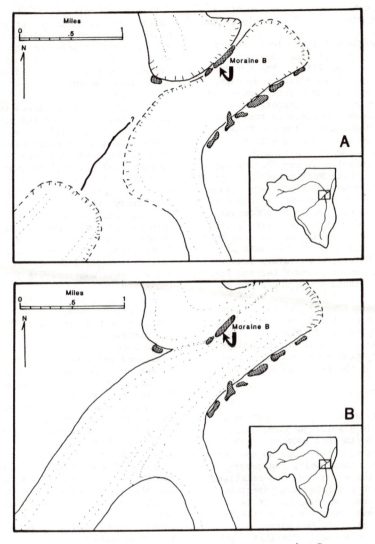

Figure 6   Possible origin of Kane advance moraine B;
A) interlobate end moraine; B) medial moraine

However, the exact origin and mode of deposition of "Moraine B" is problematic.

The provenance of "Moraine B" (Figure 5) is dominated by lithologies (24% volcanics, 55% clastics; Cotter, 1980) characteristic of North Fork Canyon, however, lithologies characteristic of both Kane Canyon (7% migmatite) and Summit Creek Canyon (4% intrusives) also occur (see Cotter, 1980, Appendix III). In addition, the position and geometry of moraine does not clearly indicate the source area.

Because "Moraine B" has a mixed provenance, it is believed to have been deposited by more than one ice-stream,
either as: A) an "interlobate" end moraine between ice eminating from Kane and North Fork Canyons (Figure 6A); or, B) a medial moraine deposited between ice streams sourced in Summit Creek and North Fork Canyons (Figure 6B). "Moraine B" has been assigned to the Kane advance (Figure 4) on the basis of surficial morphology and soil characteristics. The surficial morphology of "Moraine B" is subdued, indicating a greater period of exposure to weathering processes. In addition, the highest (oldest) glaciofluvial terrace, North Fork II terrace (Figure 3), is graded to a moraine up-valley from "Moraine B"

indicating that it is at least as old as North Fork I advance, and probably of the Kane advance.

Further evidence for the inclusion of "Moraine B" in the Kane advance was supplied by soil profiles. The two soil profiles excavated in moraines deposited by ice flowing out of North Fork Canyon (KAA and NFA, Figure 5) were calcorthids; (calcium carbonate buildup in the cca horizon). The Cca horizon in "Moraine B" (Kane advance) is extremely well-developed, and can be considered a petrocalcic horizon. In comparison, the Cca horizon in "Moraine D" (North Fork advance) is poorly developed, less well-defined, and thinner. This indicates that "Moraine B" has been exposed to the processes of pedogenesis for a longer period of time (Birkeland, 1974).

"Moraine C" (Figure 4) was assigned to the Kane advance strictly on the basis of moraine morphology, down-valley location, and its relation to younger terrace levels. No soil investigation was conducted on this moraine.

Although both Stewart (1977) and Wigley (1976) recognized two depositional events within the Copper Basin glaciation in adjacent areas, no clear relationships exist in the North Fork area to allow subdivision of the Kane advance.

6.4 The North Fork Advance

As previously discussed, two distinct, differentiable glacial advances (Kane advance and North Fork advance) were identified in the study area on the basis of morphologic, pedologic, and geographic characteristics which indicate relative age. Although absolute ages cannot be assigned to the two advances, it is assumed, because they are so distinct, that the deposits of these two advances represent two temporally distinct, climatically controlled, glacial events. Theoretically, these two glaciations were separated by a period of ice retreat and soil development.

Aside from the moraines of the older Kane advance, all glacial deposits in the area are included in the North Fork advance. All moraines of the North Fork advance are located up-valley from Kane advance moraines. Glaciofluvial deposits (including several paired terraces of differing elevation) graded to the moraines of the North Fork advance are designated "North Fork" gravels. It is unknown whether the North Fork moraines (North Fork I through North Fork IV; Figure 3) represent a single continuous deglaciation, or a series of advance and retreat events to and from successive up-valley ice margin positions.

Individual moraines of the North Fork advance cannot be differentiated on the basis of surface morphology; therefore the depositional episodes of the North Fork advance (North Fork I through North Fork IV) have been distinguished on the basis of down valley extent and their relationship to terraces.

Moss (1974) has discussed the relation of glaciofluvial terraces and glacial advances in Wyoming. Moss (1974) has found that there is a genetic relationship between moraine positions and terrace levels graded to them, and has implied synchroneity of glaciofluvial terraces and the associated moraines deposition. It is assumed, in this study, that elevated terrace gravels are glaciofluvial in origin and that they were formed during periods of glacial advance or standstill. Conversely, downcutting and the attainment of lower terrace levels is associated with deglaciation and interglacial conditions (Moss, 1974).

The geometry of North Fork I moraines (or North Fork maximum; Figures 7 and 8) indicate that during their deposition, the ice streams flowing out of Kane Creek, North Fork, and Summit Creek Canyons coalesced and interacted, but not to the extent that they did during the previous (Kane) advance (see Figures 4 and 6). During the North Fork I Advance (Figure 8), northward flowing ice from Kane Creek Canyon diverted the terminus of the northeastward flowing ice in Summit Creek Canyon slightly to the west (Figures 7 and 8) as indicated by the orientation of "Moraine E" and "Moraine F", Figure 7. These two lobes were in contact, when the ice flowing out of North Fork Canyon remained near the mouth of the canyon (Figures 7 and 8) allowing meltwater from Summit, Kane, and North Fork canyons to flow down valley.

No North Fork I terraces are present in the study area. The highest (oldest) terrace (18 m above present stream level) is graded to North Fork II moraines (Figure 3) and thus is named the "North Fork II terrace". The well-defined North Fork II terraces enable an accurate correlation of the North Fork II moraines for each of the three ice streams in the area.

North Fork II moraines have morphologic and pedologic characteristics similar to those of the North Fork I moraines. North Fork II moraines are located at the mouths of Kane Creek and North Fork Canyons, and up-valley from the North Fork I moraines in Summit Creek Canyon (Figure 9). This moraine geometry demonstrates that, unlike earlier stages of glaciation, ice streams during the deposition of the North Fork II

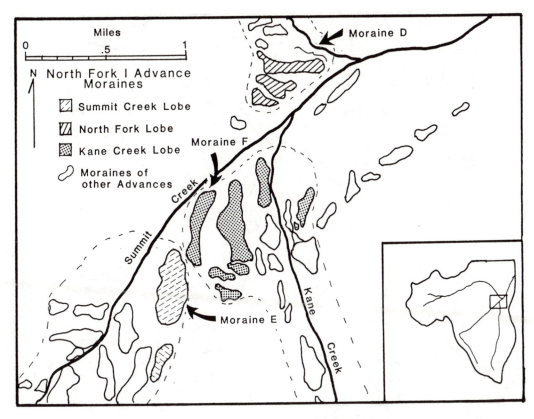

Figure 7   Provenance and location of North Fork I Moraines.
Lettered moraines (D,E and F) are discussed in text.

moraines were restricted to the principal canyons. They did not have the "expanded feet" characteristic of the previous, more extensive, episodes of deposition. Terraces and a reconstruction of the geometry of ice margins during the deposition of North Fork II moraines (Figure 10) indicate that at no time were any of the three ice streams in contact.

North Fork III moraines are located up-valley from North Fork II moraines (Figure 3). A paired terrace (North Fork III terrace; Figure 3) is graded to North Fork III moraines. This terrace is lower (5m above present stream level) and less extensive than the older North Fork II terrace. The North Fork III terrace in the three principal valleys (Figure 11), allows accurate correlation of the ice front positions at this stage of deglaciation.

The North Fork IV moraines deposited in all three major valleys represent ice margins well up-valley from those deposited during the older North Fork advances.

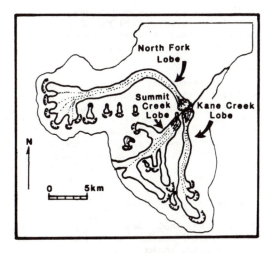

Figure 8   Reconstruction of North Fork I ice lobe geometry

437

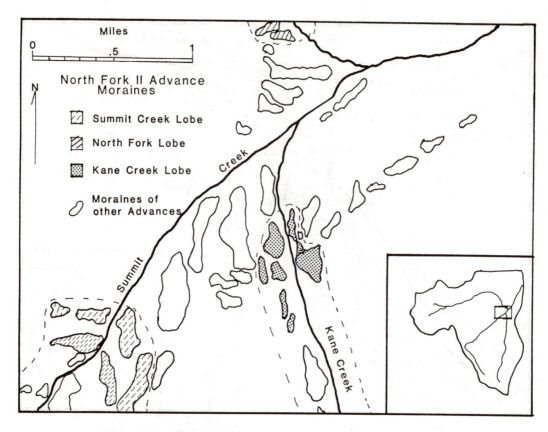

Figure 9   Provenance and location of North Fork II Moraines

Small moraines in canyons tributary to Summit Creek Canyon (Phi Kappa, Big Fall and Little Fall Canyons) indicate the complex valley glacier that occupied Summit Creek Canyon had lost most of its tributary inflow.

It is believed that the duration and timing of the deposition of North Fork IV moraines probably varied from valley to valley. Therefore, the correlation of the deposits mapped as North Fork IV moraines must be considered tenuous.   Figure 12 is an attempt to reconstruct the geometry of ice margins during the deposition of North Fork IV moraines.

6.5 Neoglacial Deposits

Deposits assigned a Neoglacial age (Cotter, 1980, Plate 1) consist entirely of alluvium, rock glacier, and rock slide deposits.   No moraines or glaciofluvial deposits of Neoglacial (post-altithermal) age were encountered.

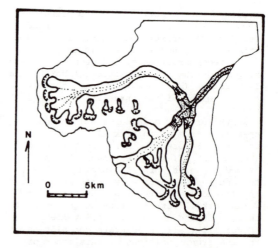

Figure 10   Reconstruction of North Fork II ice lobe geometry

438

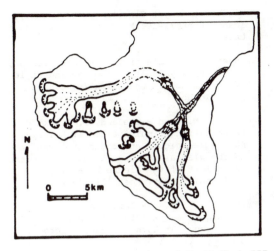

Figure 11   Reconstruction of North Fork III
ice lobe geometry

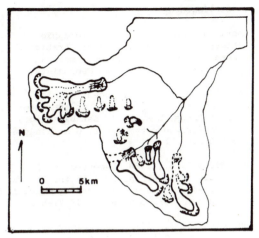

Figure 12   Reconstruction of North Fork IV
ice lobe geometry

7 CONCLUSIONS

The drainage of the North Fork of the Big Lost River has been modified by at least two major episodes of glaciation. Moraine complexes deposited during these episodes have been differentiated by relative age dating (RD) techniques (moraine morphology, pedologic characteristics, extent of glaciation, and terrace associations), and assigned to two informal, "hybrid" stratigraphic units, here named the Kane (older) and North Fork (younger) advances. The North Fork advance has been subdivided into four second order episodes, called "events", on the basis of moraine position and terrace relationships. The Kane and North Fork advances have been correlated to the Copper Basin (older) and the Potholes (younger) glaciation of the Idaho Glacial Model (Evenson et al, in press).

Moraine geometry and provenance data indicate that all moraines were deposited by three compound valley glaciers flowing out of North Fork, Kane Creek and Summit Creek Canyons. During the maximum extent of glaciation (Kane advance), the three valley glaciers coalesced to form a compound foot, or complex valley glacier (Figure 6). During the North Fork advance, maximum (N.F. I) ice flowing out of Kane Creek Canyon combined with and diverted the Summit Creek glacier to the west (Figure 8). At this time the North Fork glacier was restricted to North Fork Canyon and deposited a separate moraine near the canyon mouth (Figure 7). During subsequent advances (N.F. II, III and IV) the ice streams were not in contact with one another and each valley glacier simply deposited a series of end moraines within its respective canyon (Figures 9, 10, 11 and 12).

The development of an accurate model of the history and geometry of the Kane and North Fork advances has allowed the correlation of deposits in this study area to units in adjacent basins. Continued use of the relative age (RD) stratigraphy introduced here and in the Idaho Glacial model will ultimately lead to a unifying theory of the patterns of glaciation in this region. It will also result in the ability to more accurately correlate deposits in this portion of Idaho with deposits elsewhere in the Rocky Mountains.

8 ACKNOWLEDGMENTS

We thank Tore Vorren, Robert Stewart and James Zigmont for helpful discussions pertaining to the stratigraphic model presented here. Bobb Carson, James Parks and Beth Peters assisted with laboratory and statistical analysis and preparation of the manuscript. We are grateful to James Bloomfield, Keith Brugger and Mike Clinch for critical review. This project was supported, in part, by grants from Sigma Xi and Lehigh University.

9 Bibliography

American Commission on Stratigraphic Nomenclature, 1970, Code of Stratigraphic Nomenclature, 2nd edition, American Association of Petroleum Geologists.

Benedict, J. B., 1967, Recent glacial history of an alpine area in the Colorado Front Range, U.S.A., I. Establishing a lichen-growth curve, Jour. Glacialogy 6: 817-832.

Benedict, J. B., 1968, Recent glacial history of an Alpine area in the Colorado Front Range, U.S.A., II. Dating Glacial Deposits, Jour. Glacialogy 7: 77-87.

Benedict, J. B., 1973, Chronology of cirque glaciation, Colorado Front Range, Quat. Research 3: 584-599.

Birkeland, P. W., 1964, Pleistocene glaciation of the northern Sierra Nevada, north of Lake Tahoe, California, Jour. Geology, 72: 810-825.

Birkeland, P. W., 1973, Use of relative age dating methods in a stratigraphic study of rock glacier deposits, Mt. Sopris, Colorado, Arctic and Alpine Research 5: 401-416.

Birkeland, P. W., 1974, Pedology, Weathering, and Geomorphologic Research, Oxford Univ. Press, New York.

Birkeland, P. W., 1978, Soil development as an indication of relative age of Quaternary deposits, Baffin Island, N.W.T., Canada, Arctic and Alpine Research 10: 733-747.

Birkeland, P. W., Colman, S. M., Burke, R. M., Shroba, R. R., and Meierding, T. C., 1979, Nomenclature of alpine glacial deposits, or What's in a name?, Geology 7: 532-536.

Birkeland, P. W., Crandell, D. R., and Richmond, G. M., 1971, Status of the correlation of Quaternary stratigraphic units in the western conterminous United States, Quat. Research 1: 208-227.

Blackwelder, E., 1915, Post-Cretaceous history of the mountains of central western Wyoming, Part III, Jour. Geology 23: 307-340

Brugger, K. A., in prep., Provenance studies and ice reconstruction of the Wildhorse Canyon Glacier, Custer County, Idaho; Unpublished M. S. thesis, Lehigh University.

Brugger, K. A., Stewart, R. A. and Evenson, E. B. (this volume), Chronology and style of Glaciation in the Wildhorse Canyon Area, Idaho.

Burke, R. M., and Birkeland, P. W., 1979, Reevaluation of multiparameter Relative Dating techniques and their application to the glacial sequence along the eastern escarpment of the Sierra Nevada, California, Quat. Res. 11: 21-52.

Carroll, T., 1974, Relative age dating techniques and a late Quaternary chronology, Arikaree Cirque, Colorado, Geology 2: 321-325.

Crandell, D. R., 1967, Glaciation at Wallowa Lake, Oregon, U.S.G.S. Prof. Paper 575-C: C145-C153.

Colman, S. M. and Pierce, K. L., 1981, Weathering Rinds on Andesitic and Basaltic Stones as a Quaternary Age Indicator, Western United States, U.S.G.S. Prof. Paper 1210.

Cotter, J. F. P., 1980. The Glacial geology of the North Fork of the Big Lost River, Custer County, Idaho; unpublished M. S. thesis, Lehigh University.

Cotter, J. F. P., Clinch, J. M., Evenson, E. B. and Repsher, A. A., 1981, The glacial geology of the North Fork of the Big Lost River-Application of a new stratigraphic approach and nomenclature for the Pioneer Mts.; INQUA symposium on deposits in an alpine environment.

Currey, D. R., 1974, Probable pre-Neoglacial age of the type Temple Lake moraine, Wyoming, Arctic and Alpine Research 6: 293-300.

Dover, J. H., 1966, Bedrock geology of the Pioneer Mountains, Blaine and Custer Counties, Idaho, Ph. D. thesis, Univ. of Washington.

Dover, J. H., 1969, Bedrock geology of the Pioneer Mountains, Blaine and Custer Counties, central Idaho, Idaho Bureau of Mines and Geology Pamphlet 142.

Evenson, E. B., Cotter, J. F. P., and Clinch, J. M., In Press, Glaciation of the Pioneer Mountains: A proposed model for Idaho. In Breckenridge, R. and Bonnichsen, W. (eds), Cenozoic Geology of Idaho, Idaho Bur. Mines and Geology Pub.

Evenson, E. B., Pasquini, T. A., Stewart, R. A. and Stephens, G., 1979, Systematic provenance investigations in areas of alpine glaciation: applications to glacial geology and mineral exploration. In, Schlucter, Ch (ed.) Moraines and Varves; p. 25-42. Roterdam, Balkama.

Holmes, W. and Moss, J. H., 1955, Pleistocene geology of the southwestern Wind River Mountains, Wyoming, Geol. Soc. America Bull., 66: 629-654.

Madole, R. F., 1972, Neoglacial facies in the Colorado Front Range, Arctic and Alpine Research 4: 119-130.

Madole, R. F., 1976, Glacial geology of the Front Range, Colorado. In, Mahaney, W. C. (ed.) Quaternary Stratigraphy of North America, p. 297-318. New York, John Wiley and Sons, Inc.

Madole, R. F., 1980, Time of Pinedale deglaciation in north-central Colorado: Further considerations, Geology 8: 118-122.

Mahaney, W. C., 1972, Reinterpretation of the Neoglacial chronology in the central Colorado Front Range. In, Million, M. M. (ed.) Arctic and Mountain Environments Symposium, p. 18. Michigan, Michigan State Univ.

Mahaney, W. C., 1974, Soil Stratigraphy and genesis of Neoglacial deposits in the Arapaho and Henderson Cirques, central Colorado Front Range. In Mahaney, W. C. (ed.) Quaternary Environments - Proceedings of a Symposium; York University - Atkinson College (Toronto) Geographical Monograph No. 5: 197-240.

Mears, B., 1974, The evolution of the Rocky Mountain glacial Model. In Coates, D. R., (ed.) Glacial Geomorphology, p. 11-40. New York, New York State University Press.

Moss, J. H., 1949, Possible new glacial substage in the middle Rocky Mountains, Geol. Soc. America Bull. 60:19-72

Moss, J. H., 1951a, Early man in the Eden Valley, Univ. of Penn. Museum Monograph.

Moss, J. H., 1951b, Late glacial advance in the southern Wind River Mountains, Wyoming, Am. Jour. Sci. 249: 865-883.

Moss, J. H., 1974, The relation of river terrace formation to glaciation in the Shoshone River Basin, western Wyoming. In Coates, D. R. (ed.) Glacial Geomorphology, p. 293-315. New York, New York State Univ: Press.

Pankos, M., in prep., Provenance investigations of outwash terraces, Big Lost River Drainage, Custer County, Idaho, unpublished M. S. thesis, Lehigh University.

Pasquini, T. P., 1976, Provenance investigation of the glacial geology of the Copper Basin, Idaho, unpublished M.S. thesis, Lehigh University.

Pierce, K. L., Obradovich, J. D., and Friedman, I., 1976, Obsidian hydration dating and correlation of Bull Lake and Pinedale Glaciations near West Yellowstone, Montana: Geol. Soc. America Bull. 87: 703-710.

Repsher, A. A., 1980, Provenance of glacial deposits in the Pioneer Mountains, Blaine and Custer Counties, unpublished M. S. thesis, Lehigh University.

Repsher, A. A., Cotter, J. F. P. and Evenson, E. B., 1980, Glacial History, provenance and mineral exploration in an area of alpine glaciation Custer and Blaine Counties, Idaho, Geol. Soc. Abs. with Programs 12: 301

Richmond, G. M., 1948. Modification of Blackwelder's sequence of Pleistocene glaciation in the Wind River Mountains, Wyoming, Geol. Soc. America Bull. 59: 1400-1401.

Richmond, G. M., 1957. Three Pre-Wisconsinian glacial stages in the Rocky Mountain Region, Geol. Soc. America Bull. 68: 239-262.

Richmond, G. M., 1960a. Glaciation of the east slope of Rocky Mountain National Park, Colorado, Geol. Soc. America Bull. 71: 1371-1381.

Richmond, G. M., 1960b. Correlation of alpine and continental glacial deposits of Glacier National Park, Montana, U.S.G.S. Prof. Paper 400-B.

Richmond, G. M., 1962a. Three Pre-Bull Lake tills in the Wind River

Mountains, Wyoming. U.S.G.S. Prof. Paper 450-D.

Richmond, G. M., 1964b, Three Pre-Bull Lake tills in the Wind River Mountains, Wyoming; Reinterpreted, U.S.G.S. Prof. Paper 501-D.

Richmond, G. M., 1965, Glaciation of the Rocky Mountains. In, Wright, H. E. and Frey, D. G. (eds.) The Quaternary of the United States, p. 217-230. Princeton, N.J., Princeton Univ. Press.

Richmond, G. M., 1972, Appraisal of the future climate of the Holocene in the Rocky Mountains. Quat. Research 2: 315-322.

Richmond, G. M., 1976, Pleistocene stratigraphy and chronology in the mountains of western Wyoming. In, Mahaney, W. C., ed., Quaternary Stratigraphy of North America, p. 353-380, New York, John Wiley and Sons.

Stewart, R. A., 1977, The glacial geology of Wildhorse Canyon, Custer County, Idaho, unpublished M. S. thesis, Lehigh University.

Umpleby, S. G., Westgate, L. G. and Ross, G. P., 1930, Geology and ore deposits of the Wood River Region, Idaho, U.S.G.S. Bull. 814.

Weber, W. M., 1972, Correlation of Pleistocene glaciation in the Bitterroot Range, Montana, with fluctuations of glacial Lake Missoula, Montana, Bur. Mines and Geol. Memoir 42.

Wigley, W. C., 1976, The glacial geology of the Copper Basin, Custer Co., Idaho: A morphologic and pedogenic approach, unpublished M. S. thesis, Lehigh University.

Wigley, W. C., Pasquini, T. P., and Evenson, E. B., 1978, Glacial history of the Copper Basin, Idaho: A pedologic, provenance, and morphologic approach. In Mahaney, W. C. (ed.) Quaternary Soils; p. 265-307. England, Geol. Abstracts.

Contributions related to the field excursions:
B. Argentina

INQUA Commission on Lithology and Genesis of Quaternary Deposits:
South American Regional Meeting, Argentina 1982

JORGE RABASSA
Universidad Nacional del Comahue, Neuquén & CIC de la Provincia de Buenos Aires, La Plata, Argentina

The INQUA South American Regional Meeting on Glacigenic Deposits took place in Neuquén and San Carlos de Bariloche, Northwestern Patagonia, Argentina, between March 29 and April 6, 1982 (Figure 1). A post-Meeting "Safari" along the Patagonian Andes to Calafate, in the Southwesternmost tip of Patagonia was scheduled between April 7 and April 14 (Figure 2).

The Meeting was organized by the Department of Geography and the School of Postgraduate Studies of the Universidad Nacional del Comahue at Neuquén, and it was sponsored by the Government of the Province of Neuquén and the National Parks Survey of Argentina. The Meeting was dedicated to the memory of Carl C:zon Caldenius, a Swedish geologist who worked in Patagonia, in the fiftieth anniversary of the publication of his paramount paper (Caldenius 1932). The first two days were dedicated to the presentation of two conferences by Francisco Fidalgo (1982) and Jan Lundqvist (1982) and 20 papers at the University campus, with more than 100 Argentine investigators and students and 25 foreign scientists, coming from Austria, Bolivia, Brazil, France, New Zealand, Spain, Sweden, Switzerland, USA, USSR and West Germany. On March 31st., Prof Gerardo De Jong (Geography Department, Universidad del Comahue) conducted the group of foreign visitors on a special flight around the Province of Neuquén in one of the provincial airline planes. The aircraft toured around Volcán Lanín (3770 m a.s.l.) and its magnificient glaciers and many aspects of the geology of the Mesozoic "Neuquén Basin" were also observed.

Afterwards, the group took a regular flight to San Carlos de Bariloche from the same airport. The area of San Carlos de Bariloche was chosen for the Meeting Excursions due, not only for its beautiful scenery,

but because the glacial geology of the region has been described in the classical paper by Caldenius (1932) and later, by Flint and Fidalgo (1964). These latter authors divided the glacial deposits of this area as follows (Table 1):

TABLE 1
STRATIGRAPHY OF THE PLEISTOCENE GLACIGENIC DEPOSITS IN THE SAN CARLOS DE BARILOCHE REGION

Lithostratigraphic units	Age
NAHUEL HUAPI DRIFT II	Wisconsin
NAHUEL HUAPI DRIFT I	
EL CONDOR DRIFT	Early Wisconsin ? Pre-Wisconsin ?
PICHILEUFU DRIFT	Pre-Wisconsin

The recognition of these units was based on field studies over a large area and a relative chronology was established using weathering rates for granitic pebbles. No radiocarbon dating have been performed yet on these deposits due mainly to the lack of wood fragments in them, as Flint & Fidalgo (1964) discussed. The Wisconsin age suggested by Flint & Fidalgo (1964) was later confirmed by correlation with the Llanquihue Drift, Chile (Mercer 1976; Heusser & Flint 1977; Porter 1981) of well established Late Wisconsin age.

The Bariloche region (Figure 3) is also of high interest due to the existence of large glaciers at relatively low altitudes for their latitudinal position. These glaciers flow down the slopes of Mt. Tronador (3550 m a.s.l.), an ancient Pliocene volcano that rises well above the regional firn

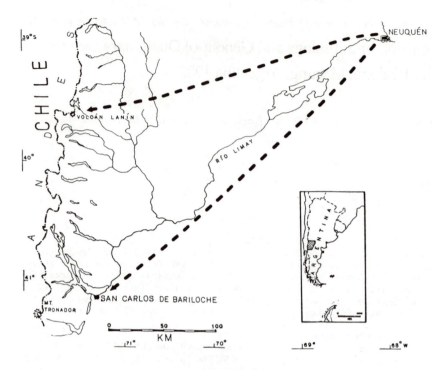

Figure 1.
Location map, showing the conference's site (Neuquén), the airplane trip to Volcán Lanín, and the Excursions' area (San Carlos de Bariloche)

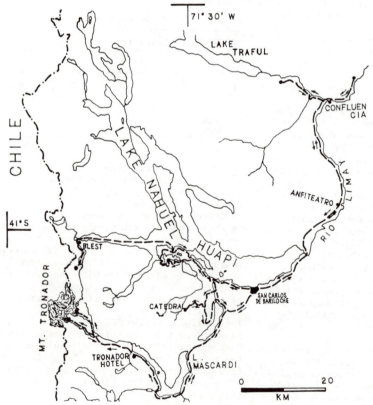

Figure 3
San Carlos de Bariloche region. Excursion N° 1 to Confluencia and Lake Traful; Excursion N° 2 to Circuito Chico and Catedral; Excursion N° 3 to Puerto Blest; Excursions N° 4a and 4b to Mount Tronador glaciers.

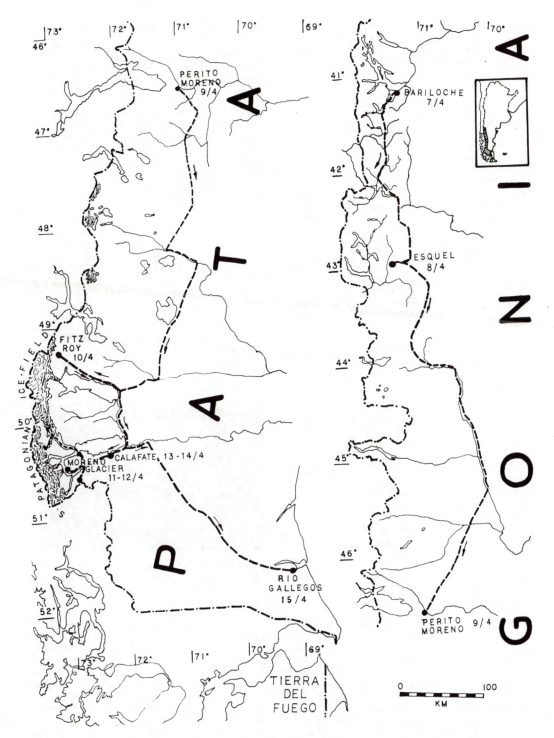

Figure 2. Post-Meeting "Safari". Itinerary map, showing the area visited and the localities where the bus stopped for overnight stay. Dotted line indicates the itinerary of the boat excursion across Lake Argentino to Upsala and Onelli glaciers.

line (1800-2000 m a.s.l., approximately;
Rabassa and others 1980). Near the terminii
of these glaciers, neoglacial deposits have
also been identified (Table 2; from Rabassa
1982, with modifications).

TABLE 2
STRATIGRAPHY OF THE HOLOCENE GLACIGENIC
DEPOSITS IN THE SAN CARLOS DE BARILOCHE
REGION

Lithostratigraphic units	Localities	Age
Sub-recent Glacigenic Deposits	Lake Alerce moraines	1945–1953
	Frías Glacier push-moraines	1945–1978
Little Ice-Age Deposits	Río Manso and Castaño Overo glaciers' moraines	Since the XIII Cent. (?)
	Cascada Alerce moraines	
	External moraines of the former Lake Témpanos (Frías Gl.)	to the XIX Cent.
Previous Neoglacial Deposits	Río Frías Valley moraines	After 6000 years ^{14}C B.P.
	Castaño Overo and Alerce valleys moraines	
Glacigenic deposits of the end and retreat of the Last Glaciation	Phase a: Lake Moreno moraines	Post-14,000 years ^{14}C B.P.
	Phase b: Puerto Blest moraines "Water-divide" drift	
Nahuel Huapi Drift		Wisconsin

With the aim of discussing the character
and extent of the glacial events corresponding to the units of Tables 1 and 2, several excursions (Rabassa 1982) were planned
and attended by more than 50 participants.
Excursion N° 1 covered the upper valley of
Río Limay (Figure 3) and the outer portions
of the Bariloche Moraine. Sedimentary deposits overlying an unweathered lapillitic
tephra which had been considered as glacigenic by Rabassa (1982) generated an active
discussion among the Commission members
and strong comments were made against my

interpretation of these units. Most of the
opinions pointed instead towards a genesis
by a volcanic mud-flow, which could have
eventually removed actual glacigenic materials from upslope. Discussion continued
on several other localities where the glacial origin of the sediments was also
questionned. Luckily (for me) the group
finally reached the "accepted" outer limit
of glaciation where till, glaciolacustrine
and glaciofluvial deposits were observed
in the Anfiteatro (Figure 4) and Bariloche
(Figure 5) moraines. The chronology of the
Pleistocene glaciations and the origin of
deformational structures in the Anfiteatro
Moraine were discussed by the participants.
On April 2nd., the itinerary of Excursion
N° 2 followed the "Circuito Chico" where
recessional depositional features of the
Nahuel-Huapi Glaciation and the erosional
glacial landscape of Lake Nahuel Huapi were
visited. Excursion N° 3 was planned as a
boat-trip across Lake Nahuel Huapi to study
the Puerto Blest Moraine (see Table 2),
interpreted as a recessional moraine of the
Frías Glacier during the end of the Nahuel
Huapi Glaciation or perhaps a readvance
during the First Neoglacial (4500 years B.P.;
Caldenius 1932, Mercer 1976, Rabassa and
others 1979).

Figure 4. Anfiteatro Moraine, El Cóndor
Glaciation. A giant ball-and-pillow structure in flow till, kame deposits and glaciolacustrine deposits.

Origin of flow tills and peat deposits
were enthusiastically discussed at Puerto
Alegre (Figure 6). Professors Calvin J
Heusser and Linda E Heusser sampled peat
beds but no significant results were obtained later through their analysis; this was
rather unexpected because plant remains
were relatively abundant in the sediment
and it will undoubtedly require a more
detailed study (Linda E Heusser, written
communication May 1982).
Finally, Excursions 4a and 4b took the
group to Cerro Tronador. Very fine weather

Figure 5. Bariloche Moraine, Nahuel Huapi Drift. A very large volcanic boulder can be observed at the edge of the moraine, near the railway. At the back, Cerro Catedral (2200 m a.s.l.; right) and Cerro Ventana (left); between them, the glacial valley of the former Tronador Glacier.

Figure 6. Puerto Alegre Moraine, Post-Nahuel Huapi Drift (Neoglacial 1?). Flow tills overlain by Holocene tephras and peat deposits.

provided good conditions to enjoy the scenery of the area. It was even warm enough to walk on the Río Manso Glacier in short sleeves. Many different aspects of glacial deposition were observed in this debris-covered, regenerated valley glacier, the largest in Argentine Northern Patagonia (Rabassa and others 1978; Figures 7 and 8). An ice-cave provided the opportunity to take a glance at the glacier sole and to reconstruct the genesis of last-winter lodgement till and observe on-going development of boulder striation. Many other features were discussed by the different groups which formed spontaneously. Friedrich Röthlisberger climbed up the glacier side, found several buried tree trunks and got back just on time for the chilled white wine that was served at the meeting point.

On the following day, the group visited Castaño Overo Glacier. Weather was very fine again and thus, the 6 Km walk through the Northern Patagonian forest was less unpleasant than expected. In this glacier (Figure 9), the participants had the possibility of looking at the Little Ice Age moraines and many recessional and ground-moraine features. Besides, comparing with the map presented by Rabassa and others (1978), the ice retreat since 1978 could be established and the corresponding deposits observed.

Figure 7. Río Manso Glacier. Aerial view by Eloy Depiante(1981) See the contrast between the clean ice of the upper glaciers and the debris-covered snout. The glacier goes well below tree limit. The snout is lying against Neoglacial 3 moraines. At the far right, Castaño Overo Glacier.

Figure 8. Río Manso Glacier. Debris-covered regenerated valley glacier. See the large crevasses in which the thin (less than 1 m) debris cover may be observed. At the back, ice-cones formed by ice and snow avalanches from the upper glaciers.

The group returned to Bariloche where on the following morning an informal meeting of the Commission was conducted by Jan Lundqvist and Ernest Müller. Most of the Argentine participants and some of the foreign visitors left the party during that day. The rest of the day was used in a trip by chair-lift to the Cerro Catedral Meteorological Station (2000 m a.s.l.), from where a general overview of Lake Nahuel Huapi and the glaciated area could be seen.

Figure 9. Castaño Overo Glacier, Mt.Tronador. A regenerated ice-cone by the avalanches from the upper glacier. On the left, see the clean trim-line in the forest, developed during the Little Ice Age. In the foreground, Little Ice Age moraines.

A reduced party of 26 participants took part of the Patagonian "Safari" (Figure 2). In fact, the "Safari" was not as unpleasant as the organizers themselves had thought. Weather remained quite acceptable and no major problems arose during the trip across one of the most uninhabited areas of the world. Even the long, endless bus journeys seemed shorter thanks to extensive discussions on the genesis of pediments and the "Patagonian Gravels" (Riggi and Fidalgo 1971) or the evolution of slopes and mass-movement deposits in volcanic mesas and plateaux. It was raining when the bus reached the Fitz Roy Hut, after three long days. Fortunately, the morning after the sky was clean and bright and the group could see Cerro Fitz Roy and Cerro Torre at sunrise (Figure 10).

Figure 10. Cerro Fitz Roy and Cerro Torre. Erosional glacial landscape developed on granitic and porphyritic rocks.

The spectacular horns and arétes of these peaks and the neighbouring ranges were observed and their genesis discussed. The trip continued to Calafate and from there to Perito Moreno Glacier (Figure 11). Two days were spent in the observation of this very active glacier, which is famous in the world because it advances rapidly traversing one of the Lake Argentino fiord-like branches and climbing up the landscape on the other side of the lake. Once the ice-dam has been built, as the glacier keeps grounded, the lake level at the Brazo Rico (the blocked arm of the lake) rises till the hydrostatic pressure of the water column (plus the seepage action through the crevasses) breaks the ice wall. When we visited the area, the glacier had already crossed the lake and was lying on the ground above lake level, enabling us to observe the acting processes of sedimentation at the ice-front (Figure 12; Figure 13).

Figure 11. Moreno Glacier, in April 1982, when the glacier had crossed the lake and climbed up the shore

Figure 12. Moreno Glacier. Depositional processes at the front of the glacier, push-moraines and ablation till accumulated as the dirty bed of the glacier melts out.

The group climbed up Cerro del Fraile on the following day, to see the Pliocene tills described by Feruglio (1944) and Mercer (1969).

The Commission members agreed on the glacigenic origin of these deposits and discussed the characteristics of the different lithological types in this important locality. The age and possible extent of this ancient glaciation were also considered.

Figure 13. Moreno Glacier, minor erosional sub-glacial features: polished surface, crescentic gouges, plucking. Ice was moving from right to left, over Cretaceous sandstones.

Finally, a boat journey was organized to get a proximal view of the calving front of Upsala Glacier, the largest glacier in Patagonia, at the Northern Branch of the Lake Argentino.

The "Safari" ended happily on April 14th., when everybody was safely on their respective flights at Río Gallegos Airport.

Saying good-by took short: the Malvinas War was starting.

REFERENCES

Caldenius, C.C. (1932). Las glaciaciones cuaternarias de la Patagonia y Tierra del Fuego. Dirección de Minas, Geología e Hidrología, Publ. N° 95, Buenos Aires. (English version published same year in Geografiska Annaler).

Feruglio, E. (1944). Estudios geológicos y glaciológicos en la región del Lago Argentino (Patagonia). Acad.Nac.Cienc. Córdoba, Boletín, 37 (1-2): 3-255.

Fidalgo, F. (1982). Glaciations in Patagonia. in: J.Rabassa, ed., INQUA Commission on Genesis and Lithology of Quaternary Deposits, South-American Regional Meeting, Excursions' Guidebook, p. 9-30, Neuquén.

Flint, R.F. & Fidalgo, F. (1964). Glacial Geology of the East Flank of the Argentine Andes between latitude 39°10'S and latitude 41°20'S. Geol.Soc.Amer.Bull., 75: 335-352.

Heusser, C.J. & Flint, R.F. (1977). Quaternary glaciations and environments of northern Isla Chiloé, Chile. Geology, 5: 305-308.

Lundqvist, J. (1982). Carl C:zon Caldenius and the Swedish work on Quaternary Geology in Argentina. This volume.

Mercer, J.H. (1976). Glacial history of Southernmost South America. Quaternary Research, 6: 125-166.

Porter, S. (1981). Quaternary glacial geology of the Chilean Lake District. Quaternary Research, 16.

Rabassa, J. (1982), editor.INQUA Commission on Genesis and Lithology of Quaternary Deposits, South-American Regional Meeting, Excursions' Guidebook, 150 pp., Neuquén, Public.Departamento de Geografía, Universidad Nacional del Comahue, 2nd. edition.

Rabassa, J.; Rubulis, S. & Suarez, J. (1978). Los glaciares del Monte Tronador, Parque Nacional Nahuel Huapi (Río Negro, Argentina). Anales Parques Nacionales, 14: 259-318, Buenos Aires.

Rabassa, J.; Rubulis, S. & Suarez, J. (1979). Rate of formation and sedimentology of 1976-1978 push-moraines, Frías Glacier, Mount Tronador (41°10'S; 71° 53'W), Argentina. in: Ch.Schluechter, ed., Moraines and Varves, A.A.Balkema, Rotterdam, p. 65-79.

Rabassa, J.; Rubulis, S. & Brandani, A. (1980).East-west and north-south snow line gradients in the northern Patagonian Andes, Argentina. World Glacier Inventory, Proc.Riederalp Workshop, Sept.1978, IAHS Publ. N° 126, p. 1-11.

Riggi, J.C. & Fidalgo, F. (1971). A review on the Rodados Patagónicos problem. Études sur le Quaternaire. Supplement au Bulletin de l'Association Française pour l'étude du Quaternaire, N° 4, VII Congress INQUA, Paris, 1969, v. 1, p. 29-36.

Carl C: zon Caldenius and the Swedish work on Quaternary geology in Argentina

JAN LUNDQVIST
University of Stockholm, Sweden

The South American Regional Meeting 1982 of the INQUA Commission on Genesis and Lithology of Quaternary Deposits has been dedicated to the memory of Dr. Carl Caldenius, Swedish and Argentinian state geologist, and the fiftieth anniversary of the publication of his monograph on the South American glacial geology. The author has kindly been invited by the organizors of the meeting to deliver a communication in honour of Caldenius.

Carl Caldenius, or Carl Carlzon which was his original name, was born in Stockholm in 1897. After finishing school he became a student of the University in Stockholm, where he especially studied Quaternary sediments in the province of Jämtland. In 1924, having adapted the name Caldenius, he defended his dissertation, which treated the sediments of Ancient Lake Ragunda in that province. The dissertation had, in spite of the great geotechnical interests of Caldenius, a purely geological content, the stratigraphy and chronology of the lake. In this work Caldenius became familiar with the geochronological clay-varve method by Gerard De Geer and an experienced co-worker of De Geer.

Contemporarily with this work for the doctor´s degree Caldenius worked as a geotechnician for the Swedish state rail-roads 1914-1922. This was an expression of his interests in construction and technics: Originally he planned to be a construction engineer.

The work with De Geer was part of the world-wide clay-varve chronological in-vestigations, which made De Geer famous - a real predecessor of the IGCP. This work would be extended also to the South American continent, partly as a result of the great interest by Dr. José Sobral, head of the Geological Survey (Dirección de Minas, Geología e Hidrología), in Buenos Aires,

Carl Caldenius in Värmland.
Photo J. Lundqvist 1952

Argentina. Discussions between Sobral and De Geer resulted in a decision that a Swedish geologist would be employed by the Geological Survey to be responsible for the geochrono-logical work in Argentina and southern Chile. For this work Caldenius was chosen. He would stay as a state geologist in Argentina for five years, 1925-1930.

During his five years in Argentina Calde-nius made several journeys in the southern parts of Patagonia, in Tierra del Fuego and the adjacent parts of Chile. The area was partly desolate, and Caldenius´ travelling in a small four-cylinder car in a more or less roadless country was an admirable per-formance - a real exploration. During these expeditions Caldenius made detailed mapping in certain areas as well as a reconnaissance of glacial deposits all over the area. This part of the work is still considered the most comprehensive as well as extensive research on glacial geology in the region. It can serve as the base for future work on

From Caldenius´ travelling in Patagonia

this topic, and the maps are still considered correct in detail.

Caldenius also permformed clay-varve investigations. He established a clay-varve
chronology, which was correlated with the
Swedish time scale of De Geer. This part of
the work can be seriously criticized - there
is a consensus of opinion that there is
little base for such teleconnections.
Actually it is rather surprising that
Caldenius devoted himself to such a speculative work, because he was a very critical
and careful scientist. The explanation is
surely to be found in the great influence
by Gerard De Geer and the fact that such a
correlation was the original aim of
Caldenius´ work.

Caldenius´ results were published mainly
in his great monograph "Las glaciaciones
cuaternarias en la Patagonia y Tierra del
Fuego" with the subheadings "Una investiga-
ción regional, estratigráfia y geocronoló-
gica. Una comparación con la escala geo-
cronológica sueca". The monograph was
published in 1932, first in Buenos Aires
and soon afterwards also in Geografiska
Annaler in Stockholm.

After leaving Argentina Caldenius extended
his geochronological work also to Australia
and New Zealand. His combined research
contributed to a better understanding of
the relationships between the glaciations
in the northern and southern hemispheres.
Not only did it comprise the Quaternary
geology, but also to some extent Permo-
-Carboniferous rhythmites.

In Sweden Caldenius worked as a "geotechnician in geology", to use his own
expression, and mainly as a state geologist
1944-1955. He was a specialist in applied
geology but took part also in the mapping
programme in Sweden. His mapping can be
characterized like his chronological work:
extremely careful and performed with greatest critical accuracy. Having retired from
his position at the Survey Caldenius continued to work as a skilful consultant and
inspiring teacher till his death in 1961.

Caldenius´ work in Argentina was a logical
continuation of the studies earlier made by
Otto Nordenskjöld. Nordenskjöld´s geographical work in Patagonia, Tierra del Fuego and
the adjacent parts of Antarctica, partly
performed in cooperation with Francisco
Moreno, also comprised important studies of
the Tehuelche formation. His work contributed to a better understanding of the true
extension of continental glaciation in
Argentina.

Thus there is an old tradition of Swedish
geological work in Argentina and adjacent
parts of the South American continent.
Several other scientists have taken part in
it, except in Quaternary geology also in
the study of volcanics, tectonics, botany,
anthropology etc. A direct successor of
Caldenius was the geographer Erik Ljungner
who started his research on the east-west
balance of the Pleistocene ice sheets of
Patagonia and Scandinavia already when
Caldenius was still working in Patagonia.
It is to be hoped that this South American-
-Swedish cooperation will continue.